Calculus

for Business, Economics, and the Social and Life Sciences

THIRD EDITION

Calculus
for Business, Economics, and the Social and Life Sciences

Laurence D. Hoffmann
Claremont McKenna College

McGraw-Hill Book Company
New York St. Louis San Francisco Auckland Bogotá Hamburg
Johannesburg London Madrid Mexico Montreal New Delhi
Panama Paris São Paulo Singapore Sydney Tokyo Toronto

CALCULUS FOR BUSINESS, ECONOMICS,
AND THE SOCIAL AND LIFE SCIENCES

1 2 3 4 5 6 7 8 9 0 DOCDOC 8 9 8 7 6 5

ISBN 0-07-029331-7

This book was set in Times Roman by Better Graphics.
The editors were Peter R. Devine and David Dunham;
the design was done by Caliber Design Planning;
the production supervisor was Diane Renda.
The cover photograph was taken by Brenda Kamen.
The drawings were done by Accurate Art, Inc.
R. R. Donnelley & Sons Company was printer and binder.

Library of Congress Cataloging-in-Publication Data

Hoffmann, Laurence D., date
    Calculus for business, economics, and the social and life sciences.

    Previous ed. published as: Calculus for the social, managerial, and life sciences.   1980.
    Includes index.
    1. Calculus. I. Title.
QA303.H569   1986   515   85-11291
ISBN 0-07-029331-7

Contents

Preface

Overview

Objectives

If you are preparing for a career in business, economics, psychology, sociology, architecture, or biology, and if you have taken high school algebra, then this book was written for you. Its primary goal is to teach you the techniques of differential and integral calculus that you are likely to encounter in undergraduate courses in your major and in your subsequent professional activities.

Applications

The text is applications-oriented. Each new concept you learn is applied to a variety of practical situations. The techniques and strategies you will need to solve applied problems are stressed. The applications are drawn from the social, managerial, and life sciences with special emphasis on business and economics.

Level of Rigor

The exposition is designed to give you a sound, intuitive understanding of the basic concepts without sacrificing mathematical accuracy. Thus, the main results are stated carefully and completely and, whenever possible, explanations are intuitive or geometric.

Problems

You learn mathematics by doing it. Each section in this text is followed by an extensive set of problems. Many involve routine computation and are designed to help you master new techniques. Others ask you to apply the new techniques to practical situations. There is a set of review problems at the end of each chapter. At the back of the book you will find the answers to the odd-numbered problems and to all the review problems.

Algebra Review

If you need to brush up on your high school algebra, there is an extensive algebra review in the appendix that includes worked examples and practice problems for you to do. You will be advised throughout the text when it might be appropriate to consult this material.

Major Features of the New Edition

While retaining the straightforward style, intuitive approach, and applications orientation of its predecessor, this edition covers more mathematics and includes more applications, particularly to business and economics.

Expanded Coverage

Among the topics that have been added are improper integrals in Chapter 7, double integrals in Chapter 10, probability density functions in Chapters 7 and 10, and infinite series in Chapter 11. The coverage of trigonometric functions has been expanded and moved from the appendix to Chapter 12. Discussions of numerical methods have also been added, including numerical integration in Chapter 7, the method of least squares in Chapter 9, and Taylor approximation and Newton's method in Chapter 11.

Applications to Business and Economics

The applications to business and economics have been augmented by discussions of elasticity of demand in Chapter 3, Section 5; consumers' surplus and willingness to spend in Chapter 6, Section 3; and indifference curves and the maximization of utility in Chapter 9, Sections 4 and 6.

Supplementary Materials

Computer Supplement

A computer supplement prepared with Granville C. Henry (of Claremont McKenna College) is available to accompany this text. In the supplement you will be introduced to using computer programs to solve calculus problems and will learn elementary programming in BASIC. In the process, you will develop an appreciation for the capabilities and limitations of both calculus and the computer.

Student's Solutions Manual

A solutions manual by Stanley M. Lukawecki of Clemson University contains complete step-by-step solutions of all the odd-numbered problems and review problems.

Instructor's Answer Manual

The answers to the even-numbered problems are contained in an answer manual available to instructors.

Acknowledgments

Many people helped with the preparation of this edition. My colleague, economist Susan Feigenbaum, offered valuable advice during the development of the new material on business and economics, for which I am particularly grateful.

Several reviewers read early versions of the manuscript. Especially helpful were the detailed comments of: Dan Anderson, University of Iowa; Bruce Edwards, University of Florida; Ronnie Goolsby, Winthrop College; Erica Jen, University of Southern California; Melvin Lax, California State University–Long Beach; Stanley Lukawecki, Clemson University; Eldon Miller, University of Mississippi; Richard Randell, University of Iowa; Anthony Shershin, Florida International University; Keith Stroyan, University of Iowa; Martin Tangora, University of Illinois at Chicago; Lee Topham, North Harris County College; Charles Votaw, Fort Hays State University; and Jonathan Weston-Dawkes, University of North Carolina.

Reviewers of the earlier editions of the book include: George Articolo, Rutgers University; Theodore J. Barth, University of California at Riverside; Barbara Lee Bleau, Pennsylvania State University; Carl Eberhardt, University of Kentucky; George Feissner, State University of New York at Cortland; Charles Frady, Georgia State University; Alexander Hahn, University of Notre Dame; Rodney T. Hansen, Montana State University; Charles Himmelberg, University of Kansas; William Heubsch, Canisius College; Roger Johnes, De Paul University; V. J. Klaussen, California State University at Fullerton; Lowell Leake, University of Cincinnati; John G. Michaels, State University of New York at Brockport; Robert A. Mills, Eastern Michigan University; Bill New, Cerritos College; J. A. Pfaltzgraff, University of North Carolina; Karen J. Schroeder, Bentley College; David Shea, University of Wisconsin; Paul Slepian, Howard University; George Springer, Indiana University; and Robert Zink, Purdue University.

Mr. Chris Hubbard checked the entire book for accuracy and worked all the new and revised problems.

My editors at McGraw-Hill have been particularly helpful, encouraging, and patient. Peter Devine and David Dunham are professional editors of the highest quality, and it has been a pleasure working with them.

Laurence D. Hoffmann

Chapter

Functions and Graphs

1 Functions

In many practical situations, the value of one quantity may depend on the value of a second. For example, the consumer demand for beef may depend on its current market price; the amount of air pollution in a metropolitan area may depend on the number of cars on the road; the value of a bottle of wine may depend on its age. Such relationships can often be represented mathematically as **functions.**

Function

A function is a rule that assigns to each object in a set A, one and only one object in a set B.

This definition is illustrated in Figure 1.1.

1

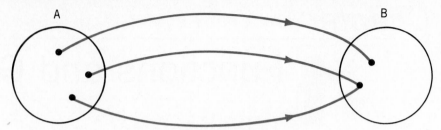

Figure 1.1 A visual representation of a function.

For most of the functions in this book, the sets A and B will be collections of real numbers. You can think of such a function as a rule that assigns "new" numbers to "old" numbers. To be called a function, the rule must have the property that it assigns one and only one "new" number to each "old" number. Here is an example.

EXAMPLE 1.1   According to a certain function, the "new" number is obtained by adding 4 to the square of the "old" number. What number does this function assign to 3?

SOLUTION
The number assigned to 3 is $3^2 + 4$, or 13.

Variables

Often you can write a function compactly by using a mathematical formula. It is traditional to let x denote the old number and y the new number, and write an equation relating x and y. For instance, you can express the function in Example 1.1 by the equation

$$y = x^2 + 4$$

The letters x and y that appear in such an equation are called **variables.** The numerical value of the variable y is determined by that of the variable x. For this reason, y is sometimes referred to as the **dependent variable** and x as the **independent variable.**

Functional Notation

There is an alternative notation for functions that is widely used and somewhat more versatile. A letter such as f is chosen to stand for the function itself, and the value that the function assigns to x is denoted by $f(x)$ instead of y. The symbol $f(x)$ is read "f of x." Using this **functional notation,** you can rewrite Example 1.1 as follows.

EXAMPLE 1.2 Find $f(3)$ if $f(x) = x^2 + 4$.

SOLUTION
$$f(3) = 3^2 + 4 = 13$$

Observe the convenience and simplicity of this notation. In Example 1.2, the compact formula $f(x) = x^2 + 4$ completely defines the function, and the simple equation $f(3) = 13$ indicates that 13 is the number that the function assigns to 3.

The use of functional notation is illustrated further in the following examples. Notice that in Example 1.3, letters other than f and x are used to denote the function and its independent variable.

EXAMPLE 1.3 If $g(t) = (t - 2)^{1/2}$, find (if possible) $g(27)$, $g(5)$, $g(2)$, and $g(1)$.

SOLUTION
Rewrite the function as $g(t) = \sqrt{t - 2}$. (If you need to brush up on fractional powers, you can consult the discussion of exponential notation in the Algebra Review at the back of the book.) Then,

$$g(27) = \sqrt{27 - 2} = \sqrt{25} = 5$$
$$g(5) = \sqrt{5 - 2} = \sqrt{3} \approx 1.7321$$

and

$$g(2) = \sqrt{2 - 2} = \sqrt{0} = 0$$

However, $g(1)$ is undefined since

$$g(1) = \sqrt{1 - 2} = \sqrt{-1}$$

and negative numbers do not have real square roots.

In the next example, two formulas are needed to define the function.

EXAMPLE 1.4 Find $f\left(-\dfrac{1}{2}\right)$, $f(1)$, and $f(2)$ if

$$f(x) = \begin{cases} \dfrac{1}{x - 1} & \text{if } x < 1 \\ 3x^2 + 1 & \text{if } x \geq 1 \end{cases}$$

SOLUTION

From the first formula,

$$f\left(-\frac{1}{2}\right) = \frac{1}{-1/2 - 1} = \frac{1}{-3/2} = -\frac{2}{3}$$

and from the second formula,

$$f(1) = 3(1)^2 + 1 = 4 \qquad \text{and} \qquad f(2) = 3(2)^2 + 1 = 13$$

The next example illustrates how functional notation is used in a practical situation. Notice that to make the algebraic formula easier to interpret, letters suggesting the relevant practical quantities are used for the function and its independent variable. (In this example, the letter C stands for "cost" and q for "quantity" manufactured.)

EXAMPLE 1.5 Suppose the total cost in dollars of manufacturing q units of a certain commodity is given by the function $C(q) = q^3 - 30q^2 + 500q + 200$.

(a) Compute the cost of manufacturing 10 units of the commodity.
(b) Compute the cost of manufacturing the 10th unit of the commodity.

SOLUTION

(a) The cost of manufacturing 10 units is the value of the total cost function when $q = 10$. That is,

$$\text{Cost of 10 units} = C(10)$$
$$= (10)^3 - 30(10)^2 + 500(10) + 200$$
$$= \$3,200$$

(b) The cost of manufacturing the 10th unit is the difference between the cost of manufacturing 10 units and the cost of manufacturing 9 units. That is,

$$\text{Cost of 10th unit} = C(10) - C(9) = 3,200 - 2,999 = \$201$$

The Domain of a Function

The set of values of the independent variable for which a function can be evaluated is called the **domain** of the function. For instance, the function $f(x) = x^2 + 4$ in Example 1.2 can be evaluated for any real number x. Thus, the domain of this function is the set of all real numbers. The domain of the function $C(q) = q^3 - 30q^2 + 500q + 200$ in Example 1.5 is also the set of all real numbers [although $C(q)$ represents total cost only for nonnegative values of q]. In the next example are two functions whose domains are restricted for algebraic reasons.

EXAMPLE 1.6 Find the domain of each of the following functions:

(a) $f(x) = \dfrac{1}{x - 3}$

(b) $g(x) = \sqrt{x - 2}$

SOLUTION

(a) Since division by any real number except zero is possible, the only value of x for which $f(x) = \dfrac{1}{x - 3}$ cannot be evaluated is $x = 3$, the value that makes the denominator of f equal to zero. Hence the domain of f consists of all real numbers except 3.

(b) Since negative numbers do not have real square roots, the only values of x for which $g(x) = \sqrt{x - 2}$ can be evaluated are those for which $x - 2$ is nonnegative, that is, for which

$$x - 2 \geq 0 \qquad \text{or} \qquad x \geq 2$$

That is, the domain of g consists of all real numbers that are greater than or equal to 2.

Composition of Functions

There are many situations in which a quantity is given as a function of one variable which, in turn, can be written as a function of a second variable. By combining the functions in an appropriate way, you can express the original quantity as a function of the second variable. This process is known as the **composition of functions.**

Composition of Functions

The composite function $g[h(x)]$ is the function formed from the two functions $g(u)$ and $h(x)$ by substituting $h(x)$ for u in the formula for $g(u)$.

The situation is illustrated in Figure 1.2.

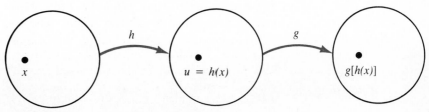

Figure 1.2 The composition of functions.

EXAMPLE 1.7 Find the composite function $g[h(x)]$ if $g(u) = u^2 + 3u + 1$ and $h(x) = x + 1$.

SOLUTION
Replace u by $x + 1$ in the formula for g to get

$$g[h(x)] = (x + 1)^2 + 3(x + 1) + 1$$
$$= (x^2 + 2x + 1) + (3x + 3) + 1$$
$$= x^2 + 5x + 5$$

The problem in Example 1.7 could have been worded more compactly as follows: Find the composite function $g(x + 1)$ where $g(u) = u^2 + 3u + 1$. The use of this compact notation is illustrated further in the next example.

EXAMPLE 1.8 Find $f(x - 1)$ if $f(x) = 3x^2 + \dfrac{1}{x} + 5$.

SOLUTION
At first glance, this problem may look confusing because the letter x appears both as the independent variable in the formula defining f and as part of the expression $x - 1$. Because of this, you may find it helpful to begin by writing the formula for f in more neutral terms, say as

$$f(\square) = 3(\square)^2 + \frac{1}{\square} + 5$$

To find $f(x - 1)$, you simply insert the expression $x - 1$ inside each box, getting

$$f(x - 1) = 3(x - 1)^2 + \frac{1}{x - 1} + 5$$

Occasionally, you will have to be able to "take apart" a given composite function $g[h(x)]$ and identify the functions $g(u)$ and $h(x)$ from which it was formed. The procedure is illustrated in the next example.

EXAMPLE 1.9 If $f(x) = \dfrac{5}{x - 2} + 4(x - 2)^3$, find functions $g(u)$ and $h(x)$ such that $f(x) = g[h(x)]$.

SOLUTION
The form of the given function is

$$f(x) = \frac{5}{\Box} + 4(\Box)^3$$

where each box contains the expression $x - 2$. Thus $f(x) = g[h(x)]$, where

$$g(u) = \frac{5}{u} + 4u^3 \qquad \text{and} \qquad h(x) = x - 2$$

Actually, in Example 1.9, there are infinitely many pairs of functions $g(u)$ and $h(x)$ that combine to give $g[h(x)] = f(x)$. [For example, $g(u) = \frac{5}{u + 1} + 4(u + 1)^3$ and $h(x) = x - 3$.] The particular pair selected in the solution to this example is the most natural one and reflects most clearly the structure of the original function $f(x)$.

An Application of Composite Functions

The next example illustrates how a composite function may arise in a practical problem.

EXAMPLE 1.10 An environmental study of a certain community suggests that the average daily level of carbon monoxide in the air will be $c(p) = 0.5p + 1$ parts per million when the population is p thousand. It is estimated that t years from now the population of the community will be $p(t) = 10 + 0.1t^2$ thousand.

(a) Express the level of carbon monoxide in the air as a function of time.
(b) When will the carbon monoxide level reach 6.8 parts per million?

SOLUTION
(a) Since the level of carbon monoxide is related to the variable p by the equation

$$c(p) = 0.5p + 1$$

and the variable p is related to the variable t by the equation

$$p(t) = 10 + 0.1t^2$$

it follows that the composite function

$$c[p(t)] = c(10 + 0.1t^2) = 0.5(10 + 0.1t^2) + 1 = 6 + 0.05t^2$$

expresses the level of carbon monoxide in the air as a function of the variable t.

(b) Set $c[p(t)]$ equal to 6.8 and solve for t to get

$$6 + 0.05t^2 = 6.8$$

$$0.05t^2 = 0.8$$

$$t^2 = \frac{0.8}{0.05} = 16$$

$$t = \sqrt{16} = 4$$

That is, 4 years from now the level of carbon monoxide will be 6.8 parts per million.

Problems

In Problems 1 through 11, compute the indicated values of the given function.

1. $f(x) = 3x^2 + 5x - 2; f(1), f(0), f(-2)$

2. $h(t) = (2t + 1)^3; h(-1), h(0), h(1)$

3. $g(x) = x + \dfrac{1}{x}; g(-1), g(1), g(2)$

4. $f(x) = \dfrac{x}{x^2 + 1}; f(2), f(0), f(-1)$

5. $h(t) = \sqrt{t^2 + 2t + 4}; h(2), h(0), h(-4)$

6. $g(u) = (u + 1)^{3/2}; g(0), g(-1), g(8)$

7. $f(t) = (2t - 1)^{-3/2}; f(1), f(5), f(13)$

8. $g(x) = 4 + |x|; g(-2), g(0), g(2)$

9. $f(x) = x - |x - 2|; f(1), f(2), f(3)$

10. $h(x) = \begin{cases} -2x + 4 & \text{if } x \le 1 \\ x^2 + 1 & \text{if } x > 1 \end{cases}; h(3), h(1), h(0), h(-3)$

11. $f(t) = \begin{cases} 3 & \text{if } t < -5 \\ t + 1 & \text{if } -5 \le t \le 5; f(-6), f(-5), f(16) \\ \sqrt{t} & \text{if } t > 5 \end{cases}$

In Problems 12 through 24, specify the domain of the given function.

12. $f(x) = x^3 - 3x^2 + 2x + 5$

13. $g(x) = \dfrac{x^2 + 5}{x + 2}$

14. $f(t) = \dfrac{t + 1}{t^2 - t - 2}$

15. $y = \sqrt{x - 5}$

16. $f(x) = \sqrt{2x - 6}$

17. $g(t) = \sqrt{t^2 + 9}$

18. $h(u) = \sqrt{u^2 - 4}$

19. $f(t) = (2t - 4)^{3/2}$

20. $y = \dfrac{x - 1}{\sqrt{x^2 + 2}}$

21. $f(x) = (x^2 - 9)^{-1/2}$

22. $h(t) = \dfrac{\sqrt{t^2 - 4}}{\sqrt{t - 4}}$

23. $g(t) = \dfrac{1}{|t - 1|}$

24. $h(x) = \sqrt{|x - 3|}$

Manufacturing cost 25. Suppose the total cost in dollars of manufacturing q units of a certain commodity is given by the function $C(q) = q^3 - 30q^2 + 400q + 500$.
(a) Compute the cost of manufacturing 20 units.
(b) Compute the cost of manufacturing the 20th unit.

Worker efficiency 26. An efficiency study of the morning shift at a certain factory indicates that an average worker who arrives on the job at 8:00 A.M. will have assembled $f(x) = -x^3 + 6x^2 + 15x$ transistor radios x hours later.
(a) How many radios will such a worker have assembled by 10:00 A.M.? (*Hint*: At 10:00 A.M., $x = 2$.)
(b) How many radios will such a worker assemble between 9:00 and 10:00 A.M.?

Temperature change 27. Suppose that t hours past midnight, the temperature in Miami was $C(t) = -\dfrac{1}{6}t^2 + 4t + 10$ degrees Celsius.
(a) What was the temperature at 2:00 P.M.?
(b) By how much did the temperature increase or decrease between 6:00 and 9:00 P.M.?

Population growth 28. It is estimated that t years from now, the population of a certain suburban community will be $P(t) = 20 - \dfrac{6}{t + 1}$ thousand.
(a) What will the population of the community be 9 years from now?
(b) By how much will the population increase during the 9th year?
(c) What will happen to the size of the population in the long run?

Experimental psychology 29. To study the rate at which animals learn, a psychology student performed an experiment in which a rat was sent repeatedly through a laboratory maze. Suppose that the time required for the rat to traverse the maze on the nth trial was approximately $f(n) = 3 + \dfrac{12}{n}$ minutes.
(a) What is the domain of the function f?
(b) For what values of n does $f(n)$ have meaning in the context of the psychology experiment?

(c) How long did it take the rat to traverse the maze on the 3rd trial?

(d) On which trial did the rat first traverse the maze in 4 minutes or less?

(e) According to the function f, what will happen to the time required for the rat to traverse the maze as the number of trials increases? Will the rat ever be able to traverse the maze in less than 3 minutes?

Poiseuille's law   30. Biologists have found that the speed of blood in an artery is a function of the distance of the blood from the artery's central axis. According to **Poiseuille's law,** the speed (in centimeters per second) of blood that is r centimeters from the central axis of an artery is given by the function $S(r) = C(R^2 - r^2)$, where C is a constant and R is the radius of the artery. Suppose that for a certain artery, $C = 1.76 \times 10^5$ centimeters and $R = 1.2 \times 10^{-2}$ centimeters.

(a) Compute the speed of the blood at the central axis of this artery.

(b) Compute the speed of the blood midway between the artery's wall and central axis.

Distribution cost   31. Suppose that the number of worker-hours required to distribute new telephone books to x percent of the households in a certain rural community is given by the function $f(x) = \dfrac{600x}{300 - x}$.

(a) What is the domain of the function f?

(b) For what values of x does $f(x)$ have a practical interpretation in this context?

(c) How many worker-hours were required to distribute new telephone books to the first 50 percent of the households?

(d) How many worker-hours were required to distribute new telephone books to the entire community?

(e) What percentage of the households in the community had received new telephone books by the time 150 worker-hours had been expended?

Immunization   32. Suppose that during a nationwide program to immunize the population against a certain form of influenza, public health officials found that the cost of inoculating x percent of the population was approximately $f(x) = \dfrac{150x}{200 - x}$ million dollars.

(a) What is the domain of the function f?

(b) For what values of x does $f(x)$ have a practical interpretation in this context?

(c) What was the cost of inoculating the first 50 percent of the population?

(d) What was the cost of inoculating the second 50 percent of the population?

(e) What percentage of the population had been inoculated by the time 37.5 million dollars had been spent?

Speed of a moving object

33. A ball has been dropped from the top of a building. Its height (in feet) after t seconds is given by the function $H(t) = -16t^2 + 256$.
   (a) How high will the ball be after 2 seconds?
   (b) How far will the ball travel during the 3rd second?
   (c) How tall is the building?
   (d) When will the ball hit the ground?

In Problems 34 through 41, find the composite function $g[h(x)]$.

34. $g(u) = u^2 + 4$, $h(x) = x - 1$

35. $g(u) = 3u^2 + 2u - 6$, $h(x) = x + 2$

36. $g(u) = (2u + 10)^2$, $h(x) = x - 5$

37. $g(u) = (u - 1)^3 + 2u^2$, $h(x) = x + 1$

38. $g(u) = \dfrac{1}{u}$, $h(x) = x^2 + x - 2$

39. $g(u) = \dfrac{1}{u^2}$, $h(x) = x - 1$

40. $g(u) = u^2$, $h(x) = \dfrac{1}{x - 1}$

41. $g(u) = \sqrt{u + 1}$, $h(x) = x^2 - 1$

In Problems 42 through 49, find the indicated composite function.

42. $f(x + 1)$ where $f(x) = x^2 + 5$

43. $f(x - 2)$ where $f(x) = 2x^2 - 3x + 1$

44. $f(x + 3)$ where $f(x) = (2x - 6)^2$

45. $f(x - 1)$ where $f(x) = (x + 1)^5 - 3x^2$

46. $f\left(\dfrac{1}{x}\right)$ where $f(x) = 3x + \dfrac{2}{x}$

47. $f(x^2 + 3x - 1)$ where $f(x) = \sqrt{x}$

48. $f(x^2 - 2x + 9)$ where $f(x) = 2x - 20$

49. $f(x + 1)$ where $f(x) = \dfrac{x - 1}{x}$

In Problems 50 through 55, find functions $h(x)$ and $g(u)$ such that $f(x) = g[h(x)]$.

50. $f(x) = (x^5 - 3x^2 + 12)^3$

51. $f(x) = \sqrt{3x - 5}$

52. $f(x) = (x - 1)^2 + 2(x - 1) + 3$

53. $f(x) = \dfrac{1}{x^2 + 1}$

54. $f(x) = \sqrt{x + 4} - \dfrac{1}{(x + 4)^3}$

55. $f(x) = \sqrt{x + 3} - \dfrac{1}{(x + 4)^3}$

Air pollution

56. An environmental study of a certain suburban community suggests that the average daily level of carbon monoxide in the air will be $c(p) = 0.4p + 1$ parts per million when the population is p thousand. It is estimated that t years from now the population of the community will be $p(t) = 8 + 0.2t^2$ thousand.
    (a) Express the level of carbon monoxide in the air as a function of time.
    (b) What will the carbon monoxide level be 2 years from now?
    (c) When will the carbon monoxide level reach 6.2 parts per million?

Manufacturing cost

57. At a certain factory, the total cost of manufacturing q units during the daily production run is $C(q) = q^2 + q + 900$ dollars. On a typical workday, $q(t) = 25t$ units are manufactured during the first t hours of a production run.
    (a) Express the total manufacturing cost as a function of t.
    (b) How much will have been spent on production by the end of the 3rd hour?
    (c) When will the total manufacturing cost reach \$11,000?

Consumer demand

58. An importer of Brazilian coffee estimates that local consumers will buy approximately $Q(p) = \dfrac{4{,}374}{p^2}$ kilograms of the coffee per week when the price is p dollars per kilogram. It is estimated that t weeks from now the price of Brazilian coffee will be $p(t) = 0.04t^2 + 0.2t + 12$ dollars per kilogram.
    (a) Express the weekly consumer demand for the coffee as a function of t.
    (b) How many kilograms of the coffee will consumers be buying from the importer 10 weeks from now?
    (c) When will the demand for the coffee be 30.375 kilograms?

2 Graphs

Graphs have visual impact. They also reveal information that may not be evident from verbal or algebraic descriptions. Two graphs depicting practical relationships are shown in Figure 2.1.

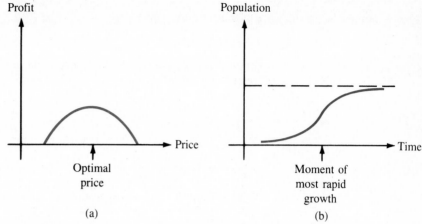

Figure 2.1 (a) A profit function. (b) Bounded population growth.

The graph in Figure 2.1a describes the effect that the market price of a commodity has on the manufacturer's total profit. According to the graph, profit will be small if the market price is either very low or very high. (Can you explain why this might occur?) The fact that the graph has a peak suggests that there is an optimal selling price at which the manufacturer's profit will be greatest.

The graph in Figure 2.1b represents population growth when environmental factors impose an upper bound on the possible size of the population. It indicates that the rate of population growth increases at first and then decreases as the size of the population gets closer and closer to the upper bound.

The Graph of a Function

To represent a function $y = f(x)$ geometrically as a graph, it is traditional to use a rectangular coordinate system on which units for the independent variable x are marked on the horizontal axis and units for the dependent variable y on the vertical axis.

The Graph of a Function

> The graph of a function f consists of all points whose coordinates (x, y) satisfy the equation $y = f(x)$.

In Chapter 3 you will see efficient techniques involving calculus that you can use to draw accurate graphs of functions. For many functions, however, you can make a fairly good sketch by the elementary method of **plotting points.** This method is illustrated in the following examples.

EXAMPLE 2.1   Graph the function $f(x) = x^2$.

SOLUTION

Begin by computing $f(x)$ for several convenient values of x and summarize the results in a table.

x	-2	-1	$-\frac{1}{2}$	0	$\frac{1}{2}$	1	2
$f(x)$	4	1	$\frac{1}{4}$	0	$\frac{1}{4}$	1	4

Then plot the corresponding points $(x, f(x))$ and connect them by a smooth curve. The resulting graph is shown in Figure 2.2.

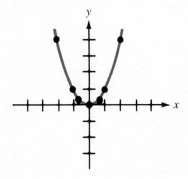

Figure 2.2 The graph of the function $y = x^2$.

The next example illustrates how to sketch the graph of a function defined by more than one formula.

EXAMPLE 2.2   Graph the function

$$f(x) = \begin{cases} 2x & \text{if } 0 \leq x < 1 \\ \dfrac{2}{x} & \text{if } 1 \leq x < 4 \\ 3 & \text{if } x \geq 4 \end{cases}$$

SOLUTION

When making a table of values for this function, remember to use the formula that is appropriate for the particular value of x. Using the formula $f(x) = 2x$ when $0 \leq x < 1$, the formula $f(x) = \dfrac{2}{x}$ when $1 \leq x < 4$, and the formula $f(x) = 3$ when $x \geq 4$, you can compile the following table:

x	0	$\frac{1}{2}$	1	2	3	4	5	6
$f(x)$	0	1	2	1	$\frac{2}{3}$	3	3	3

Now plot the corresponding points $(x, f(x))$ and draw the graph as in Figure 2.3. Notice that the pieces for $0 \leq x < 1$ and $1 \leq x < 4$ are connected to each other at the point $(1, 2)$, but that the piece for $x \geq 4$ is separated from the rest of the graph.

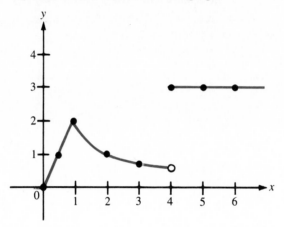

Figure 2.3 The graph for Example 2.2.

The x and y Intercepts

The points (if any) at which a graph crosses the x and y axes are called the **x and y intercepts,** respectively. The y intercept is the point on the graph whose x coordinate is zero, and the x intercepts are the points on the graph whose y coordinates are zero.

How to Find the x and y Intercepts

> To find the y intercept (if any) of $y = f(x)$, set x equal to zero and compute y.
>
> To find the x intercepts (if any) of $y = f(x)$, set $f(x)$ equal to zero and solve for x.

Here is an example.

EXAMPLE 2.3   Graph the function $f(x) = x^3 - x^2 - 6x$. Include all the x and y intercepts.

SOLUTION

The y intercept is $f(0) = 0$.

To find the x intercepts, set $f(x)$ equal to zero and solve for x by factoring the equation as follows:

$$0 = x^3 - x^2 - 6x$$
$$0 = x(x^2 - x - 6)$$
$$0 = x(x - 3)(x + 2)$$
$$x = 0 \qquad x = 3 \qquad \text{or} \qquad x = -2$$

It follows that the x intercepts are $(0, 0)$, $(3, 0)$, and $(-2, 0)$.

Now make a table of values (including the intercepts) and plot the corresponding points $(x, f(x))$.

x	-3	-2	-1	0	1	2	3	4
$f(x)$	-18	0	4	0	-6	-8	0	24

The graph is shown in Figure 2.4.

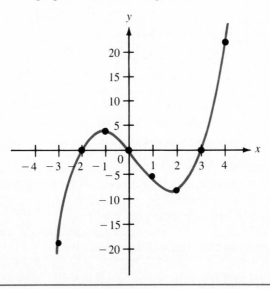

Figure 2.4 The graph of the function $y = x^3 - x^2 - 6x$.

Discontinuities

A function whose graph is unbroken is said to be **continuous.** For example, the functions sketched in Figures 2.2 and 2.4 are continuous. On the other hand, the function sketched in Figure 2.3 is not continuous because its graph breaks into two separate pieces when $x = 4$.

A gap or break in the graph of a function is called a **discontinuity.** For the functions you will encounter in this text, discontinuities may arise in one of the following two ways.

Two Common Types of Discontinuities

> A function defined in several pieces will have discontinuities if the graphs of the individual pieces are not connected to each other.
>
> A function defined as a quotient will have a discontinuity whenever its denominator is zero.

In Figure 2.3 accompanying Example 2.2, you saw a discontinuity of the first type. The piece of the graph for the interval $x \geq 4$ was not connected to the rest of the graph, creating a discontinuity when $x = 4$.

The second type of discontinuity is illustrated in the next example. The function is defined as a quotient, and since division by zero is impossible, a discontinuity occurs when the denominator is zero.

EXAMPLE 2.4   Graph the function $f(x) = \dfrac{x^2 + x - 2}{x - 2}$.

SOLUTION

Since the denominator is zero when $x = 2$, the function is undefined and will have a discontinuity for this value of x. To remind yourself that there is no point on the graph whose x coordinate is 2, you might begin your sketch by drawing a broken vertical line at $x = 2$. (See Figure 2.5.) Your graph should not cross this line.

The y intercept is $f(0) = \dfrac{-2}{-2} = 1$. To find the x intercepts, put the function in factored form.

$$f(x) = \frac{(x + 2)(x - 1)}{x - 2}$$

From the factored numerator you see that $f(x)$ is zero when $x = -2$ and $x = 1$. Hence the x intercepts are $(-2, 0)$ and $(1, 0)$.

Now make a table of values. To find out what the graph looks like near the discontinuity at $x = 2$, you should include some values of x that are close to 2.

x	-50	-4	-2	-1	0	1	$\frac{3}{2}$	2	$\frac{5}{2}$	3	4	5	6	50
$f(x)$	-47.1	-1.7	0	0.7	1	0	-3.5		13.5	10	9	9.3	10	53.1

Now plot the corresponding points $(x, f(x))$ and draw the graph as shown in Figure 2.5. Don't forget that there should be a break in the graph when $x = 2$.

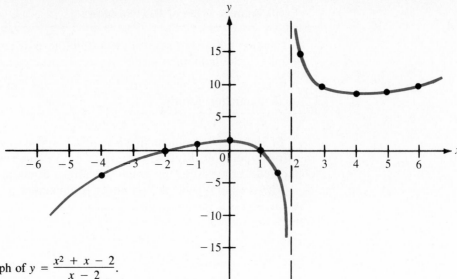

Figure 2.5 The graph of $y = \dfrac{x^2 + x - 2}{x - 2}$.

Polynomials

A **polynomial** is a function of the form

$$f(x) = a_0 + a_1x + a_2x^2 + \cdots + a_nx^n$$

where n is a nonnegative integer and $a_0, a_1, \ldots, a_n$ are constants. If $a_n \neq 0$, the integer n is said to be the **degree** of the polynomial. For example, the function $f(x) = 3x^5 - 6x^2 + 7$ is a polynomial of degree 5. Polynomials are continuous functions. It can be shown that the graph of a polynomial of degree n is an unbroken curve that crosses the x axis no more than n times. To illustrate some of the possibilities, the graphs of three polynomials of degree 3 are shown in Figure 2.6.

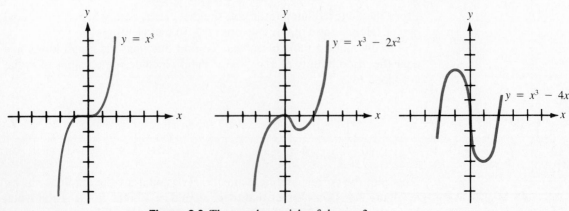

Figure 2.6 Three polynomials of degree 3.

Many profit functions in economics are polynomials. Here is an example.

EXAMPLE 2.5   A manufacturer can produce radios at a cost of $10 apiece and estimates that if they are sold for x dollars apiece, consumers will buy approximately $80 - x$ radios each month. Express the manufacturer's monthly profit as a function of the price x, graph this function, and estimate the price at which the manufacturer's profit will be greatest.

SOLUTION
Begin by stating the desired relationship in words.

Profit = (number of radios sold)(profit per radio)

Now replace the words by algebraic expressions. You know that

Number of radios sold = $80 - x$

Moreover, since the radios are produced at a cost of $10 apiece and sold for x dollars apiece, it follows that

Profit per radio = $x - 10$

Let $P(x)$ denote the profit and conclude that

$$P(x) = (80 - x)(x - 10)$$

(Notice that the profit function is factored. Resist the temptation to multiply it out. It is already in its most convenient form.)
To sketch this function, make a table of representative values and plot the corresponding points $(x, P(x))$ as shown in Figure 2.7.

x	10	20	30	40	50	60	70	80
$P(x)$	0	600	1,000	1,200	1,200	1,000	600	0

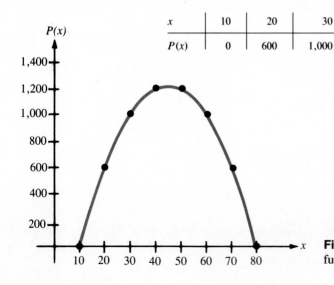

Figure 2.7 The profit function $P(x) = (80 - x)(x - 10)$.

Notice that the x intercepts of the profit function are (10, 0) and (80, 0). Can you explain these x intercepts in economic terms?

The graph suggests that the price at which the manufacturer's profit will be greatest is approximately \$45. In Chapter 3 you will learn how to use calculus to find this optimal price exactly.

Rational Functions

The quotient of two polynomials is called a **rational function.** For example, the function $f(x) = \dfrac{x^2 + x - 2}{x - 2}$ in Example 2.4 is rational. So is the function $f(x) = 1 + \dfrac{1}{x}$ since it can be rewritten as $f(x) = \dfrac{x + 1}{x}$. Since division by zero is impossible, a rational function has a discontinuity whenever its denominator is zero. Many cost functions in economics are rational. Here is an example.

EXAMPLE 2.6   For each shipment of raw materials, a manufacturer must pay an ordering fee to cover handling and transportation. After they are received, the raw materials must be stored until needed and storage costs result. If each shipment of raw materials is large, ordering costs will be low because few shipments are required, but storage costs will be high. If each shipment is small, ordering costs will be high because many shipments will be required, but storage costs will drop. A manufacturer estimates that if each shipment contains x units, the total cost of obtaining and storing the year's supply of raw materials will be $C(x) = x + \dfrac{160{,}000}{x}$ dollars. Sketch the relevant portion of this cost function and estimate the optimal shipment size.

SOLUTION

$C(x)$ is a rational function with a discontinuity when $x = 0$ and represents cost for nonnegative values of x. Compile a table for some representative nonnegative values of x and plot the corresponding points to get the graph shown in Figure 2.8.

x	100	200	300	400	500	600	700	800
$C(x)$	1,700	1,000	833	800	820	867	929	1,000

The graph indicates that total cost will be high if shipments are very small or very large and that the optimal shipment size is approximately 400 units.

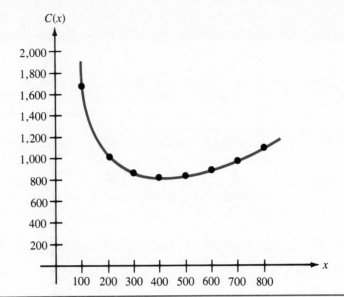

Figure 2.8 The cost function $C(x) = x + \dfrac{160,000}{x}$.

Problems

In Problems 1 through 24, sketch the graph of the given function. Include all x and y intercepts. (You may use a calculator to help with the computations.)

1. $f(x) = x$

2. $f(x) = x^2$

3. $f(x) = x^3$

4. $f(x) = x^4$

5. $f(x) = \dfrac{1}{x}$

6. $f(x) = \dfrac{1}{x^2}$

7. $f(x) = \dfrac{1}{x^3}$

8. $f(x) = \sqrt{x}$

9. $f(x) = 2x - 1$

10. $f(x) = 2 - 3x$

11. $f(x) = -x^2$

12. $f(x) = -\dfrac{1}{x^2}$

13. $f(x) = \begin{cases} x - 1 & \text{if } x \le 0 \\ x + 1 & \text{if } x > 0 \end{cases}$

14. $f(x) = \begin{cases} x^2 - 1 & \text{if } x \le 2 \\ 3 & \text{if } x > 2 \end{cases}$

15. $f(x) = (x - 1)(x + 2)$

16. $f(x) = (x + 2)(x + 1)$

17. $f(x) = x^2 - x - 6$

18. $f(x) = x^3 - 4x$

19. $f(x) = \dfrac{1}{x - 2}$

20. $f(x) = \dfrac{1}{(x + 2)^2}$

21. $f(x) = \dfrac{x^2 - 2x}{x + 1}$

22. $f(x) = \dfrac{x^2 - 2x - 3}{x - 1}$

23. $f(x) = x + \dfrac{1}{x}$

24. $f(x) = x - \dfrac{1}{x}$

Manufacturing cost

25. A manufacturer can produce cassette tape recorders at a cost of $20 apiece. It is estimated that if the tape recorders are sold for x dollars apiece, consumers will buy $120 - x$ of them a month. Express the manufacturer's monthly profit as a function of price, graph this function, and use the graph to estimate the optimal selling price.

Retail sales

26. A bookstore can obtain an atlas from the publisher at a cost of $5 per copy and estimates that if it sells the atlas for x dollars per copy, approximately $20(22 - x)$ copies will be sold each month. Express the bookstore's monthly profit from the sale of the atlas as a function of price, graph this function, and use the graph to estimate the optimal selling price.

Consumer expenditure

27. The consumer demand for a certain commodity is $D(p) = -200p + 12,000$ units per month when the market price is p dollars per unit.
(a) Graph this demand function.
(b) Express consumers' total monthly expenditure for the commodity as a function of p. (The total monthly expenditure is the total amount of money consumers spend each month on the commodity.)
(c) Graph the total monthly expenditure function.
(d) Discuss the economic significance of the p intercepts of the expenditure function.
(e) Use the graph in part (c) to estimate the market price that generates the greatest consumer expenditure.

Speed of a moving object

28. If an object is thrown vertically upward from the ground with an initial speed of 160 feet per second, its height (in feet) t seconds later is given by the function $H(t) = -16t^2 + 160t$.
(a) Graph the function $H(t)$.
(b) Use the graph in part (a) to determine when the object will hit the ground.
(c) Use the graph in part (a) to estimate how high the object will rise.

Distribution cost

29. Suppose that the number of worker-hours required to distribute new telephone books to x percent of the households in a certain rural community is given by the function $f(x) = \dfrac{600x}{300 - x}$. Sketch this function and specify what portion of the graph is relevant to the practical situation under consideration.

Immunization

30. Suppose that during a nationwide program to immunize the population against a certain form of influenza, public health officials found

that the cost of inoculating x percent of the population was approximately $f(x) = \dfrac{150x}{200 - x}$ million dollars. Sketch this function and specify what portion of the graph is relevant to the practical situation under consideration.

Experimental psychology

31. To study the rate at which animals learn, a psychology student performed an experiment in which a rat was sent repeatedly through a laboratory maze. Suppose that the time required for the rat to traverse the maze on the nth trial was approximately $f(n) = 3 + \dfrac{12}{n}$ minutes.

   (a) Graph the function $f(n)$.

   (b) What portion of the graph is relevant to the practical situation under consideration?

   (c) What happens to the graph as n increases without bound? Interpret your answer in practical terms.

Inventory cost

32. A manufacturer estimates that if each shipment of raw materials contains x units, the total cost of obtaining and storing the year's supply of raw materials will be $C(x) = 2x + \dfrac{80,000}{x}$ dollars. Sketch the relevant portion of the graph of this cost function and estimate the optimal shipment size.

Production cost

33. A manufacturer estimates that if x machines are used, the cost of a production run will be $C(x) = 20x + \dfrac{2,000}{x}$ dollars. Sketch the relevant portion of this cost function and estimate how many machines the manufacturer should use to minimize cost.

Average cost

34. Suppose the total cost of manufacturing x units of a certain commodity is $C(x) = x^2 + 6x + 19$ dollars. Express the average cost per unit as a function of the number of units produced and, on the same set of axes, sketch the total cost and average cost functions. (*Hint:* Average cost is total cost divided by the number of units produced.)

Average cost

35. Suppose the total cost (in dollars) of manufacturing x units is given by the function $C(x) = x^2 + 4x + 16$. Express the average cost per unit as a function of the number of units produced and, on the same set of axes, sketch the total cost and average cost functions.

36. (a) Graph the functions $y = x^2$ and $y = x^2 + 3$. How are the graphs related?

   (b) Without further computation, graph the function $y = x^2 - 5$.

   (c) Suppose $g(x) = f(x) + c$, where c is a constant. How are the graphs of f and g related? Explain.

37. (a) Graph the functions $y = x^2$ and $y = -x^2$. How are the graphs related?

(b) Suppose $g(x) = -f(x)$. How are the graphs of f and g related? Explain.

38. Graph the function $y = \dfrac{1}{x}$. Then, without further computation, graph the function $y = 3 - \dfrac{1}{x}$. (*Hint:* Combine the rules you discovered in Problems 36c and 37b.)

39. (a) Graph the functions $y = x^2$ and $y = (x - 2)^2$. How are the graphs related?
    (b) Without further computation, graph the function $y = (x + 1)^2$.
    (c) Suppose $g(x) = f(x - c)$, where c is a constant. How are the graphs of f and g related? Explain.

40. Graph the function $y = \dfrac{1}{x}$. Then, without further computation, graph the function $y = 3 - \dfrac{1}{x + 2}$.

Distance formula 41. Show that the distance d between the two points (x_1, y_1) and (x_2, y_2) is given by the formula

$$d = \sqrt{(x_2 - x_1)^2 + (y_2 - y_1)^2}$$

(*Hint:* Apply the pythagorean theorem to the right triangle whose hypotenuse is the line segment joining the two points.)

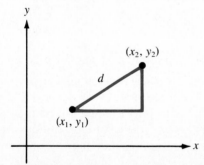

42. Compute the distance between the two given points using the formula in Problem 41.
    (a) $(1, 0)$ and $(0, 1)$
    (b) $(5, -1)$ and $(2, 3)$
    (c) $(2, 6)$ and $(2, -1)$

3 Linear Functions

In many practical situations, the rate at which one quantity changes with respect to another is constant. Here is a simple example from economics.

EXAMPLE 3.1   A manufacturer's total cost consists of a fixed overhead of $200 plus production costs of $50 per unit. Express the total cost as a function of the number of units produced and draw the graph.

SOLUTION
Let x denote the number of units produced and $C(x)$ the corresponding total cost. Then,

Total cost = (cost per unit)(number of units) + overhead

where

Cost per unit = 50
Number of units = x
Overhead = 200

Hence,

$C(x) = 50x + 200$

The graph of this cost function is sketched in Figure 3.1.

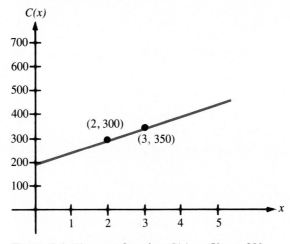

Figure 3.1 The cost function $C(x) = 50x + 200$.

The total cost in Example 3.1 increases at a constant rate of $50 per unit. As a result, its graph in Figure 3.1 is a straight line, increasing in height by 50 units for each 1-unit increase in x.

In general, a function whose value changes at a constant rate with respect to its independent variable is said to be a **linear function.** This is because the graph of such a function is a straight line. In algebraic terms, a linear function is a function of the form

$f(x) = a_0 + a_1 x$

where a_0 and a_1 are constants. For example, the functions $f(x) = \frac{3}{2} + 2x$, $f(x) = -5x$, and $f(x) = 12$ are all linear. Linear functions are traditionally written in the form

$$y = mx + b$$

where m and b are constants. This standard notation will be used in the discussion that follows.

Linear Functions

A linear function is a function that changes at a constant rate with respect to its independent variable.

The graph of a linear function is a straight line.

The equation of a linear function can be written in the form

$$y = mx + b$$

where m and b are constants.

To work with linear functions, you will need to know the following things about straight lines.

The Slope of a Line

The **slope** of a line is the amount by which the y coordinate of a point on the line changes when the x coordinate is increased by 1. You can compute the slope of a nonvertical line if you know any two of its points. Suppose (x_1, y_1) and (x_2, y_2) lie on a line as indicated in Figure 3.2.

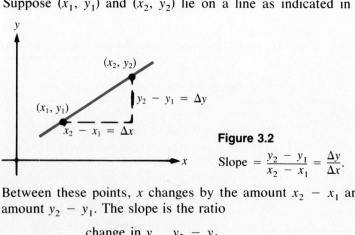

Figure 3.2

$$\text{Slope} = \frac{y_2 - y_1}{x_2 - x_1} = \frac{\Delta y}{\Delta x}.$$

Between these points, x changes by the amount $x_2 - x_1$ and y by the amount $y_2 - y_1$. The slope is the ratio

$$\text{Slope} = \frac{\text{change in } y}{\text{change in } x} = \frac{y_2 - y_1}{x_2 - x_1}$$

It is sometimes convenient to use the symbol $\triangle y$ instead of $y_2 - y_1$ to denote the change in y. The symbol $\triangle y$ is read "delta y." Similarly, the symbol $\triangle x$ is used to denote the change $x_2 - x_1$.

The Slope of a Line

> The slope of the nonvertical line passing through the points (x_1, y_1) and (x_2, y_2) is given by the formula
>
> $$\text{Slope} = \frac{\triangle y}{\triangle x} = \frac{y_2 - y_1}{x_2 - x_1}$$

The use of this formula is illustrated in the following example.

EXAMPLE 3.2   Find the slope of the line joining the points $(-2, 5)$ and $(3, -1)$.

SOLUTION

$$\text{Slope} = \frac{\triangle y}{\triangle x} = \frac{-1 - 5}{3 - (-2)} = -\frac{6}{5}$$

The situation is illustrated in Figure 3.3.

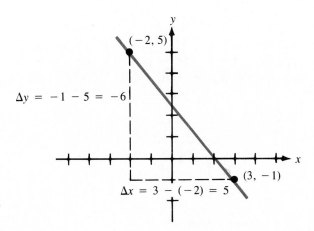

Figure 3.3 The line joining $(-2, 5)$ and $(3, -1)$.

The sign and magnitude of the slope of a line indicate the line's direction and steepness, respectively. The slope is positive if the height of the line increases as x increases and is negative if the height decreases as x increases. The absolute value of the slope is large if the slant of the line is severe and small if the slant of the line is gradual. The situation is illustrated in Figure 3.4.

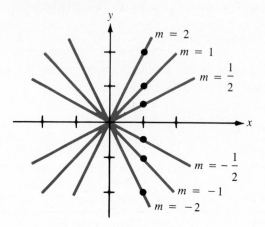

Figure 3.4 The direction and steepness of a line.

Horizontal and Vertical Lines

Horizontal and vertical lines (Figure 3.5a and 3.5b) have particularly simple equations. The y coordinates of all the points on a horizontal line are the same. Hence, a horizontal line is the graph of a linear function of the form $y = b$, where b is a constant. The slope of a horizontal line is zero, since changes in x produce no changes in y.

The x coordinates of all the points on a vertical line are equal. Hence, vertical lines are characterized by equations of the form $x = c$, where c is a constant. The slope of a vertical line is undefined. This is because only the y coordinates of points on a vertical line can change, and so the denominator of the quotient $\dfrac{\text{change in } y}{\text{change in } x}$ is zero.

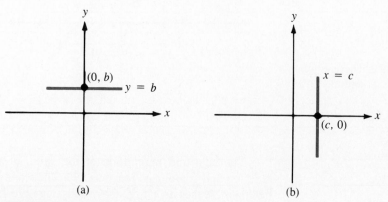

(a)

(b)

Figure 3.5 Horizontal and vertical lines.

The Slope-Intercept Form of the Equation of a Line

The constants m and b in the equation $y = mx + b$ of a nonvertical line have geometric interpretations. The coefficient m is the slope of the line. To see this, suppose that (x_1, y_1) and (x_2, y_2) are two points on the line $y = mx + b$. Then, $y_1 = mx_1 + b$ and $y_2 = mx_2 + b$ and so

$$\text{Slope} = \frac{y_2 - y_1}{x_2 - x_1} = \frac{(mx_2 + b) - (mx_1 + b)}{x_2 - x_1}$$

$$= \frac{mx_2 - mx_1}{x_2 - x_1} = \frac{m(x_2 - x_1)}{x_2 - x_1} = m$$

The constant b in the equation $y = mx + b$ is the value of y corresponding to $x = 0$. Hence, b is the height at which the line $y = mx + b$ crosses the y axis, and the corresponding point $(0, b)$ is the y intercept of the line. The situation is illustrated in Figure 3.6.

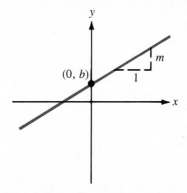

Figure 3.6 The slope and y intercept of the line $y = mx + b$.

Because the constants m and b in the equation $y = mx + b$ correspond to the slope and y intercept, respectively, this form of the equation of a line is known as the **slope-intercept form.**

The Slope-Intercept Form of the Equation of a Line

The equation

$$y = mx + b$$

is the equation of the line whose slope is m and whose y intercept is $(0, b)$.

The slope-intercept form of the equation of a line is particularly useful when geometric information about a line (such as its slope or y intercept) is to be determined from the line's algebraic representation. Here is a typical example.

EXAMPLE 3.3   Find the slope and y intercept of the line $3y + 2x = 6$ and draw the graph.

SOLUTION
The first step is to put the equation $3y + 2x = 6$ in slope-intercept form $y = mx + b$. To do this, solve for y to get

$$3y = -2x + 6 \quad \text{or} \quad y = -\frac{2}{3}x + 2$$

It follows that the slope is $-\frac{2}{3}$ and the y intercept is $(0, 2)$.

To graph a linear function, plot two of its points and draw a straight line through them. In this case, you already know one point, the y intercept $(0, 2)$. A convenient choice for the x coordinate of the second point is $x = 3$, since the corresponding y coordinate is $y = -\frac{2}{3}(3) + 2 = 0$. Draw a line through the points $(0, 2)$ and $(3, 0)$ to obtain the graph shown in Figure 3.7.

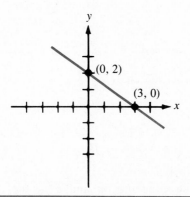

Figure 3.7 The line $3y + 2x = 6$.

The Point-Slope Form of the Equation of a Line

Geometric information about a line can be obtained readily from the slope-intercept formula $y = mx + b$. There is another form of the equation of a line, however, that is usually more efficient for problems in which a line's geometric properties are known and the goal is to find the equation of the line.

The Point-Slope Form of the Equation of a Line

The equation

$$y - y_0 = m(x - x_0)$$

is an equation of the line that passes through the point (x_0, y_0) and that has slope equal to m.

The point-slope form of the equation of a line is simply the formula for slope in disguise. To see this, suppose that the point (x, y) lies on the line that passes through a given point (x_0, y_0) and that has slope m. Using the points (x, y) and (x_0, y_0) to compute the slope, you get

$$\frac{y - y_0}{x - x_0} = m$$

which you can put in point-slope form

$$y - y_0 = m(x - x_0)$$

by simply multiplying both sides by $x - x_0$.

The use of the point-slope form of the equation of a line is illustrated in the next two examples.

EXAMPLE 3.4   Find an equation of the line that passes through the point $(5, 1)$ and whose slope is equal to $\frac{1}{2}$.

SOLUTION

Use the formula $y - y_0 = m(x - x_0)$ with $(x_0, y_0) = (5, 1)$ and $m = \frac{1}{2}$ to get

$$y - 1 = \frac{1}{2}(x - 5)$$

which you can rewrite as

$$y = \frac{1}{2}x - \frac{3}{2}$$

The graph is shown in Figure 3.8.

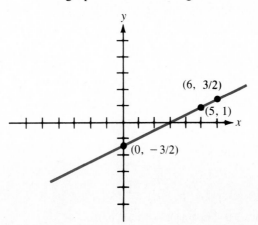

Figure 3.8 The line $y = \frac{1}{2}x - \frac{3}{2}$.

Instead of the point-slope form, the slope-intercept form could have been used to solve the problem in Example 3.4. For practice, solve the problem this way. Notice that the solution based on the point-slope formula is more efficient.

The next example illustrates how you can use the point-slope form to find an equation of a line that passes through two given points.

EXAMPLE 3.5 Find an equation of the line that passes through the points $(3, -2)$ and $(1, 6)$.

SOLUTION
First compute the slope

$$m = \frac{6 - (-2)}{1 - 3} = \frac{8}{-2} = -4$$

Then use the point-slope formula with $(1, 6)$ as the given point (x_0, y_0) to get

$$y - 6 = -4(x - 1) \quad \text{or} \quad y = -4x + 10$$

Convince yourself that the resulting equation would have been the same if you had chosen $(3, -2)$ to be the given point (x_0, y_0).

The graph is shown in Figure 3.9.

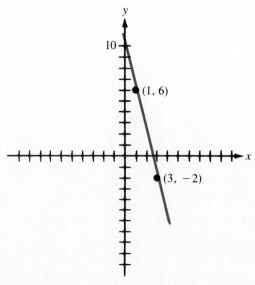

Figure 3.9 The line $y = -4x + 10$.

Practical Applications

If the rate of change of one quantity with respect to a second quantity is constant, the function relating the quantities must be linear. The constant rate of change is the slope of the corresponding line. The next two examples illustrate techniques you can use to find the appropriate linear functions in such situations.

EXAMPLE 3.6   Since the beginning of the year, the price of a loaf of whole-wheat bread at a local supermarket has been rising at a constant rate of 2 cents per month. By November first, the price had reached 64 cents per loaf. Express the price of the bread as a function of time and determine the price at the beginning of the year.

SOLUTION

Let x denote the number of months that have elapsed since the first of the year and y the price of a loaf of bread. Since y changes at a constant rate with respect to x, the function relating y to x must be linear and its graph a straight line. Since the price y increases by 2 each time x increases by 1, the slope of the line must be 2. The fact that the price was 64 cents on November first (10 months after the first of the year) implies that the line passes through the point (10, 64). To write an equation defining y as a function of x, use the point-slope formula $y - y_0 = m(x - x_0)$ with

$$m = 2 \quad \text{and} \quad (x_0, y_0) = (10, 64)$$

to get

$$y - 64 = 2(x - 10) \quad \text{or} \quad y = 2x + 44$$

The corresponding line is shown in Figure 3.10. Notice that the y intercept is (0, 44), which implies that the price of the bread at the beginning of the year was 44 cents per loaf.

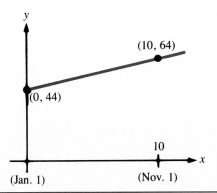

Figure 3.10 The rising price of bread: $y = 2x + 44$.

EXAMPLE 3.7 The average SAT scores of incoming students at an eastern liberal arts college have been declining at a constant rate in recent years. In 1980, the average SAT score was 582 while in 1985 it was only 552.

(a) Express the average SAT score as a function of time.
(b) If the trend continues, what will the average SAT score of incoming students be in 1990?
(c) If the trend continues, when will the average SAT score be 534?

SOLUTION
(a) Let x denote the number of years since 1980 and y the average SAT score of incoming students. Since y changes at a constant rate with respect to x, the function relating y to x must be linear. Since $y = 582$ when $x = 0$ and $y = 552$ when $x = 5$, the corresponding straight line must pass through the points (0, 582) and (5, 552). The slope of this line is

$$m = \frac{582 - 552}{0 - 5} = -6$$

Since one of the given points happens to be the y intercept (0, 582), you can use the slope-intercept form to conclude immediately that

$$y = -6x + 582$$

The corresponding line is shown in Figure 3.11.

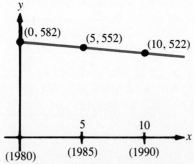

Figure 3.11 Declining SAT scores: $y = -6x + 582$.

(b) To predict the average SAT score in 1990, compute y when $x = 10$ to get

$$y = -6(10) + 582 = 522$$

(c) Set y equal to 534 and solve for x to get

$$534 = -6x + 582 \qquad 6x = 48 \qquad \text{or} \qquad x = 8$$

Thus the SAT score will be 534 in 1988, 8 years after 1980.

Problems

In Problems 1 through 5, find the slope (if possible) of the line that passes through the given pair of points.

1. $(2, -3)$ and $(0, 4)$
2. $(-1, 2)$ and $(2, 5)$
3. $(2, 0)$ and $(0, 2)$
4. $(5, -1)$ and $(-2, -1)$
5. $(2, 6)$ and $(2, -4)$

In Problems 6 through 16, find the slope and y intercept (if they exist) of the given line and draw a graph.

6. $y = 3x$
7. $y = 5x + 2$
8. $y = 3x - 6$
9. $x + y = 2$
10. $3x + 2y = 6$
11. $2x - 4y = 12$
12. $5y - 3x = 4$
13. $4x = 2y + 6$
14. $\dfrac{x}{2} + \dfrac{y}{5} = 1$
15. $y = 2$
16. $x = -3$

In Problems 17 through 27, write an equation for the line with the given properties.

17. Through $(2, 0)$ with slope 1
18. Through $(-1, 2)$ with slope $\dfrac{2}{3}$
19. Through $(5, -2)$ with slope $-\dfrac{1}{2}$
20. Through $(0, 0)$ with slope 5
21. Through $(2, 5)$ and parallel to the x axis
22. Through $(2, 5)$ and parallel to the y axis
23. Through $(1, 0)$ and $(0, 1)$
24. Through $(2, 5)$ and $(1, -2)$
25. Through $(-2, 3)$ and $(0, 5)$
26. Through $(1, 5)$ and $(3, 5)$
27. Through $(1, 5)$ and $(1, -4)$

Manufacturing cost

28. A manufacturer's total cost consists of a fixed overhead of $5,000 plus production costs of $60 per unit. Express the total cost as a function of the number of units produced and draw the graph.

Car rental   29. A certain car rental agency charges $20 per day plus 14 cents per mile.
(a) Express the cost of renting a car from this agency for 1 day as a function of the number of miles driven and draw the graph.
(b) How much does it cost to rent a car for a 1-day trip of 50 miles?
(c) How many miles were driven if the daily rental cost was $45.20?

Course registration   30. Students at a state college may preregister for their fall classes by mail during the summer. Those who do not preregister must register in person in September. The registrar can process 35 students per hour during the September registration period. Suppose that after 4 hours in September, a total of 360 students (including those who preregistered) have been registered.
(a) Express the number of students registered as a function of time and draw the graph.
(b) How many students were registered after 3 hours?
(c) How many students preregistered during the summer?

Membership fees   31. Membership in a swimming club costs $150 for the 12-week summer season. If a member joins after the start of the season, the fee is prorated; that is, it is reduced linearly.
(a) Express the membership fee as a function of the number of weeks that have elapsed by the time the membership is purchased and draw the graph.
(b) Compute the cost of a membership that is purchased 5 weeks after the start of the season.

Linear depreciation   32. A doctor owns $1,500 worth of medical books which, for tax purposes, are assumed to depreciate linearly to zero over a 10-year period. That is, the value of the books decreases at a constant rate so that it is equal to zero at the end of 10 years. Express the value of the books as a function of time and draw the graph.

Linear depreciation   33. A manufacturer buys $20,000 worth of machinery that depreciates linearly so that its trade-in value after 10 years will be $1,000.
(a) Express the value of the machinery as a function of its age and draw the graph.
(b) Compute the value of the machinery after 4 years.

Water consumption   34. Since the beginning of the month, a local reservoir has been losing water at a constant rate. On the 12th of the month, the reservoir held 200 million gallons of water and on the 21st, it held only 164 million gallons.
(a) Express the amount of water in the reservoir as a function of time and draw the graph.
(b) How much water was in the reservoir on the 8th of the month?

Car pooling   35. To encourage motorists to form car pools, the transit authority in a certain metropolitan area has been offering a special reduced rate at toll bridges for vehicles containing four or more persons. When the program began 30 days ago, 157 vehicles qualified for the reduced rate during the morning rush hour. Since then, the number of vehicles qualifying has been increasing at a constant rate and today, 247 vehicles qualified.

(a) Express the number of vehicles qualifying each morning for the reduced rate as a function of time and draw the graph.

(b) If the trend continues, how many vehicles will qualify during the morning rush hour 14 days from now?

Metric conversion   36. (a) Temperature measured in degrees Fahrenheit is a linear function of temperature measured in degrees Celsius. Use the facts that $0°$ Celsius is equal to $32°$ Fahrenheit and $100°$ Celsius is equal to $212°$ Fahrenheit to write an equation for this linear function.

(b) Use the function you obtained in part (a) to convert $15°$ Celsius to Fahrenheit.

(c) Convert $68°$ Fahrenheit to Celsius.

Appreciation of assets   37. The value of a certain rare book doubles every 10 years. The book was originally worth \$3.

(a) How much is the book worth when it is 30 years old? When it is 40 years old?

(b) Is the relationship between the value of the book and its age linear? Explain.

Parallel lines   38. What is the relationship between the slopes of parallel lines? Explain your answer. Using this relationship, write equations for the lines with the following properties:

(a) Through $(1, 3)$ and parallel to the line $4x + 2y = 7$.

(b) Through $(0, 2)$ and parallel to the line $2y - 3x = 5$.

(c) Through $(-2, 5)$ and parallel to the line through the points $(1, 2)$ and $(6, -1)$.

Perpendicular lines   39. Show that if a line L_1 with slope m_1 is perpendicular to a line L_2 with slope m_2, then $m_1 = -\dfrac{1}{m_2}$. (*Hint:* Find expressions for the slopes of the perpendicular lines L_1 and L_2 in the accompanying figure. Then apply the pythagorean theorem, together with the distance formula from Section 2, Problem 41, to the right triangle OAB to obtain the desired relationship between the slopes.)

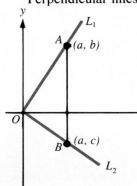

40. Using the result from Problem 39, write equations for the lines with the following properties:

(a) Through $(-1, 3)$ and perpendicular to the line $4x + 2y = 7$.

(b) Through $(0, 0)$ and perpendicular to the line $2y - 3x = 5$.

(c) Through $(2, 1)$ and perpendicular to the line joining $(0, 3)$ and $(2, -1)$.

4 Intersections of Graphs

Sometimes it is necessary to determine when two functions are equal. This is the case, for example, when an economist wants to compute the market price at which the consumer demand for a commodity will be equal to its supply. It occurs when a manufacturer seeks to determine how many units must be sold before revenue exceeds cost. And it occurs when a political analyst attempts to predict how long it will take for the popularity of a certain challenger to reach that of the incumbent.

In geometric terms, the values of x for which two functions $f(x)$ and $g(x)$ are equal are the x coordinates of the points at which their graphs intersect. The situation is illustrated in Figure 4.1.

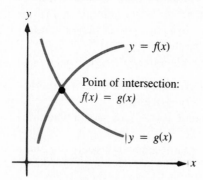

Figure 4.1 The intersection of two graphs.

To find the points of intersection algebraically, you set $f(x)$ equal to $g(x)$ and solve for x. Here are three examples illustrating some of the algebraic techniques.

EXAMPLE 4.1   Where do the lines $y = 2x + 1$ and $y = -x + 4$ intersect?

SOLUTION
Solve the equation

$$2x + 1 = -x + 4$$

to get

$$3x = 3 \quad \text{or} \quad x = 1$$

To find the corresponding value of y, substitute $x = 1$ into either of the original equations $y = 2x + 1$ or $y = -x + 4$. You will get $y = 3$ from

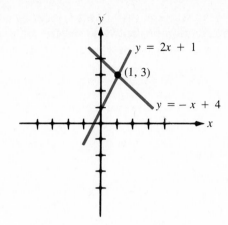

Figure 4.2 The lines $y = 2x + 1$ and $y = -x + 4$.

which you can conclude that $(1, 3)$ is the point of intersection. The graphs are shown in Figure 4.2.

EXAMPLE 4.2   Find the points of intersection of the graphs of the functions $f(x) = 2x$ and $g(x) = x^2$.

SOLUTION
Set $f(x)$ equal to $g(x)$ and solve to get

$$2x = x^2$$
$$x^2 - 2x = 0$$
$$x(x - 2) = 0$$
$$x = 0 \quad \text{or} \quad x = 2$$

Now substitute these values of x into either equation $y = f(x)$ or $y = g(x)$ and you will find that $y = 0$ when $x = 0$ and $y = 4$ when $x = 2$. It follows that the points of intersection are $(0, 0)$ and $(2, 4)$, as shown in Figure 4.3.

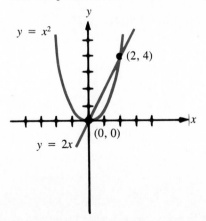

Figure 4.3 The line $y = 2x$ and the curve $y = x^2$.

In the next example, the **quadratic formula** will be needed to find the points of intersection of the given graphs. (A review of the use of this formula is in Section A of the appendix at the back of the book.)

EXAMPLE 4.3 Find the points of intersection of the line $y = 3x + 2$ and the curve $y = x^2$.

SOLUTION
Rewrite the equation

$$x^2 = 3x + 2$$

as

$$x^2 - 3x - 2 = 0$$

Since the expression $x^2 - 3x - 2$ has no obvious factors, use the quadratic formula to get

$$x = \frac{-(-3) \pm \sqrt{(-3)^2 - 4(1)(-2)}}{2(1)} = \frac{3 \pm \sqrt{17}}{2}$$

It follows that the x coordinates of the points of intersection are

$$x = \frac{3 + \sqrt{17}}{2} \simeq 3.56 \quad \text{and} \quad x = \frac{3 - \sqrt{17}}{2} \simeq -0.56$$

(These computations were done with a calculator, and the answers were rounded off to two decimal places.)

Computing the corresponding y coordinates from the equation $y = x^2$ you find that the points of intersection are approximately (3.56, 12.67) and (−0.56, 0.31). (Due to round-off approximations, you will get slightly different values for the y coordinates if you use the equation $y = 3x + 2$.)

The graphs are shown in Figure 4.4.

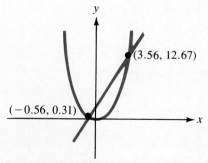

Figure 4.4 The line $y = 3x + 2$ and the curve $y = x^2$.

Break-Even Analysis

Intersections of graphs arise in business in the context of **break-even analysis.** In a typical situation, a manufacturer wishes to determine how many units of a certain commodity have to be sold for total revenue to equal total cost. Suppose that x denotes the number of units manufactured and sold, and let $C(x)$ and $R(x)$ be the corresponding total cost and total revenue, respectively. A pair of (linear) cost and revenue curves is sketched in Figure 4.5.

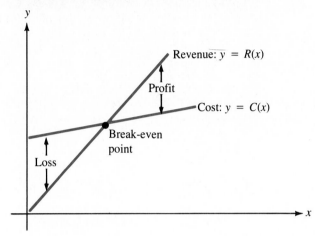

Figure 4.5 Linear cost and revenue curves.

Because of fixed overhead costs, the total cost curve is initially higher than the total revenue curve. Hence, at low levels of production, the manufacturer suffers a loss. At higher levels of production, however, the total revenue curve is the higher one and the manufacturer realizes a profit. The point at which the two curves cross is called the **break-even point,** because when total revenue equals total cost, the manufacturer breaks even, experiencing neither a profit nor a loss. Here is an example.

EXAMPLE 4.4   A manufacturer can sell a certain product for $110 per unit. Total cost consists of a fixed overhead of $7,500 plus production costs of $60 per unit.

(a) How many units must the manufacturer sell to break even?
(b) What is the manufacturer's profit or loss if 100 units are sold?
(c) How many units must the manufacturer sell to realize a profit of $1,250?

SOLUTION

Let x denote the number of units manufactured and sold. Then the total revenue is given by the function

$$R(x) = 110x$$

and the total cost by the function

$$C(x) = 7,500 + 60x$$

(a) To find the break-even point, set $R(x)$ equal to $C(x)$ and solve, getting

$$110x = 7,500 + 60x$$
$$50x = 7,500$$
$$x = 150$$

It follows that the manufacturer will have to sell 150 units to break even. The situation is illustrated in Figure 4.6.

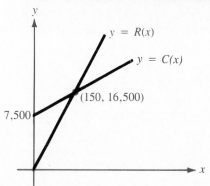

Figure 4.6 $R(x) = 110x$ and $C(x) = 7,500 + 60x$.

(b) The profit $P(x)$ is revenue minus cost. Hence,

$$P(x) = R(x) - C(x) = 110x - (7,500 + 60x) = 50x - 7,500$$

The profit from the sale of 100 units is

$$P(100) = 5,000 - 7,500 = -2,500$$

The minus sign indicates a negative profit (or loss), which was expected since 100 units is less than the break-even level of 150 units. It follows that the manufacturer will lose $2,500 if 100 units are sold.

(c) To determine the number of units that must be sold to generate a profit of $1,250, set the formula for profit $P(x)$ equal to 1,250 and solve for x. You get

$$50x - 7,500 = 1,250$$
$$50x = 8,750$$
$$x = \frac{8,750}{50} = 175$$

from which you can conclude that 175 units must be sold to generate the desired profit.

The next example illustrates how break-even analysis can be used as a tool for decision-making.

EXAMPLE 4.5   A certain car rental agency charges $14 plus 15 cents per mile. A second agency charges $20 plus 5 cents per mile. Which agency offers the better deal?

SOLUTION
The answer depends on the number of miles the car is driven. For short trips, the first agency charges less than the second, but for long trips, the second charges less than the first. You can use break-even analysis to determine the number of miles for which the two agencies charge the same amount.

Suppose a car is to be driven x miles. Then the first agency will charge

$$C_1(x) = 14 + 0.15x \quad \text{dollars}$$

and the second will charge

$$C_2(x) = 20 + 0.05x \quad \text{dollars}$$

If you set these expressions equal to each other and solve, you get

$$14 + 0.15x = 20 + 0.05x$$
$$0.1x = 6 \quad \text{or} \quad x = 60$$

This implies that the two agencies charge the same amount if the car is driven 60 miles. For shorter distances, the first agency offers the better deal and for longer distances, the second agency does. The situation is illustrated in Figure 4.7.

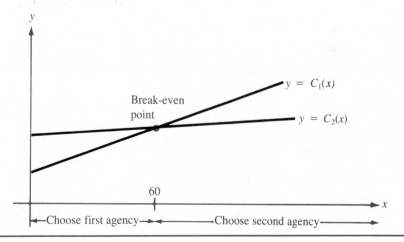

Figure 4.7 Car rental costs at competing agencies.

Market Equilibrium

An important economic application involving intersections of graphs arises in connection with the **law of supply and demand**. In this context, we think of the market price p of a commodity as determining the number of units of the commodity that manufacturers are willing to supply as well as the number of units that consumers are willing to buy. In most cases, manufacturers' supply $S(p)$ increases and consumers' demand $D(p)$ decreases as the market price p increases. A pair of supply and demand curves is sketched in Figure 4.8. (The letter q used to label the vertical axis stands for "quantity.")

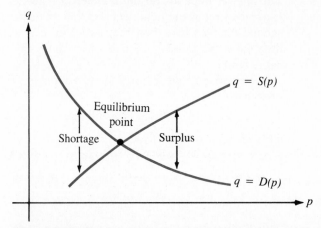

Figure 4.8 Market equilibrium: The intersection of supply and demand.

(Actually, an economist's graph of these functions would not look quite like the one in Figure 4.8. For technical reasons, when dealing with supply and demand curves, economists usually break with mathematical tradition and use the horizontal axis for the dependent variable q and the vertical axis for the independent variable p.)

The point of intersection of the supply and demand curves is called the point of **market equilibrium.** The p coordinate of this point (the **equilibrium price**) is the market price at which supply equals demand, that is, the market price at which there will be neither a surplus nor a shortage of the commodity.

The law of supply and demand asserts that in a situation of pure competition, a commodity will tend to be sold at its equilibrium price. If the commodity is sold for more than the equilibrium price, there will be an unsold surplus on the market and retailers will tend to lower their prices. On the other hand, if the commodity is sold for less than the equilibrium price, the demand will exceed the supply and retailers will be inclined to raise their prices.

Here is an example.

EXAMPLE 4.6 Find the equilibrium price and the corresponding number of units supplied and demanded if the supply function for a certain commodity is $S(p) = p^2 + 3p - 70$ and the demand function is $D(p) = 410 - p$.

SOLUTION

Set $S(p)$ equal to $D(p)$ and solve for p to get

$$p^2 + 3p - 70 = 410 - p$$
$$p^2 + 4p - 480 = 0$$
$$(p - 20)(p + 24) = 0$$
$$p = 20 \quad \text{or} \quad p = -24$$

Since only positive values of p are meaningful in this practical problem, you can conclude that the equilibrium price is \$20. Since the corresponding supply and demand are equal, use the simpler demand equation to compute this quantity to get

$$D(20) = 410 - 20 = 390$$

Hence, 390 units are supplied and demanded when the market is in equilibrium.

The supply and demand curves are sketched in Figure 4.9. Notice that the supply curve crosses the p axis when $p = 7$. (Verify this.) What is the economic interpretation of this fact?

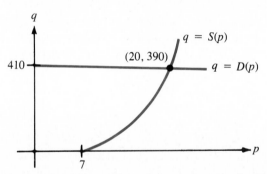

Figure 4.9 The supply and demand curves for Example 4.6.

Problems

In Problems 1 through 17, find the points of intersection (if any) of the given pair of curves and draw the graphs.

1. $y = 3x + 5$ and $y = -x + 3$

2. $y = 5x - 14$ and $y = 4 - x$

3. $y = 3x + 8$ and $y = 3x - 2$

4. $y = x^2$ and $y = 6 - x$

5. $y = x^2 - x$ and $y = x - 1$

6. $y = x^3 - 6x^2$ and $y = -x^2$

7. $y = x^3$ and $y = x^2$

8. $y = x^3$ and $y = -x^3$

9. $y = x^2 + 2$ and $y = x$

10. $3y - 2x = 5$ and $y + 3x = 9$

11. $2x - 3y = -8$ and $3x - 5y = -13$

12. $y = \dfrac{1}{x}$ and $y = x^2$

13. $y = \dfrac{1}{x^2}$ and $y = 4$

14. $y = \dfrac{1}{x}$ and $y = \dfrac{1}{x^2}$

15. $y = \dfrac{1}{x^2}$ and $y = -x^2$

16. $y = x^2$ and $y = 2x + 2$

17. $y = x^2 - 2x$ and $y = x - 1$

Break-even analysis   18. A furniture manufacturer can sell dining-room tables for $70 apiece. The manufacturer's total cost consists of a fixed overhead of $8,000 plus production costs of $30 per table.
(a) How many tables must the manufacturer sell to break even?
(b) How many tables must the manufacturer sell to make a profit of $6,000?
(c) What will be the manufacturer's profit or loss if 150 tables are sold?
(d) On the same set of axes, graph the manufacturer's total revenue and total cost functions. Explain how the overhead can be read from the graph.

Break-even analysis   19. During the summer, a group of students builds kayaks in a converted garage. The rental for the garage is $600 for the summer, and the materials needed to build a kayak cost $25. The kayaks can be sold for $175 apiece.
(a) How many kayaks must the students sell to break even?
(b) How many kayaks must the students sell to make a profit of $450?

Checking accounts   20. The charge for maintaining a checking account at a certain bank is $2 per month plus 5 cents for each check that is written. A competing bank charges $1 per month plus 9 cents per check. Find a criterion for deciding which bank offers the better deal.

Membership fees   21. Membership in a private tennis club costs $500 per year and entitles the member to use the courts for a fee of $1 per hour. At a competing club, membership costs $440 per year and the charge for the use of the courts is $1.75 per hour. If only financial considerations are to be taken into account, how should a tennis player choose which club to join?

Property tax   22. Under the provisions of a proposed property tax bill, a homeowner will pay $100 plus 8 percent of the assessed value of the house. Under the provisions of a competing bill, the homeowner will pay $1,900 plus 2 percent of the assessed value. If only financial considerations are taken into account, how should a homeowner decide which bill to support?

Supply and demand   23. The supply and demand functions for a certain commodity are $S(p) = 4p + 200$ and $D(p) = -3p + 480$, respectively. Find the equilibrium price and the corresponding number of units supplied and demanded, and draw the supply and demand curves on the same set of axes.

Supply and demand   24. When electric blenders are sold for p dollars apiece, manufacturers will supply $\frac{p^2}{10}$ blenders to local retailers while the local demand will be $60 - p$ blenders. At what market price will the manufacturers' supply of electric blenders be equal to the consumers' demand for the blenders? How many blenders will be sold at this price?

Supply and demand   25. The supply and demand functions for a certain commodity are $S(p) = p - 10$ and $D(p) = \frac{5,600}{p}$, respectively.

(a) Find the equilibrium price and the corresponding number of units supplied and demanded.

(b) Draw the supply and demand curves on the same set of axes.

(c) Where does the supply curve cross the p axis? What is the economic significance of this point?

Supply and demand   26. Suppose that the supply and demand functions for a certain commodity are $S(p) = ap + b$ and $D(p) = cp + d$, respectively.

(a) What can you say about the signs of the coefficients a, b, c, and d if the supply and demand curves are oriented as shown in the accompanying figure?

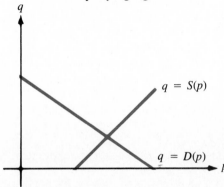

(b) Express the equilibrium price in terms of the coefficients a, b, c, and d.

(c) Use your answer in part (b) to determine what happens to the equilibrium price as a increases.

(d) Use your answer in part (b) to determine what happens to the equilibrium price as d increases.

Spy story   27. The hero of a popular spy story has escaped from the headquarters of an international diamond smuggling ring in the tiny Mediterranean country of Azusa. Our hero, driving a stolen milk truck at 72 kilometers per hour, has a 40-minute head start on his pursuers who are chasing him in a Ferrari going 168 kilometers per hour. The distance from the smugglers' headquarters to the border, and freedom, is 83.8 kilometers. Will our hero make it?

Air travel   28. Two jets bound for Los Angeles leave New York 30 minutes apart. The first travels 550 miles per hour, while the second goes 650 miles per hour. At what time will the second plane pass the first?

5 Functional Models

A mathematical representation of a practical situation is called a **mathematical model**. In preceding sections, you saw models representing such quantities as manufacturing cost, air pollution levels, population size, supply, and demand. In this section, you will see examples illustrating some of the techniques you can use to build mathematical models of your own.

A Profit Function

In the following example, profit is expressed as a function of the price at which a product is sold.

EXAMPLE 5.1   A manufacturer can produce radios at a cost of $2 apiece. The radios have been selling for $5 apiece, and, at this price, consumers have been buying 4,000 radios a month. The manufacturer is planning to raise the price of the radios and estimates that for each $1 increase in the price, 400 fewer radios will be sold each month. Express the manufacturer's monthly profit as a function of the price at which the radios are sold.

SOLUTION
Begin by stating the desired relationship in words.

   Profit = (number of radios sold)(profit per radio)

Since the goal is to express profit as a function of price, the independent variable is price and the dependent variable is profit. Let x denote the price at which the radios will be sold and $P(x)$ the corresponding profit.

Next, express the number of radios sold in terms of the variable x. You know that 4,000 radios are sold each month when the price is $5 and that 400 fewer will be sold each month for each $1 increase in the price. Thus,

Number of radios sold = 4,000 − 400(number of $1 increases)

The number of $1 increases in the price is the difference $x - 5$ between the new and old selling prices. Hence,

$$\begin{aligned}
\text{Number of radios sold} &= 4{,}000 - 400(x - 5) \\
&= 400[10 - (x - 5)] \\
&= 400(15 - x)
\end{aligned}$$

The profit per radio is simply the difference between the selling price x and the cost $2. That is,

Profit per radio = $x - 2$

If you now substitute the algebraic expressions for the number of radios sold and the profit per radio into the verbal equation with which you began, you will get

$$P(x) = 400(15 - x)(x - 2)$$

The graph of this factored polynomial is sketched in Figure 5.1. Actually, only the portion of the graph for $x \geq 5$ is relevant to the original problem as stated. (Can you give a practical interpretation of the portion between $x = 2$ and $x = 5$?) Notice that the profit function reaches a maximum for some value of x near $x = 8$. In Chapter 3, you will learn how to use calculus to find this optimal selling price.

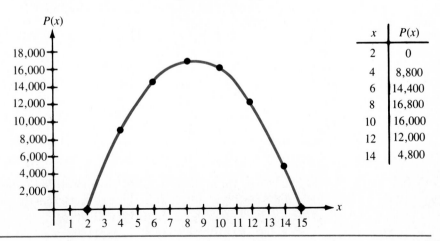

x	$P(x)$
2	0
4	8,800
6	14,400
8	16,800
10	16,000
12	12,000
14	4,800

Figure 5.1 The profit function $P(x) = 400(15 - x)(x - 2)$.

Elimination of Variables

In the next two examples, the quantity you are seeking is expressed most naturally in terms of two variables. You will have to eliminate one of these before you can write the quantity as a function of a single variable.

EXAMPLE 5.2 The highway department is planning to build a picnic area for motorists along a major highway. It is to be rectangular with an area of 5,000 square yards and is to be fenced off on the three sides not adjacent to the highway. Express the number of yards of fencing required as a function of the length of the unfenced side.

SOLUTION
It is natural to start by introducing two variables, say x and y, to denote the length of the sides of the picnic area (Figure 5.2) and to express the number of yards F of required fencing in terms of these two variables:

$$F = x + 2y$$

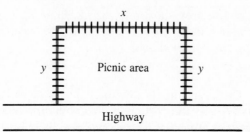

Figure 5.2 Rectangular picnic area.

Since the goal is to express the number of yards of fencing as a function of x alone, you must find a way to express y in terms of x. To do this, use the fact that the area is to be 5,000 square yards and write

$$xy = 5,000$$

Solve this equation for y

$$y = \frac{5,000}{x}$$

and substitute the resulting expression for y into the formula for F to get

$$F(x) = x + 2\left(\frac{5,000}{x}\right) = x + \frac{10,000}{x}$$

A graph of the relevant portion of this rational function is sketched in Figure 5.3. Notice that there is some length x for which the amount of required fencing is minimal. In Chapter 3 you will compute this optimal value of x using calculus.

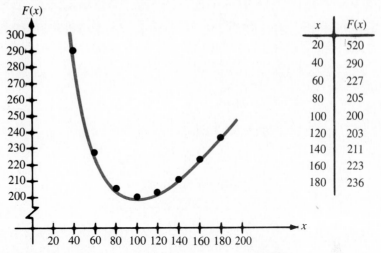

x	$F(x)$
20	520
40	290
60	227
80	205
100	200
120	203
140	211
160	223
180	236

Figure 5.3 The length of fencing: $F(x) = x + \dfrac{10,000}{x}$.

EXAMPLE 5.3   A cylindrical can is to have capacity (volume) of 24π cubic inches. The cost of the material used for the top and bottom of the can is 3 cents per square inch, and the cost of the material used for the curved side is 2 cents per square inch. Express the cost of constructing the can as a function of its radius.

SOLUTION

Let r denote the radius of the circular top and bottom, h the height of the can, and C the cost (in cents) of constructing the can. Then,

$$C = \text{cost of top} + \text{cost of bottom} + \text{cost of side}$$

where, for each component of cost,

$$\text{Cost} = (\text{cost per square inch})(\text{number of square inches})$$
$$= (\text{cost per square inch})(\text{area})$$

The area of the circular top (or bottom) is πr^2, and the cost per square inch of the top (or bottom) is 3 cents. Hence,

$$\text{Cost of top} = 3\pi r^2 \quad \text{and} \quad \text{Cost of bottom} = 3\pi r^2$$

To find the area of the curved side, imagine the top and bottom of the can removed and the side cut and spread out to form a rectangle as shown in Figure 5.4.

The height of the rectangle is the height h of the can. The length of the rectangle is the circumference $2\pi r$ of the circular top (or bottom) of the can. Hence, the area of the rectangle (or curved side) is $2\pi rh$ square inches. Since the cost of the side is 2 cents per square inch, it follows that

Figure 5.4 Cylindrical can for Example 5.3.

Cost of side $= 2(2\pi rh) = 4\pi rh$

Putting it all together,

$$C = 3\pi r^2 + 3\pi r^2 + 4\pi rh = 6\pi r^2 + 4\pi rh$$

Since the goal is to express the cost as a function of the radius alone, you must find a way to express the height h in terms of r. To do this, use the fact that the volume $V = \pi r^2 h$ is to be 24π. That is, set $\pi r^2 h$ equal to 24π and solve for h to get

$$\pi r^2 h = 24\pi \qquad \text{or} \qquad h = \frac{24}{r^2}$$

Now substitute this expression for h into the formula for C to get

$$C(r) = 6\pi r^2 + 4\pi r\left(\frac{24}{r^2}\right)$$

or

$$C(r) = 6\pi r^2 + \frac{96\pi}{r}$$

A graph of the relevant portion of this cost function is sketched in Figure 5.5. Notice that there is some radius r for which the cost is

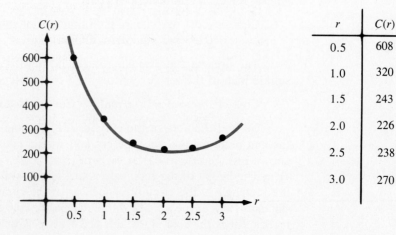

Figure 5.5 The cost function:

$$C(r) = 6\pi r^2 + \frac{96\pi}{r}.$$

r	$C(r)$
0.5	608
1.0	320
1.5	243
2.0	226
2.5	238
3.0	270

minimal. In Chapter 3 you will learn how to find this optimal radius using calculus.

Functions Involving Multiple Formulas

In the next example, you will need three formulas to define the desired function.

EXAMPLE 5.4   During the 1977 drought, residents of Marin County, California, were faced with a severe water shortage. To discourage excessive use of water, the County Water District initiated drastic rate increases. The monthly rate for a family of four was $1.22 per 100 cubic feet of water for the first 1,200 cubic feet, $10 per 100 cubic feet for the next 1,200 cubic feet, and $50 per 100 cubic feet thereafter. Express the monthly water bill for a family of four as a function of the amount of water used.

SOLUTION
Let x denote the number of hundred-cubic-feet units of water used by the family during the month and $C(x)$ the corresponding cost in dollars. If $0 \leq x \leq 12$, the cost is simply the cost per unit times the number of units used:

$$C(x) = 1.22x$$

If $12 < x \leq 24$, each of the first 12 units costs $1.22, and so the total cost of these 12 units is $1.22(12) = 14.64$ dollars. Each of the remaining $x - 12$ units costs $10, and hence the total cost of these units is $10(x - 12)$ dollars. The cost of all x units is the sum

$$C(x) = 14.64 + 10(x - 12) = 10x - 105.36$$

If $x > 24$, the cost of the first 12 units is $1.22(12) = 14.64$ dollars, the cost of the next 12 units is $10(12) = 120$ dollars, and the cost of the remaining $x - 24$ units is $50(x - 24)$ dollars. The cost of all x units is the sum

$$C(x) = 14.64 + 120 + 50(x - 24) = 50x - 1,065.36$$

Combining these three formulas you get

$$C(x) = \begin{cases} 1.22x & \text{if } 0 \leq x \leq 12 \\ 10x - 105.36 & \text{if } 12 < x \leq 24 \\ 50x - 1,065.36 & \text{if } x > 24 \end{cases}$$

The graph of this function is shown in Figure 5.6. Notice that the graph consists of three line segments, each one steeper than the preceding one. What aspect of the practical situation is reflected by the increasing steepness of the lines?

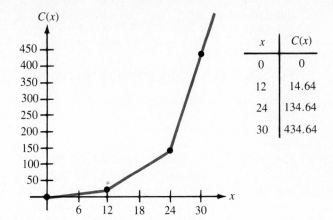

x	$C(x)$
0	0
12	14.64
24	134.64
30	434.64

Figure 5.6 The cost of water in Marin County.

Proportionality

The following concepts of proportionality are used frequently in the construction of mathematical models.

Direct Proportionality

To say that Q is proportional (or directly proportional) to x means that there is a constant k for which $Q = kx$.

Inverse Proportionality

To say that Q is inversely proportional to x means that there is a constant k for which $Q = \dfrac{k}{x}$.

Joint Proportionality

To say that Q is jointly proportional to x and y means that Q is directly proportional to the product of x and y; that is, there is a constant k for which $Q = kxy$.

Here is an example from biology.

EXAMPLE 5.5 When environmental factors impose an upper bound on its size, population grows at a rate that is jointly proportional to its current size and the difference between its current size and the upper bound. Express the rate of population growth as a function of the size of the population.

SOLUTION

Let p denote the size of the population, $R(p)$ the corresponding rate of population growth, and b the upper bound placed on the population by the environment. Then,

Difference between population and bound $= b - p$

and so

$$R(p) = kp(b - p)$$

where k is the constant of proportionality.

A graph of this factored polynomial is sketched in Figure 5.7. In Chapter 3, you will use calculus to compute the population size for which the rate of population growth is greatest.

Figure 5.7 The rate of bounded population growth:
$R(p) = kp(b - p)$.

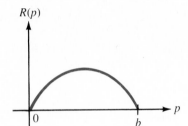

Problems

Retail sales

1. A bookstore can obtain a certain gift book from the publisher at a cost of $3 per book. The bookstore has been offering the book at the price of $15 per copy and, at this price, has been selling 200 copies a month. The bookstore is planning to lower its price to stimulate sales and estimates that for each $1 reduction in the price, 20 more books will be sold each month. Express the bookstore's monthly profit from the sale of this book as a function of the selling price, draw the graph, and estimate the optimal selling price.

Retail sales

2. A manufacturer has been selling lamps at the price of $6 per lamp, and, at this price, consumers have been buying 3,000 lamps a month. The manufacturer wishes to raise the price and estimates that for each $1 increase in the price, 1,000 fewer lamps will be sold each month. The manufacturer can produce the lamps at a cost of $4 per lamp. Express the manufacturer's monthly profit as a function of the price at which the lamps are sold, draw the graph, and estimate the optimal selling price.

Transportation costs

3. A bus company is willing to charter buses only to groups of 35 or more people. If a group contains exactly 35 people, each person pays

$60. In larger groups, everybody's fare is reduced by 50 cents for each person in excess of 35. Express the bus company's revenue as a function of the size of the group, draw the graph, and estimate the size of the group that will maximize the revenue.

Agricultural yield

4. A Florida citrus grower estimates that if 60 orange trees are planted, the average yield per tree will be 400 oranges. The average yield will decrease by 4 oranges per tree for each additional tree planted on the same acreage. Express the grower's total yield as a function of the number of additional trees planted, draw the graph, and estimate the total number of trees the grower should plant to maximize yield.

Harvesting

5. Farmers can get $2 per bushel for their potatoes on July first, and after that, the price drops by 2 cents per bushel per day. On July first, a farmer has 80 bushels of potatoes in the field and estimates that the crop is increasing at the rate of 1 bushel per day. Express the farmer's revenue from the sale of the potatoes as a function of the time at which the crop is harvested, draw the graph, and estimate when the farmer should harvest the potatoes to maximize revenue.

Recycling

6. To raise money, a service club has been collecting used bottles that it plans to deliver to a local glass company for recycling. Since the project began 80 days ago, the club has collected 24,000 pounds of glass for which the glass company currently offers 1 cent per pound. However, since bottles are accumulating faster than they can be recycled, the company plans to reduce by 1 cent each day the price it will pay for 100 pounds of used glass. Assuming that the club can continue to collect bottles at the same rate and that transportation costs make more than one trip to the glass company unfeasible, express the club's revenue from its recycling project as a function of the number of additional days the project runs. Draw the graph and estimate when the club should conclude the project and deliver the bottles to maximize its revenue.

Fencing

7. A city recreation department plans to build a rectangular playground 3,600 square meters in area. The playground is to be surrounded by a fence. Express the length of the fencing as a function of the length of one of the sides of the playground, draw the graph, and estimate the dimensions of the playground requiring the least amount of fencing.

Area

8. Express the area of a rectangular field whose perimeter is 320 meters as a function of the length of one of its sides. Draw the graph and estimate the dimensions of the field of maximum area.

Construction cost

9. A closed box with a square base is to have a volume of 250 cubic meters. The material for the top and bottom of the box costs $2 per square meter, and the material for the sides costs $1 per square meter. Express the construction cost of the box as a function of the length of its base.

Construction cost   10. An open box with a square base is to be built for $48. The sides of the box will cost $3 per square meter, and the base will cost $4 per square meter. Express the volume of the box as a function of the length of its base.

Volume   11. An open box is to be made from a square piece of cardboard, 18 inches by 18 inches, by removing a small square from each corner and folding up the flaps to form the sides. Express the volume of the resulting box as a function of the length x of a side of the removed squares. Draw the graph and estimate the value of x for which the volume of the resulting box is greatest.

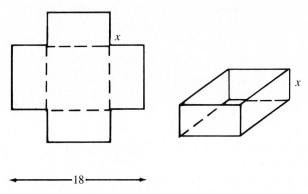

Packaging   12. A beer can holds 12 fluid ounces, which is approximately 6.89π cubic inches. Express the surface area of the can as a function of its radius. (Recall that the volume of a cylinder of radius r and height h is $\pi r^2 h$, the circumference of a circle of radius r is $2\pi r$, and the area of a circle of radius r is πr^2.)

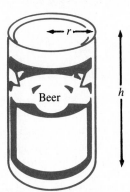

Packaging   13. A cylindrical can is to hold 4π cubic inches of frozen orange juice. The cost per square inch of constructing the metal top and bottom is twice the cost per square inch of constructing the cardboard side. Express the cost of constructing the can as a function of its radius if the cost of the side is 0.02 cent per square inch.

Volume 14. A cylindrical can with no top has been made from 27π square inches of metal. Express the volume of the can as a function of its radius.

Admission fees 15. A local natural history museum charges admission to groups according to the following policy: Groups of fewer than 50 people are charged a rate of $1.50 per person, while groups of 50 people or more are charged a reduced rate of $1 per person.
(a) Express the amount a group will be charged for admission as a function of its size and draw the graph.
(b) How much money will a group of 49 people save in admission costs if it can recruit 1 additional member?

Discounts 16. A record club offers the following special sale: If 5 records are bought at the full price of $6 apiece, additional records can then be bought at half price. There is a limit of 9 records per customer. Express the cost of the records as a function of the number bought and draw the graph.

Postal rates 17. There was a time when the postal rate for letters was 20 cents for the first ounce or fraction thereof and 17 cents for each additional ounce or fraction thereof. Express the cost of sending a letter as a function of its weight x for $0 < x \leq 4$ and draw the graph.

Telegram rates 18. In 1984, the rate for interstate telegrams was $8.45 for 10 words or less plus 45 cents for each additional word. Express the cost of sending a telegram as a function of its length and draw the graph.

Income tax 19. The following table is taken from the 1983 federal income tax rate schedule for single taxpayers:

If the taxable income is:		The income tax is:	
Over	but not over		of the excess over
15,000	18,200	2,097 + 24%	15,000
18,200	23,500	2,865 + 28%	18,200
23,500	28,800	4,349 + 32%	23,500

(a) Express an individual's income tax as a function of the taxable income x for $15,000 < x \leq 28,800$ and draw the graph.
(b) Your graph in part (a) should consist of three line segments. Compute the slope of each segment. What happens to these slopes as the taxable income increases? Explain the behavior of the slopes in practical terms.

Transportation cost 20. A bus company has adopted the following pricing policy for groups wishing to charter its buses: Groups containing no more than 40 people will be charged a fixed amount of $2,400 (40 times $60). In groups containing between 40 and 80 people, everyone will pay $60 minus 50 cents for each person in excess of 40. The company's lowest fare of $40 per person will be offered to groups that have 80 members or more. Express the bus company's revenue as a function of the size of the group and draw the graph.

Population growth   21. In the absence of environmental constraints, population grows at a rate proportional to its size. Express the rate of population growth as a function of the size of the population.

Radioactive decay   22. A sample of radium decays at a rate proportional to the amount of radium remaining. Express the rate of decay of the sample as a function of the amount remaining.

Temperature change   23. The rate at which the temperature of an object changes is proportional to the difference between its own temperature and the temperature of the surrounding medium. Express this rate as a function of the temperature of the object.

The spread of an epidemic   24. The rate at which an epidemic spreads through a community is jointly proportional to the number of people who have caught the disease and the number who have not. Express this rate as a function of the number of people who have caught the disease.

Political corruption   25. The rate at which people are implicated in a government scandal is jointly proportional to the number of people already implicated and the number of people involved who have not yet been implicated. Express this rate as a function of the number of people who have been implicated.

Production cost   26. At a certain factory, setup cost is directly proportional to the number of machines used, and operating cost is inversely proportional to the number of machines used. Express the total cost as a function of the number of machines used.

Transportation cost   27. A truck is hired to transport goods from a factory to a warehouse. The driver's wages are figured by the hour and so are inversely proportional to the speed at which the truck is driven. The cost of gasoline is directly proportional to the speed. Express the total cost of operating the truck as a function of the speed at which it is driven.

Distance   28. A car traveling east at 80 kilometers per hour and a truck traveling south at 60 kilometers per hour start at the same intersection. Express the distance between them as a function of time. (*Hint:* Use the pythagorean theorem.)

Distance   29. A truck is 300 miles due east of a car and is traveling west at the constant speed of 30 miles per hour. Meanwhile, the car is going north at the constant speed of 60 miles per hour. Express the distance between the car and truck as a function of time.

Installation cost   30. A cable is to be run from a power plant on one side of a river 900 meters wide to a factory on the other side, 3,000 meters downstream. The cable will be run in a straight line from the power plant to some point *P* on the opposite bank, and then along the bank to the factory.

The cost of running the cable across the water is $5 per meter, while the cost over land is $4 per meter. Let x be the distance from P to the point directly across the river from the power plant, and express the cost of installing the cable as a function of x.

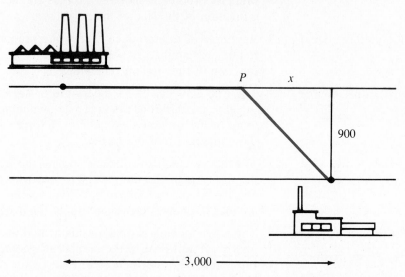

Poster design   **31.** A rectangular poster contains 25 square centimeters of print surrounded by margins of 2 centimeters on each side and 4 centimeters on the top and bottom. Express the total area of the poster (printing plus margins) as a function of the width of the printed portion.

Production cost   **32.** A company has received an order from the city recreation department to manufacture 8,000 Styrofoam kickboards for its summer swimming program. The company owns several machines, each of which can produce 30 kickboards an hour. The cost of setting up the machines to produce these particular kickboards is $20 per machine. Once the machines have been set up, the operation is fully automated and can be overseen by a single production supervisor earning $4.80 per hour. Express the cost of producing the 8,000 kickboards as a function of the number of machines used, draw the graph, and estimate the number of machines the company should use to minimize cost.

Chapter Summary and Review Problems

Important Terms, Symbols, and Formulas

Function
Independent and dependent variables
Functional notation: $f(x)$
Domain of a function

Composition of functions: $g[h(x)]$
Graph of a function: the points $(x, f(x))$
x and y intercepts
Continuous function
Discontinuity
Polynomial
Rational function
Linear function; constant rate of change
Slope: $m = \dfrac{\Delta y}{\Delta x} = \dfrac{y_2 - y_1}{x_2 - x_1}$
Slope-intercept formula: $y = mx + b$
Point-slope formula: $y - y_0 = m(x - x_0)$
Intersection of graphs; break-even analysis
Market equilibrium; law of supply and demand
Direct proportionality: $Q = kx$
Inverse proportionality: $Q = \dfrac{k}{x}$
Joint proportionality: $Q = kxy$

Review Problems

1. Specify the domain of each of the following functions:
   (a) $f(x) = x^2 - 2x + 6$
   (b) $f(x) = \dfrac{x - 3}{x^2 + x - 2}$
   (c) $f(x) = \sqrt{x^2 - 9}$

2. As advances in technology result in the production of increasingly powerful and compact calculators, the price of calculators currently on the market drops. Suppose that x months from now, the price of a certain model will be $P(x) = 40 + \dfrac{30}{x + 1}$ dollars.
   (a) What will the price be 5 months from now?
   (b) By how much will the price drop during the 5th month?
   (c) When will the price be $43?
   (d) What will happen to the price in the long run?

3. Find the composite function $g[h(x)]$.
   (a) $g(u) = u^2 + 2u + 1$, $h(x) = 1 - x$
   (b) $g(u) = \dfrac{1}{2u + 1}$, $h(x) = x + 2$
   (c) $g(u) = \sqrt{1 - u}$, $h(x) = 2x + 4$

4. (a) Find $f(x - 2)$ if $f(x) = x^2 - x + 4$.
   (b) Find $f(x^2 + 1)$ if $f(x) = \sqrt{x} + \dfrac{2}{x - 1}$.
   (c) Find $f(x + 1) - f(x)$ if $f(x) = x^2$.

5. Find functions $h(x)$ and $g(u)$ such that $f(x) = g[h(x)]$
   (a) $f(x) = (x^2 + 3x + 4)^5$
   (b) $f(x) = (3x + 1)^2 + \dfrac{5}{2(3x + 2)^3}$

6. An environmental study of a certain community suggests that the average daily level of smog in the air will be $Q(p) = \sqrt{0.5p + 19.4}$ units when the population is p thousand. It is estimated that t years from now, the population will be $p(t) = 8 + 0.2t^2$ thousand.
   (a) Express the level of smog in the air as a function of time.
   (b) What will the smog level be 3 years from now?
   (c) When will the smog level reach 5 units?

7. Find the value of c for which the curve $y = 3x^2 - 2x + c$ passes through the point $(2, 4)$.

8. Graph the following functions:

   (a) $f(x) = x^2 + 2x - 8$

   (b) $f(x) = \begin{cases} 1/x^2 & \text{if } x < 0 \\ x & \text{if } 0 \le x < 3 \\ 4 & \text{if } x \ge 3 \end{cases}$

   (c) $f(x) = \dfrac{-2}{x - 3}$

   (d) $f(x) = x + \dfrac{2}{x}$

   (e) $f(x) = \dfrac{x^2 - 1}{x - 3}$

9. The consumer demand for a certain commodity is $D(p) = -50p + 800$ units per month when the market price is p dollars per unit.
   (a) Graph this demand function.
   (b) Express consumers' total monthly expenditure for the commodity as a function of p and draw the graph.
   (c) Use the graph in part (b) to estimate the market price at which the total expenditure for the commodity is greatest.

10. A private college in the southwest has launched a fund-raising campaign. Suppose that college officials estimate that it will take $f(x) = \dfrac{10x}{150 - x}$ weeks to reach x percent of their goal.
    (a) Sketch the relevant portion of the graph of this function.
    (b) How long will it take to reach 50 percent of the campaign's goal?
    (c) How long will it take to reach 100 percent of the goal?

11. Find the slope and y intercept of the given line and draw the graph.
    (a) $y = 3x + 2$
    (b) $5x - 4y = 20$
    (c) $2y + 3x = 0$
    (d) $\dfrac{x}{3} + \dfrac{y}{2} = 4$

12. Find the equation of the line with slope 5 and y intercept $(0, -4)$.

13. Find the equation of the line that passes through $(1, 3)$ and has slope -2.

14. Find the equation of the line through the points (2, 4) and (1, −3).

15. Since the beginning of the year, the price of unleaded gasoline has been increasing at a constant rate of 2 cents per gallon per month. By June first, the price had reached $1.03 per gallon.
    (a) Express the price of unleaded gasoline as a function of time and draw the graph.
    (b) What was the price at the beginning of the year?
    (c) What will the price be on October first?

16. The circulation of a newspaper is increasing at a constant rate. Three months ago the circulation was 3,200. Today it is 4,400.
    (a) Express the circulation as a function of time and draw the graph.
    (b) What will the circulation be 2 months from now?

17. Find the points of intersection (if any) of the given pair of curves and draw the graphs.
    (a) $y = -3x + 5$ and $y = 2x - 10$
    (b) $y = x + 7$ and $y = -2 + x$
    (c) $y = x^2 - 1$ and $y = 1 - x^2$
    (d) $y = x^2$ and $y = 15 - 2x$
    (e) $y = \dfrac{24}{x^2}$ and $y = 3x$

18. One plumber charges $25 plus $16 per half hour. A second charges $31 plus $14 per half hour. Find a criterion for deciding which plumber to call if only financial considerations are to be taken into account.

19. A manufacturer can sell a certain product for $80 per unit. Total cost consists of a fixed overhead of $4,500 plus production costs of $50 per unit.
    (a) How many units must the manufacturer sell to break even?
    (b) What is the manufacturer's profit or loss if 200 units are sold?
    (c) How many units must the manufacturer sell to realize a profit of $900?

20. A manufacturer can produce bookcases at a cost of $10 apiece. Sales figures indicate that if the bookcases are sold for x dollars apiece, approximately $50 - x$ will be sold each month. Express the manufacturer's monthly profit as a function of the selling price x, draw the graph, and estimate the optimal selling price.

21. A retailer can obtain cameras from the manufacturer at a cost of $50 apiece. The retailer has been selling the cameras at the price of $80 apiece, and, at this price, consumers have been buying 40 cameras a month. The retailer is planning to lower the price to stimulate sales and estimates that for each $5 reduction in the price, 10 more cameras will be sold each month. Express the retailer's monthly profit from

the sale of the cameras as a function of the selling price. Draw the graph and estimate the optimal selling price.

22. A cylindrical can with no top is to be constructed for 80 cents. The cost of the material used for the bottom is 3 cents per square inch, and the cost of the material used for the curved side is 2 cents per square inch. Express the volume of the can as a function of its radius.

23. A manufacturing firm has received an order to make 400,000 souvenir medals commemorating the 15th anniversary of the landing of Apollo 11 on the moon. The firm owns several machines, each of which can produce 200 medals per hour. The cost of setting up the machines to produce the medals is $80 per machine, and the total operating cost is $5.76 per hour. Express the cost of producing the 400,000 medals as a function of the number of machines used. Draw the graph and estimate the number of machines the firm should use to minimize cost.

24. The following table is taken from the 1983 federal income tax rate schedule for unmarried heads of households:

If the taxable income is:		The income tax is:	
Over	but not over		of the excess over
15,000	18,200	2,000 + 21%	15,000
18,200	23,500	2,672 + 25%	18,200
23,500	28,800	3,997 + 29%	23,500

Express an individual's income tax as a function of the taxable income x for $15,000 < x \leq 28,800$ and draw the graph.

25. Psychologists believe that when a person is asked to recall a set of facts, the rate at which the facts are recalled is proportional to the number of relevant facts in the subject's memory that have not yet been recalled. Express the recall rate as a function of the number of facts that have been recalled.

Chapter

2 | Differentiation: Basic Concepts

1   The Derivative

Differentiation is a mathematical technique of exceptional power and versatility. It is one of the two central concepts in the branch of mathematics called **calculus** and has a variety of applications including curve sketching, the optimization of functions, and the analysis of rates of change. To set the stage, here are brief introductory discussions of two of the types of practical problems that can be solved using calculus.

Optimization Problems

A typical problem to which calculus can be applied is the profit maximization problem you saw in Example 5.1 of Chapter 1. Recall that in that problem, a manufacturer's monthly profit from the sale of radios was $P(x) = 400(15 - x)(x - 2)$ dollars when the radios were sold for x dollars

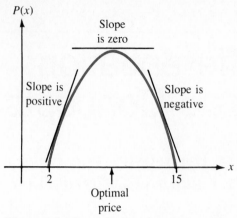

Figure 1.1 The profit function $P(x) = 400(15 - x)(x - 2)$.

apiece. The graph of this profit function, which is reproduced in Figure 1.1, suggests that there is an optimal selling price x at which the manufacturer's profit will be greatest. In geometric terms, the optimal price is the x coordinate of the peak of the graph.

In this relatively simple example, the peak can be characterized in terms of lines that are **tangent** to the graph. In particular, the peak is the only point on the graph at which the tangent line is horizontal, that is, at which the slope of the tangent is zero. To the left of the peak, the slope of the tangent is positive. To the right of the peak, the slope is negative. But just at the peak itself, the curve "levels off" and the slope of its tangent is zero. (The complete solution of this optimization problem using these ideas will be given a little later in this section.)

Rate of Change

Calculus is also one of the techniques used to find the rate of change of a function. Recall from Chapter 1, Section 3, that the rate of change of a linear function with respect to its independent variable is equal to the steepness or slope of its straight-line graph. Moreover, this steepness or rate of change is constant.

If the function under consideration does not happen to be linear, its rate of change with respect to its independent variable is still the steepness of its graph, measured in this case by the slope of the line that is tangent to the graph at the point in question. Since the graph is not a straight line, its steepness, or rate of change, is not constant but varies from point to point. (These ideas are of fundamental importance and will be developed much more carefully in Section 3.) The situation is illustrated in Figure 1.2.

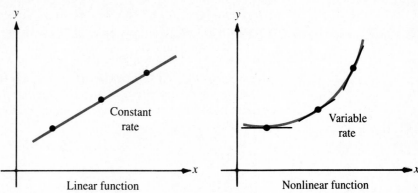

Figure 1.2 Rate of change = slope (of tangent).

The Slope of a Tangent

The preceding discussion suggests that you could solve optimization problems and compute rates of change if you had a procedure for finding the slope of the tangent to a curve at a given point. Such a procedure will now be developed. Throughout the development, you may rely on your intuitive understanding that the tangent to a curve at a point is the line through the point that indicates the direction of the curve.

The goal is to solve the following general problem: Given a point $(x, f(x))$ on the graph of a function f, find the slope of the line that is tangent to the graph at this point. The situation is illustrated in Figure 1.3.

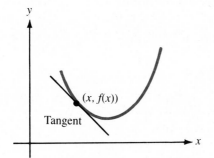

Figure 1.3 A tangent to the curve $y = f(x)$.

In Chapter 1, Section 3, you saw that the slope of the line passing through two points (x_1, y_1) and (x_2, y_2) is given by the formula

$$\text{Slope} = \frac{\Delta y}{\Delta x} = \frac{y_2 - y_1}{x_2 - x_1}$$

Unfortunately, in the present situation, you know only one point on the tangent line, namely the point of tangency $(x, f(x))$. Hence, direct computation of the slope is impossible and you are forced to try an indirect approach.

The strategy is to approximate the tangent by other lines whose slopes can be computed directly. In particular, consider lines joining the given point $(x, f(x))$ to neighboring points on the graph of f. These lines, shown in Figure 1.4, are called **secants** and are good approximations to the tangent, provided the neighboring point is close to the given point.

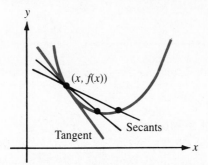

Figure 1.4 Secants approximating a tangent.

In fact, you can make the slope of the secant as close as you like to the slope of the tangent by choosing the neighboring point sufficiently close to the given point $(x, f(x))$. This suggests that you should be able to determine the slope of the tangent itself by first computing the slopes of related secants and then studying the behavior of these slopes as the neighboring points get closer and closer to the given point.

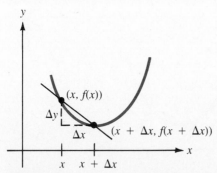

Figure 1.5 A secant through the curve $y = f(x)$.

To compute the slope of a secant, first label the coordinates of the neighboring point as indicated in Figure 1.5. In particular, let Δx denote the change in the x coordinate between the given point $(x, f(x))$ and the neighboring point. The x coordinate of the neighboring point is $x + \Delta x$, and since the point lies on the graph of f, its y coordinate is $f(x + \Delta x)$. Since the change in the y coordinate is $\Delta y = f(x + \Delta x) - f(x)$, it follows that

$$\text{Slope of secant} = \frac{\Delta y}{\Delta x} = \frac{f(x + \Delta x) - f(x)}{x + \Delta x - x} = \frac{f(x + \Delta x) - f(x)}{\Delta x}$$

Remember that this quotient is not the slope of the tangent but only an approximation to it. If Δx is small, however, the neighboring point $(x + \Delta x, f(x + \Delta x))$ is close to the given point $(x, f(x))$, and the approximation is a good one. In fact, the slope of the actual tangent is the number that this quotient approaches as Δx approaches zero.

A calculation illustrating this procedure is performed in the following example.

EXAMPLE 1.1   Find the slope of the line that is tangent to the graph of the function $f(x) = x^2$ at the point $(2, 4)$.

SOLUTION
A sketch of f showing the given point $(2, 4)$ and a related secant is drawn in Figure 1.6.

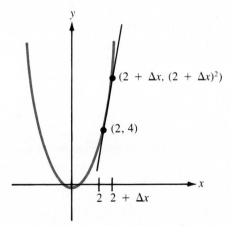

Figure 1.6 The curve $y = x^2$ and a secant through $(2, 4)$.

Since the x coordinate of the given point is 2, it follows that the x coordinate of the neighboring point is $2 + \Delta x$, and the y coordinate of this point is $(2 + \Delta x)^2$. Hence,

$$\text{Slope of secant} = \frac{(2 + \Delta x)^2 - 4}{\Delta x}$$

Your goal is to find the number that this quotient approaches as Δx approaches zero. Before you can do this, you must rewrite the quotient in simpler form. (Do you see what would happen if you let Δx become zero in the numerator and denominator of the quotient in its present form?) To simplify the quotient, expand the term $(2 + \Delta x)^2$, rewrite the numerator, and then divide numerator and denominator by Δx as follows:

$$\text{Slope of secant} = \frac{(2 + \Delta x)^2 - 4}{\Delta x}$$

$$= \frac{4 + 4\Delta x + (\Delta x)^2 - 4}{\Delta x}$$

$$= \frac{4\Delta x + (\Delta x)^2}{\Delta x}$$

$$= \frac{\Delta x(4 + \Delta x)}{\Delta x}$$

$$= 4 + \Delta x$$

Now you can let Δx approach zero. Since the slope $4 + \Delta x$ of the secant approaches 4 as Δx approaches zero, it follows that at the given point (2, 4), the slope of the tangent must be 4.

In the preceding example, you found the slope of the tangent to the curve $y = x^2$ at a particular point (2, 4). In the next example, you will perform the same calculation again, this time representing the given point algebraically as (x, x^2). The result will be a formula into which you can substitute any value of x to calculate the slope of the tangent to the curve at the point (x, x^2).

EXAMPLE 1.2 Derive a formula expressing the slope of the tangent to the curve $y = x^2$ as a function of the x coordinate of the point of tangency.

SOLUTION

Represent the point of tangency as (x, x^2) and the neighboring point as $(x + \Delta x, (x + \Delta x)^2)$ as shown in Figure 1.7. Then,

$$\text{Slope of secant} = \frac{(x + \Delta x)^2 - x^2}{\Delta x}$$

$$= \frac{x^2 + 2x \Delta x + (\Delta x)^2 - x^2}{\Delta x}$$

$$= \frac{2x \Delta x + (\Delta x)^2}{\Delta x}$$

$$= \frac{\Delta x(2x + \Delta x)}{\Delta x}$$

$$= 2x + \Delta x$$

Figure 1.7 The curve $y = x^2$ and a secant through (x, x^2).

Since the slope $2x + \Delta x$ of the secant approaches $2x$ as Δx approaches zero, it follows that at the point (x, x^2), the slope of the tangent is $2x$.

For example, at the point $(2, 4)$, $x = 2$, and so the slope of the tangent is $2(2) = 4$, as you found in Example 1.1.

The Derivative

In Example 1.2, you started with a function f and derived a related function that expressed the slope of its tangent in terms of the x coordinate of the point of tangency. This derived function is known as the **derivative** of f and is frequently denoted by the symbol f', which is read "f prime." In Example 1.2 you discovered that the derivative of x^2 is $2x$; that is, you found that if $f(x) = x^2$, then $f'(x) = 2x$.

Here is a summary of the situation.

Geometric Interpretation of the Derivative

> The derivative $f'(x)$ expresses the slope of the tangent to the curve $y = f(x)$ as a function of the x coordinate of the point of tangency.

How to Compute the Derivative of $f(x)$

> Step 1. Form the difference quotient (the slope of a secant).
>
> $$\frac{f(x + \Delta x) - f(x)}{\Delta x}$$
>
> Step 2. Simplify the difference quotient algebraically.
> Step 3. Let Δx approach zero in the simplified difference quotient.

Limit Notation

The fact that the difference quotient $\dfrac{f(x + \Delta x) - f(x)}{\Delta x}$ approaches the derivative $f'(x)$ as Δx approaches zero is traditionally expressed more compactly using **limit notation** as follows.

The Derivative

$$f'(x) = \lim_{\Delta x \to 0} \frac{f(x + \Delta x) - f(x)}{\Delta x}$$

The symbol lim is an abbreviation of the word "limit." The expression $\lim_{\Delta x \to 0}$ is read "the limit as Δx approaches zero" and indicates that Δx is to get closer and closer to zero in the formula that follows. The complete equation defining the derivative is read "f prime of x equals the limit as Δx approaches zero of $\dfrac{f(x + \Delta x) - f(x)}{\Delta x}$."

The use of limit notation in the computation of a derivative is illustrated in the next example. Also illustrated is the geometric interpretation of the derivative as the slope of the tangent.

EXAMPLE 1.3   (a) Find the derivative of the function $f(x) = \dfrac{1}{x}$.

(b) Find the equation of the line that is tangent to the graph of the function $f(x) = \dfrac{1}{x}$ when $x = 2$.

SOLUTION

(a) $f'(x) = \displaystyle\lim_{\Delta x \to 0} \frac{f(x + \Delta x) - f(x)}{\Delta x}$ definition of derivative

$= \displaystyle\lim_{\Delta x \to 0} \frac{1/(x + \Delta x) - 1/x}{\Delta x}$ since $f(x) = 1/x$

$= \displaystyle\lim_{\Delta x \to 0} \frac{1/(x + \Delta x) - 1/x}{\Delta x} \cdot \frac{x(x + \Delta x)}{x(x + \Delta x)}$

$= \displaystyle\lim_{\Delta x \to 0} \frac{x - (x + \Delta x)}{\Delta x[x(x + \Delta x)]}$

$= \displaystyle\lim_{\Delta x \to 0} \frac{-\Delta x}{\Delta x[x(x + \Delta x)]}$

$= \displaystyle\lim_{\Delta x \to 0} \frac{-1}{x(x + \Delta x)}$

$$= -\frac{1}{x^2} \qquad \text{after } \Delta x \to 0$$

Notice the algebraic technique used in the third line of the computation that led to the simplification of the difference quotient. Notice also the form of the computation, in particular that the expression $\lim_{\Delta x \to 0}$ is written on each line until after the limit has actually been performed in the last line.

(b) To find the slope of the tangent when $x = 2$, compute $f'(2)$.

$$\text{Slope of tangent} = f'(2) = -\frac{1}{4}$$

To find the y coordinate of the point of tangency, compute $f(2)$.

$$y = f(2) = \frac{1}{2}$$

Now use the point-slope formula $y - y_0 = m(x - x_0)$ with $m = -\frac{1}{4}$ and $(x_0, y_0) = \left(2, \frac{1}{2}\right)$ to conclude that the equation of the tangent is

$$y - \frac{1}{2} = -\frac{1}{4}(x - 2) \qquad \text{or} \qquad y = -\frac{1}{4}x + 1$$

The situation is illustrated in Figure 1.8.

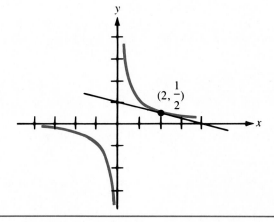

Figure 1.8 The curve $y = \frac{1}{x}$ and the tangent when $x = 2$.

Derivative Notation

Symbols other than f' are often used to denote the derivative. For example, if y rather than $f(x)$ is used to denote the function itself, the symbol $\frac{dy}{dx}$

$\left(\text{suggesting slope } \dfrac{\Delta y}{\Delta x} \right)$ is frequently used instead of $f'(x)$. Hence, instead of the statement

> If $f(x) = x^2$, then $f'(x) = 2x$

you could write

> If $y = x^2$, then $\dfrac{dy}{dx} = 2x$

The symbol $\dfrac{dy}{dx}$ is read "the derivative of y with respect to x." Sometimes the two notations are combined as in the statement

> If $f(x) = x^2$, then $\dfrac{df}{dx} = 2x$

By omitting reference to y and f altogether, you can condense these statements and write

$$\frac{d}{dx}(x^2) = 2x$$

which is read "the derivative with respect to x of x^2 equals $2x$."

The Maximization of Profit

To illustrate how the derivative can be applied to practical situations, here, as promised, is the solution to the profit maximization problem that was discussed at the beginning of the section. A more thorough treatment of the use of calculus to solve optimization problems will be given in Chapter 3.

EXAMPLE 1.4   Suppose a manufacturer's profit from the sale of radios is given by the function $P(x) = 400(15 - x)(x - 2)$, where x is the price at which the radios are sold. Find the optimal selling price.

SOLUTION
For reference, the graph of this profit function is sketched once again in Figure 1.9.

Your goal is to find the value of x for which the profit $P(x)$ is greatest. This is the value of x for which the slope of the tangent is zero. Since the slope of the tangent is given by the derivative, begin by computing $P'(x)$. In this case, it is easier to work with the unfactored form of the profit function.

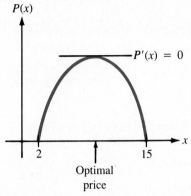

Figure 1.9 The profit function $P(x) = 400(15 - x)(x - 2)$.

$$P(x) = -400x^2 + 6{,}800x - 12{,}000$$

From the definition of the derivative,

$$P'(x) = \lim_{\Delta x \to 0} \frac{P(x + \Delta x) - P(x)}{\Delta x} =$$

$$\lim_{\Delta x \to 0} \frac{-400(x + \Delta x)^2 + 6{,}800(x + \Delta x) - 12{,}000 - (-400x^2 + 6{,}800x - 12{,}000)}{\Delta x}$$

$$= \lim_{\Delta x \to 0} \frac{-400(\Delta x)^2 - 800x\,\Delta x + 6{,}800\Delta x}{\Delta x}$$

$$= \lim_{\Delta x \to 0} (-400\Delta x - 800x + 6{,}800)$$

$$= -800x + 6{,}800$$

To find the value of x for which the slope of the tangent is zero, set the derivative equal to zero and solve the resulting equation for x as follows:

$$P'(x) = 0$$

$$-800x + 6{,}800 = 0$$

$$800x = 6{,}800$$

$$x = \frac{6{,}800}{800} = 8.5$$

It follows that $x = 8.5$ is the x coordinate of the peak of the graph and that the optimal selling price is $8.50 per radio.

Limits

As you have seen, the derivative is the value that a certain difference quotient approaches as the variable Δx approaches zero. In general, mathematicians use the word **limit** to denote the value that a function approaches as its variable approaches a specific number. Limits play a central role in modern mathematics and form the basis for a rigorous development of calculus.

In this text, you will need only an intuitive understanding of the limit concept and familiarity with the use of limit notation. However, for those wishing greater insight into this important theoretical topic, there is a more detailed discussion of limits in Section B of the appendix.

Differentiability and Continuity

Not all functions have a derivative for every value of x. Three functions that do not have derivatives when $x = 0$ are sketched in Figure 1.10.

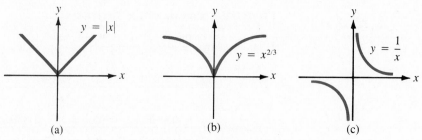

Figure 1.10 Three functions that are not differentiable at $x = 0$.

At the point $(0, 0)$ of the function $f(x) = |x|$ in Figure 1.10a, the tangent line cannot be uniquely determined. As a result, the derivative, which gives the slope of the tangent, cannot be defined for $x = 0$. The function $f(x) = x^{2/3}$ in Figure 1.10b has a vertical tangent when $x = 0$. Since the slope of a vertical line is undefined, this function has no derivative when $x = 0$. The function $f(x) = \dfrac{1}{x}$ in Figure 1.10c has no derivative at $x = 0$ because the function itself is undefined for this value of x.

A function that has a derivative when $x = a$ is said to be **differentiable at $x = a$**. A function that is differentiable at all values of x in its domain is said to be a **differentiable function**. The graphs of differentiable functions must be "smooth." They cannot have corners or cusps as the graphs in Figure 1.10a and 1.10b do. Most of the functions you will encounter in this text will be differentiable at most points. It can be shown, for example, that polynomials are differentiable everywhere and that rational functions fail to have derivatives only at those values of x that make their denominators zero.

Recall from Chapter 1, Section 2, that a function whose graph is an unbroken curve is said to be **continuous.** (A more formal definition is given in the appendix.) It can be shown that every differentiable function must be continuous, although, as the function $f(x) = |x|$ in Figure 1.10a indicates, not every continuous function is differentiable.

Problems

In Problems 1 through 7, compute the derivative of the given function and find the slope of the line that is tangent to its graph for the specified value of x.

1. $f(x) = 5x - 3; x = 2$

2. $f(x) = x^2 - 1; x = -1$

3. $f(x) = 2x^2 - 3x + 5; x = 0$

4. $f(x) = x^3 - 1; x = 2$

5. $f(x) = \dfrac{2}{x}; x = \dfrac{1}{2}$

6. $f(x) = \dfrac{1}{x^2}; x = 2$

7. $f(x) = \sqrt{x}; x = 9$

In Problems 8 through 11, compute the derivative of the given function and find the equation of the line that is tangent to its graph for the specified value of x.

8. $f(x) = x^2 + x + 1; x = 2$

9. $f(x) = x^3 - x; x = -2$

10. $f(x) = \dfrac{3}{x^2}; x = \dfrac{1}{2}$

11. $f(x) = 2\sqrt{x}; x = 4$

12. Suppose $f(x) = x^2$.
    (a) Compute the slope of the secant joining the points on the graph of f whose x coordinates are $x = -2$ and $x = -1.9$.
    (b) Use calculus to compute the slope of the line that is tangent to the graph when $x = -2$ and compare this slope to your answer in part (a).

13. Suppose $f(x) = x^3$.
    (a) Compute the slope of the secant joining the points on the graph of f whose x coordinates are $x = 1$ and $x = 1.1$.
    (b) Use calculus to compute the slope of the line that is tangent to the graph when $x = 1$ and compare this slope with your answer in part (a).

14. (a) Find the derivative of the linear function $f(x) = 3x - 2$.
    (b) Find the equation of the tangent to the graph of this function at the point $(-1, -5)$.
    (c) Explain how the answers to parts (a) and (b) could have been obtained from geometric considerations with no calculation whatsoever.

15. Sketch the graph of the function $y = x^2 - 3x$ and use calculus to find its lowest point.

16. Sketch the graph of the function $y = 1 - x^2$ and use calculus to find its highest point.

17. Sketch the graph of the function $y = x^3 - x^2$. Determine the values of x for which the derivative is zero. What happens to the graph at the corresponding points?

Maximization of profit   18. A manufacturer can produce tape recorders at a cost of $20 apiece. It is estimated that if the tape recorders are sold for x dollars apiece, consumers will buy $120 - x$ of them a month. Use calculus to determine the price at which the manufacturer's profit will be the greatest.

19. What can you conclude about the graph of a function between $x = a$ and $x = b$ if its derivative is positive whenever $a \leq x \leq b$?

20. Sketch a graph of a function f whose derivative has all of the following properties:
    (a) $f'(x) > 0$ when $x < 1$ and when $x > 5$
    (b) $f'(x) < 0$ when $1 < x < 5$
    (c) $f'(1) = 0$ and $f'(5) = 0$

21. (a) Find the derivatives of the functions $y = x^2$ and $y = x^2 - 3$ and account geometrically for their similarity.
    (b) Without further computation, find the derivative of the function $y = x^2 + 5$.

22. (a) Find the derivative of the function $y = x^2 + 3x$.
    (b) Find the derivatives of the functions $y = x^2$ and $y = 3x$ separately.
    (c) How is the answer in part (a) related to the answers in part (b)?
    (d) In general, if $f(x) = g(x) + h(x)$, what would you guess is the relationship between the derivative of f and the derivatives of g and h?

23. (a) Compute the derivatives of the functions $y = x^2$ and $y = x^3$.
    (b) Examine your answers in part (a). Can you detect a pattern? What do you think is the derivative of $y = x^4$? How about the derivative of $y = x^{27}$?

2   Techniques of Differentiation

In Section 1, you learned how to find the derivative of a function by letting Δx approach zero in the expression for the slope of a secant. For even the simplest functions, this process is tedious and time-consuming. In this section, you will see some rules that simplify the process. Justification of some of these rules will be given at the end of the section, after you have had a chance to practice using them.

The Derivative of a Power Function

A **power function** is a function of the form $f(x) = x^n$, where n is a real number. For example, $f(x) = x^2$, $f(x) = x^{-3}$, and $f(x) = x^{1/2}$ are all power functions. So are $f(x) = \dfrac{1}{x^2}$ and $f(x) = \sqrt[3]{x}$ since they can be rewritten as $f(x) = x^{-2}$ and $f(x) = x^{1/3}$, respectively. Here is a simple rule you can use to find the derivative of any power function. The proof of this rule will be given in Chapter 4.

The Power Rule

For any number n,

$$\frac{d}{dx}(x^n) = nx^{n-1}$$

That is, to find the derivative of x^n, reduce the power of x by 1 and multiply by the original power.

According to this rule, the derivative of x^2 is $2x^1$ or $2x$, which agrees with the result in Example 1.2. Similarly, the derivative of x^3 is $3x^2$, which should agree with the result you obtained in Problem 23 of the preceding section. Here are a few more calculations illustrating the use of the power rule.

EXAMPLE 2.1   Differentiate (find the derivative of) each of the following functions:

(a) $y = x^{27}$

(b) $y = \dfrac{1}{x^{27}}$

(c) $y = \sqrt{x}$

(d) $y = \dfrac{1}{\sqrt{x}}$

SOLUTION

In each case, use exponential notation to write the function as a power function and then apply the general rule. (You can find a review of exponential notation in Section A of the appendix at the back of the book.)

(a) $\dfrac{d}{dx}(x^{27}) = 27x^{27-1} = 27x^{26}$

(b) $\dfrac{d}{dx}\left(\dfrac{1}{x^{27}}\right) = \dfrac{d}{dx}(x^{-27}) = -27x^{-27-1} = -27x^{-28} = -\dfrac{27}{x^{28}}$

(c) $\dfrac{d}{dx}(\sqrt{x}) = \dfrac{d}{dx}(x^{1/2}) = \dfrac{1}{2}x^{1/2-1} = \dfrac{1}{2}x^{-1/2} = \dfrac{1}{2\sqrt{x}}$

(d) $\dfrac{d}{dx}\left(\dfrac{1}{\sqrt{x}}\right) = \dfrac{d}{dx}(x^{-1/2}) = -\dfrac{1}{2}x^{-1/2-1} = -\dfrac{1}{2}x^{-3/2} = -\dfrac{1}{2\sqrt{x^3}}$

The Derivative of a Constant

The derivative of any constant function is zero. This is because the graph of a constant function $y = c$ is a horizontal line and its slope is zero. Thus, for example, if $f(x) = 5$, then $f'(x) = 0$.

The Derivative of a Constant

> For any constant c,
>
> $$\dfrac{d}{dx}(c) = 0$$
>
> That is, the derivative of a constant is zero.

The Derivative of a Constant Times a Function

The next rule expresses the fact that the curve $y = cf(x)$ is c times as steep as the curve $y = f(x)$.

The Constant Multiple Rule

> For any constant c,
>
> $$\dfrac{d}{dx}(cf) = c\dfrac{df}{dx}$$
>
> That is, the derivative of a constant times a function is equal to the constant times the derivative of the function.

EXAMPLE 2.2 Differentiate the function $y = 3x^5$.

SOLUTION

You already know that $\dfrac{d}{dx}(x^5) = 5x^4$. Combining this with the constant multiple rule you get

$$\dfrac{d}{dx}(3x^5) = 3\dfrac{d}{dx}(x^5) = 3(5x^4) = 15x^4$$

The Derivative of a Sum

The next rule states that a sum can be differentiated term by term.

The Sum Rule

$$\frac{d}{dx}(f + g) = \frac{df}{dx} + \frac{dg}{dx}$$

That is, the derivative of a sum is the sum of the individual derivatives.

EXAMPLE 2.3 Differentiate the function $y = x^2 + 3x^5$.

SOLUTION

You know that $\frac{d}{dx}(x^2) = 2x$ and that $\frac{d}{dx}(3x^5) = 15x^4$. According to the sum rule, you simply add these derivatives to get the derivative of the sum $x^2 + 3x^5$. That is,

$$\frac{d}{dx}(x^2 + 3x^5) = \frac{d}{dx}(x^2) + \frac{d}{dx}(3x^5) = 2x + 15x^4$$

By combining the sum rule with the power and constant multiple rules, you can differentiate any polynomial. Here is an example.

EXAMPLE 2.4 Differentiate the polynomial $y = 5x^3 - 4x^2 + 12x - 8$.

SOLUTION
Differentiate this sum term by term to get

$$\frac{dy}{dx} = \frac{d}{dx}(5x^3) + \frac{d}{dx}(-4x^2) + \frac{d}{dx}(12x) + \frac{d}{dx}(-8)$$

$$= 15x^2 - 8x^1 + 12x^0 + 0$$

$$= 15x^2 - 8x + 12$$

The Derivative of a Product

Suppose you wanted to differentiate the product $y = x^2(3x + 1)$. You might be tempted to differentiate the factors x^2 and $3x + 1$ separately, and then multiply your answers. That is, since $\frac{d}{dx}(x^2) = 2x$ and $\frac{d}{dx}(3x + 1) = 3$, you might suspect that $\dfrac{dy}{dx} = 6x$. However, this answer is wrong. To see this, rewrite the function as $y = 3x^3 + x^2$ and observe that the derivative is $9x^2 + 2x$ and not $6x$. The derivative of a product is *not* the product of the individual derivatives. Here is the correct formula for the derivative of a product.

The Product Rule

$$\frac{d}{dx}(fg) = f\frac{dg}{dx} + g\frac{df}{dx}$$

That is, the derivative of a product is the first factor times the derivative of the second plus the second factor times the derivative of the first.

The use of this rule is illustrated in the next example.

EXAMPLE 2.5   Differentiate the function $y = x^2(3x + 1)$.

SOLUTION
According to the product rule,

$$\frac{d}{dx}[x^2(3x + 1)] = x^2\frac{d}{dx}(3x + 1) + (3x + 1)\frac{d}{dx}(x^2)$$

$$= x^2(3) + (3x + 1)(2x)$$

$$= 9x^2 + 2x$$

which is precisely the result that was obtained when the product was multiplied out and differentiated as a sum.

The Derivative of a Quotient

The derivative of a quotient is not the quotient of the individual derivatives. Here is the correct rule.

The Derivative of a Quotient

$$\frac{d}{dx}\left(\frac{f}{g}\right) = \frac{g\dfrac{df}{dx} - f\dfrac{dg}{dx}}{g^2}$$

The quotient rule is probably the most complicated formula you have had to learn so far in this book. Here is one way to remember it. The numerator resembles the product rule except that it contains a minus sign, which makes the order in which the terms are written important. Begin by squaring the denominator (since this is easy to do) and then, while still thinking of the original denominator, copy it in the numerator. This gets you started with the proper term in the numerator, and you can easily write down the rest thinking of the product rule. Don't forget to insert the minus sign, without which the rule would not have been so hard to remember in the first place!

Using the quotient rule, you can now differentiate any rational function. Here is an example.

EXAMPLE 2.6   Differentiate the rational function $y = \dfrac{x^2 + 2x - 21}{x - 3}$.

SOLUTION
According to the quotient rule,

$$\frac{dy}{dx} = \frac{(x - 3)\dfrac{d}{dx}(x^2 + 2x - 21) - (x^2 + 2x - 21)\dfrac{d}{dx}(x - 3)}{(x - 3)^2}$$

$$= \frac{(x - 3)(2x + 2) - (x^2 + 2x - 21)(1)}{(x - 3)^2}$$

$$= \frac{2x^2 - 4x - 6 - x^2 - 2x + 21}{(x - 3)^2}$$

$$= \frac{x^2 - 6x + 15}{(x - 3)^2}$$

A Word of Advice

The quotient rule is somewhat cumbersome, so don't use it unnecessarily. Consider the following example.

EXAMPLE 2.7   Differentiate the function $y = \dfrac{2}{3x^2} - \dfrac{x}{3} + \dfrac{4}{5} + \dfrac{x + 1}{x}$.

SOLUTION
Don't use the quotient rule! Instead, rewrite the function as

$$y = \frac{2}{3}x^{-2} - \frac{1}{3}x + \frac{4}{5} + 1 + x^{-1}$$

and then apply the power rule term by term to get

$$\frac{dy}{dx} = \frac{2}{3}(-2x^{-3}) - \frac{1}{3} + 0 + 0 + (-1)x^{-2}$$

$$= -\frac{4}{3}x^{-3} - \frac{1}{3} - x^{-2}$$

$$= -\frac{4}{3x^3} - \frac{1}{3} - \frac{1}{x^2}$$

Derivation of the Sum Rule

The rules you have seen in this section can all be derived from the definition of the derivative. Rigorous derivations of these rules involve technical properties of limits and are discussed in the appendix at the back of the book. However, even without these technical details, you should be able to get a feel for how the rules are derived. To see why the sum rule is true, for example, consider a secant through the graph of the function $f + g$ as shown in Figure 2.1.

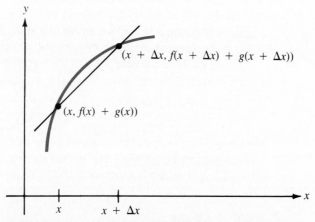

Figure 2.1 A secant through the curve $y = f(x) + g(x)$.

Begin with the expression for the slope of this secant and then rewrite it as the sum of two quotients, one involving f and the other involving g as follows:

$$\text{Slope of secant} = \frac{[f(x + \Delta x) + g(x + \Delta x)] - [f(x) + g(x)]}{\Delta x}$$

$$= \frac{f(x + \Delta x) - f(x)}{\Delta x} + \frac{g(x + \Delta x) - g(x)}{\Delta x}$$

To get the derivative of $f + g$, let Δx approach zero in this expression. Since

$$\lim_{\Delta x \to 0} \frac{f(x + \Delta x) - f(x)}{\Delta x} = \frac{df}{dx}$$

and

$$\lim_{\Delta x \to 0} \frac{g(x + \Delta x) - g(x)}{\Delta x} = \frac{dg}{dx}$$

it follows that

$$\frac{d}{dx}(f + g) = \frac{df}{dx} + \frac{dg}{dx}$$

and the sum rule is proved.

Derivation of the Product Rule

To show that $\frac{d}{dx}(fg) = f\frac{dg}{dx} + g\frac{df}{dx}$, begin with the appropriate difference quotient and rewrite it using the algebraic "trick" of subtracting and then adding the quantity $f(x + \Delta x)g(x)$ in the numerator as follows:

$$\text{Slope of secant} = \frac{f(x + \Delta x)g(x + \Delta x) - f(x)g(x)}{\Delta x}$$

$$= \frac{f(x + \Delta x)g(x + \Delta x) - f(x + \Delta x)g(x)}{\Delta x}$$

$$+ \frac{f(x + \Delta x)g(x) - f(x)g(x)}{\Delta x}$$

$$= f(x + \Delta x)\frac{g(x + \Delta x) - g(x)}{\Delta x}$$

$$+ g(x)\frac{f(x + \Delta x) - f(x)}{\Delta x}$$

Now let Δx approach zero. Since

$$\lim_{\Delta x \to 0} \frac{f(x + \Delta x) - f(x)}{\Delta x} = \frac{df}{dx}$$

$$\lim_{\Delta x \to 0} \frac{g(x + \Delta x) - g(x)}{\Delta x} = \frac{dg}{dx}$$

and

$$\lim_{\Delta x \to 0} f(x + \Delta x) = f(x)$$

it follows that

$$\frac{d}{dx}(fg) = f\frac{dg}{dx} + g\frac{df}{dx}$$

and the product rule is proved.

Problems

In Problems 1 through 25, differentiate the given function. Do as much of the computation as possible in your head and simplify your answers.

1. $y = x^2 + 2x + 3$

2. $y = 3x^5 - 4x^3 + 9x - 6$

3. $f(x) = x^9 - 5x^8 + x + 12$

4. $f(x) = \frac{1}{4}x^8 - \frac{1}{2}x^6 - x + 2$

5. $y = \frac{1}{x} + \frac{1}{x^2} - \frac{1}{\sqrt{x}}$

6. $y = \frac{3}{x} - \frac{2}{x^2} + \frac{2}{3x^3}$

7. $f(x) = \sqrt{x^3} + \frac{1}{\sqrt{x^3}}$

8. $f(x) = 2\sqrt{x^3} + \frac{4}{\sqrt{x}} - \sqrt{2}$

9. $y = -\frac{x^2}{16} + \frac{2}{x} - x^{3/2} + \frac{1}{3x^2} + \frac{x}{3}$

10. $y = -\frac{2}{x^2} + x^{2/3} + \frac{1}{2\sqrt{x}} + \frac{x^2}{4} + \sqrt{5} + \frac{x + 2}{3}$

11. $f(x) = (2x + 1)(3x - 2)$ (Use the product rule.)

12. $f(x) = (x^2 - 5)(1 - 2x)$ (Use the product rule.)

13. $y = 10(3x + 1)(1 - 5x)$ (Use the product rule.)

14. $y = 400(15 - x^2)(3x - 2)$ (Use the product rule.)

15. $f(x) = \frac{1}{3}(x^5 - 2x^3 + 1)$

16. $f(x) = -3(5x^3 - 2x + 5)$

17. $y = \frac{x + 1}{x - 2}$

18. $y = \frac{2x - 3}{5x + 4}$

19. $f(x) = \frac{x}{x^2 - 2}$

20. $f(x) = \frac{1}{x - 2}$

21. $y = \frac{3}{x + 5}$

22. $y = \frac{x^2 + 1}{1 - x^2}$

23. $f(x) = \frac{x^2 - 3x + 2}{2x^2 + 5x - 1}$

24. $f(x) = \frac{x^2 + 2x + 1}{3}$

25. $y = (2x + 1)(x - 3)(1 - 4x)$

In Problems 26 through 29, find the equation of the line that is tangent to the graph of the given function at the specified point.

26. $y = x^5 - 3x^3 - 5x + 2$; $(1, -5)$

27. $y = (x^2 + 1)(1 - x^3)$; $(1, 0)$

28. $f(x) = \frac{x + 1}{x - 1}$; $(0, -1)$

29. $f(x) = 1 - \dfrac{1}{x} + \dfrac{2}{\sqrt{x}}; \left(4, \dfrac{7}{4}\right)$

In Problems 30 through 33, find the equation of the line that is tangent to the graph of the given function at the point $(x, f(x))$ for the specified value of x.

30. $f(x) = x^4 - 3x^3 + 2x^2 - 6; x = 2$

31. $f(x) = x - \dfrac{1}{x^2}; x = 1$

32. $f(x) = \dfrac{x^2 + 2}{x^2 - 2}; x = -1$

33. $f(x) = (x^3 - 2x^2 + 3x - 1)(x^5 - 4x^2 + 2); x = 0$

34. (a) Differentiate the function $y = 2x^2 - 5x - 3$.
    (b) Now factor the function in part (a) as $y = (2x + 1)(x - 3)$ and differentiate using the product rule. Show that the two answers are the same.

35. (a) Use the quotient rule to differentiate the function $y = \dfrac{2x - 3}{x^3}$.
    (b) Rewrite the function as $y = x^{-3}(2x - 3)$ and differentiate using the product rule.
    (c) Rewrite the function as $y = 2x^{-2} - 3x^{-3}$ and differentiate.
    (d) Show that your answers to parts (a), (b), and (c) are the same.

36. The product rule tells you how to differentiate the product of any two functions, while the constant multiple rule tells you how to differentiate products in which one of the factors is constant. Show that the two rules are consistent. In particular, use the product rule to show that $\dfrac{d}{dx}(cf) = c\dfrac{df}{dx}$ if c is a constant.

37. Sketch the graph of the function $f(x) = x^2 - 4x - 5$ and use calculus to determine its lowest point.

38. Sketch the graph of the function $f(x) = 3 - 2x - x^2$ and use calculus to determine its highest point.

39. Find numbers a and b such that the lowest point on the graph of the function $f(x) = ax^2 + bx$ is $(3, -8)$.

40. Find numbers a, b, and c such that the graph of the function $f(x) = ax^2 + bx + c$ will have x intercepts at $(0, 0)$ and $(5, 0)$, and a tangent with slope 1 when $x = 2$.

41. Find the equations of all the tangents to the graph of the function $f(x) = x^2 - 4x + 25$ that pass through the origin $(0, 0)$.

42. Find all the points (x, y) on the graph of the function $y = 4x^2$ with the property that the tangent to the graph at (x, y) passes through the point $(2, 0)$.

Consumer expenditure

43. The consumer demand for a certain commodity is $D(p) = -200p + 12{,}000$ units per month when the market price is p dollars per unit.
    (a) Express consumers' total monthly expenditure for the commodity as a function of p and draw the graph.
    (b) Use calculus to determine the market price for which the consumer expenditure is greatest.

44. Show that $\dfrac{d}{dx}(fgh) = fg\dfrac{dh}{dx} + fh\dfrac{dg}{dx} + gh\dfrac{df}{dx}$. (*Hint:* Apply the product rule twice.)

45. Derive the quotient rule. {*Hint:* Show that the difference quotient is

$$\frac{1}{\Delta x}\left[\frac{f(x + \Delta x)}{g(x + \Delta x)} - \frac{f(x)}{g(x)}\right] = \frac{g(x)f(x + \Delta x) - f(x)g(x + \Delta x)}{g(x + \Delta x)g(x)\Delta x}$$

Before letting Δx approach zero, rewrite this quotient using the trick of subtracting and adding $g(x)f(x)$ in the numerator.}

3 Rate of Change and Marginal Analysis

In this section, you will see how the derivative can be interpreted as a rate of change. Viewed in this way, derivatives may represent such quantities as the rate at which population grows, the speed of a moving object, a manufacturer's marginal cost, the rate of inflation, or the rate at which natural resources are being depleted.

At the beginning of this chapter, you saw a brief discussion suggesting the relationship between rates of change and derivatives. The reasoning was that the rate of change of a function with respect to its independent variable is equal to the steepness of its graph, which is measured by the slope of its tangent at the point in question. Since the slope of the tangent is given by the derivative of the function, it follows that the rate of change is equal to the derivative.

The purpose of this section is to make the relationship between derivatives and rates of change more precise. The discussion begins with a familiar practical situation, which will serve as a model for the general case.

Average and Instantaneous Speed

Imagine that a car is moving along a straight road, and that $D(t)$ is its distance from its starting point after t hours. Suppose you want to determine the speed of the car at a particular time t but do not have access to the car's speedometer. Here's what you can do.

First record the position of the car at time t, and again at some later time $t + \Delta t$. That is, determine $D(t)$ and $D(t + \Delta t)$. Then compute the **average speed** of the car between the times t and $t + \Delta t$ as follows:

$$\text{Average speed} = \frac{\text{change in distance}}{\text{change in time}} = \frac{D(t + \Delta t) - D(t)}{\Delta t}$$

The situation is illustrated in Figure 3.1.

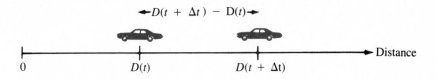

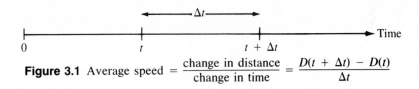

Figure 3.1 Average speed $= \dfrac{\text{change in distance}}{\text{change in time}} = \dfrac{D(t + \Delta t) - D(t)}{\Delta t}$

Since the speed of the car may fluctuate during the time interval from t to $t + \Delta t$, it is unlikely that this average speed is equal to the **instantaneous speed** (the speed shown on the speedometer) at time t. However, if Δt is small, the possibility of drastic changes in speed is small, and the average speed may be a fairly good approximation to the instantaneous speed. Indeed, you can find the instantaneous speed at time t by letting Δt approach zero in the expression for the average speed. That is,

$$\text{Instantaneous speed} = \lim_{\Delta t \to 0} \frac{D(t + \Delta t) - D(t)}{\Delta t}$$

This limit expression for instantaneous speed should look familiar. It is exactly the formula defining the derivative $D'(t)$ of the function $D(t)$. Thus, the instantaneous speed (or rate of change of distance) is equal to the derivative of distance.

Average and Instantaneous Rate of Change

These ideas can be extended to more general situations. Suppose that y is a function of x, say $y = f(x)$. Corresponding to a change from x to $x + \Delta x$, the variable y changes by an amount $\Delta y = f(x + \Delta x) - f(x)$. The resulting **average rate of change of y with respect to x** is the difference quotient

$$\text{Average rate of change} = \frac{\text{change in } y}{\text{change in } x} = \frac{f(x + \Delta x) - f(x)}{\Delta x}$$

As the interval over which you are averaging becomes shorter (that is, as Δx approaches zero), the average rate of change approaches what you would intuitively call the **instantaneous rate of change of y with respect to x,** and the difference quotient approaches the derivative $f'(x)$ or $\dfrac{dy}{dx}$. That is,

$$\text{Instantaneous rate of change} = \lim_{\Delta x \to 0} \frac{f(x + \Delta x) - f(x)}{\Delta x}$$

$$= f'(x) = \frac{dy}{dx}$$

Instantaneous Rate of Change

> If $y = f(x)$, the instantaneous rate of change of y with respect to x is given by the derivative of f. That is,
>
> $$\text{Rate of change} = f'(x) = \frac{dy}{dx}$$

EXAMPLE 3.1   It is estimated that x months from now, the population of a certain community will be $P(x) = x^2 + 20x + 8{,}000$.

(a) At what rate will the population be changing with respect to time 15 months from now?

(b) By how much will the population actually change during the 16th month?

SOLUTION

(a) The rate of change of the population with respect to time is the derivative of the population function. That is,

$$\text{Rate of change} = P'(x) = 2x + 20$$

The rate of change of the population 15 months from now will be

$$P'(15) = 2(15) + 20 = 50 \text{ people per month}$$

(b) The actual change in the population during the 16th month is the difference between the population at the end of 16 months and the population at the end of 15 months. That is,

$$\text{Change in population} = P(16) - P(15) = 8{,}576 - 8{,}525$$
$$= 51 \text{ people}$$

The reason for the difference in Example 3.1 between the actual change in population during the 16th month in part (b) and the monthly rate of change at the beginning of that month in part (a) is that the rate of change of the population varied during the month. The instantaneous rate of change in part (a) can be thought of as the change in population that would occur during the 16th month if the rate of change of the population remained constant.

Approximation

In Example 3.1, the rate of change of population at the beginning of the month in part (a) was close to, but not equal to, the actual change in population during the month in part (b). In geometric terms, the difference between these two quantities is the difference between the slope of a tangent and the slope of a nearby secant. In particular, the derivative $P'(15)$ in part (a) is the slope of the line that is tangent to the population curve $P(x)$ when $x = 15$. The difference $P(16) - P(15)$ in part (b) is the slope

$$\frac{\text{Change in } P}{\text{Change in } x} = \frac{P(15 + 1) - P(15)}{1}$$

of the secant joining the two points on the curve whose x coordinates are 15 and 16. The situation is illustrated in Figure 3.2.

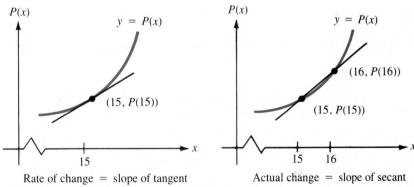

Rate of change = slope of tangent   Actual change = slope of secant

Figure 3.2 The relationship between the rate of change and the actual change.

The answers to parts (a) and (b) of Example 3.1 were almost equal because the points $(15, P(15))$ and $(16, P(16))$ are close together and lie on a portion of the curve that does not bend very much. For such points, the slope of the secant is a good approximation to the slope of the tangent. Because the similarity of the answers in parts (a) and (b) of Example 3.1 is typical, and because it is usually easier to compute the derivative for one value of x than to compute and subtract values of the function for two values of x, one often uses the derivative of a function to approximate the change in the function caused by a 1-unit increase in its variable.

Approximation Formula

$f'(x) \simeq$ change in f caused by a 1-unit increase in x

Here is an example.

EXAMPLE 3.2   It is estimated that the weekly output at a certain plant is $Q(x) = -x^3 + 60x^2 + 1,200x$ units, where x is the number of workers employed at the plant. Currently 30 workers are employed. Use calculus to estimate the change in the weekly output that will result from the addition of 1 more worker to the work force.

SOLUTION

The rate of change of the output Q with respect to the number x of workers is the derivative

$$Q'(x) = -3x^2 + 120x + 1,200$$

For any value of x, this derivative is an approximation to the number of additional units that will be produced each week from the hiring of 1 more worker. Hence,

$$\text{Change in output} \simeq Q'(30) = -3(30)^2 + 120(30) + 1,200$$
$$= 2,100 \text{ units}$$

For practice, compute the change in output exactly and compare your answer to the approximation. Is the approximation a good one?

Marginal Analysis in Economics

In economics, the rate of change of the total production cost with respect to the number of units produced is called the **marginal cost**. It is measured in dollars per unit and is often a good approximation to the cost of producing 1 additional unit.

Marginal Cost

If $C(q)$ is the total cost of manufacturing q units, then
$$\text{Marginal cost} = C'(q)$$

> The marginal cost is approximately the cost of producing 1 additional unit. That is,
>
> $$C'(q) \simeq \text{cost of producing the } (q + 1)\text{st unit}$$

EXAMPLE 3.3   Suppose the total cost of manufacturing q units of a certain commodity is $C(q) = 3q^2 + 5q + 10$.

(a) Derive a formula for the marginal cost.
(b) Use the marginal cost to approximate the cost of producing the 51st unit.
(c) Compute the actual cost of producing the 51st unit.

SOLUTION
(a) The marginal cost is the derivative $C'(q) = 6q + 5$.
(b) The 51st unit is the $(q + 1)$st unit when $q = 50$. Hence,

$$\text{Cost of 51st unit} \simeq C'(50) = 6(50) + 5 = \$305$$

(c) The actual cost of producing the 51st unit is the difference between the cost of producing 51 units and the cost of producing 50 units. That is,

$$\text{Cost of 51st unit} = C(51) - C(50) = 8{,}068 - 7{,}760 = \$308$$

In the two preceding examples, the derivative of a function was used to estimate the change in the function produced by a 1-unit increase in the size of its variable. In economics, the term **marginal analysis** is used to denote this approximation procedure.

Percentage Rate of Change

In many practical situations, the rate of change of a quantity is not as significant as its **percentage rate of change**. For example, a yearly rate of change in population of 500 people in a city of 5 million would be negligible, while the same rate of change could have enormous impact in a town of 2,000. The percentage rate of change compares the rate of change of a quantity with the size of that quantity:

$$\text{Percentage rate of change} = 100 \frac{\text{rate of change of quantity}}{\text{size of quantity}}$$

For example, a rate of change of 500 people per year in the population of a city of 5 million amounts to a percentage rate of change of only $\frac{100(500)}{5{,}000{,}000} = 0.01$ percent of the population per year. On the other hand, the same rate of change in a town of 2,000 is equal to a percentage rate of change of $\dfrac{100(500)}{2,000} = 25$ percent of the population per year.

Here is the formula for percentage rate of change written in terms of the derivative.

Percentage Rate of Change

> If $y = f(x)$, the percentage rate of change of y with respect to x is given by the formula
>
> $$\text{Percentage rate of change} = 100\frac{f'(x)}{f(x)} = 100\frac{dy/dx}{y}$$

EXAMPLE 3.4 The gross national product (GNP) of a certain country was $N(t) = t^2 + 5t + 100$ billion dollars t years after 1980.

(a) At what rate was the GNP changing with respect to time in 1985?

(b) At what percentage rate was the GNP changing with respect to time in 1985?

SOLUTION

(a) The rate of change of the GNP is the derivative $N'(t) = 2t + 5$. The rate of change in 1985 was $N'(5) = 2(5) + 5 = 15$ billion dollars per year.

(b) The percentage rate of change of the GNP in 1985 was

$$100\frac{N'(5)}{N(5)} = 100\frac{15}{150} = 10 \text{ percent per year}$$

Problems

Newspaper circulation

1. It is estimated that t years from now, the circulation of a local newspaper will be $C(t) = 100t^2 + 400t + 5,000$.

   (a) Derive an expression for the rate at which the circulation will be changing with respect to time t years from now.

   (b) At what rate will the circulation be changing with respect to time 5 years from now? Will the circulation be increasing or decreasing at that time?

   (c) By how much will the circulation actually change during the 6th year?

Speed of a moving object

2. An object moves along a straight line so that after t minutes, its distance from its starting point is $D(t) = 10t + \dfrac{5}{t + 1}$ meters.

(a) At what speed is the object moving at the end of 4 minutes?

(b) How far does the object actually travel during the 5th minute?

Worker efficiency

3. An efficiency study of the morning shift at a certain factory indicates that an average worker who arrives on the job at 8:00 A.M. will have assembled $f(x) = -x^3 + 6x^2 + 15x$ transistor radios x hours later.

(a) Derive a formula for the rate at which the worker will be assembling radios after x hours.

(b) At what rate will the worker be assembling radios at 9:00 A.M.?

(c) How many radios will the worker actually assemble between 9:00 and 10:00 A.M.?

Air pollution

4. An environmental study of a certain suburban community suggests that t years from now, the average level of carbon monoxide in the air will be $Q(t) = 0.05t^2 + 0.1t + 3.4$ parts per million.

(a) At what rate will the carbon monoxide level be changing with respect to time 1 year from now?

(b) By how much will the carbon monoxide level change this year?

(c) By how much will the carbon monoxide level change over the next 2 years?

Population growth

5. It is estimated that t years from now, the population of a certain suburban community will be $P(t) = 20 - \dfrac{6}{t + 1}$ thousand.

(a) Derive a formula for the rate at which the population will be changing with respect to time t years from now.

(b) At what rate will the population be growing 1 year from now?

(c) By how much will the population actually increase during the 2nd year?

(d) At what rate will the population be growing 9 years from now?

(e) What will happen to the rate of population growth in the long run?

SAT scores

6. It is estimated that x years from now, the average SAT score of the incoming students at an eastern liberal arts college will be $f(x) = -6x + 582$.

(a) Derive an expression for the rate at which the average SAT score will be changing with respect to time x years from now.

(b) What is the significance of the fact that the expression in part (a) is a constant? What is the significance of the fact that the constant in part (a) is negative?

Distance

7. Two cars leave an intersection at the same time. One travels east at a constant speed of 60 kilometers per hour, while the other goes north at a constant speed of 80 kilometers per hour. Find an expression for the rate at which the distance between the cars is changing with respect to time.

Marginal analysis

8. It is estimated that the weekly output at a certain plant is $Q(x) =$

$-x^2 + 2,100x$ units, where x is the number of workers employed at the plant. Currently there are 60 workers employed at the plant.

(a) Use calculus to estimate the effect that 1 additional worker will have on the weekly output.

(b) Compute the actual change in the weekly output that will result if 1 additional worker is hired.

Marginal analysis

9. Suppose the total cost in dollars of manufacturing q units is $C(q) = 3q^2 + q + 500$.

(a) Use marginal analysis to estimate the cost of manufacturing the 41st unit.

(b) Compute the actual cost of manufacturing the 41st unit.

Marginal analysis

10. A manufacturer's total cost is $C(q) = 0.1q^3 - 0.5q^2 + 500q + 200$ dollars, where q is the number of units produced.

(a) Use marginal analysis to estimate the cost of manufacturing the 4th unit.

(b) Compute the actual cost of manufacturing the 4th unit.

Marginal analysis

11. A manufacturer's total monthly revenue is $R(q) = 240q + 0.05q^2$ dollars when q units are produced and sold during the month. Currently, the manufacturer is producing 80 units a month and is planning to increase the monthly output by 1 unit.

(a) Use marginal analysis to estimate the additional revenue that will be generated by the production and sale of the 81st unit.

(b) Use the revenue function to compute the actual additional revenue that will be generated by the production and sale of the 81st unit.

Marginal analysis

12. At a certain factory, the daily output is $Q(L) = 360L^{1/3}$ units, where L is the size of the labor force measured in worker-hours. Currently 1,000 worker-hours of labor are used each day. Use marginal analysis to estimate the effect that 1 additional worker-hour of labor will have on the daily output.

Marginal analysis

13. At a certain factory, the daily output is $Q(K) = 600K^{1/2}$ units, where K denotes the capital investment measured in units of $1,000. The current capital investment is $900,000. Use marginal analysis to estimate the effect that an additional capital investment of $1,000 will have on the daily output.

Marginal analysis

14. At a certain factory, the daily output is $Q = 3,000K^{1/2}L^{1/3}$ units, where K denotes the firm's capital investment measured in units of $1,000 and L denotes the size of the labor force measured in worker-hours. Suppose that the current capital investment is $400,000 and that 1,331 worker-hours of labor are used each day. Use marginal analysis to estimate the effect that an additional capital investment of $1,000 will have on the daily output if the size of the labor force is not changed.

Population growth 15. It is projected that x months from now, the population of a certain town will be $P(x) = 2x + 4x^{3/2} + 5,000$.
(a) At what rate will the population be changing with respect to time 9 months from now?
(b) At what percentage rate will the population be changing with respect to time 9 months from now?

Annual earnings 16. The gross annual earnings of a certain company were $A(t) = 0.1t^2 + 10t + 20$ thousand dollars t years after its formation in 1980.
(a) At what rate were the gross annual earnings of the company growing with respect to time in 1984?
(b) At what percentage rate were the gross annual earnings growing with respect to time in 1984?

Property tax 17. Records indicate that x years after 1980, the average property tax on a three-bedroom home in a certain community was $T(x) = 20x^2 + 40x + 600$ dollars.
(a) At what rate was the property tax increasing with respect to time in 1986?
(b) At what percentage rate was the property tax increasing with respect to time in 1986?

Population growth 18. It is estimated that t years from now, the population of a certain town will be $P(t) = t^2 + 200t + 10,000$.
(a) Express the percentage rate of change of the population as a function of t, simplify this function algebraically, and draw its graph.
(b) What will happen to the percentage rate of change of the population in the long run?

Salary increases 19. Your starting salary will be $24,000 and you will get a raise of $2,000 each year.
(a) Express the percentage rate of change of your salary as a function of time and draw the graph.
(b) At what percentage rate will your salary be increasing after 1 year?
(c) What will happen to the percentage rate of change of your salary in the long run?

Gross national product 20. The gross national product (GNP) of a certain country is growing at a constant rate. In 1980, the GNP was 125 billion dollars and in 1982, the GNP was 155 billion dollars. At what percentage rate was the GNP growing in 1985?

Free fall If an object is dropped or thrown vertically, its height (in feet) after t seconds is $H(t) = -16t^2 + S_0t + H_0$, where S_0 is the initial speed of the object and H_0 is its initial height. Use this formula to solve Problems 21 through 24.

21. A stone is dropped (with initial speed zero) from the top of a building 144 feet high.
    (a) When will the stone hit the ground? [That is, for what value of t is $H(t)$ equal to zero?]
    (b) With what speed will the stone hit the ground?

22. A ball is thrown vertically upward from the ground ($H_0 = 0$) with an initial speed of 160 feet per second ($S_0 = 160$).
    (a) When will the ball hit the ground?
    (b) With what speed will the ball hit the ground?
    (c) When will the ball reach its maximum height? (*Hint:* The speed of the ball will be zero when the ball reaches its maximum height.)
    (d) How high will the ball rise?

23. You are standing on the top of a building and throw a ball vertically upward. After 2 seconds, the ball passes you on the way down, and 2 seconds after that it hits the ground below.
    (a) What was the initial speed of the ball?
    (b) How high is the building?
    (c) What is the speed of the ball when it passes you on the way down?
    (d) What is the speed of the ball as it hits the ground?

24. A ball is thrown vertically upward from the ground with a certain initial speed S_0.
    (a) Derive a formula for the time at which the ball hits the ground.
    (b) Use the result of part (a) to prove that the ball will be falling at a speed of S_0 feet per second when it hits the ground.

Linear functions   25. Use calculus to prove that if y is a linear function of x, the rate of change of y with respect to x is constant.

Linear functions   26. If y is a linear function of x, what will happen to the percentage rate of change of y with respect to x as x increases without bound? Explain.

Manufacturing cost   27. Suppose the total manufacturing cost C at a certain factory is a function of the number q of units produced, which, in turn, is a function of the number t of hours during which the factory has been operating.

(a) What quantity is represented by the derivative $\dfrac{dC}{dq}$? In what units is this quantity measured?

(b) What quantity is represented by the derivative $\dfrac{dq}{dt}$? In what units is this quantity measured?

(c) What quantity is represented by the product $\dfrac{dC}{dq}\dfrac{dq}{dt}$? In what units is this quantity measured?

4   Approximation by Differentials

In the preceding section, you learned that the derivative of a function is its rate of change and that this rate of change is often a good approximation to the change in the function resulting from a 1-unit increase in its variable. But what if the change in the variable is something other than a 1-unit increase? In this section, the approximation procedure will be generalized, and you will be able to use calculus to estimate how a function is affected by *any* small change in the size of its variable.

The procedure is based on the fact that if y is a function of x, then

$$\text{Change in } y \simeq \left(\begin{array}{c}\text{rate of change of } y \\ \text{with respect to } x\end{array}\right)(\text{change in } x)$$

If the rate of change of y with respect to x happens to be constant, the approximation is exact. If the rate of change is not constant but the change in x is small, large variations in the rate of change are unlikely, and hence the approximation will be a good one.

Since the rate of change is given by the derivative, the approximation formula can be written more compactly as follows.

Approximation Formula

If $y = f(x)$ and Δx is a small change in x, then the corresponding change in y is

$$\Delta y \simeq \frac{dy}{dx}\Delta x$$

or, in functional notation, the corresponding change in f is

$$\Delta f = f(x + \Delta x) - f(x) \simeq f'(x)\Delta x$$

That is, the change in the function is approximately the derivative of the function times the change in its variable.

Notice that if you divide both sides of the approximation formula
$$f(x + \Delta x) - f(x) \simeq f'(x)\Delta x$$
by Δx you get
$$\frac{f(x + \Delta x) - f(x)}{\Delta x} \simeq f'(x)$$
which shows that the approximation formula is simply a restatement of the fact that the difference quotient is close to the derivative when Δx is small.

The use of the approximation formula is illustrated in the following example.

EXAMPLE 4.1   Suppose the total cost in dollars of manufacturing q units of a certain commodity is $C(q) = 3q^2 + 5q + 10$. If the current level of production is 40 units, estimate how the total cost will change if 40.5 units are produced.

SOLUTION

In this problem, the current value of the variable is $q = 40$ and the change in the variable is $\Delta q = 0.5$. By the approximation formula, the corresponding change in cost is

$$\Delta C = C(40.5) - C(40) \simeq C'(40)\,\Delta q = C'(40)(0.5)$$

Since

$$C'(q) = 6q + 5 \quad \text{and} \quad C'(40) = 6(40) + 5 = 245$$

it follows that

$$\Delta C \simeq C'(40)(0.5) = 245(0.5) = \$122.50$$

For practice, compute the actual change in cost caused by the increase in the level of production from 40 to 40.5 and compare your answer with the approximation. Is the approximation a good one?

In the next example, the approximation formula is used to estimate the maximum error in a calculation that is based on figures obtained by imperfect measurement.

EXAMPLE 4.2   You measure the side of a cube to be 12 centimeters long and conclude that the volume of the cube is $12^3 = 1{,}728$ cubic centimeters. If your measurement of the side is accurate to within 2 percent, approximately how accurate is your calculation of the volume?

SOLUTION

The volume of the cube is $V(x) = x^3$, where x is the length of a side. The error you make in computing the volume if you take the length of the side to be 12 when it is really $12 + \Delta x$ is

$$\Delta V = V(12 + \Delta x) - V(12) \simeq V'(12)\Delta x$$

Your measurement of the side can be off by as much as 2 percent, that is, by as much as $0.02(12) = 0.24$ centimeter in either direction. Hence the maximum error in your measurement of the side is $\Delta x = \pm 0.24$, and the corresponding maximum error in your calculation of the volume is

$$\text{Maximum error in volume} = \Delta V \simeq V'(12)(\pm 0.24)$$

Since

$$V'(x) = 3x^2 \quad \text{and} \quad V'(12) = 3(12)^2 = 432$$

it follows that

$$\text{Maximum error in volume} \simeq 432(\pm 0.24) = \pm 103.68$$

This says that, at worst, your calculation of the volume as 1,728 cubic centimeters is off by approximately 103.68 cubic centimeters.

In the next example, the desired change in the function is given, and the goal is to estimate the necessary corresponding change in the variable.

EXAMPLE 4.3   The daily output at a certain factory is $Q(L) = 900L^{1/3}$ units, where L denotes the size of the labor force measured in worker-hours. Currently, 1,000 worker-hours of labor are used each day. Use calculus to estimate the number of additional worker-hours of labor that will be needed to increase daily output by 15 units.

SOLUTION
Solve for ΔL using the approximation formula

$$\Delta Q \simeq Q'(L)\Delta L$$

with

$$\Delta Q = 15 \quad L = 1,000 \quad \text{and} \quad Q'(L) = 300L^{-2/3}$$

to get

$$15 \simeq 300(1,000)^{-2/3} \, \Delta L$$

or

$$\Delta L \simeq \frac{15}{300}(1,000)^{2/3} = \frac{15}{300}(10)^2 = 5 \text{ worker-hours}$$

Approximation of Percentage Change

The **percentage change** of a quantity expresses the change in the quantity as a percentage of its size prior to the change. In particular,

$$\text{Percentage change} = 100 \frac{\text{change in quantity}}{\text{size of quantity}}$$

This formula can be combined with the approximation formula and written in functional notation as follows.

Approximation Formula for Percentage Change

> If Δx is a (small) change in x, the corresponding percentage change in the function $f(x)$ is
>
> $$\text{Percentage change in } f = 100\frac{\Delta f}{f(x)} \simeq 100\frac{f'(x)\Delta x}{f(x)}$$

EXAMPLE 4.4 The GNP of a certain country was $N(t) = t^2 + 5t + 200$ billion dollars t years after 1980. Use calculus to estimate the percentage change in the GNP during the first quarter of 1984.

Use the formula

$$\text{Percentage change in } N \simeq 100\frac{N'(t)\Delta t}{N(t)}$$

with

$$t = 4 \qquad \Delta t = 0.25 \qquad \text{and} \qquad N'(t) = 2t + 5$$

to get

$$\text{Percentage change in } N \simeq 100\frac{N'(4)0.25}{N(4)}$$

$$= 100\frac{[2(4) + 5](0.25)}{(4)^2 + 5(4) + 200}$$

$$\simeq 1.38 \text{ percent}$$

The next example illustrates how the percentage change can sometimes be estimated even though the numerical value of the variable is not known.

EXAMPLE 4.5 At a certain factory, the daily output is $Q(K) = 4{,}000K^{1/2}$ units, where K denotes the firm's capital investment. Use calculus to estimate the percentage increase in output that will result from a 1 percent increase in capital investment.

SOLUTION
The derivative is $Q'(K) = 2{,}000K^{-1/2}$. The fact that K increases by 1 percent means that $\Delta K = 0.01K$. Hence,

$$\text{Percentage change in } Q \simeq 100\frac{Q'(K)\Delta K}{Q(K)}$$

$$= 100\frac{2{,}000K^{-1/2}(0.01K)}{4{,}000K^{1/2}}$$

$$= \frac{2{,}000K^{1/2}}{4{,}000K^{1/2}} \qquad \text{since } K^{-1/2}K = K^{1/2}$$

$$= \frac{2{,}000}{4{,}000} = 0.5 \text{ percent}$$

Differentials

The expression $f'(x)\Delta x$ on the right-hand side of the approximation formula $\Delta f \simeq f'(x)\Delta x$ is sometimes called the **differential** of f and is denoted by df. Similarly, the expression $\frac{dy}{dx}\Delta x$ on the right-hand side of the other form of the approximation formula $\Delta y \simeq \frac{dy}{dx}\Delta x$ is known as the differential of y and is denoted by dy. Thus, if Δx is small,

$$\Delta y \simeq dy \qquad \text{where} \qquad dy = \frac{dy}{dx}\Delta x$$

Geometric Interpretation

The approximation of Δy by its differential dy has a simple geometric interpretation that is illustrated in Figure 4.1.

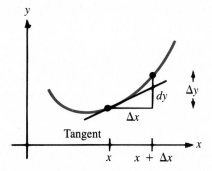

Figure 4.1 Approximation of Δy by the differential dy.

Since the slope of the tangent is $\frac{dy}{dx}$, the differential $dy = \frac{dy}{dx}\Delta x$ is the change in the height of the tangent corresponding to a change from x to $x + \Delta x$. On the other hand, Δy is the change in the height of the curve corresponding to this change in x. Hence, approximating Δy by the differential dy is the same as approximating the change in height of a curve by the change in height of its tangent. If Δx is small, this approximation is a good one.

Problems

In each of the following problems, use calculus to obtain the required estimate.

Manufacturing

1. A manufacturer's total cost is $C(q) = 0.1q^3 - 0.5q^2 + 500q + 200$ dollars when the level of production is q units. The current level of production is 4 units, and the manufacturer is planning to increase this to 4.1 units. Estimate how the total cost will change as a result.

Manufacturing

2. A manufacturer's total monthly revenue is $R(q) = 240q + 0.05q^2$ dollars when q units are produced during the month. Currently the manufacturer is producing 80 units a month and is planning to decrease the monthly output by 0.65 unit. Estimate how the total monthly revenue will change as a result.

Newspaper circulation

3. It is projected that t years from now, the circulation of a local newspaper will be $C(t) = 100t^2 + 400t + 5,000$. Estimate the amount by which the circulation will increase during the next 6 months. (*Hint:* The current value of the variable is $t = 0$.)

Population growth

4. It is projected that t years from now, the population of a certain suburban community will be $P(t) = 20 - \dfrac{6}{t + 1}$ thousand. By approximately how much will the population increase during the next quarter-year?

Air pollution

5. An environmental study of a certain community suggests that t years from now, the average level of carbon monoxide in the air will be $Q(t) = 0.05t^2 + 0.1t + 3.4$ parts per million. By approximately how much will the carbon monoxide level change during the coming 6 months?

Efficiency

6. An efficiency study of the morning shift at a certain factory indicates that an average worker who arrives on the job at 8:00 A.M. will have assembled $f(x) = -x^3 + 6x^2 + 15x$ transistor radios x hours later. Approximately how many radios will the worker assemble between 9:00 and 9:15 A.M.?

Production

7. At a certain factory, the daily output is $Q(K) = 600K^{1/2}$ units, where K denotes the capital investment measured in units of $1,000. The current capital investment is $900,000. Estimate the effect that an additional capital investment of $800 will have on the daily output.

Production

8. At a certain factory, the daily output is $Q(L) = 60,000L^{1/3}$ units, where L denotes the size of the labor force measured in worker-hours. Currently 1,000 worker-hours of labor are used each day. Estimate the effect on output that will be produced if the labor force is cut to 940 worker-hours.

Circulation of blood

9. The speed of blood flowing along the central axis of a certain artery is

$S(R) = 1.8 \times 10^5 R^2$ centimeters per second, where R is the radius of the artery. A medical researcher measures the radius of the artery to be 1.2×10^{-2} centimeter and makes an error of 5×10^{-4} centimeter. Estimate the amount by which the calculated value of the speed of the blood will differ from the true speed if the incorrect value of the radius is used in the formula.

Area 10. You measure the radius of a circle to be 12 centimeters and use the formula $A = \pi r^2$ to calculate the area. If your measurement of the radius is accurate to within 3 percent, approximately how accurate is your calculation of the area?

Volume 11. You measure the radius of a sphere to be 6 inches and use the formula $V = \frac{4}{3}\pi r^3$ to calculate the volume. If your measurement of the radius is accurate to within 1 percent, approximately how accurate is your calculation of the volume?

Production 12. The daily output at a certain factory is $Q(L) = 300L^{2/3}$ units, where L denotes the size of the labor force measured in worker-hours. Currently 512 worker-hours of labor are used each day. Estimate the number of additional worker-hours of labor that will be needed to increase daily output by 12.5 units.

Manufacturing 13. A manufacturer's total cost is $C(q) = \frac{1}{6}q^3 + 642q + 400$ dollars when q units are produced. The current level of production is 4 units. Estimate the amount by which the manufacturer should decrease production to reduce the total cost by $130.

Property tax 14. Records indicate that x years after 1980, the average property tax on a three-bedroom home in a certain community was $T(x) = 60x^{3/2} + 40x + 1,200$ dollars. Estimate the percentage by which the property tax increased during the first half of 1984.

Annual earnings 15. The gross annual earnings of a company were $A(t) = 0.1t^2 + 10t + 20$ thousand dollars t years after its formation in 1980. Estimate the percentage change in the gross annual earnings during the third quarter of 1984.

Volume 16. Estimate what will happen to the volume of a cube if the length of each side is decreased by 2 percent. Express your answer as a percentage and verify that your answer is consistent with the calculation in Example 4.2.

Area 17. Estimate what will happen to the area of a circle if the radius is increased by 1 percent. Express your answer as a percentage.

Volume 18. Estimate what will happen to the volume of a sphere if the radius is increased by 1 percent. Express your answer as a percentage.

Circulation of blood 19. According to Poiseuille's law, the speed of blood flowing along the central axis of an artery of radius R is $S(R) = cR^2$, where c is a constant. Estimate the percentage error you will make in the calculation of $S(R)$ from this formula if you make a 1 percent error in the measurement of R.

Production 20. The output at a certain factory is $Q(K) = 400K^{1/2}$ units, where K denotes the firm's capital investment. Estimate the percentage increase in output that will result from a 1 percent increase in capital investment.

Production 21. The output at a certain factory is $Q(L) = 9,000L^{1/3}$ units, where L denotes the size of the labor force. Estimate the percentage increase in output that will result from a 1.5 percent increase in the size of the labor force.

Production 22. The output at a certain factory is $Q = 600K^{1/2}L^{1/3}$ units, where K denotes the capital investment and L the size of the labor force. Estimate the percentage increase in output that will result from a 2 percent increase in the size of the labor force if capital investment is not changed.

Production 23. At a certain factory, the daily output is $Q(K) = 1,200K^{1/2}$ units, where K denotes the firm's capital investment. Estimate the percentage increase in capital investment that is needed to produce a 1.2 percent increase in output.

Volume 24. Estimate the largest percentage error you can allow in the measurement of the radius of a sphere if you want the error in the calculation of its volume using the formula $V = \frac{4}{3}\pi r^3$ to be no greater than 8 percent.

Volume 25. A soccer ball made of leather $\frac{1}{8}$ inch thick has an inner diameter of $8\frac{1}{2}$ inches. Estimate the volume of its leather shell. (*Hint:* Think of the volume of the shell as a certain change ΔV in volume.)

Volume 26. A melon in the form of a sphere has a rind $\frac{1}{5}$ inch thick and an inner diameter of 8 inches. Estimate what percentage of the total volume of the melon is rind.

5 The Chain Rule

In many practical situations, a quantity is given as a function of one variable which, in turn, can be thought of as a function of a second variable. In such cases, the rate of change of the quantity with respect to the second variable is equal to the rate of change of the quantity with respect to the first variable times the rate of change of the first variable with respect to the second.

For example, suppose the total manufacturing cost at a certain factory is a function of the number of units produced which, in turn, is a function of the number of hours during which the factory has been operating. Let C, q, and t denote the cost (in dollars), the number of units, and the number of hours, respectively. Then,

$$\frac{dC}{dq} = \frac{\text{rate of change of cost}}{\text{with respect to output}} \qquad \text{(dollars per unit)}$$

and

$$\frac{dq}{dt} = \frac{\text{rate of change of output}}{\text{with respect to time}} \qquad \text{(units per hour)}$$

The product of these two rates is the rate of change of cost with respect to time. That is,

$$\frac{dC}{dq} \frac{dq}{dt} = \frac{\text{rate of change of cost}}{\text{with respect to time}} \qquad \text{(dollars per hour)}$$

Since the rate of change of cost with respect to time is also given by the derivative $\frac{dC}{dt}$, it follows that

$$\frac{dC}{dt} = \frac{dC}{dq} \frac{dq}{dt}$$

This formula is a special case of an important rule called the **chain rule**.

The Chain Rule

Suppose y is a function of u and u is a function of x. Then y can be regarded as a function of x and

$$\frac{dy}{dx} = \frac{dy}{du} \frac{du}{dx}$$

That is, the derivative of y with respect to x is the derivative of y with respect to u times the derivative of u with respect to x.

Notice that one way to remember the chain rule is to pretend that the derivatives $\frac{dy}{du}$ and $\frac{du}{dx}$ are quotients and cancel du, reducing the expression $\frac{dy}{du} \frac{du}{dx}$ on the right-hand side of the equation to the expression $\frac{dy}{dx}$ on the left-hand side.

Here are two examples illustrating the use of the chain rule.

EXAMPLE 5.1   Find $\dfrac{dy}{dx}$ if $y = u^3 - 3u^2 + 1$ and $u = x^2 + 2$.

SOLUTION
Since

$$\frac{dy}{du} = 3u^2 - 6u \qquad \text{and} \qquad \frac{du}{dx} = 2x$$

it follows that

$$\frac{dy}{dx} = \frac{dy}{du}\frac{du}{dx} = (3u^2 - 6u)(2x)$$

Notice that this derivative is expressed in terms of the variables x and u. Since you are thinking of y as a function of x, it is more natural to express $\dfrac{dy}{dx}$ in terms of x alone. To do this, substitute $x^2 + 2$ for u in the expression for $\dfrac{dy}{dx}$ and simplify the answer as follows.

$$\frac{dy}{dx} = [3(x^2 + 2)^2 - 6(x^2 + 2)](2x)$$

$$= 6x(x^2 + 2)[(x^2 + 2) - 2]$$

$$= 6x(x^2 + 2)(x^2) = 6x^3(x^2 + 2)$$

For practice, check this answer by first substituting $u = x^2 + 2$ into the original expression for y and then differentiating with respect to x.

In the next example, you will see how to use the chain rule to calculate a derivative for a particular value of the independent variable.

EXAMPLE 5.2   Find $\dfrac{dy}{dx}$ when $x = 1$ if $y = \dfrac{u}{u + 1}$ and $u = 3x^2 - 1$.

SOLUTION
Since

$$\frac{dy}{du} = \frac{(u + 1)(1) - u(1)}{(u + 1)^2} = \frac{1}{(u + 1)^2} \qquad \text{by the quotient rule}$$

and

$$\frac{du}{dx} = 6x$$

it follows that

$$\frac{dy}{dx} = \frac{dy}{du}\frac{du}{dx} = \left[\frac{1}{(u+1)^2}\right](6x) = \frac{6x}{(u+1)^2}$$

The goal is to evaluate this derivative when $x = 1$. One way to do this is to replace u by its algebraic formula as in Example 5.1 and then evaluate the resulting expression when $x = 1$. However, it is easier to substitute numbers than algebraic expressions, and so it is preferable to compute the numerical value of u first and then substitute. In particular, when $x = 1$, the original formula $u = 3x^2 - 1$ gives $u = 3(1)^2 - 1 = 2$. Now substitute $x = 1$ and $u = 2$ in the formula for $\frac{dy}{dx}$ to conclude that when $x = 1$,

$$\frac{dy}{dx} = \frac{6(1)}{(2+1)^2} = \frac{6}{9} = \frac{2}{3}$$

Composite Functions

Recall from Chapter 1, Section 1, that the composite function $g[h(x)]$ is the function formed from functions $g(u)$ and $h(x)$ by substituting $h(x)$ for u in the formula for $g(u)$. The chain rule is actually a rule for differentiating composite functions and can be rewritten using functional notation as follows.

The Chain Rule in Functional Notation

If $g(u)$ and $h(x)$ are differentiable functions,

$$\frac{d}{dx}g[h(x)] = g'[h(x)]h'(x)$$

To see that this is nothing more than a restatement of the previous version of the chain rule, suppose that $y = g[h(x)]$. Then,

$$y = g(u) \qquad \text{where} \qquad u = h(x)$$

and, by the chain rule,

$$\frac{dy}{dx} = \frac{dy}{du}\frac{du}{dx} = g'(u)h'(x) = g'[h(x)]h'(x)$$

The use of this form of the chain rule is illustrated in the next example.

EXAMPLE 5.3 Differentiate the function $f(x) = \sqrt{x^2 + 3x + 2}$.

SOLUTION

Think of $f(x)$ as the composite function $g[h(x)]$, where

$$g(u) = \sqrt{u} = u^{1/2} \qquad \text{and} \qquad h(x) = x^2 + 3x + 2$$

Then,

$$g'(u) = \frac{1}{2}u^{-1/2} \qquad \text{and} \qquad h'(x) = 2x + 3$$

and, by the chain rule,

$$f'(x) = g'[h(x)]h'(x)$$

$$= \frac{1}{2}(x^2 + 3x + 2)^{-1/2}(2x + 3)$$

$$= \frac{2x + 3}{2\sqrt{x^2 + 3x + 2}}$$

The Chain Rule for Powers

In Section 2, you learned the rule

$$\frac{d}{dx}(x^n) = nx^{n-1}$$

for differentiating power functions. There is a closely related rule, which is a special case of the chain rule, that you can use to differentiate functions of the form $[h(x)]^n$, that is, functions that are powers of other functions. According to this rule, you begin by computing $n[h(x)]^{n-1}$ and then multiply this expression by the derivative of h.

The Chain Rule for Powers

For any real number n and differentiable function h,

$$\frac{d}{dx}[h(x)]^n = n[h(x)]^{n-1}\frac{d}{dx}[h(x)]$$

To derive the chain rule for powers, think of $[h(x)]^n$ as the composite function

$$[h(x)]^n = g[h(x)] \qquad \text{where} \qquad g(u) = u^n$$

Then,

$$g'(u) = nu^{n-1} \qquad \text{and} \qquad h'(x) = \frac{d}{dx}[h(x)]$$

and, by the chain rule,

$$\frac{d}{dx}[h(x)]^n = \frac{d}{dx}g[h(x)] = g'[h(x)]h'(x) = n[h(x)]^{n-1}\frac{d}{dx}[h(x)]$$

The use of the chain rule for powers is illustrated in the following examples.

EXAMPLE 5.4   Differentiate the function $f(x) = (2x^4 - x)^3$.

SOLUTION
One way to do this problem is to expand the function and rewrite it as

$$f(x) = 8x^{12} - 12x^9 + 6x^6 - x^3$$

and then differentiate this polynomial term by term to get

$$f'(x) = 96x^{11} - 108x^8 + 36x^5 - 3x^2$$

But see how much easier it is to use the chain rule for powers. According to this rule,

$$f'(x) = 3(2x^4 - x)^2\frac{d}{dx}(2x^4 - x) = 3(2x^4 - x)^2(8x^3 - 1)$$

Not only is this method easier, but the answer even comes out in factored form!

In the next example, the solution to Example 5.3 is written more compactly with the aid of the chain rule for powers.

EXAMPLE 5.5   Differentiate the function $f(x) = \sqrt{x^2 + 3x + 2}$.

SOLUTION
Rewrite the function as $f(x) = (x^2 + 3x + 2)^{1/2}$ and apply the chain rule for powers to get

$$f'(x) = \frac{1}{2}(x^2 + 3x + 2)^{-1/2}\frac{d}{dx}(x^2 + 3x + 2)$$

$$= \frac{1}{2}(x^2 + 3x + 2)^{-1/2}(2x + 3)$$

$$= \frac{2x + 3}{2\sqrt{x^2 + 3x + 2}}$$

EXAMPLE 5.6 Differentiate the function $f(x) = \dfrac{1}{(2x + 3)^5}$.

SOLUTION

Do *not* use the quotient rule! It's much easier to rewrite the function as

$$f(x) = (2x + 3)^{-5}$$

and apply the chain rule for powers to get

$$f'(x) = -5(2x + 3)^{-6}\frac{d}{dx}(2x + 3) = -5(2x + 3)^{-6}(2) = -\frac{10}{(2x + 3)^6}$$

The chain rule is often used in combination with the other rules you learned in Section 2. The next example involves the product rule.

EXAMPLE 5.7 Differentiate the function $f(x) = (3x + 1)^4(2x - 1)^5$ and simplify your answer.

SOLUTION

First apply the product rule to get

$$f'(x) = (3x + 1)^4\frac{d}{dx}[(2x - 1)^5] + (2x - 1)^5\frac{d}{dx}[(3x + 1)^4]$$

Continue by applying the chain rule for powers twice to get

$$f'(x) = (3x + 1)^4[5(2x - 1)^4(2)] + (2x - 1)^5[4(3x + 1)^3(3)]$$
$$= 10(3x + 1)^4(2x - 1)^4 + 12(2x - 1)^5(3x + 1)^3$$

Finally, simplify your answer by factoring to get

$$f'(x) = 2(3x + 1)^3(2x - 1)^4[5(3x + 1) + 6(2x - 1)]$$
$$= 2(3x + 1)^3(2x - 1)^4[15x + 5 + 12x - 6]$$
$$= 2(3x + 1)^3(2x - 1)^4(27x - 1)$$

EXAMPLE 5.8 Differentiate the function $f(x) = \sqrt{\dfrac{x + 1}{x - 1}}$ and simplify your answer.

SOLUTION

First rewrite the function as

$$f(x) = \left(\frac{x + 1}{x - 1}\right)^{1/2}$$

and then apply the chain rule for powers to get

$$f'(x) = \frac{1}{2}\left(\frac{x+1}{x-1}\right)^{-1/2}\frac{d}{dx}\left(\frac{x+1}{x-1}\right)$$

Now use the quotient rule to get

$$\frac{d}{dx}\left(\frac{x+1}{x-1}\right) = \frac{(x-1)(1) - (x+1)(1)}{(x-1)^2} = -\frac{2}{(x-1)^2}$$

and substitute the result into the equation for $f'(x)$ to get

$$f'(x) = \frac{1}{2}\left(\frac{x+1}{x-1}\right)^{-1/2}\left[-\frac{2}{(x-1)^2}\right]$$

$$= -\frac{(x+1)^{-1/2}}{(x-1)^{-1/2}}\frac{1}{(x-1)^2}$$

$$= -\frac{(x+1)^{-1/2}}{(x-1)^{3/2}}$$

$$= -\frac{1}{(x+1)^{1/2}(x-1)^{3/2}}$$

Applications of the Chain Rule

In many practical problems, a quantity is given as a function of one variable which, in turn, can be written as a function of a second variable, and the goal is to find the rate of change of the original quantity with respect to the second variable. Such problems can be solved by means of the chain rule. Here is an example.

EXAMPLE 5.9 An environmental study of a certain suburban community suggests that the average daily level of carbon monoxide in the air will be $c(p) = \sqrt{0.5p^2 + 17}$ parts per million when the population is p thousand. It is estimated that t years from now, the population of the community will be $p(t) = 3.1 + 0.1t^2$ thousand. At what rate will the carbon monoxide level be changing with respect to time 3 years from now?

SOLUTION

The goal is to find $\frac{dc}{dt}$ when $t = 3$. Since

$$\frac{dc}{dp} = \frac{1}{2}(0.5p^2 + 17)^{-1/2}[0.5(2p)] = \frac{1}{2}p(0.5p^2 + 17)^{-1/2}$$

and

$$\frac{dp}{dt} = 0.2t$$

it follows from the chain rule that

$$\frac{dc}{dt} = \frac{dc}{dp}\frac{dp}{dt} = \frac{1}{2}p(0.5p^2 + 17)^{-1/2}(0.2t) = \frac{0.1pt}{\sqrt{0.5p^2 + 17}}$$

When $t = 3$,

$$p = p(3) = 3.1 + 0.1(3)^2 = 4$$

and so

$$\frac{dc}{dt} = \frac{0.1(4)3}{\sqrt{0.5(4)^2 + 17}}$$

$$= \frac{1.2}{\sqrt{25}} = \frac{1.2}{5} = 0.24 \text{ part per million per year}$$

Problems

In Problems 1 through 10, use the chain rule to compute the derivative $\dfrac{dy}{dx}$ and simplify your answer.

1. $y = u^2 + 1, u = 3x - 2$

2. $y = 2u^2 - u + 5, u = 1 - x^2$

3. $y = \sqrt{u}, u = x^2 + 2x - 3$

4. $y = u^2 + 2u - 3, u = \sqrt{x}$

5. $y = \dfrac{1}{u^2}, u = x^2 + 1$

6. $y = \dfrac{1}{u}, u = 3x^2 + 5$

7. $y = \dfrac{1}{\sqrt{u}}, u = x^2 - 9$

8. $y = u^2 + u - 2, u = \dfrac{1}{x}$

9. $y = \dfrac{1}{u - 1}, u = x^2$

10. $y = u^2, u = \dfrac{1}{x - 1}$

In Problems 11 through 16, use the chain rule to compute the derivative $\dfrac{dy}{dx}$ for the given value of x.

11. $y = 3u^4 - 4u + 5, u = x^3 - 2x - 5; x = 2$

12. $y = u^5 - 3u^2 + 6u - 5, u = x^2 - 1; x = 1$

13. $y = \sqrt{u}, u = x^2 - 2x + 6; x = 3$

14. $y = 3u^2 - 6u + 2, u = \dfrac{1}{x^2}; x = \dfrac{1}{3}$

15. $y = \dfrac{1}{u}, u = 3 - \dfrac{1}{x^2}; x = \dfrac{1}{2}$

16. $y = \dfrac{1}{u + 1}$, $u = x^3 - 2x + 5$; $x = 0$

In Problems 17 through 35, differentiate the given function and simplify your answer.

17. $f(x) = (2x + 1)^4$

18. $f(x) = \sqrt{5x^6 - 12}$

19. $f(x) = (x^5 - 4x^3 - 7)^8$

20. $f(x) = (3x^4 - 7x^2 + 9)^5$

21. $f(x) = \dfrac{1}{5x^2 - 6x + 2}$

22. $f(x) = \dfrac{2}{(6x^2 + 5x + 1)^2}$

23. $f(x) = \dfrac{1}{\sqrt{4x^2 + 1}}$

24. $f(x) = \dfrac{1}{\sqrt{5x^3 + 2}}$

25. $f(x) = \dfrac{3}{(1 - x^2)^4}$

26. $f(x) = \dfrac{2}{3(5x^4 + 1)^2}$

27. $f(x) = (1 + \sqrt{3x})^5$

28. $f(x) = \sqrt{1 + \dfrac{1}{3x}}$

29. $f(x) = (x + 2)^3(2x - 1)^5$

30. $f(x) = 2(3x + 1)^4(5x - 3)^2$

31. $f(x) = \sqrt{\dfrac{3x + 1}{2x - 1}}$

32. $f(x) = \left(\dfrac{x + 2}{2 - x}\right)^3$

33. $f(x) = \dfrac{(x + 1)^5}{(1 - x)^4}$

34. $f(x) = \dfrac{(1 - 2x)^2}{(3x + 1)^3}$

35. $f(x) = \dfrac{3x + 1}{\sqrt{1 - 4x}}$

In Problems 36 through 39, find an equation of the line that is tangent to the graph of f for the given value of x.

36. $f(x) = (3x^2 + 1)^2$; $x = -1$

37. $f(x) = (x^2 - 3)^5(2x - 1)^3$; $x = 2$

38. $f(x) = \dfrac{1}{(2x - 1)^6}$; $x = 1$

39. $f(x) = \left(\dfrac{x + 1}{x - 1}\right)^3$; $x = 3$

40. Differentiate the function $f(x) = (3x + 5)^2$ by two different methods, first using the chain rule and then the product rule. Show that the two answers are the same.

Annual earnings 41. The gross annual earnings of a certain company were $f(t) = \sqrt{10t^2 + t + 236}$ thousand dollars t years after its formation in January 1980.

(a) At what rate were the gross annual earnings of the company growing in January 1984?

(b) At what percentage rate were the gross annual earnings growing in January 1984?

Manufacturing cost 42. At a certain factory, the total cost of manufacturing q units during the daily production run is $C(q) = 0.2q^2 + q + 900$ dollars. From experience it has been determined that approximately $q(t) = t^2 + 100t$ units are manufactured during the first t hours of a production run. Compute the rate at which the total manufacturing cost is changing with respect to time 1 hour after production commences.

Air pollution 43. It is estimated that t years from now, the population of a certain suburban community will be $p(t) = 20 - \dfrac{6}{t+1}$ thousand. An environmental study indicates that the average daily level of carbon monoxide in the air will be $c(p) = 0.5\sqrt{p^2 + p + 58}$ parts per million when the population is p thousand. Find the rate at which the level of carbon monoxide will be changing with respect to time 2 years from now.

Consumer demand 44. When electric blenders are sold for p dollars apiece, local consumers will buy $D(p) = \dfrac{8{,}000}{p}$ blenders a month. It is estimated that t months from now, the price of the blenders will be $p(t) = 0.04t^{3/2} + 15$ dollars. Compute the rate at which the monthly demand for the blenders will be changing with respect to time 25 months from now. Will the demand be increasing or decreasing?

Consumer demand 45. An importer of Brazilian coffee estimates that local consumers will buy approximately $D(p) = \dfrac{4{,}374}{p^2}$ pounds of the coffee per week when the price is p dollars per pound. It is estimated that t weeks from now, the price of Brazilian coffee will be $p(t) = 0.02t^2 + 0.1t + 6$ dollars per pound. At what rate will the weekly demand for the coffee be changing with respect to time 10 weeks from now? Will the demand be increasing or decreasing?

Consumer demand 46. When a certain commodity is sold for p dollars per unit, consumers will buy $D(p) = \dfrac{40{,}000}{p}$ units per month. It is estimated that t months from now, the price of the commodity will be $p(t) = 0.4t^{3/2} + 6.8$ dollars per unit. At what percentage rate will the monthly demand for the commodity be changing with respect to time 4 months from now?

Air pollution 47. It is estimated that t years from now, the population of a certain community will be $p(t) = 12 - \dfrac{6}{t+1}$ thousand. An environmental study indicates that the average daily level of carbon monoxide in the air will be $c(p) = 0.6\sqrt{p^2 + 2p + 24}$ units when the population is p thousand. At what percentage rate will the level of carbon monoxide be changing with respect to time 2 years from now?

48. Suppose $L(x)$ is a function with the property that $L'(x) = \dfrac{1}{x}$. Use the chain rule to find the derivatives of the following functions and simplify your answers:

    (a) $f(x) = L(x^2)$

    (b) $f(x) = L\left(\dfrac{1}{x}\right)$

    (c) $f(x) = L\left(\dfrac{2}{3\sqrt{x}}\right)$

    (d) $f(x) = L\left(\dfrac{2x + 1}{1 - x}\right)$

49. Prove the chain rule for powers for $n = 2$ by using the product rule to compute $\dfrac{dy}{dx}$ if $y = [h(x)]^2$.

50. Prove the chain rule for powers for $n = 3$ by using the product rule and the result of Problem 49 to compute $\dfrac{dy}{dx}$ if $y = [h(x)]^3$. {*Hint:* Begin by writing y as $h(x)[h(x)]^2$.}

6 Implicit Differentiation

Explicit and Implicit Functions

The functions you have worked with so far have all been given by equations of the form $y = f(x)$ in which the dependent variable y on the left is given explicitly by an expression on the right involving the independent variable. A function in this form is said to be in **explicit form.** For example, the functions

$$y = x^2 + 3x + 1 \qquad y = \frac{x^3 + 1}{2x - 3} \qquad \text{and} \qquad y = \sqrt{1 - x^2}$$

are all functions in explicit form.

Sometimes, practical problems will lead to equations in which the function y is not written explicitly in terms of the independent variable x, equations such as

$$x^2y^3 - 6 = 5y^3 + x \qquad \text{and} \qquad x^2y + 2y^3 = 3x + 2y$$

for example. Since it has not been solved for y, such an equation is said to **define y implicitly as a function of x** and the function y is said to be in **implicit form.**

Differentiation of Functions in Implicit Form

Suppose you have an equation that defines y implicitly as a function of x, and you want to find the derivative $\frac{dy}{dx}$. For instance, you may be interested in the slope of a line that is tangent to the graph of the equation at a particular point. One approach might be to solve the equation for y explicitly and then differentiate using the techniques you already know. Unfortunately, it is not always possible to find y explicitly. For example, there is no obvious way to solve for y in the equation $x^2y + 2y^3 = 3x + 2y$. Moreover, even when you can solve for y explicitly, the resulting formula is often complicated and unpleasant to differentiate. For example, the equation

$$x^2y^3 - 6 = 5y^3 + x$$

can be solved for y to give

$$x^2y^3 - 5y^3 = x + 6$$

$$y^3(x^2 - 5) = x + 6$$

$$y = \left(\frac{x + 6}{x^2 - 5}\right)^{1/3}$$

The computation of $\frac{dy}{dx}$ for this function in explicit form would be tedious, involving both the chain rule and the quotient rule.

Fortunately, there is a simple technique based on the chain rule that you can use to find $\frac{dy}{dx}$ without first solving for y explicitly. This technique is known as **implicit differentiation**. It consists of differentiating both sides of the (unsolved) equation with respect to x and then solving algebraically for $\frac{dy}{dx}$. Here is an example illustrating the technique.

EXAMPLE 6.1   Find $\frac{dy}{dx}$ if $x^2y + 2y^3 = 3x + 2y$.

SOLUTION
You are going to differentiate both sides of the given equation with respect to x. So that you won't forget that y is actually a function of x, temporarily replace y by the symbol $f(x)$ and begin by rewriting the equation as

$$x^2f(x) + 2[f(x)]^3 = 3x + 2f(x)$$

Now differentiate both sides of this equation term by term with respect to x. By the product rule,

$$\frac{d}{dx}[x^2 f(x)] = x^2 f'(x) + 2x f(x)$$

By the chain rule for powers,

$$\frac{d}{dx}\{2[f(x)]^3\} = 6[f(x)]^2 f'(x)$$

and by the constant multiple rule,

$$\frac{d}{dx}(3x) = 3 \qquad \text{and} \qquad \frac{d}{dx}[2f(x)] = 2f'(x)$$

Putting it all together you get

$$x^2 f'(x) + 2x f(x) + 6[f(x)]^2 f'(x) = 3 + 2f'(x)$$

Since $f(x) = y$ and $f'(x) = \dfrac{dy}{dx}$, you can now rewrite this equation as

$$x^2 \frac{dy}{dx} + 2xy + 6y^2 \frac{dy}{dx} = 3 + 2\frac{dy}{dx}$$

Finally, solve this equation for $\dfrac{dy}{dx}$ to get

$$x^2 \frac{dy}{dx} + 6y^2 \frac{dy}{dx} - 2\frac{dy}{dx} = 3 - 2xy$$

$$(x^2 + 6y^2 - 2)\frac{dy}{dx} = 3 - 2xy$$

$$\frac{dy}{dx} = \frac{3 - 2xy}{x^2 + 6y^2 - 2}$$

Notice that the formula for $\dfrac{dy}{dx}$ contains both the independent variable x and the dependent variable y. This is usual when derivatives are computed implicitly.

So that you would not forget to use the chain rule for powers when first learning implicit differentiation, it was suggested in Example 6.1 that you temporarily replace y by $f(x)$. As soon as you feel comfortable with the technique, try to leave out this unnecessary step and differentiate the given equation directly. Just keep in mind that y is really a function of x and remember to use the chain rule when it is appropriate. Here's what the solution to Example 6.1 looks like without the unnecessary substitution of $f(x)$.

EXAMPLE 6.2   Find $\dfrac{dy}{dx}$ if $x^2y + 2y^3 = 3x + 2y$.

SOLUTION

Differentiate both sides of the equation as it stands with respect to x. Remember that y is really a function of x and that you will have to use the chain rule to differentiate powers of y. In particular,

$$x^2\frac{dy}{dx} + 2xy + 6y^2\frac{dy}{dx} = 3 + 2\frac{dy}{dx}$$

Now solve for $\dfrac{dy}{dx}$ as before to get

$$x^2\frac{dy}{dx} + 6y^2\frac{dy}{dx} - 2\frac{dy}{dx} = 3 - 2xy$$

$$(x^2 + 6y^2 - 2)\frac{dy}{dx} = 3 - 2xy$$

$$\frac{dy}{dx} = \frac{3 - 2xy}{x^2 + 6y^2 - 2}$$

Here is an outline of the procedure.

Implicit Differentiation

Suppose an equation defines y implicitly as a function of x. To find $\dfrac{dy}{dx}$:

1. Differentiate both sides of the equation with respect to x. Remember that y is really a function of x and use the chain rule when differentiating terms containing y.

2. Solve the differentiated equation algebraically for $\dfrac{dy}{dx}$.

The technique is further illustrated in the next example.

EXAMPLE 6.3   Find $\dfrac{dy}{dx}$ if $(x^2 - 3y^2)^4 = x^2y^3$.

SOLUTION

Differentiate both sides of the equation with respect to x using the chain rule for powers on the left and the product rule on the right to get

$$4(x^2 - 3y^2)^3 \frac{d}{dx}(x^2 - 3y^2) = x^2 \frac{d}{dx}(y^3) + y^3 \frac{d}{dx}(x^2)$$

$$4(x^2 - 3y^2)^3 \left(2x - 6y\frac{dy}{dx}\right) = x^2 \left(3y^2 \frac{dy}{dx}\right) + y^3(2x)$$

$$4(x^2 - 3y^2)^3 \left(2x - 6y\frac{dy}{dx}\right) = 3x^2y^2 \frac{dy}{dx} + 2xy^3$$

Now solve for $\frac{dy}{dx}$. Whatever you do, *don't* expand $(x^2 - 3y^2)^3$! Instead, use the distributive law on the left-hand side to get

$$4(x^2 - 3y^2)^3(2x) - 4(x^2 - 3y^2)^3 \left(6y\frac{dy}{dx}\right) = 3x^2y^2 \frac{dy}{dx} + 2xy^3$$

or

$$8x(x^2 - 3y^2)^3 - 24y(x^2 - 3y^2)^3 \frac{dy}{dx} = 3x^2y^2 \frac{dy}{dx} + 2xy^3$$

Now bring all the terms containing $\frac{dy}{dx}$ to the left and all the others to the right to get

$$-24y(x^2 - 3y^2)^3 \frac{dy}{dx} - 3x^2y^2 \frac{dy}{dx} = 2xy^3 - 8x(x^2 - 3y^2)^3$$

Finally, factor out $\frac{dy}{dx}$ on the left and divide to get

$$[-24y(x^2 - 3y^2)^3 - 3x^2y^2]\frac{dy}{dx} = 2xy^3 - 8x(x^2 - 3y^2)^3$$

or

$$\frac{dy}{dx} = \frac{2xy^3 - 8x(x^2 - 3y^2)^3}{-24y(x^2 - 3y^2)^3 - 3x^2y^2}$$

Computing Slopes of Tangents by Implicit Differentiation

In the next example, you will see how to use implicit differentiation to find the slope of a tangent.

EXAMPLE 6.4 Find the slope of the line that is tangent to the curve $x^2y^3 - 6 = 5y^3 + x$ when $x = 2$.

SOLUTION
Differentiate both sides of the equation with respect to x (using the product rule on the left-hand side) to get

$$3x^2y^2\frac{dy}{dx} + 2xy^3 = 15y^2\frac{dy}{dx} + 1$$

and solve for $\frac{dy}{dx}$ to get

$$(3x^2y^2 - 15y^2)\frac{dy}{dx} = 1 - 2xy^3$$

or

$$\frac{dy}{dx} = \frac{1 - 2xy^3}{3x^2y^2 - 15y^2}$$

The desired slope is the value of this derivative when $x = 2$. Before you can compute this value, you have to find the value of y that corresponds to $x = 2$. To do this, substitute $x = 2$ into the original equation and solve getting

$$4y^3 - 6 = 5y^3 + 2$$
$$y^3 = -8$$
$$y = -2$$

Now substitute $x = 2$ and $y = -2$ into the formula for $\frac{dy}{dx}$ to conclude that

$$\text{Slope of tangent} = \frac{dy}{dx} = \frac{1 - 2(2)(-2)^3}{3(2)^2(-2)^2 - 15(-2)^2} = -\frac{11}{4}$$

Application to Economics

Implicit differentiation is used in economics in both practical and theoretical work. In Chapter 3, Section 5, you will see it used to derive certain theoretical relationships. A more practical application of implicit differentiation is given in the next example, which is actually a preview of the discussion in Chapter 9, Section 4, of level curves of functions of two variables.

EXAMPLE 6.5 Suppose the output at a certain factory is $Q = 2x^3 + x^2y + y^3$ units, where x is the number of hours of skilled labor used and y the number of hours of unskilled labor. The current labor force consists of 30 hours of skilled labor and 20 hours of unskilled labor. Use calculus to estimate the change in unskilled labor y that should be made to offset a 1-hour increase in skilled labor x so that output will be maintained at its current level.

SOLUTION

The current level of output is the value of Q when $x = 30$ and $y = 20$. That is,

$$Q = 2(30)^3 + (30)^2(20) + (20)^3 = 80,000 \text{ units}$$

If output is to be maintained at this level, the relationship between skilled labor x and unskilled labor y is given by the equation

$$80,000 = 2x^3 + x^2y + y^3$$

which defines y implicitly as a function of x.

The goal is to estimate the change in y that corresponds to a 1-unit increase in x when x and y are related by this equation. As you saw in Section 3, the change in y caused by a 1-unit increase in x can be approximated by the derivative $\dfrac{dy}{dx}$. To find this derivative, use implicit differentiation. (Remember that the derivative of the constant 80,000 on the left-hand side is zero.)

$$0 = 6x^2 + x^2\frac{dy}{dx} + y\frac{d}{dx}(x^2) + 3y^2\frac{dy}{dx}$$

$$0 = 6x^2 + x^2\frac{dy}{dx} + 2xy + 3y^2\frac{dy}{dx}$$

$$-(x^2 + 3y^2)\frac{dy}{dx} = 6x^2 + 2xy$$

$$\frac{dy}{dx} = -\frac{6x^2 + 2xy}{x^2 + 3y^2}$$

Now evaluate this derivative when $x = 30$ and $y = 20$ to conclude that

$$\text{Change in } y \simeq \frac{dy}{dx} = -\frac{6(30)^2 + 2(30)(20)}{(30)^2 + 3(20)^2} \simeq -3.14 \text{ hours}$$

That is, to maintain the current level of output unskilled labor should be decreased by approximately 3.14 hours to offset a 1-hour increase in skilled labor.

Did you notice that the solution to Example 6.5 contains an unnecessary step? The calculation of the current output to be 80,000 units could be omitted and the equation relating x and y written as

$$C = 2x^3 + x^2y + y^3$$

where C is a constant that stands for the (unspecified) current level of output. Since the derivative of any constant is zero, the differentiated equations will be the same.

Problems

In Problems 1 through 10, find $\dfrac{dy}{dx}$ by implicit differentiation.

1. $x^2 + y^2 = 25$

2. $x^2 + y = x^3 + y^2$

3. $x^3 + y^3 = xy$

4. $5x - x^2y^3 = 2y$

5. $y^2 + 2xy^2 - 3x + 1 = 0$

6. $\dfrac{1}{x} + \dfrac{1}{y} = 1$

7. $(2x + y)^3 = x$

8. $(x - 2y)^2 = y$

9. $(x^2 + 3y^2)^5 = 2xy$

10. $(3xy^2 + 1)^4 = 2x - 3y$

In Problems 11 through 17, find the slope of the line that is tangent to the given curve at the specified value of x.

11. $x^2 = y^3$; $x = 8$

12. $\dfrac{1}{x} - \dfrac{1}{y} = 2$; $x = \dfrac{1}{4}$

13. $xy = 2$; $x = 2$

14. $x^2y^3 - 2xy = 6x + y + 1$; $x = 0$

15. $(1 - x + y)^3 = x + 7$; $x = 1$

16. $(x^2 - 2y)^3 = 2xy^2 + 64$; $x = 0$

17. $(2xy^3 + 1)^3 = 2x - y^3$; $x = 0$

In Problems 18 through 21, find $\dfrac{dy}{dx}$ in two ways: by implicit differentiation of the given equation, and by differentiation of an explicit formula for y. In each case, show that the two answers are really the same.

18. $x^2 + y^3 = 12$

19. $xy + 2y = x^2$

20. $x + \dfrac{1}{y} = 5$

21. $xy - x = y + 2$

Manufacturing 22. The output at a certain plant is $Q = 0.08x^2 + 0.12xy + 0.03y^2$ units per day, where x is the number of hours of skilled labor used and y the number of hours of unskilled labor used. Currently, 80 hours of skilled labor and 200 hours of unskilled labor are used each day. Use calculus to estimate the change in unskilled labor that should be made to offset a 1-hour increase in skilled labor so that output will be maintained at its current level.

Manufacturing 23. At a certain factory, output Q is related to inputs x and y by the equation $Q = 2x^3 + 3x^2y^2 + (1 + y)^3$. If the current levels of input are $x = 30$ and $y = 20$, use calculus to estimate the change in input y

that should be made to offset a decrease of 0.8 unit in input x so that output will be maintained at its current level.

7 Higher-Order Derivatives

The Rate of Change of a Rate of Change

This section is about the rate of change of the rate of change of a quantity. Such rates arise in a variety of situations. For example, the acceleration of a car is the rate of change with respect to time of its speed, which is itself the rate of change with respect to time of its distance. If distance is measured in miles and time in hours, the speed (or rate of change of distance) is measured in miles per hour, and the acceleration (or rate of change of speed) is measured in miles per hour per hour.

Statements about the rate of change of a rate of change are used frequently in economics. In inflationary times, for example, you may hear a government economist assure the nation that although the inflation rate is increasing, the rate at which it is doing so is decreasing. That is, prices are still going up but not as quickly as they were before.

The Second Derivative

The rate of change with respect to x of a function $f(x)$ is given by its derivative $f'(x)$. Similarly, the rate of change of $f'(x)$ is given by its derivative, that is, by the derivative of the derivative of the original function. The derivative of the derivative of a function $f(x)$ is known as the **second derivative** of the function and is denoted by the symbol $f''(x)$, which is read "f double prime of x." The derivative $f'(x)$ is sometimes called the **first derivative** to distinguish it from the second derivative $f''(x)$.

If the function is denoted by y instead of $f(x)$, the symbol $\dfrac{d^2y}{dx^2}$ (which is read "the second derivative of y with respect to x") is often used instead of $f''(x)$ to denote the second derivative.

The Second Derivative

> The second derivative of a function is the derivative of its derivative. If $y = f(x)$, the second derivative is denoted by
>
> $$\frac{d^2y}{dx^2} \quad \text{or} \quad f''(x)$$
>
> The second derivative gives the rate of change of the rate of change of the original function.

Computation of the Second Derivative

You don't have to use any new rules to find the second derivative of a function. Just find the first derivative and then differentiate again.

EXAMPLE 7.1   Find the second derivative of the function $f(x) = 5x^4 - 3x^2 - 3x + 7$.

SOLUTION
Compute the first derivative

$$f'(x) = 20x^3 - 6x - 3$$

and then differentiate again to get

$$f''(x) = 60x^2 - 6$$

EXAMPLE 7.2   Find the second derivative of the function $y = (x^2 + 1)^5$.

SOLUTION
Compute the first derivative using the chain rule to get

$$\frac{dy}{dx} = 5(x^2 + 1)^4(2x) = 10x(x^2 + 1)^4$$

Then differentiate again using the product rule to get

$$\frac{d^2y}{dx^2} = 10x[4(x^2 + 1)^3(2x)] + 10(x^2 + 1)^4$$

$$= 80x^2(x^2 + 1)^3 + 10(x^2 + 1)^4$$

$$= 10(x^2 + 1)^3[8x^2 + (x^2 + 1)]$$

$$= 10(x^2 + 1)^3(9x^2 + 1)$$

EXAMPLE 7.3   Find the second derivative of the function $f(x) = \dfrac{3x - 2}{(x - 1)^2}$.

SOLUTION
By the quotient rule,

$$f'(x) = \frac{(x - 1)^2(3) - (3x - 2)[2(x - 1)(1)]}{(x - 1)^4}$$

$$= \frac{(x - 1)[3(x - 1) - 2(3x - 2)]}{(x - 1)^4}$$

$$= \frac{3x - 3 - 6x + 4}{(x - 1)^3}$$

$$= \frac{1 - 3x}{(x - 1)^3}$$

By the quotient rule again,

$$f''(x) = \frac{(x - 1)^3(-3) - (1 - 3x)[3(x - 1)^2(1)]}{(x - 1)^6}$$

$$= \frac{-3(x - 1)^2[(x - 1) + (1 - 3x)]}{(x - 1)^6}$$

$$= \frac{-3(-2x)}{(x - 1)^4} = \frac{6x}{(x - 1)^4}$$

A Word of Advice

Before computing the second derivative of a function, always take the time to simplify the first derivative as much as possible. The more complicated the form of the first derivative is, the more tedious the computation of the second derivative will be.

The Second Derivative by Implicit Differentiation

The next example illustrates how to find second derivatives by implicit differentiation.

EXAMPLE 7.4 Use implicit differentiation to find $\frac{d^2y}{dx^2}$ if $4x^2 - 2y^2 = 9$. Express your answer in terms of x and y and simplify.

SOLUTION
First differentiate both sides of the equation implicitly with respect to x to get

$$8x - 4y\frac{dy}{dx} = 0 \quad \text{or} \quad \frac{dy}{dx} = \frac{8x}{4y} = \frac{2x}{y}$$

Now differentiate implicitly with respect to x again, this time using the quotient rule.

$$\frac{d^2y}{dx^2} = \frac{d}{dx}\left(\frac{2x}{y}\right)$$

$$= \frac{y\frac{d}{dx}(2x) - 2x\left(\frac{dy}{dx}\right)}{y^2}$$

$$= \frac{2y - 2x\frac{dy}{dx}}{y^2}$$

Use the fact that $\dfrac{dy}{dx} = \dfrac{2x}{y}$ to replace the symbol $\dfrac{dy}{dx}$ so that the second derivative will be expressed in terms of just x and y.

$$\frac{d^2y}{dx^2} = \frac{2y - 2x\left(\dfrac{2x}{y}\right)}{y^2}$$

Simplify the resulting quotient by multiplying numerator and denominator by y to get

$$\frac{d^2y}{dx^2} = \frac{2y^2 - 4x^2}{y^3}$$

Finally, notice from the original equation that

$$2y^2 - 4x^2 = -(4x^2 - 2y^2) = -9$$

and make this substitution in the numerator to get

$$\frac{d^2y}{dx^2} = -\frac{9}{y^3}$$

Applications of the Second Derivative

The second derivative will be used in Chapter 3, Section 2, to obtain information about the shapes of graphs. In Sections 3 and 4 of that chapter, the second derivative will appear again, this time in the solution of optimization problems. Here is a more elementary application illustrating the interpretation of the second derivative as the rate of change of a rate of change.

EXAMPLE 7.5 An efficiency study of the morning shift at a certain factory indicates that an average worker who arrives on the job at 8:00 A.M. will have produced $Q(t) = -t^3 + 6t^2 + 24t$ units t hours later.

(a) Compute the worker's rate of production at 11:00 A.M.
(b) At what rate is the worker's rate of production changing with respect to time at 11:00 A.M.?
(c) Use calculus to estimate the change in the worker's rate of production between 11:00 and 11:10 A.M.
(d) Compute the actual change in the worker's rate of production between 11:00 and 11:10 A.M.

SOLUTION
(a) The worker's rate of production is the first derivative

$$Q'(t) = -3t^2 + 12t + 24$$

of the output function $Q(t)$. At 11:00 A.M., $t = 3$ and the rate of production is

$$Q'(3) = -3(3)^2 + 12(3) + 24 = 33 \text{ units per hour}$$

(b) The rate of change of the rate of production is the second derivative

$$Q''(t) = -6t + 12$$

of the output function. At 11:00 A.M., this rate is

$$Q''(3) = -6(3) + 12 = -6 \text{ units per hour per hour}$$

The minus sign indicates that the worker's rate of production is decreasing; that is, the worker is slowing down. The rate of this decrease in efficiency at 11:00 A.M. is 6 units per hour per hour.

(c) To estimate the change in the production rate $Q'(t)$ due to a change in t of $\Delta t = \dfrac{1}{6}$ hour, apply the approximation formula from Section 4 to the function $Q'(t)$ to get

Change in rate of production $= \Delta Q' \simeq Q''(t)\,\Delta t$

Evaluate this expression when $t = 3$ and $\Delta t = \dfrac{1}{6}$ to conclude that

$$\begin{matrix}\text{Change in rate} \\ \text{of production}\end{matrix} \simeq Q''(3)\Delta t = -6\left(\frac{1}{6}\right) = -1 \text{ unit per hour}$$

That is, the worker's rate of production (which was 33 units per hour at 11:00 A.M.) will decrease by approximately 1 unit per hour (to approximately 32 units per hour) during the subsequent 10 minutes.

(d) The actual change in the worker's rate of production between 11:00 and 11:10 A.M. is the difference between the values of the rate $Q'(t)$ when $t = 3$ and when $t = 3\dfrac{1}{6} = \dfrac{19}{6}$. That is,

$$\begin{matrix}\text{Actual change in} \\ \text{rate of production}\end{matrix} = Q'\left(\frac{19}{6}\right) - Q'(3)$$

$$= \left[-3\left(\frac{19}{6}\right)^2 + 12\left(\frac{19}{6}\right) + 24\right]$$

$$- [-3(3)^2 + 12(3) + 24]$$

$$\simeq 31.92 - 33 = -1.08 \text{ units per hour}$$

That is, by 11:10 A.M., the worker's rate of production, which was 33 units per hour at 11:00 A.M., will actually have decreased by 1.08 units per hour to 31.92 units per hour.

Higher-Order Derivatives

If you differentiate the second derivative $f''(x)$ of a function $f(x)$ one more time, you get the third derivative $f'''(x)$. Differentiate again and you get the fourth derivative, which is denoted by $f^{(4)}(x)$ since the prime-notation $f''''(x)$ begins to get cumbersome. In general, the derivative obtained from $f(x)$ after n successive differentiations is called the **nth derivative** or **derivative of order n.**

The *n*th Derivative

> For any positive integer n, the nth derivative of a function is obtained from the function by differentiating successively n times. If the original function is $y = f(x)$, the nth derivative is denoted by
>
> $$\frac{d^n y}{dx^n} \quad \text{or} \quad f^{(n)}(x)$$

EXAMPLE 7.6 Find the 5th derivative of each of the following functions:

(a) $f(x) = 4x^3 + 5x^2 + 6x - 1$

(b) $y = \dfrac{1}{x}$

SOLUTION

(a) $f'(x) = 12x^2 + 10x + 6$

$\quad f''(x) = 24x + 10$

$\quad f'''(x) = 24$

$\quad f^{(4)}(x) = 0$

$\quad f^{(5)}(x) = 0$

(b) $\dfrac{dy}{dx} = \dfrac{d}{dx}(x^{-1}) = -x^{-2} = -\dfrac{1}{x^2}$

$\dfrac{d^2 y}{dx^2} = \dfrac{d}{dx}(-x^{-2}) = 2x^{-3} = \dfrac{2}{x^3}$

$\dfrac{d^3 y}{dx^3} = \dfrac{d}{dx}(2x^{-3}) = -6x^{-4} = -\dfrac{6}{x^4}$

$\dfrac{d^4 y}{dx^4} = \dfrac{d}{dx}(-6x^{-4}) = 24x^{-5} = \dfrac{24}{x^5}$

$\dfrac{d^5 y}{dx^5} = \dfrac{d}{dx}(24x^{-5}) = -120x^{-6} = -\dfrac{120}{x^6}$

Derivatives of higher order will play a central role in Chapter 11, Section 3, in which a powerful technique for approximating functions will be developed.

Problems

In Problems 1 through 16, find the second derivative of the given function. In each case, use the appropriate notation for the second derivative and simplify your answer. (Don't forget to simplify the first derivative as much as possible before computing the second derivative.)

1. $f(x) = 5x^{10} - 6x^5 - 27x + 4$

2. $f(x) = \frac{2}{5}x^5 - 4x^3 + 9x^2 - 6x - 2$

3. $y = 5\sqrt{x} + \frac{3}{x^2} + \frac{1}{3\sqrt{x}} + \frac{1}{2}$

4. $y = \frac{2}{3x} - \sqrt{2x} + \sqrt{2}x - \frac{1}{6\sqrt{x}}$

5. $f(x) = (3x + 1)^5$

6. $f(x) = \frac{2}{5x + 1}$

7. $y = (x^2 + 5)^8$

8. $y = (1 - 2x^3)^4$

9. $f(x) = \sqrt{1 + x^2}$

10. $f(x) = \frac{1}{(3x^2 - 1)^2}$

11. $y = \frac{2}{1 + x^2}$

12. $y = \frac{x}{(x + 1)^2}$

13. $f(x) = x(2x + 1)^4$ (Use the product rule.)

14. $f(x) = 2x(x + 4)^3$ (Use the product rule.)

15. $y = \left(\frac{x}{x + 1}\right)^2$

16. $y = \frac{(x - 2)^3}{x^2}$

In Problems 17 through 19, use implicit differentiation to find $\frac{d^2y}{dx^2}$. Express your answer in terms of x and y and simplify.

17. $2y^2 - 5x^2 = 3$ 18. $2x^2 - 3y^2 = 7$

19. $ax^2 + by^2 = 1$, where a and b are constants.

Worker efficiency 20. An efficiency study of the morning shift at a certain factory indicates that an average worker who arrives on the job at 8:00 A.M. will have produced $Q(t) = -t^3 + 8t^2 + 15t$ units t hours later.
(a) Compute the worker's rate of production at 9:00 A.M.
(b) At what rate is the worker's rate of production changing with respect to time at 9:00 A.M.?
(c) Use calculus to estimate the change in the worker's rate of production between 9:00 and 9:15 A.M.
(d) Compute the actual change in the worker's rate of production between 9:00 and 9:15 A.M.

Inflation 21. It is projected that t months from now, the average price per unit for goods in a certain sector of the economy will be $P(t) = -t^3 + 7t^2 + 200t + 300$ dollars.
(a) At what rate will the price per unit be increasing with respect to time 5 months from now?
(b) At what rate will the rate of price increase be changing with respect to time 5 months from now?
(c) Use calculus to estimate the change in the rate of price increase during the first half of the 6th month.
(d) Compute the actual change in the rate of price increase during the first half of the 6th month.

Population growth 22. Suppose that a 5-year projection of population trends suggests that t years from now, the population of a certain community will be $P(t) = -t^3 + 9t^2 + 48t + 200$ thousand.
(a) At what rate will the population be growing 3 years from now?
(b) At what rate will the rate of population growth be changing with respect to time 3 years from now?
(c) Use calculus to estimate the change in the rate of population growth during the first month of the 4th year.
(d) Compute the actual change in the rate of population growth during the first month of the 4th year.

Acceleration The **acceleration** of a moving object is the rate of change of its speed with respect to time. Use this concept in Problems 23 through 25.

23. An object moves along a straight line so that after t seconds, its distance from its starting point is $D(t) = t^3 - 12t^2 + 100t + 12$ meters. Find the acceleration of the object after 3 seconds. Is the object slowing down or speeding up at this time?

24. If an object is dropped or thrown vertically, its height (in feet) after t seconds is $H(t) = -16t^2 + S_0 t + H_0$, where S_0 is the initial speed of the object and H_0 its initial height.
(a) Derive an expression for the acceleration of the object.
(b) How does the acceleration vary with time?
(c) What is the significance of the fact that the answer to part (a) is negative?

25. After t hours of an 8-hour trip, a car has gone $D(t) = 64t + \frac{10}{3}t^2 - \frac{2}{9}t^3$ kilometers.
   (a) Derive a formula expressing the acceleration of the car as a function of time.
   (b) At what rate is the speed of the car changing with respect to time at the end of 6 hours? Is the speed increasing or decreasing at this time?
   (c) By how much does the speed of the car actually change during the 7th hour?

26. Find $f^{(4)}(x)$ if $f(x) = x^5 - 2x^4 + x^3 - 3x^2 + 5x - 6$.

27. Find $\frac{d^3y}{dx^3}$ if $y = \sqrt{x} - \frac{1}{2x} + \frac{x}{\sqrt{2}}$.

28. Find $f'''(x)$ if $f(x) = \frac{1}{\sqrt{3x}} - \frac{2}{x^2} + \sqrt{2}$.

Chapter Summary and Review Problems

Important Terms, Symbols, and Formulas

Secant line; tangent line

Derivative: $f'(x) = \lim\limits_{\Delta x \to 0} \dfrac{f(x + \Delta x) - f(x)}{\Delta x}$

Geometric interpretation: derivative = slope of the tangent

Limits, continuity, differentiable functions

Power rule: $\dfrac{d}{dx}(x^n) = nx^{n-1}$

Derivative of a constant: $\dfrac{d}{dx}(c) = 0$

Constant multiple rule: $\dfrac{d}{dx}(cf) = c\dfrac{df}{dx}$

Sum rule: $\dfrac{d}{dx}(f + g) = \dfrac{df}{dx} + \dfrac{dg}{dx}$

Product rule: $\dfrac{d}{dx}(fg) = f\dfrac{dg}{dx} + g\dfrac{df}{dx}$

Quotient rule: $\dfrac{d}{dx}\left(\dfrac{f}{g}\right) = \dfrac{g\dfrac{df}{dx} - f\dfrac{dg}{dx}}{g^2}$

Average rate of change $= \dfrac{f(x + \Delta x) - f(x)}{\Delta x}$

Instantaneous rate of change $= f'(x) = \dfrac{dy}{dx}$

Approximation: $f'(x) \simeq$ change in f due to a 1-unit increase in x

Marginal cost $=$ derivative of cost $\simeq$ cost of 1 additional unit

Marginal analysis

Percentage rate of change of y with respect to $x = 100\dfrac{dy/dx}{y}$

Approximation formula: $\Delta y \simeq \dfrac{dy}{dx}\Delta x$; $\Delta f = f(x + \Delta x) - f(x) \simeq f'(x)\Delta x$

Percentage change in $f = 100\dfrac{\Delta f}{f} \simeq 100\dfrac{f'(x)\Delta x}{f(x)}$

Differential: $dy = \dfrac{dy}{dx}\Delta x =$ change in y along the tangent line

Chain rule: $\dfrac{dy}{dx} = \dfrac{dy}{du}\dfrac{du}{dx}$; $\dfrac{d}{dx}g[h(x)] = g'[h(x)]h'(x)$

Chain rule for powers: $\dfrac{d}{dx}[h(x)]^n = n[h(x)]^{n-1}h'(x)$

Implicit differentiation

Second derivative: $\dfrac{d^2y}{dx^2} = f''(x) =$ rate of change of rate of change

nth derivative: $\dfrac{d^ny}{dx^n} = f^{(n)}(x)$

Review Problems

1. Use the definition to find the derivative of the given function.

   (a) $f(x) = x^2 - 3x + 1$

   (b) $f(x) = \dfrac{1}{x - 2}$

2. Differentiate the following functions using the rules. Simplify your answers.

   (a) $f(x) = 6x^4 - 7x^3 + 2x + \sqrt{2}$

   (b) $f(x) = x^3 - \dfrac{1}{3x^5} + 2\sqrt{x} - \dfrac{3}{x} + \dfrac{1 - 2x}{x^3}$

   (c) $y = \dfrac{2 - x^2}{3x^2 + 1}$

(d) $y = (2x + 5)^3(x + 1)^2$

(e) $f(x) = (5x^4 - 3x^2 + 2x + 1)^{10}$

(f) $f(x) = \sqrt{x^2 + 1}$

(g) $y = \left(x + \dfrac{1}{x}\right)^2 - \dfrac{5}{\sqrt{3x}}$

(h) $y = \left(\dfrac{x + 1}{1 - x}\right)^2$

(i) $f(x) = (3x + 1)\sqrt{6x + 5}$

(j) $f(x) = \dfrac{(3x + 1)^3}{(1 - 3x)^4}$

(k) $y = \sqrt{\dfrac{1 - 2x}{3x + 2}}$

3. Find the equation of the line that is tangent to the graph of f at the point $(x, f(x))$ for the given value of x.

   (a) $f(x) = x^2 - 3x + 2; x = 1$

   (b) $f(x) = \dfrac{4}{x - 3}; x = 1$

   (c) $f(x) = \dfrac{x}{x^2 + 1}; x = 0$

   (d) $f(x) = \sqrt{x^2 + 5}; x = -2$

4. After x weeks, the number of people using a new rapid transit system was approximately $N(x) = 6x^3 + 500x + 8,000$.
   (a) At what rate was the use of the system changing with respect to time after 8 weeks?
   (b) By how much did the use of the system change during the 8th week?

5. It is estimated that the weekly output at a certain plant is $Q(x) = 50x^2 + 9,000x$ units, where x is the number of workers employed at the plant. Currently there are 30 workers employed at the plant.
   (a) Use calculus to estimate the change in the weekly output that will result from the addition of 1 worker to the force.
   (b) Compute the actual change in output that will result from the addition of 1 worker.

6. It is projected that t months from now, the population of a certain town will be $P(t) = 3t + 5t^{3/2} + 6,000$. At what percentage rate will the population be changing with respect to time 4 months from now?

7. At a certain factory, the daily output is $Q(L) = 20,000L^{1/2}$ units, where L denotes the size of the labor force measured in worker-

hours. Currently 900 worker-hours of labor are used each day. Use calculus to estimate the effect on output that will be produced if the labor force is cut to 885 worker-hours.

8. The gross national product of a certain country was $N(t) = t^2 + 6t + 300$ billion dollars t years after 1980. Use calculus to estimate the percentage change in the GNP during the second quarter of 1984.

9. The level of air pollution in a certain city is proportional to the square of the population. Use calculus to estimate the percentage by which the air-pollution level will increase if the population increases by 5 percent.

10. The output at a certain factory is $Q(L) = 600L^{2/3}$ units, where L is the size of the labor force. The manufacturer wishes to increase output by 1 percent. Use calculus to estimate the percentage increase in labor that will be required.

11. Use the chain rule to find $\dfrac{dy}{dx}$.

    (a) $y = 5u^2 + u - 1; u = 3x + 1$

    (b) $y = \dfrac{1}{u^2}; u = 2x + 3$

12. Use the chain rule to find $\dfrac{dy}{dx}$ for the given value of x.

    (a) $y = u^3 - 4u^2 + 5u + 2, u = x^2 + 1; x = 1$

    (b) $y = \sqrt{u}, u = x^2 + 2x - 4; x = 2$

13. At a certain factory, approximately $q(t) = t^2 + 50t$ units are manufactured during the first t hours of a production run, and the total cost of manufacturing q units is $C(q) = 0.1q^2 + 10q + 400$ dollars. Find the rate at which the manufacturing cost is changing with respect to time 2 hours after production commences.

14. It is estimated that t years from now, the population of a certain suburban community will be $p(t) = 10 - \dfrac{20}{(t + 1)^2}$ thousand. An environmental study indicates that the average daily level of carbon monoxide in the air will be $c(p) = 0.8\sqrt{p^2 + p + 139}$ units when the population is p thousand. At what percentage rate will the level of carbon monoxide be changing with respect to time 1 year from now?

15. Find $\dfrac{dy}{dx}$ by implicit differentiation.

    (a) $5x + 3y = 12$ (b) $x^2y = 1$

    (c) $(2x + 3y)^5 = x + 1$ (d) $(1 - 2xy^3)^5 = x + 4y$

16. Use implicit differentiation to find the slope of the line that is tangent to the given curve for the specified value of x.

(a) $xy^3 = 8; x = 1$

(b) $x^2y - 2xy^3 + 6 = 2x + 2y; x = 0$

17. The output Q at a certain factory is related to inputs x and y by the equation $Q = x^3 + 2xy^2 + 2y^3$. If the current levels of input are $x = 10$ and $y = 20$, use calculus to estimate the change in input y that should be made to offset an increase of 0.5 in input x so that output will be maintained at its current level.

18. Find the second derivative of each of the following functions:

(a) $f(x) = 6x^5 - 4x^3 + 5x^2 - 2x + \dfrac{1}{x}$

(b) $y = (3x^2 + 2)^4$

(c) $f(x) = \dfrac{x - 1}{(x + 1)^2}$

19. Use implicit differentiation to find $\dfrac{d^2y}{dx^2}$ if $3x^2 - 2y^2 = 6$.

20. An efficiency study of the morning shift at a certain factory indicates that an average worker who arrives on the job at 8:00 A.M. will have produced $Q(t) = -t^3 + 9t^2 + 12t$ units t hours later.
(a) Compute the worker's rate of production at 9:00 A.M.
(b) At what rate is the worker's rate of production changing with respect to time at 9:00 A.M.?
(c) Use calculus to estimate the change in the worker's rate of production between 9:00 and 9:06 A.M.
(d) Compute the actual change in the worker's rate of production between 9:00 and 9:06 A.M.

21. Find the 4th derivative of each of the following functions:

(a) $y = 2x^5 + 5x^4 - 2x + \dfrac{1}{x}$

(b) $f(x) = \sqrt{3x} + \dfrac{3}{2x^2}$

Chapter

3 Differentiation: Further Topics

1   Increase and Decrease; Relative Extrema

In Example 1.4 of Chapter 2, Section 1, we used calculus to maximize a profit function like the one shown in Figure 1.1. In particular we observed that the maximum value corresponded to the unique point on the graph at

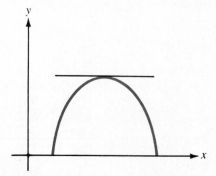

Figure 1.1 A profit function.

which the slope of the tangent was zero, and we therefore set the derivative equal to zero and solved for x.

The simplicity of this example is misleading. In general, not every point at which the derivative of a function is zero is a peak of its graph. Two functions whose derivatives are zero when $x = 0$ are sketched in Figure 1.2. Both have horizontal tangents at $(0, 0)$, but the function $y = x^2$ in Figure 1.2a reaches its *lowest* point at $(0, 0)$, while the function $y = x^3$ in Figure 1.2b has neither a maximum nor a minimum at this point.

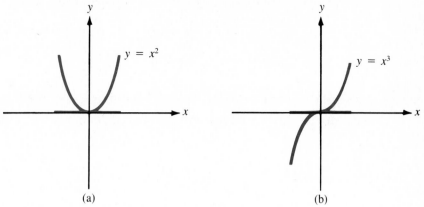

(a)   (b)

Figure 1.2 Two functions with horizontal tangents when $x = 0$.

The situation is further complicated by the existence of functions that have maxima or minima at points at which the derivative is not even defined. Two such functions are sketched in Figure 1.3.

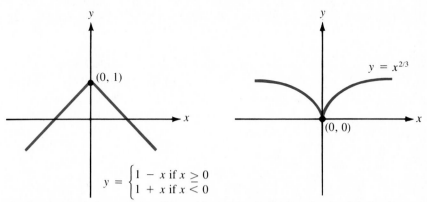

$$y = \begin{cases} 1 - x \text{ if } x \geq 0 \\ 1 + x \text{ if } x < 0 \end{cases}$$

Figure 1.3 Two functions with extrema where the derivative is undefined.

In this section, you will learn a systematic procedure you can use to locate and identify maxima and minima of differentiable functions. In the process, you will also see how to use derivatives to help you sketch the graphs of functions.

Relative Maxima and Minima

A **relative maximum** of a function is a peak, a point on the graph of the function that is higher than any neighboring point on the graph. A **relative minimum** is the bottom of a valley, a point on the graph that is lower than any neighboring point. (The relative maxima and minima of a function are sometimes called **relative extrema**.) The function sketched in Figure 1.4 has a relative maximum at $x = b$ and relative minima at $x = a$ and $x = c$. Notice that a relative maximum need not be the highest point on a graph. It is maximal only *relative to* neighboring points. Similarly, a relative minimum need not be the lowest point on the graph.

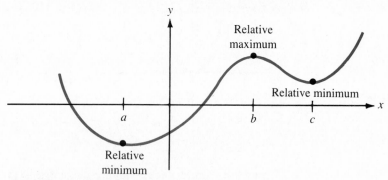

Figure 1.4 Relative maxima and minima.

Increase and Decrease of Functions

A function is said to be **increasing** if its graph is rising as x increases, and **decreasing** if its graph is falling as x increases. The function in Figure 1.5 is increasing for $a < x < b$ and for $x > c$. It is decreasing for $x < a$ and for $b < x < c$.

If you know the intervals on which a function is increasing and decreasing, you can easily identify its relative maxima and minima. A relative maximum occurs when the function stops increasing and starts

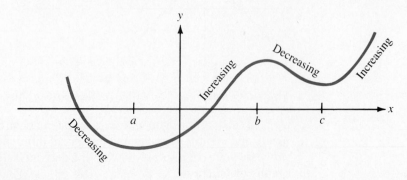

Figure 1.5 Increase and decrease of a function.

decreasing. In Figure 1.5, this happens when $x = b$. A relative minimum occurs when the function stops decreasing and starts increasing. In Figure 1.5, this happens when $x = a$ and $x = c$.

The Sign of the Derivative

You can find out where a differentiable function is increasing or decreasing by checking the sign of its (first) derivative. This is because the derivative is the slope of the tangent. When the derivative is positive, the slope of the tangent is positive and the function is increasing. When the derivative is negative, the slope of the tangent is negative and the function is decreasing. The situation is illustrated in Figure 1.6.

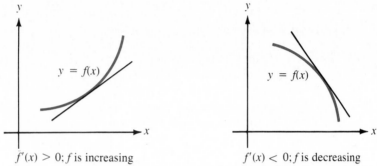

$f'(x) > 0; f$ is increasing $f'(x) < 0; f$ is decreasing

Figure 1.6 The geometric significance of the sign of the first derivative.

Here is a more precise statement of the situation.

Derivative Test for Intervals of Increase and Decrease

If $f'(x) > 0$ on the interval $a < x < b$, then f is increasing on this interval.

If $f'(x) < 0$ on the interval $a < x < b$, then f is decreasing on this interval.

Critical Points

Since a function is increasing when its derivative is positive and decreasing when its derivative is negative, the only points at which the function can have a relative maximum or minimum are those at which its derivative is either zero or undefined. A point in the domain of a function at which the derivative is zero or undefined is said to be a **critical point** of the function.

Three functions with critical points at which the derivative is zero are shown in Figure 1.7. In each case, the line that is tangent to the graph

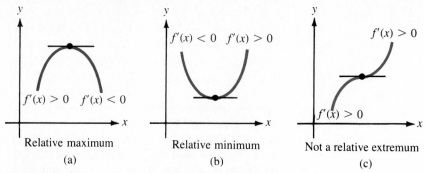

Figure 1.7 Three critical points at which $f'(x) = 0$.

at the critical point is horizontal because the derivative (or slope of the tangent) is zero.

Three functions with critical points at which the derivative is undefined are shown in Figure 1.8. In Figure 1.8b and Figure 1.8c, the tangent line is vertical and hence the derivative (or slope of the tangent) is undefined. In Figure 1.8a, the derivative is undefined at the critical point because the function does not even have a (uniquely defined) tangent line at this point.

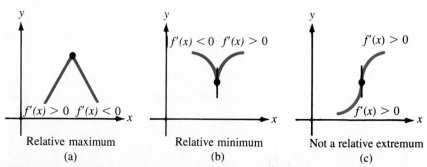

Figure 1.8 Three critical points at which $f'(x)$ is undefined.

Critical Points

A critical point of a function is a point in the domain of the function at which either

(1) the derivative is zero

or

(2) the derivative is undefined.

The critical points are the possible relative maxima and minima of the function.

Classification of Critical Points

Every relative extremum is a critical point. However, as you saw in Figures 1.7c and 1.8c, not every critical point is necessarily a relative extremum. If the derivative is positive to the left of a critical point and negative to the right of it, the graph changes from increasing to decreasing and the critical point is a relative maximum (Figures 1.7a and 1.8a). If the derivative is negative to the left of a critical point and positive to the right of it, the graph changes from decreasing to increasing and the critical point is a relative minimum (Figures 1.7b and 1.8b). If the sign of the derivative is the same on both sides of the critical point, the direction of the graph does not change and the critical point is neither a relative maximum nor a relative minimum (Figures 1.7c and 1.8c).

Curve Sketching Using the First Derivative

The preceding observations suggest the following general procedure you can use to sketch functions and find their intervals of increase and decrease and their relative extrema.

How to Use the First Derivative to Graph a Function

Step 1. Compute the derivative $f'(x)$ and put it in factored form if possible.

Step 2. Find the values of x for which the derivative is zero or undefined. For each of these values of x that is in the domain of f, substitute x into the function $f(x)$ to get the y coordinate of the corresponding critical point.

Step 3. Plot the critical points on the graph. These are the only points at which relative extrema can possibly occur.

Step 4. Determine where the function is increasing or decreasing by checking the sign of the derivative on the intervals whose endpoints are the values of x from Step 2.

Step 5. Sketch the graph so that it increases on the intervals on which the derivative is positive, decreases on the intervals on which the derivative is negative, and has a horizontal tangent where the derivative is zero.

Here are some examples illustrating the technique.

Graphs of Polynomials

EXAMPLE 1.1   Find the intervals of increase and decrease and the relative extrema of the function $f(x) = 2x^3 + 3x^2 - 12x - 7$ and sketch the graph.

SOLUTION

Begin by computing and factoring the derivative

$$f'(x) = 6x^2 + 6x - 12 = 6(x^2 + x - 2) = 6(x + 2)(x - 1)$$

From the factored form of the derivative you can see that $f'(x) = 0$ when $x = -2$ and when $x = 1$. The corresponding critical points are $(-2, f(-2)) = (-2, 13)$ and $(1, f(1)) = (1, -14)$. Begin the sketch (Figure 1.9a) by plotting these critical points. (To help you remember that the graph should have horizontal tangents at these points, you can draw a short horizontal line segment through each.)

To find the intervals of increase and decrease of the function, check the sign of the derivative for $x < -2$, for $-2 < x < 1$, and for $x > 1$.

On the interval $x < -2$ (at $x = -3$, for example), both factors $x + 2$ and $x - 1$ are negative. Hence the derivative $f'(x) = 6(x + 2)(x - 1)$ is positive and the function is increasing on this interval.

On the interval $-2 < x < 1$ (at $x = 0$, for example), the factor $x + 2$ is positive while the factor $x - 1$ is still negative. Hence the derivative is negative and the function is decreasing on this interval.

Finally, on the interval $x > 1$ (at $x = 2$, for example), both factors $x + 2$ and $x - 1$ are positive. Hence the derivative is positive and the function is increasing on this interval.

These observations are summarized in the following table.

Interval	(x + 2)	(x − 1)	Sign of f′	Increase or decrease of f
$x < -2$	−	−	+	Increasing
$-2 < x < 1$	+	−	−	Decreasing
$x > 1$	+	+	+	Increasing

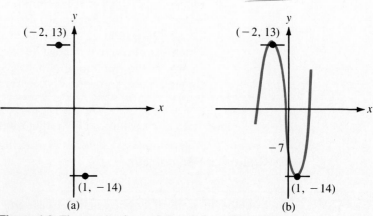

(a) (b)

Figure 1.9 The graph of $y = 2x^3 + 3x^2 - 12x - 7$.

Complete the sketch in Figure 1.9b by drawing the curve so that it increases for $x < -2$, decreases for $-2 < x < 1$, increases again for $x > 1$, and levels off at the critical points. Notice that the function has a relative maximum at the critical point $(-2, 13)$ and a relative minimum at the critical point $(1, -14)$. Notice also that the y intercept is $(0, f(0)) = (0, -7)$.

EXAMPLE 1.2   Find the intervals of increase and decrease and the relative extrema of the function $f(x) = x^4 + 8x^3 + 18x^2 - 8$ and sketch the graph.

SOLUTION
The derivative is

$$f'(x) = 4x^3 + 24x^2 + 36x = 4x(x^2 + 6x + 9) = 4x(x + 3)^2$$

which is zero when $x = 0$ and $x = -3$. The corresponding critical points are $(0, f(0)) = (0, -8)$ and $(-3, f(-3)) = (-3, 19)$.

To find the intervals of increase and decrease of the function, check the sign of the derivative for $x < -3$, $-3 < x < 0$, and $x > 0$.

Interval	x	$(x + 3)^2$	Sign of f'	Increase or decrease of f
$x < -3$	$-$	$+$	$-$	Decreasing
$-3 < x < 0$	$-$	$+$	$-$	Decreasing
$x > 0$	$+$	$+$	$+$	Increasing

Sketch the graph using this information as shown in Figure 1.10. Notice that since the graph is decreasing on both sides of the critical point $(-3, 19)$, this point is neither a relative maximum nor a relative minimum. On the other hand, the critical point $(0, -8)$ is a relative minimum.

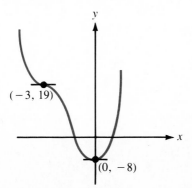

Figure 1.10 The graph of $y = x^4 + 8x^3 + 18x^2 - 8$.

Graphs of Rational Functions

EXAMPLE 1.3   Find the intervals of increase and decrease and the relative extrema of the rational function $f(x) = \dfrac{x^2}{x - 2}$ and sketch the graph.

SOLUTION

Since the denominator of $f(x)$ is zero when $x = 2$, the function is undefined and has a discontinuity for this value of x. To remind yourself that there is no point on the graph whose x coordinate is 2, begin your sketch by drawing a broken vertical line at $x = 2$. Your graph should not cross this line.

By the quotient rule, the derivative is

$$f'(x) = \frac{(x - 2)(2x) - x^2(1)}{(x - 2)^2} = \frac{2x^2 - 4x - x^2}{(x - 2)^2} = \frac{x(x - 4)}{(x - 2)^2}$$

which is zero when its numerator is zero, that is, when $x = 0$ and $x = 4$. The corresponding critical points are $(0, f(0)) = (0, 0)$ and $(4, f(4)) = (4, 8)$. The derivative is also undefined at $x = 2$, but this is not a critical point since $x = 2$ is not in the domain of the function.

To identify the intervals of increase and decrease, check the sign of the derivative on the intervals determined by the critical points. Since $f'(x)$ is undefined at $x = 2$, its sign for $0 < x < 2$ could be different from its sign for $2 < x < 4$, and so you will have to check these two intervals separately.

Interval	x	$(x - 4)$	$(x - 2)^2$	Sign of f'	Increase or decrease of f
$x < 0$	$-$	$-$	$+$	$+$	Increasing
$0 < x < 2$	$+$	$-$	$+$	$-$	Decreasing
$2 < x < 4$	$+$	$-$	$+$	$-$	Decreasing
$x > 4$	$+$	$+$	$+$	$+$	Increasing

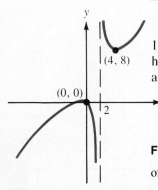

Now sketch the graph using this information as shown in Figure 1.11. Don't forget that the graph has a discontinuity at $x = 2$. Notice that f has a relative maximum at the critical point $(0, 0)$ and a relative minimum at the critical point $(4, 8)$.

Figure 1.11 The graph of $y = \dfrac{x^2}{x - 2}$.

Critical Points at Which $f'(x)$ Is Undefined

In the next example, the relative extremum occurs at a critical point at which the derivative is not defined.

EXAMPLE 1.4   Find the intervals of increase and decrease and the relative extrema of the function $f(x) = x^{2/3}$, and draw the graph.

SOLUTION
The derivative is

$$f'(x) = \frac{2}{3}x^{-1/3} = \frac{2}{3\sqrt[3]{x}}$$

which is never zero, but is undefined when $x = 0$, which is in the domain of f. Hence, the corresponding point $(0, f(0)) = (0, 0)$ is the only critical point.

To find the intervals of increase and decrease, check the sign of the derivative for $x < 0$ and $x > 0$.

Interval	Sign of f'	Increase or decrease of f
$x < 0$	$-$	Decreasing
$x > 0$	$+$	Increasing

The graph is sketched in Figure 1.12. Notice that the tangent to the graph at the relative minimum $(0, 0)$ is vertical and its slope is undefined. This corresponds to the fact that the derivative is undefined when $x = 0$.

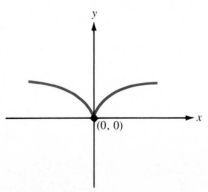

Figure 1.12 The graph of $y = x^{2/3}$.

The Use of the Quadratic Formula

In the next example, you will need the quadratic formula to find the critical points of the given function. You may want to use a calculator to help with the computations.

EXAMPLE 1.5   Find the intervals of increase and decrease and the relative extrema of the function $f(x) = x^3 + 2x^2 - x + 1$ and sketch the graph.

SOLUTION
The derivative is

$$f'(x) = 3x^2 + 4x - 1$$

which has no obvious factors. According to the quadratic formula, $3x^2 + 4x - 1 = 0$ when

$$x = \frac{-4 \pm \sqrt{4^2 - 4(3)(-1)}}{2(3)} = \frac{-4 \pm \sqrt{28}}{6}$$

Hence, the x coordinates of the critical points are

$$x = \frac{-4 + \sqrt{28}}{6} \approx 0.22 \quad \text{and} \quad x = \frac{-4 - \sqrt{28}}{6} \approx -1.55$$

and the corresponding y coordinates are

$$y \approx f(0.22) \approx 0.89 \quad \text{and} \quad y \approx f(-1.55) \approx 3.63$$

For each of the intervals $x < -1.55$, $-1.55 < x < 0.22$, and $x > 0.22$, use any convenient value of x in the interval to check the sign of the derivative. The values $x = -2$, $x = 0$, and $x = 1$ would be reasonable choices.

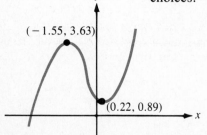

Interval	Sign of f'	Increase or decrease of f
$x < -1.55$	$+$	Increasing
$-1.55 < x < 0.22$	$-$	Decreasing
$x > 0.22$	$+$	Increasing

The graph is sketched in Figure 1.13. Notice that the critical point $(-1.55, 3.63)$ is a relative maximum and that the critical point $(0.22, 0.89)$ is a relative minimum. Notice also that the y intercept is $(0, f(0)) = (0, 1)$.

Figure 1.13 The graph of $y = x^3 + 2x^2 - x + 1$.

Problems

In Problems 1 through 3, specify the intervals on which the derivative of the given function is positive and on which it is negative.

1.

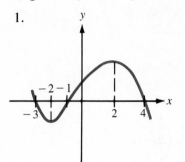

2.

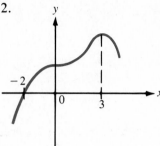

3.

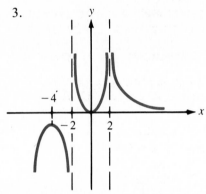

In Problems 4 through 30, find the intervals of increase and decrease and the relative extrema of the given function and sketch the graph.

4. $f(x) = x^2 - 4x + 5$

5. $f(x) = x^3 + 3x^2 + 1$

6. $f(x) = x^3 - 3x - 4$

7. $f(x) = \frac{1}{3}x^3 - 9x + 2$

8. $f(x) = x^5 - 5x^4 + 100$

9. $f(x) = 3x^5 - 5x^3$

10. $f(x) = 3x^4 - 8x^3 + 6x^2 + 2$

11. $f(x) = 324x - 72x^2 + 4x^3$

12. $f(x) = 2x^3 + 6x^2 + 6x + 5$

13. $f(x) = 10x^6 + 24x^5 + 15x^4 + 3$

14. $f(x) = (x - 1)^5$

15. $f(x) = 3 - (x + 1)^3$

16. $f(x) = (x^2 - 1)^5$

17. $f(x) = (x^2 - 1)^4$

18. $f(x) = (x^3 - 1)^4$

19. $f(x) = \dfrac{x^2}{x - 1}$

20. $f(x) = \dfrac{x^2}{x + 2}$

21. $f(x) = \dfrac{x^2 - 3x}{x + 1}$

22. $f(x) = \dfrac{1}{x^2 - 9}$

23. $f(x) = x + \dfrac{1}{x}$

24. $f(x) = 2x + \dfrac{18}{x} + 1$

25. $f(x) = 6x^2 + \dfrac{12{,}000}{x}$

26. $f(x) = 1 + x^{1/3}$

27. $f(x) = x^{3/5}$

28. $f(x) = 2 + (x - 1)^{2/3}$

29. $f(x) = x^3 - 2x^2 - 3x + 2$

30. $f(x) = x^3 - 3x^2 + 2x + 1$

31. Sketch a graph of a function that has all of the following properties:
    (a) $f'(x) > 0$ when $x < -5$ and when $x > 1$
    (b) $f'(x) < 0$ when $-5 < x < 1$
    (c) $f(-5) = 4$ and $f(1) = -1$

32. Sketch a graph of a function that has all of the following properties:
    (a) $f'(x) < 0$ when $x < -1$
    (b) $f'(x) > 0$ when $-1 < x < 3$ and when $x > 3$
    (c) $f'(-1) = 0$ and $f'(3) = 0$

33. Sketch a graph of a function that has all of the following properties:
    (a) $f'(x) > 0$ when $x > 2$
    (b) $f'(x) < 0$ when $x < 0$ and when $0 < x < 2$
    (c) $x = 0$ is not in the domain of f

34. Sketch a graph of a function that has all of the following properties:
    (a) $f'(x) > 0$ when $-1 < x < 3$ and when $x > 6$
    (b) $f'(x) < 0$ when $x < -1$ and when $3 < x < 6$
    (c) $f'(-1) = 0$ and $f'(6) = 0$
    (d) $f'(x)$ is undefined when $x = 3$

35. Find constants a, b, and c such that the graph of the function $f(x) = ax^2 + bx + c$ has a relative maximum at $(5, 12)$ and crosses the y axis at $(0, 3)$.

36. Use calculus to prove that the relative extremum of the quadratic function $y = ax^2 + bx + c$ occurs when $x = -\dfrac{b}{2a}$.

37. Use calculus to prove that the relative extremum of the quadratic function $y = (x - p)(x - q)$ occurs midway between its x intercepts.

2 Concavity: Curve Sketching; The Second-Derivative Test

In the preceding section, you saw that the sign of the first derivative of a function tells you where its graph is increasing and where it is decreasing. In this section, you will see that the sign of the second derivative also gives useful information about the shape of the graph. By way of introduction, here is a brief description of a situation from industry that can be analyzed with the aid of the second derivative.

The Efficiency of a Worker

The number of units that a factory worker can produce in t hours is often given by a function $Q(t)$ like the one whose graph is shown in Figure 2.1.

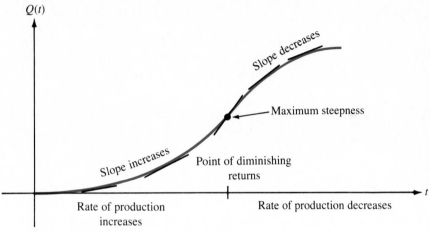

Figure 2.1 The output of a factory worker.

Notice that at first, the graph is not very steep. The steepness increases, however, until the graph reaches a point of maximum steepness, after which the steepness begins to decrease. This reflects the fact that at first, the worker's rate of production is low. The rate of production increases, however, as the worker settles into a routine, and continues to increase until the worker is performing at maximum efficiency, after which fatigue sets in and the rate of production begins to decrease. The moment of maximum efficiency is known in economics as the **point of diminishing returns.**

The behavior of the graph of this production function on either side of the point of diminishing returns can be described in terms of its tangent lines. To the left of this point, the slope of the tangent increases as t increases. To the right of this point, the slope of the tangent decreases as t increases. It is this increase and decrease of slopes that we shall examine in this section with the aid of the second derivative. (We shall return to the questions of worker efficiency and diminishing returns in Section 4 in the discussion of optimization.)

Concavity

The following notions of **concavity** are used to describe the increase and decrease of the slope of the tangent to a curve.

Concavity

> A curve is said to be concave upward if the slope of its tangent increases as it moves along the curve from left to right.
>
> A curve is said to be concave downward if the slope of its tangent decreases as it moves along the curve from left to right.

In Figure 2.1, for example, the production curve was concave upward to the left of the point of diminishing returns and concave downward to the right of this point. The notions of concavity are illustrated further in Figure 2.2, in which the curve is concave upward to the left of $x = a$ and concave downward to the right of $x = a$.

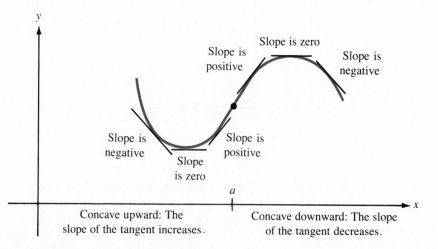

Figure 2.2 Concavity and the slope of the tangent.

The Sign of the Second Derivative

There is a simple characterization of concavity in terms of the sign of the second derivative. It is based on the fact (established in the preceding section) that a quantity increases when its derivative is positive and decreases when its derivative is negative. The second derivative comes into the picture when this fact is applied to the first derivative (or slope of the tangent). Here is the argument.

Suppose the second derivative f'' is positive on an interval. This implies that the first derivative f' must be increasing on the interval. But f' is the slope of the tangent. Hence the slope of the tangent is increasing, and so the graph of f is concave upward on the interval. On the other hand, if f'' is negative on an interval, then f' is decreasing. This implies that the slope of the tangent is decreasing, and so the graph of f is concave downward on the interval.

Here is a summary of the situation.

Test for Concavity

> If $f''(x) > 0$ on the interval $a < x < b$, then f is concave upward on this interval.
> If $f''(x) < 0$ on the interval $a < x < b$, then f is concave downward on this interval.

Inflection Points

A point in the domain of a function at which the concavity of the function changes is called an **inflection point.** The production function in Figure 2.1 has an inflection point at the point of diminishing returns. The function in Figure 2.2 has an inflection point when $x = a$. If the second derivative of a function is defined at an inflection point, its value there must be zero. Inflection points can also occur at points in the domain of the function where the second derivative is undefined.

Second-Order Critical Points

Points in the domain of a function at which the second derivative is zero or undefined are called **second-order critical points.** (The points in the domain of a function at which the first derivative is zero or undefined are sometimes called **first-order critical points.**) Second-order critical points are to inflection points as first-order critical points are to relative extrema. In particular, every inflection point is a second-order critical point, but not every second-order critical point is necessarily an inflection point.

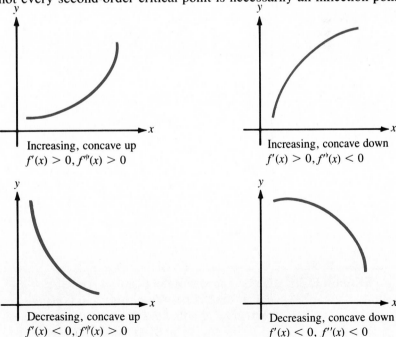

Increasing, concave up
$f'(x) > 0, f'''(x) > 0$

Increasing, concave down
$f'(x) > 0, f'''(x) < 0$

Decreasing, concave up
$f'(x) < 0, f'''(x) > 0$

Decreasing, concave down
$f'(x) < 0, f''(x) < 0$

Figure 2.3 Possible combinations of increase and decrease and concavity.

The Shape of a Curve

Do not confuse the concavity of a curve with its increase or decrease. A curve that is concave upward on an interval may be increasing or decreasing on that interval. Similarly, a curve that is concave downward may be increasing or decreasing. The four possibilities are illustrated in Figure 2.3.

Curve Sketching Using the First and Second Derivatives

The preceding observations can be combined with the techniques developed in Section 1 of this chapter to get the following procedure for obtaining detailed graphs of functions using calculus.

How to Use Calculus to Graph a Function

Step 1. Compute the first derivative $f'(x)$, factor it if possible, find the first-order critical points, and plot them on the graph.

Step 2. Compute the second derivative $f''(x)$, factor it if possible, find the second-order critical points, and plot them on the graph.

Step 3. Use the x coordinates of the first- and second-order critical points (and of any discontinuities) to divide the x axis into a collection of intervals. Check the signs of the first and second derivatives on each of these intervals.

Step 4. Draw the graph on each interval according to the following table:

Sign of f'	Sign of f''	Increase or decrease of f	Concavity of f	Shape of f
+	+	Increasing	C-up	
+	−	Increasing	C-down	
−	+	Decreasing	C-up	
−	−	Decreasing	C-down	

Here are two examples.

EXAMPLE 2.1 Determine where the function $f(x) = x^4 + 8x^3 + 18x^2 - 8$ is increasing, decreasing, concave upward, and concave downward. Find the relative extrema and inflection points and draw the graph.

SOLUTION

The first derivative is

$$f'(x) = 4x^3 + 24x^2 + 36x = 4x(x^2 + 6x + 9) = 4x(x + 3)^2$$

which is zero when $x = 0$ and $x = -3$. The corresponding first-order critical points are $(0, f(0)) = (0, -8)$ and $(-3, f(-3)) = (-3, 19)$.

The second derivative (computed from the unfactored form of the first derivative) is

$$f''(x) = 12x^2 + 48x + 36 = 12(x^2 + 4x + 3) = 12(x + 3)(x + 1)$$

which is zero when $x = -3$ and $x = -1$. The corresponding second-order critical points are $(-3, f(-3)) = (-3, 19)$ and $(-1, f(-1)) = (-1, 3)$.

Plot these critical points and check the signs of $f'(x)$ and $f''(x)$ on each of the intervals defined by their x coordinates.

Interval	$f'(x)$	$f''(x)$	Increase or decrease of f	Concavity of f	Shape of f
$x < -3$	−	+	Decreasing	C-up	
$-3 < x < -1$	−	−	Decreasing	C-down	
$-1 < x < 0$	−	+	Decreasing	C-up	
$x > 0$	+	+	Increasing	C-up	

Draw the graph as shown in Figure 2.4 by connecting the critical points with a curve of appropriate shape on each interval.

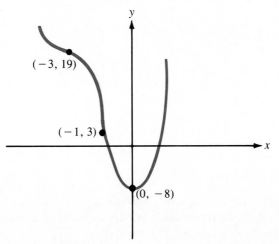

Figure 2.4 The graph of $y = x^4 + 8x^3 + 18x^2 - 8$.

Notice that the first-order critical point $(0, -8)$ is a relative minimum, while the first-order critical point $(-3, 19)$ is neither a relative minimum nor a relative maximum, and that both of the second-order critical points $(-3, 19)$ and $(-1, 3)$ are inflection points.

EXAMPLE 2.2   Determine where the function $f(x) = \dfrac{x}{(x + 1)^2}$ is increasing, decreasing, concave upward, and concave downward. Find the relative extrema and inflection points and draw the graph.

SOLUTION
Since the denominator of $f(x)$ is zero when $x = -1$, the function is undefined and has a discontinuity for this value of x. To remind yourself of this, begin your sketch by drawing a broken vertical line at $x = -1$. Your graph should not cross this line.

By the quotient rule, the first derivative is

$$f'(x) = \frac{(x + 1)^2(1) - x[2(x + 1)(1)]}{(x + 1)^4}$$

$$= \frac{(x + 1)[(x + 1) - 2x]}{(x + 1)^4}$$

$$= \frac{1 - x}{(x + 1)^3}$$

which is zero when $x = 1$. The corresponding first-order critical point is $(1, f(1)) = \left(1, \dfrac{1}{4}\right)$. (The derivative is also undefined at $x = -1$, but this is not a critical point since it is not in the domain of the function.)

By the quotient rule again,

$$f''(x) = \frac{(x + 1)^3(-1) - (1 - x)[3(x + 1)^2(1)]}{(x + 1)^6}$$

$$= \frac{(x + 1)^2[-(x + 1) - 3(1 - x)]}{(x + 1)^6}$$

$$= \frac{-x - 1 - 3 + 3x}{(x + 1)^4} = \frac{2x - 4}{(x + 1)^4} = \frac{2(x - 2)}{(x + 1)^4}$$

which is zero when $x = 2$. The corresponding second-order critical point is $(2, f(2)) = \left(2, \dfrac{2}{9}\right)$.

Plot these critical points and check the signs of $f'(x)$ and $f''(x)$ on each of the intervals defined by the x coordinates of the critical points and the discontinuity.

Interval	f'(x)	f"(x)	Increase or decrease of f	Concavity of f	Shape of f
$x < -1$	$-$	$-$	Decreasing	C-down	⌒
$-1 < x < 1$	$+$	$-$	Increasing	C-down	⌒
$1 < x < 2$	$-$	$-$	Decreasing	C-down	⌒
$x > 2$	$-$	$+$	Decreasing	C-up	⌣

Draw the graph as shown in Figure 2.5 using a curve of appropriate shape on each interval. Don't forget that the graph has a discontinuity at $x = -1$.

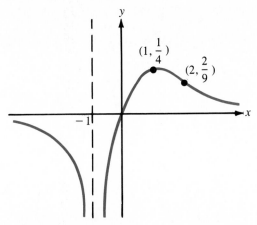

Figure 2.5 The graph of $y = \dfrac{x}{(x + 1)^2}$.

Notice that the first-order critical point $\left(1, \dfrac{1}{4}\right)$ is a relative maximum and that the second-order critical point $\left(2, \dfrac{2}{9}\right)$ is an inflection point.

Notice also that the y intercept is $(0, f(0)) = (0, 0)$. The x intercepts of a rational function occur when the numerator is zero, which happens in this case only when $x = 0$. Hence $(0, 0)$ is the only x intercept and so the graph crosses the x axis only at this point.

The Second-Derivative Test

The second derivative can be used to classify the first-order critical points of a function as relative maxima or relative minima. Here is a statement of the procedure, which is known as the **second-derivative test.**

The Second-Derivative Test

Suppose $f'(a) = 0$.

If $f''(a) > 0$, then f has a relative minimum at $x = a$.
If $f''(a) < 0$, then f has a relative maximum at $x = a$.

However, if $f''(a) = 0$, the test is inconclusive and f may have a relative maximum, a relative minimum, or no relative extremum at all at $x = a$.

To see why the second-derivative test works, look at Figure 2.6, which shows the four possibilities that can occur when $f'(a) = 0$.

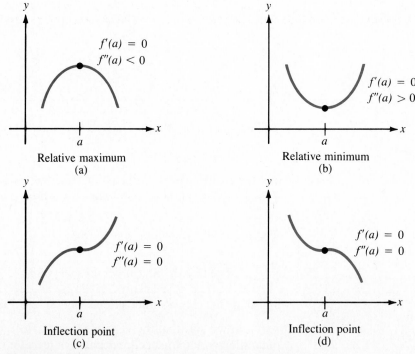

Figure 2.6 Behavior of a graph when the first derivative is zero.

Figure 2.6a suggests that at a relative maximum, f must be concave downward and so $f''(a) \leq 0$. Figure 2.6b suggests that at a relative minimum, f must be concave upward and so $f''(a) \geq 0$. On the other hand, Figures 2.6c and 2.6d suggest that if a point at which $f'(a) = 0$ is not a relative extremum, it must be an inflection point and so $f''(a)$, if it is defined, must be zero. It follows that if $f'(a) = 0$ and $f''(a) < 0$, the corresponding critical point must be a relative maximum, while if

$f'(a) = 0$ and $f''(a) > 0$, the corresponding critical point must be a relative minimum.

The use of the second-derivative test is illustrated in the following example.

EXAMPLE 2.3   Use the second-derivative test to find the relative maxima and minima of the function $f(x) = 2x^3 + 3x^2 - 12x - 7$.

SOLUTION
Since the first derivative

$$f'(x) = 6x^2 + 6x - 12 = 6(x + 2)(x - 1)$$

is zero when $x = -2$ and $x = 1$, the corresponding points $(-2, 13)$ and $(1, -14)$ are the first-order critical points of f. To test these points, compute the second derivative

$$f''(x) = 12x + 6$$

and evaluate it at $x = -2$ and $x = 1$. Since

$$f''(-2) = -18 < 0$$

it follows that the critical point $(-2, 13)$ is a relative maximum, and since

$$f''(1) = 18 > 0$$

it follows that the critical point $(1, -14)$ is a relative minimum. For reference, the graph of f is sketched in Figure 2.7.

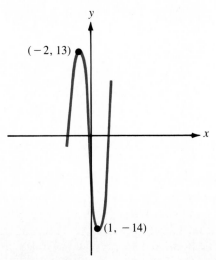

Figure 2.7 The graph of $y = 2x^3 + 3x^2 - 12x - 7$.

The function in Example 2.3 is the same one you analyzed in Example 1.1 of the preceding section using only the first derivative. Notice the relative ease with which you can now identify the extrema. Using the second-derivative test, you check the sign of $f''(x)$ only at the critical point itself. Using the first derivative, you had to check the sign of $f'(x)$ over entire intervals on either side of the critical point.

There are, however, some limitations of the second-derivative test. For some functions, the work involved in computing the second derivative is time-consuming and may diminish the appeal of the test. Moreover, the test applies only to critical points at which the derivative is zero and not to those at which the derivative is undefined. Finally, if both $f'(a)$ and $f''(a)$ are zero, the second derivative test tells you nothing whatsoever about the nature of the critical point. This is illustrated in Figure 2.8, which shows the graphs of three functions whose first and second derivatives are both zero when $x = 0$. When it is impractical or impossible to apply the second-derivative test, you can always return to the methods of the preceding section and use the first derivative to check the increase or decrease of the function on either side of the critical point.

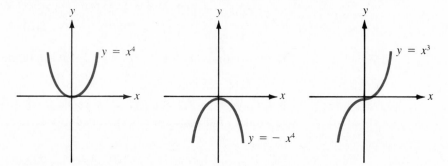

Figure 2.8 Three functions whose first and second derivatives are zero at $x = 0$.

Problems

In Problems 1 and 2, determine where the second derivative of the function is positive and where it is negative.

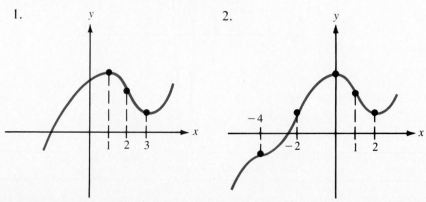

In Problems 3 through 25, determine where the given function is increasing, decreasing, concave upward, and concave downward. Find the relative extrema and inflection points and draw the graph.

3. $f(x) = \dfrac{1}{3}x^3 - 9x + 2$

4. $f(x) = x^3 + 3x^2 + 1$

5. $f(x) = x^4 - 4x^3 + 10$

6. $f(x) = x^3 - 3x^2 + 3x + 1$

7. $f(x) = (x - 2)^3$

8. $f(x) = x^5 - 5x$

9. $f(x) = (x^2 - 5)^3$

10. $f(x) = (x - 2)^4$

11. $f(x) = x + \dfrac{1}{x}$

12. $f(x) = (x^2 - 3)^2$

13. $f(x) = \dfrac{x^2}{x - 3}$

14. $f(x) = 1 + 2x + \dfrac{18}{x}$

15. $f(x) = (x + 1)^{1/3}$

16. $f(x) = \dfrac{x^2 - 3x}{x + 1}$

17. $f(x) = (x + 1)^{4/3}$

18. $f(x) = (x + 1)^{2/3}$

19. $f(x) = \sqrt{x^2 + 1}$

20. $f(x) = (x + 1)^{5/3}$

21. $f(x) = 2x(x + 4)^3$

22. $f(x) = \left(\dfrac{x}{x + 1}\right)^2$

23. $f(x) = \dfrac{2}{1 + x^2}$

24. $f(x) = \dfrac{x + 1}{(x - 1)^2}$

25. $f(x) = \dfrac{(x - 2)^3}{x^2}$

26. Sketch the graph of a function that has all of the following properties:
    (a) $f'(x) > 0$ when $x < -1$ and when $x > 3$
    (b) $f'(x) < 0$ when $-1 < x < 3$
    (c) $f''(x) < 0$ when $x < 2$
    (d) $f''(x) > 0$ when $x > 2$

27. Sketch the graph of a function that has all of the following properties:
    (a) $f'(x) > 0$ when $x < 2$ and when $2 < x < 5$
    (b) $f'(x) < 0$ when $x > 5$
    (c) $f'(2) = 0$
    (d) $f''(x) < 0$ when $x < 2$ and when $4 < x < 7$
    (e) $f''(x) > 0$ when $2 < x < 4$ and when $x > 7$

28. Sketch the graph of a function that has all of the following properties:
    (a) $f'(x) > 0$ when $x < 1$
    (b) $f'(x) < 0$ when $x > 1$
    (c) $f''(x) > 0$ when $x < 1$ and when $x > 1$
    What can you say about the first derivative of f when $x = 1$?

29. The first derivative of a certain function is $f'(x) = x^2 - 4x$.

    (a) On what intervals is f increasing? Decreasing?

    (b) On what intervals is f concave upward? Concave downward?

    (c) Find the x coordinates of the relative extrema and inflection points of f.

30. The first derivative of a certain function is $f'(x) = x^2 - 2x - 8$.

    (a) On what intervals is f increasing? Decreasing?

    (b) On what intervals is f concave upward? Concave downward?

    (c) Find the x coordinates of the relative extrema and inflection points of f.

In Problems 31 through 33, the *first derivative* of a certain function is sketched. In each case, determine where the *function* itself is increasing, decreasing, concave upward, and concave downward, and find the x coordinates of its relative extrema and inflection points.

31.

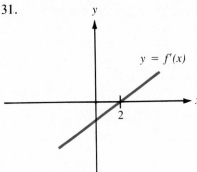

32.

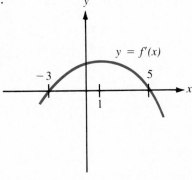

33.

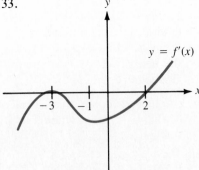

34. Use calculus to show that the graph of the quadratic function $y = ax^2 + bx + c$ is concave upward if a is positive and concave downward if a is negative.

In Problems 35 through 40, use the second-derivative test to find the relative maxima and minima of the given function.

35. $f(x) = x^3 + 3x^2 + 1$ 36. $f(x) = x^4 - 2x^2 + 3$

37. $f(x) = (x^2 - 9)^2$ 38. $f(x) = x + \dfrac{1}{x}$

39. $f(x) = 2x + 1 + \dfrac{18}{x}$ 40. $f(x) = \dfrac{x^2}{x - 2}$

3   Absolute Maxima and Minima

In most practical optimization problems, the goal is to find the **absolute maximum** or **absolute minimum** of a particular function on some relevant interval. The absolute maximum of a function on an interval is the largest value of the function on the interval. The absolute minimum is the smallest value. Absolute extrema often coincide with relative extrema, but not always. For example, in Figure 3.1, the absolute maximum and relative maximum on the interval $a \leq x \leq b$ are the same. The absolute minimum, on the other hand, occurs at the endpoint $x = a$, which is not a relative minimum.

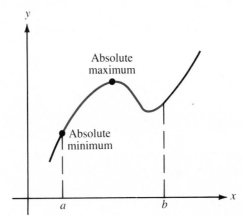

Figure 3.1 Absolute extrema.

In this section, you will learn how to find absolute extrema of functions on intervals. Particular attention will be given to two special cases, which include virtually all the practical optimization problems you are likely to encounter. In Sections 4 and 5, these optimization techniques will be applied to a variety of practical situations.

Absolute Extrema on Closed Intervals

A **closed interval** is an interval of the form $a \leq x \leq b$, that is, an interval that contains both of its endpoints. It can be shown that a function that is

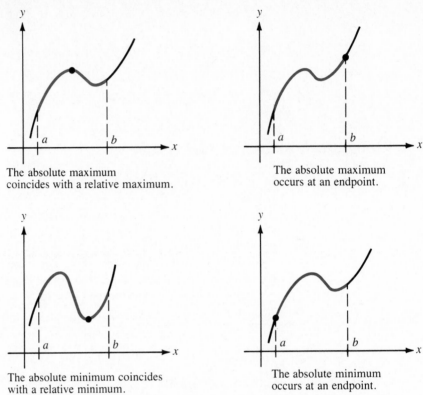

The absolute maximum
coincides with a relative maximum.

The absolute maximum
occurs at an endpoint.

The absolute minimum coincides
with a relative minimum.

The absolute minimum
occurs at an endpoint.

Figure 3.2 Absolute extrema of a continuous function on a closed interval.

continuous on a closed interval attains an absolute maximum and an absolute minimum on that interval. An absolute extremum can occur either at a relative extremum in the interval or at an endpoint $x = a$ or $x = b$. The possibilities are illustrated in Figure 3.2. These observations suggest the following simple procedure for locating and identifying absolute extrema of continuous functions on closed intervals.

How to Find the Absolute Extrema of a Continuous Function f on a Closed Interval $a \le x \le b$

Step 1. Find the x coordinates of all the first-order critical points of f in the interval $a \le x \le b$.

Step 2. Compute $f(x)$ at these critical points and at the endpoints $x = a$ and $x = b$.

Step 3. Select the largest and smallest values of $f(x)$ obtained in Step 2. These are the absolute maximum and absolute minimum, respectively.

The procedure is illustrated in the following example.

EXAMPLE 3.1   Find the absolute maximum and absolute minimum of the function $f(x) = 2x^3 + 3x^2 - 12x - 7$ on the interval $-3 \leq x \leq 0$.

SOLUTION
From the derivative

$$f'(x) = 6x^2 + 6x - 12 = 6(x + 2)(x - 1)$$

you see that the first-order critical points occur when $x = -2$ and $x = 1$. Of these, only $x = -2$ lies in the interval $-3 \leq x \leq 0$. Compute $f(x)$ at $x = -2$ and at the endpoints $x = -3$ and $x = 0$.

$$f(-2) = 13 \qquad f(-3) = 2 \qquad f(0) = -7$$

Compare these values to conclude that the absolute maximum of f on the interval $-3 \leq x \leq 0$ is $f(-2) = 13$ and the absolute minimum is $f(0) = -7$.

Notice that you did not have to classify the critical points or draw the graph to locate the absolute extrema. The sketch in Figure 3.3 is presented only for the sake of illustration.

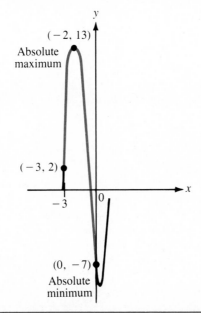

Figure 3.3 The absolute extrema of $y = 2x^3 + 3x^2 - 12x - 7$ on $-3 \leq x \leq 0$.

Applications

Here are two applications of this technique.

EXAMPLE 3.2   For several weeks, the highway department has been recording the speed of freeway traffic flowing past a certain downtown exit. The data suggest that between the hours of 1:00 and 6:00 P.M. on a normal weekday, the speed of the traffic at the exit is approximately $S(t) = t^3 - 10.5t^2 + 30t + 20$ miles per hour, where t is the number of hours past noon. At what time between 1:00 and 6:00 P.M. is the traffic moving the fastest, and at what time is it moving the slowest?

SOLUTION

The goal is to find the absolute maximum and absolute minimum of the function $S(t)$ on the interval $1 \leq t \leq 6$. From the derivative

$$S'(t) = 3t^2 - 21t + 30 = 3(t^2 - 7t + 10) = 3(t - 2)(t - 5)$$

you get the t coordinates $t = 2$ and $t = 5$ of the first-order critical points, both of which lie in the interval $1 \leq t \leq 6$.

Now compute $S(t)$ for these values of t and at the endpoints $t = 1$ and $t = 6$ to get

$$S(1) = 40.5 \qquad S(2) = 46 \qquad S(5) = 32.5 \qquad S(6) = 38$$

Since the largest of these values is $S(2) = 46$ and the smallest is $S(5) = 32.5$, you can conclude that the traffic is moving fastest at 2:00 P.M. when its speed is 46 miles per hour and slowest at 5:00 P.M. when its speed is 32.5 miles per hour.

For reference, the graph of S is sketched in Figure 3.4.

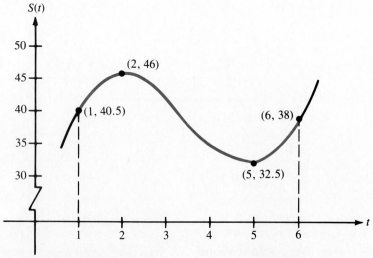

Figure 3.4 Traffic speed: $S(t) = t^3 - 10.5t^2 + 30t + 20$.

The next example comes from biology.

EXAMPLE 3.3   When you cough, the radius of your trachea (windpipe) decreases, affecting the speed of the air in the trachea. If r_0 is the normal radius of the trachea, the relationship between the speed S of the air and the radius r of the trachea during a cough is given by a function of the form $S(r) = ar^2(r_0 - r)$, where a is a positive constant. Find the radius r for which the speed of the air is greatest.

SOLUTION

The radius r of the contracted trachea cannot be greater than the normal radius r_0 nor less than zero. Hence, the goal is to find the absolute maximum of $S(r)$ on the interval $0 \leq r \leq r_0$.

First differentiate $S(r)$ with respect to r using the product rule and factor the derivative as follows. (Note that a and r_0 are constants.)

$$S'(r) = -ar^2 + (r_0 - r)(2ar) = ar[-r + 2(r_0 - r)] = ar(2r_0 - 3r)$$

Then set the factored derivative equal to zero and solve to get the r coordinates of the first-order critical points.

$$ar(2r_0 - 3r) = 0$$

$$r = 0 \quad \text{or} \quad r = \frac{2}{3}r_0$$

Both of these values of r lie in the interval $0 \leq r \leq r_0$, and one is actually an endpoint of the interval. Compute $S(r)$ for these two values of r and at the other endpoint $r = r_0$ to get

$$S(0) = 0 \qquad S\left(\frac{2}{3}r_0\right) = \frac{4a}{27}r_0^3 \qquad S(r_0) = 0$$

Compare these values and conclude that the speed of the air is greatest when the radius of the contracted trachea is $\frac{2}{3}r_0$, that is, when it is two-thirds the radius of the uncontracted trachea.

A graph of the function $S(r)$ is sketched in Figure 3.5. Notice that the r intercepts of the graph are obvious from the factored function $S(r) = ar^2(r_0 - r)$. Notice also that the graph was drawn so that it has a horizontal tangent when $r = 0$, reflecting the fact that $S'(0) = 0$.

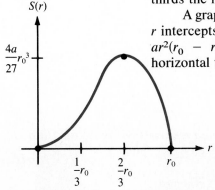

Figure 3.5 The speed of air during a cough: $S(r) = ar^2(r_0 - r)$.

Absolute Extrema on Intervals That Are Not Closed

When the interval on which you wish to maximize or minimize a continuous function is not of the form $a \leq x \leq b$, the procedure illustrated in the preceding examples no longer applies. This is because there is no longer any guarantee that the function actually has an absolute maximum or minimum on the interval in question. On the other hand, if an absolute extremum does exist and the function is continuous on the interval, the absolute extremum will still occur at a relative extremum or endpoint contained in the interval. Two of the possibilities for functions on unbounded intervals are illustrated in Figure 3.6.

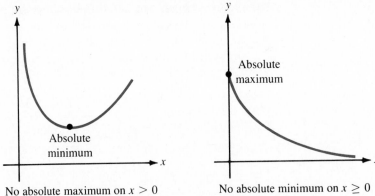

No absolute maximum on $x > 0$ No absolute minimum on $x \geq 0$

Figure 3.6 Extrema of functions on unbounded intervals.

To find the absolute extrema of a continuous function on an interval that is not of the form $a \leq x \leq b$, you still evaluate the function at all the critical points and endpoints that are contained in the interval. However, before you can draw any final conclusions, you must find out if the function actually has relative extrema on the interval. One way to do this is to use the first derivative to determine where the function is increasing and where it is decreasing and sketch the graph. The technique is illustrated in the next example.

EXAMPLE 3.4   Find the absolute maximum and absolute minimum (if any) of the function $f(x) = x^2 + \dfrac{16}{x}$ on the interval $x > 0$.

SOLUTION
The function is continuous on the interval $x > 0$ since its only discontinuity occurs at $x = 0$. The derivative is

$$f'(x) = 2x - \frac{16}{x^2} = \frac{2x^3 - 16}{x^2} = \frac{2(x^3 - 8)}{x^2}$$

which is zero when

$$x^3 - 8 = 0 \qquad x^3 = 8 \qquad \text{or} \qquad x = 2$$

Since the interval has no right-hand endpoint and does not contain its left-hand endpoint at $x = 0$, the only possible absolute extremum is

$$f(2) = (2)^2 + \frac{16}{2} = 12$$

at the first-order critical point.

Since $f'(x) < 0$ for $0 < x < 2$ and $f'(x) > 0$ for $x > 2$, the graph of f is decreasing for $0 < x < 2$ and increasing for $x > 2$, as shown in Figure 3.7. It follows that $f(2) = 12$ is the absolute minimum of f on the interval $x > 0$ and that there is no absolute maximum.

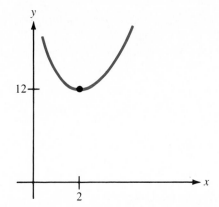

Figure 3.7 The function $y = x^2 + \dfrac{16}{x}$ on the interval $x > 0$.

Absolute Extrema on Intervals Containing Only One Critical Point

In many practical optimization problems, the interval with which you are working will contain only one first-order critical point of the function. When this happens, you can use the second-derivative test from Section 2 to identify its absolute extremum, even though this test is really a test for only relative extrema. The reason is that, in this special case, every *relative* maximum or minimum is necessarily also an *absolute* maximum or minimum.

To see why, look at Figure 3.8, in which the curve is eventually lower than its relative minimum at $x = c_1$. What allows this to happen is the existence of the relative maximum (i.e., another first-order critical point) at $x = c_2$, where the curve "turns around" and starts down again. On an interval that does not contain this additional first-order critical point, the relative minimum at $x = c_1$ would be the absolute minimum. Hence, on such an interval, the second-derivative test, which identifies relative maxima and minima, also identifies absolute maxima and minima. Here is a more precise statement of the situation.

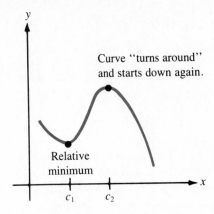

Curve "turns around" and starts down again.

Relative minimum

c_1 c_2

Figure 3.8 The relative minimum is not the absolute minimum because of the effect of another critical point.

The Second-Derivative Test for Absolute Extrema

Suppose that f is continuous on an interval on which $x = a$ is the only first-order critical point and that $f'(a) = 0$.

If $f''(a) > 0$, then $f(a)$ is the absolute minimum of f on the interval.
If $f''(a) < 0$, then $f(a)$ is the absolute maximum of f on the interval.

The second-derivative test for absolute extrema can be applied to any interval, whether closed or not. The only requirement is that the function be continuous and have only one critical point on the interval.

Here are two examples.

EXAMPLE 3.5 Suppose that when q units of a certain commodity are produced, the total manufacturing cost is $C(q) = 3q^2 + 5q + 75$ dollars. At what level of production will the average cost per unit be smallest?

SOLUTION
The average cost per unit is the total cost divided by the number of units produced. Thus, if q units are produced, the average cost is

$$A(q) = \frac{C(q)}{q} = \frac{3q^2 + 5q + 75}{q} = 3q + 5 + \frac{75}{q}$$

dollars per unit. Since only positive values of q are meaningful in this context, the goal is to find the absolute minimum of the function $A(q)$ on the unbounded interval $q > 0$.

The derivative of A is

$$A'(q) = 3 - \frac{75}{q^2}$$

which is zero when

$$3 - \frac{75}{q^2} = 0 \qquad q^2 = 25 \qquad \text{or} \qquad q = \pm 5$$

Of these two values of q, only $q = 5$ is in the relevant interval $q > 0$. Since $q = 5$ is the only first-order critical point in this interval, and since $A(q) = 3q + 5 + \frac{75}{q}$ is continuous on the interval (its only discontinuity occurring at $q = 0$, which is not in the interval), you can apply the second-derivative test for absolute extrema.

The second derivative is $A''(q) = \frac{150}{q^3}$. Since $A''(5) = \frac{150}{(5)^3} > 0$, it follows that $A(q)$ has an absolute minimum at $q = 5$. That is, the average cost per unit is minimal when 5 units are produced and is equal to $A(5) = 35$ dollars per unit at this level of production.

For reference, the relevant portion of the graph of the average cost function is sketched in Figure 3.9. Notice that it is decreasing for $0 < q < 5$ and increasing for $q > 5$ and has a relative and absolute minimum when $q = 5$.

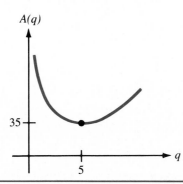

Figure 3.9 The average cost: $A(q) = 3q + 5 + \frac{75}{q}$.

Applied Optimality Conditions

In the next example, the goal is not to find the absolute maximum or minimum of a particular function, but rather to prove that when the absolute extremum occurs, a certain condition must be satisfied.

EXAMPLE 3.6 Suppose that at a certain factory, setup cost is directly proportional to the number of machines used, and operating cost is inversely proportional to the number of machines used. Show that when the total cost is minimal, the setup cost is equal to the operating cost.

SOLUTION

Let x be the number of machines used and $C(x)$ the corresponding total cost. Since

Total cost = setup cost + operating cost

and

Setup cost = $k_1 x$ and   Operating cost = $\dfrac{k_2}{x}$

where k_1 and k_2 are positive constants of proportionality, it follows that

$$C(x) = k_1 x + \frac{k_2}{x}$$

Since only positive values of x are meaningful in this context, the goal is to minimize $C(x)$ on the interval $x > 0$.

The derivative of C is

$$C'(x) = k_1 - \frac{k_2}{x^2}$$

which is zero when

$$k_1 - \frac{k_2}{x^2} = 0 \quad \text{or} \quad k_1 = \frac{k_2}{x^2}$$

At this point in the calculation, notice that if you multiply both sides of the equation by x, you will get

$$k_1 x = \frac{k_2}{x}$$

But,

$$k_1 x = \text{setup cost} \quad \text{and} \quad \frac{k_2}{x} = \text{operating cost}$$

and so this equation says

Setup cost = operating cost

which is exactly what you were trying to show!

All that remains is to verify that the function $C(x)$ actually has an absolute minimum when its derivative is zero. To do this, solve the equation $C'(x) = 0$ for x to get

$$k_1 - \frac{k_2}{x^2} = 0 \quad x^2 = \frac{k_2}{k_1} \quad \text{or} \quad x = \pm \sqrt{\frac{k_2}{k_1}}$$

Only the positive square root is in the relevant interval $x > 0$. Since this is the only first-order critical point in the interval and since the function

$C(x) = k_1x + \dfrac{k_2}{x}$ is continuous on this interval (its only discontinuity occurring at $x = 0$), you can apply the second-derivative test for absolute extrema. In particular, the second derivative is $C''(x) = \dfrac{2k_2}{x^3}$, which is clearly positive when $x > 0$ (since k_2 is a positive constant). It follows that the point at which the derivative is zero is indeed the absolute minimum you were seeking.

For reference, the graph of the cost function is sketched in Figure 3.10.

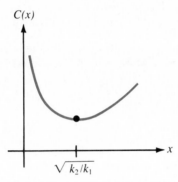

Figure 3.10 Total cost: $C(x) = k_1x + \dfrac{k_2}{x}$.

Problems

In Problems 1 through 16, find the absolute maximum and absolute minimum (if any) of the given function on the specified interval.

1. $f(x) = x^2 + 4x + 5$; $-3 \le x \le 1$

2. $f(x) = x^3 + 3x^2 + 1$; $-3 \le x \le 2$

3. $f(x) = \dfrac{1}{3}x^3 - 9x + 2$; $0 \le x \le 2$

4. $f(x) = x^5 - 5x^4 + 1$; $0 \le x \le 5$

5. $f(x) = 3x^5 - 5x^3$; $-2 \le x \le 0$

6. $f(x) = 10x^6 + 24x^5 + 15x^4 + 3$; $-1 \le x \le 1$

7. $f(x) = (x^2 - 4)^5$; $-3 \le x \le 2$

8. $f(x) = \dfrac{x^2}{x - 1}$; $-2 \le x \le -\dfrac{1}{2}$

9. $f(x) = x + \dfrac{1}{x}$; $\dfrac{1}{2} \le x \le 3$

10. $f(x) = \dfrac{1}{x^2 - 9}; \; 0 \le x \le 2$ 11. $f(x) = x + \dfrac{1}{x}; \; x > 0$

12. $f(x) = 2x + \dfrac{32}{x}; \; x > 0$ 13. $f(x) = \dfrac{1}{x}; \; x > 0$

14. $f(x) = \dfrac{1}{x^2}; \; x > 0$ 15. $f(x) = \dfrac{1}{x + 1}; \; x \ge 0$

16. $f(x) = \dfrac{1}{(x + 1)^2}; \; x \ge 0$

In Problems 17 through 25, solve the practical optimization problem and use one of the techniques from this section to verify that you have actually found the desired absolute extremum.

Group membership   17. Suppose that x years after its founding in 1970, a certain national consumers' association had a membership of $f(x) = 100(2x^3 - 45x^2 + 264x)$.
   (a) At what time between 1970 and 1984 was the membership of the association largest? What was the membership at that time?
   (b) At what time between 1971 and 1984 was the membership of the association smallest? What was the membership at that time?

Broadcasting   18. An all-news radio station has made a survey of the listening habits of local residents between the hours of 5:00 P.M. and midnight. The survey indicates that the percentage of the local adult population that is tuned in to the station x hours after 5:00 P.M. is $f(x) = \dfrac{1}{8}(-2x^3 + 27x^2 - 108x + 240)$.
   (a) At what time between 5:00 P.M. and midnight are the most people listening to the station? What percentage of the population is listening at this time?
   (b) At what time between 5:00 P.M. and midnight are the fewest people listening? What percentage of the population is listening at this time?

Profit   19. A manufacturer can produce radios at a cost of $5 apiece and estimates that if they are sold for x dollars apiece, consumers will buy $20 - x$ radios a day. At what price should the manufacturer sell the radios to maximize profit?

Consumer expenditure   20. The demand function for a certain commodity is $D(p) = 160 - 2p$, where p is the price at which the commodity is sold. At what price is the total consumer expenditure for the commodity greatest?

Circulation   21. Poiseuille's law asserts that the speed of blood that is r centimeters from the central axis of an artery of radius R is $S(r) = c(R^2 - r^2)$, where c is a positive constant. Where is the speed of the blood greatest?

Respiration

22. Biologists define the flow F of air in the trachea by the formula $F = SA$, where S is the speed of the air and A the area of a cross section of the trachea.

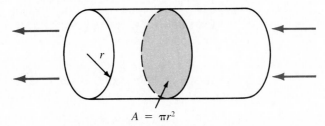

$$A = \pi r^2$$

(a) Assume that the cross section of the trachea is circular and use the formula from Example 3.3 for the speed of air in the trachea during a cough to express the flow F as a function of the radius r.

(b) Find the radius r for which the flow is greatest.

Population growth

23. When environmental factors impose an upper bound on its possible size, population grows at a rate that is jointly proportional to its current size and the difference between its current size and the upper bound. (Recall from Chapter 1, Section 5, that joint proportionality means direct proportionality to the product of the two quantities.) Show that the rate of population growth will be greatest when the population has climbed to 50 percent of its upper bound.

The spread of an epidemic

24. The rate at which an epidemic spreads through a community is jointly proportional to the number of people who have caught the disease and the number who have not. Show that the epidemic is spreading most rapidly when half the people have caught the disease.

Transportation cost

25. A truck is hired to transport goods from a factory to a warehouse. The driver's wages are figured by the hour and so are inversely proportional to the speed at which the truck is driven. The amount of gasoline used is directly proportional to the speed at which the truck is driven, and the price of gasoline remains constant during the trip. Show that the total cost is smallest at the speed for which the driver's wages are equal to the cost of the gasoline used.

Microeconomics

26. An economic law states that under normal circumstances, profit is maximized when marginal revenue equals marginal cost. (Recall that marginal revenue and marginal cost are the derivatives of total revenue and total cost, respectively.) Use the theory of extrema to show why this law is true. (You may assume without proof that the critical point is actually the desired absolute maximum.)

Average cost

27. Suppose the total cost in dollars of manufacturing q units is given by the function $C(q) = 3q^2 + q + 48$.

(a) Express the average manufacturing cost per unit as a function of q.

(b) For what value of q is the average cost the smallest?

(c) For what value of q is the average cost equal to the marginal cost? Compare this value with your answer in part (b) and look at the general economic law stated in Problem 28.

(d) On the same set of axes, graph the total cost, marginal cost, and average cost functions.

Microeconomics 28. An economic law states that under normal circumstances, average cost is smallest when it equals marginal cost. Derive this law using calculus. [*Hint:* If $C(q)$ represents the total cost of manufacturing q units, then average cost is $A(q) = \dfrac{C(q)}{q}$. Assume that average cost will be smallest when $A'(q) = 0$. Use the quotient rule to compute $A'(q)$; then set $A'(q)$ equal to zero, and the economic law will emerge.]

4   Practical Optimization Problems

In this section, you will learn how to combine the techniques of model building from Chapter 1, Section 5, with the optimization techniques of Chapter 3, Section 3, to solve practical optimization problems.

The first step in solving such a problem is to decide precisely what you are to optimize. Once you have identified this quantity, choose a letter to represent it. You may find it helpful to choose a letter that is closely related to the quantity, such as R for revenue or A for area.

Your goal is to represent the quantity to be optimized as a function of some other variable so that you can apply calculus. It is usually a good idea to express the desired function in words before trying to represent it mathematically.

Once the function has been expressed in words, the next step is to choose an appropriate variable. Often, the choice is obvious. Sometimes you will be faced with a choice among several natural variables. When this happens, think ahead and try to choose the variable that leads to the simplest functional representation. In some problems, the quantity to be optimized is expressed most naturally in terms of two variables. If so, you will have to find a way to write one of these variables in terms of the other.

The next step is to express the quantity to be optimized as a function of the variable you have chosen. In most problems, the function has a practical interpretation only when the variable lies in a certain interval. Once you have written the function and identified the appropriate interval, the difficult part is done and the rest is routine. To complete the solution, simply apply the techniques from the preceding section to optimize your function on the specified interval.

Here are several examples to illustrate the procedure.

EXAMPLE 4.1 A manufacturer can produce radios at a cost of $2 apiece. The radios have been selling for $5 apiece, and, at this price, consumers have been buying 4,000 radios a month. The manufacturer is planning to raise the price of the radios and estimates that for each $1 increase in the price, 400 fewer radios will be sold each month. At what price should the manufacturer sell the radios to maximize profit?

SOLUTION

Let x denote the new price at which the radios will be sold and $P(x)$ the corresponding profit. The goal is to maximize the profit. As in Example 5.1 of Chapter 1, Section 5, begin by stating the formula for profit in words.

Profit = (number of radios sold)(profit per radio)

Since 4,000 radios are sold each month when the price is $5 and 400 fewer will be sold each month for each $1 increase in the price, it follows that

Number of radios sold = 4,000 − 400(number of $1 increases)

The number of $1 increases in the price is the difference $x - 5$ between the new and old selling prices. Hence,

$$\text{Number of radios sold} = 4,000 - 400(x - 5)$$
$$= 400[10 - (x - 5)]$$
$$= 400(15 - x)$$

The profit per radio is simply the difference between the selling price x and the cost $2. That is,

Profit per radio = $x - 2$

Putting it all together,

$$P(x) = 400(15 - x)(x - 2)$$

The goal is to find the absolute maximum of the profit function $P(x)$. You have some leeway in the selection of the relevant interval for this problem. Since the new price x is to be at least as high as the old price of $5, a reasonable choice is the interval $x \geq 5$. On the other hand, the number of radios sold is $400(15 - x)$, which will drop to zero when $x = 15$. If you assume that the manufacturer will not price the radios so high that no one will buy them, you can restrict the optimization problem to the closed interval $5 \leq x \leq 15$. Of the two choices, this closed interval is probably preferable since you have a special technique for maximizing functions on closed intervals.

To find the critical points, compute the derivative (using the product and constant multiple rules) to get

$$P'(x) = 400[(15 - x)(1) + (x - 2)(-1)]$$
$$= 400(15 - x - x + 2) = 400(17 - 2x)$$

which is zero when

$$17 - 2x = 0 \quad \text{or} \quad x = 8.5$$

Comparing the values of the profit function

$$P(5) = 12{,}000 \quad P(8.5) = 16{,}900 \quad \text{and} \quad P(15) = 0$$

at this critical point and at the endpoints of the interval, you can conclude that the maximum possible profit is $16,900, which will be generated if the radios are sold for $8.50 apiece.

For reference, the graph of the profit function is sketched in Figure 4.1.

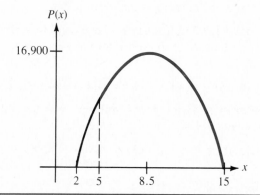

Figure 4.1 The profit function $P(x) = 400(15 - x)(x - 2)$.

An Alternative Solution to the Profit Maximization Problem

Since the number of radios sold in Example 4.1 is described in terms of the number of $1 increases in price, you may prefer to let the variable x denote the number of $1 increases in price rather than the new price itself. If you make this choice, you should find that the corresponding profit function is

$$P(x) = 400(10 - x)(3 + x)$$

and that the relevant interval is $0 \le x \le 10$. The absolute maximum in this case will occur when $x = 3.5$, that is, when the old price is increased from $5 by $3.50 to a new price of $8.50. For practice, work out the details of the solution in this case.

Elimination of Variables

In the next example, the quantity to be optimized is expressed most naturally in terms of two variables. Before you can apply calculus, you will have to write one of these variables in terms of the other to get a function of just one variable.

EXAMPLE 4.2   The highway department is planning to build a picnic area for motorists along a major highway. It is to be rectangular with an area of 5,000 square yards and is to be fenced off on the three sides not adjacent to the highway. What is the least amount of fencing that will be needed to complete the job?

SOLUTION

As in Example 5.2 of Chapter 1, Section 5, label the sides of the picnic area as indicated in Figure 4.2 and let F denote the amount of fencing required. Then,

$$F = x + 2y$$

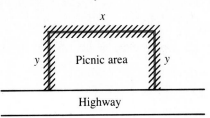

Figure 4.2 Rectangular picnic area.

The fact that the area is to be 5,000 tells you that

$$xy = 5,000 \quad \text{or} \quad y = \frac{5,000}{x}$$

To rewrite F in terms of the single variable x, substitute this expression for y into the formula for F to get

$$F(x) = x + 2\left(\frac{5,000}{x}\right) = x + \frac{10,000}{x}$$

Since $F(x)$ has a practical interpretation for any positive value of x, your goal is to find the absolute minimum of $F(x)$ on the interval $x > 0$. To find the critical points, set the derivative

$$F'(x) = 1 - \frac{10,000}{x^2}$$

equal to zero and solve for x, getting

$$1 - \frac{10,000}{x^2} = 0 \quad x^2 = 10,000 \quad \text{or} \quad x = \pm 100$$

Only the positive value $x = 100$ lies in the interval $x > 0$. Since this is the only critical point in the interval, you can apply the second-derivative test for absolute extrema. In particular, the second derivative is $F''(x) = \frac{20,000}{x^3}$, which is positive when $x > 0$. Hence the critical point at $x = 100$ is the absolute minimum of F on the interval. That is, the least amount of fencing needed to complete the job is $F(100) = 200$ yards. For reference, the graph of $F(x)$ is sketched in Figure 4.3.

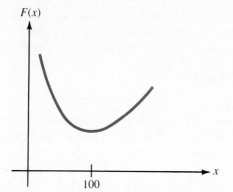

Figure 4.3 The graph of $F(x) = x + \dfrac{10,000}{x}$ for $x > 0$.

Optimization Problems Involving the Pythagorean Theorem

In the next example, the function to be optimized is obtained geometrically with the aid of the pythagorean theorem.

EXAMPLE 4.3   A cable is to be run from a power plant on one side of a river 900 meters wide to a factory on the other side, 3,000 meters downstream. The cost of running the cable under the water is $5 per meter while the cost over land is $4 per meter. What is the most economical route over which to run the cable?

SOLUTION
To help you visualize the situation, begin by drawing a diagram as shown in Figure 4.4. (Notice that in drawing the diagram in Figure 4.4, we have already assumed that the cable should be run in a *straight line* from the power plant to some point P on the opposite bank. Do you see why this assumption is justified?)

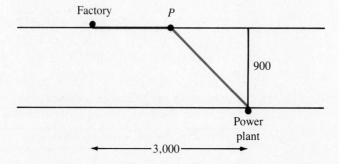

Figure 4.4 Relative positions of factory, river, and power plant.

The goal is to minimize the cost of installing the cable. Let C denote this cost and represent C as follows:

$$C = 5(\text{number of meters of cable under water})$$
$$+ 4(\text{number of meters of cable over land})$$

Since you wish to describe the optimal route over which to run the cable, it will be convenient to choose a variable in terms of which you can easily locate the point P. Two reasonable choices for the variable x are illustrated in Figure 4.5.

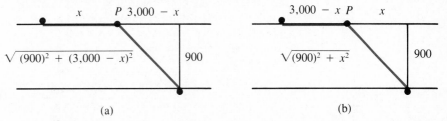

Figure 4.5 Two choices for the variable x.

(a) (b)

Before plunging into the calculations, take a minute to decide which choice of variables is more advantageous. In Figure 4.5a, the distance across the water from the power plant to the point P is (by the pythagorean theorem) $\sqrt{(900)^2 + (3{,}000 - x)^2}$, and the corresponding total cost function is

$$C(x) = 5\sqrt{(900)^2 + (3{,}000 - x)^2} + 4x$$

In Figure 4.5b, the distance across the water is $\sqrt{(900)^2 + x^2}$, and the total cost function is

$$C(x) = 5\sqrt{(900)^2 + x^2} + 4(3{,}000 - x)$$

The second of these functions is the more attractive since the term $3{,}000 - x$ is merely multiplied by 4, while in the first function it is squared and appears under the radical. Hence, you should choose x as in Figure 4.5b and work with the total cost function

$$C(x) = 5\sqrt{(900)^2 + x^2} + 4(3{,}000 - x)$$

Since the distances x and $3{,}000 - x$ cannot be negative, the relevant interval is $0 \leq x \leq 3{,}000$, and your goal is to find the absolute minimum of the function $C(x)$ on this closed interval.

To find the critical points, compute the derivative

$$C'(x) = \frac{5}{2}[(900)^2 + x^2]^{-1/2}(2x) - 4 = \frac{5x}{\sqrt{(900)^2 + x^2}} - 4$$

and set it equal to zero to get

$$\frac{5x}{\sqrt{(900)^2 + x^2}} - 4 = 0 \quad \text{or} \quad \sqrt{(900)^2 + x^2} = \frac{5}{4}x$$

Square both sides of this equation and solve for x to get

$$(900)^2 + x^2 = \frac{25}{16}x^2$$

$$x^2 = \frac{16}{9}(900)^2$$

$$x = \pm\frac{4}{3}(900) = \pm 1,200$$

Since only the positive value $x = 1,200$ is in the interval $0 \le x \le 3,000$, compute $C(x)$ at this critical point and at the endpoints $x = 0$ and $x = 3,000$. Since

$$C(0) = 5\sqrt{(900)^2 + 0} + 4(3,000 - 0) = 16,500$$

$$C(1,200) = 5\sqrt{(900)^2 + (1,200)^2} + 4(3,000 - 1,200) = 14,700$$

$$C(3,000) = 5\sqrt{(900)^2 + (3,000)^2} + 4(3,000 - 3,000) \simeq 15,660$$

it follows that the minimal installation cost is $14,700, which will occur if the cable reaches the opposite bank 1,200 meters downstream from the power plant.

Cylinder Problems

The next example involves the volume and surface area of a cylindrical can. For reference, here is a list of the formulas from geometry that you will need to solve problems of this type. In these formulas, r denotes the radius and h the height.

Area of circular top (or bottom) $= \pi r^2$

Area of curved side $= 2\pi rh$

Volume of cylinder $=$ (area of base)(height) $= \pi r^2 h$

To see why the area of the curved side is $2\pi rh$, imagine the top and bottom of the cylinder removed and the side cut and spread out to form a rectangle as in Figure 4.6. The height of the rectangle is the height h of the cylinder, and the length of the rectangle is the circumference $2\pi r$ of the

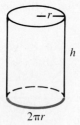
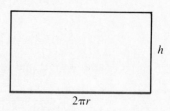

Figure 4.6 Cylindrical can and its curved side spread out.

circular top (or bottom). Hence, the area of the rectangle (or curved side) is $2\pi rh$.

EXAMPLE 4.4   A cylindrical can is to be constructed to hold a fixed volume of liquid. The cost of the material used for the top and bottom of the can is 3 cents per square inch, and the cost of the material used for the curved side is 2 cents per square inch. Use calculus to derive a simple relationship between the radius and height of the can that is the least costly to construct.

SOLUTION
Let r denote the radius, h the height, C the cost (in cents), and V the (fixed) volume. The goal is to minimize cost:

Cost = cost of top + cost of bottom + cost of side

where, for each component of the cost,

Cost = (cost per square inch)(area)

Hence,

Cost of top = cost of bottom = $3\pi r^2$

and

Cost of side = $2(2\pi rh)$ = $4\pi rh$

and the total cost is

$$C = 3\pi r^2 + 3\pi r^2 + 4\pi rh = 6\pi r^2 + 4\pi rh$$

Before you can apply calculus, you must write the cost in terms of just one variable. To do this, use the fact that the can is to have a fixed volume V and solve the equation

$$V = \pi r^2 h$$

for h to get

$$h = \frac{V}{\pi r^2} \quad \text{(where } V \text{ is a constant.)}$$

Substitute this expression for h into the formula for C to get

$$C(r) = 6\pi r^2 + 4\pi r\left(\frac{V}{\pi r^2}\right) = 6\pi r^2 + \frac{4V}{r}$$

Since the radius can be any positive number, the goal is to find the absolute minimum of $C(r)$ on the interval $r > 0$.
The derivative is

$$C'(r) = 12\pi r - \frac{4V}{r^2} \quad \text{Remember that } V \text{ is a constant.}$$

which is zero when

$$12\pi r = \frac{4V}{r^2} \qquad \text{or} \qquad 3r = \frac{V}{\pi r^2}$$

But

$$\frac{V}{\pi r^2} = h$$

and so this equation says

$$3r = h$$

That is, assuming that the cost function is minimized when its derivative is zero, the height of the least expensive can must be 3 times its radius.

Finally, to verify that this actually corresponds to the absolute minimum, convince yourself that $C(r)$ has only one first-order critical point, at $r = \sqrt[3]{\dfrac{V}{3\pi}}$, on the interval $r > 0$ and apply the second-derivative test for absolute extrema. Since the second derivative

$$C''(r) = 12\pi + \frac{8V}{r^3}$$

is clearly positive when $r > 0$, it follows that the critical point that led to the relationship $3r = h$ corresponds to the absolute minimum of the cost function on the interval $r > 0$.

For reference, the relevant portion of the graph of $C(r)$ is sketched in Figure 4.7.

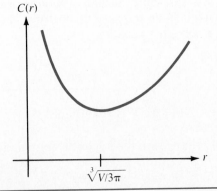

Figure 4.7 The cost function $C(r) = 6\pi r^2 + \dfrac{4V}{r}$ for $r > 0$.

The Maximum Efficiency of a Worker

In the next example, we return to the questions of worker efficiency and diminishing returns, which were introduced in Section 2 of this chapter. Our goal will be to maximize a worker's rate of production. The moment at which this maximum efficiency occurs is known in economics as the

point of diminishing returns. As usual, the strategy will be to set the first derivative of the function that is to be maximized equal to zero. In this case, the function to be maximized is the rate of change or first derivative of the worker's output function. Hence, it will be the second derivative of the output function that will be set equal to zero to find the point of diminishing returns.

EXAMPLE 4.5 An efficiency study of the morning shift at a factory indicates that an average worker who arrives on the job at 8:00 A.M. will have produced $Q(t) = -t^3 + 9t^2 + 12t$ units t hours later. At what time during the morning is the worker performing most efficiently?

SOLUTION
The worker's rate of production is the derivative

$$R(t) = Q'(t) = -3t^2 + 18t + 12$$

of the output function $Q(t)$. Assuming that the morning shift runs from 8:00 A.M. until noon, the goal is to find the absolute maximum of the function $R(t)$ on the closed interval $0 \leq t \leq 4$. The derivative of R is

$$R'(t) = Q''(t) = -6t + 18$$

which is zero when $t = 3$. Comparing the values

$$R(0) = 12 \qquad R(3) = 39 \qquad R(4) = 36$$

of the production-rate function at the critical point and endpoints, you can conclude that the rate of production will be greatest and the worker performing most efficiently when $t = 3$, that is, at 11:00 A.M. when the worker is performing at the rate of 39 units per hour.

The graphs of the output function $Q(t)$ and its derivative, the rate of production $R(t)$, are sketched in Figure 4.8. Notice that the production curve is steepest and the rate of production greatest when $t = 3$.

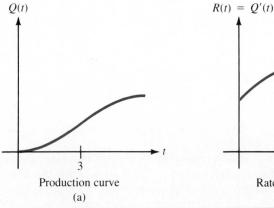

Production curve
(a)

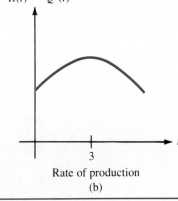

Rate of production
(b)

Figure 4.8 A production curve and corresponding rate of production.

Optimization of Discrete Functions

In the next example, the function to be maximized has practical meaning only when its independent variable is a whole number. However, the optimization procedure leads to a fractional value of this variable, and additional analysis is needed to obtain a meaningful solution.

EXAMPLE 4.6 A bus company will charter a bus that holds 50 people to groups of 35 or more. If a group contains exactly 35 people, each person pays $60. In larger groups, everybody's fare is reduced by $1 for each person in excess of 35. Determine the size of the group for which the bus company's revenue will be greatest.

SOLUTION

Let R denote the bus company's revenue. Then,

R = (number of people in the group)(fare per person)

You could let x denote the total number of people in the group, but it is slightly more convenient to let x denote the number of people in excess of 35. Then,

Number of people in the group = $35 + x$

and

Fare per person = $60 - x$

and the revenue function is

$R(x) = (35 + x)(60 - x)$

Since x represents the number of people in excess of 35, and since the total size of the group must be between 35 and 50, it follows that x must be in the interval $0 \leq x \leq 15$. The goal is to find the absolute maximum of $R(x)$ on this closed interval.

The derivative is

$R'(x) = (35 + x)(-1) + (60 - x)(1) = 25 - 2x$

which is zero when $x = 12.5$. Since

$R(0) = 2,100 \qquad R(12.5) = 2,256.25 \qquad R(15) = 2,250$

it follows that the absolute maximum of $R(x)$ on the interval $0 \leq x \leq 15$ occurs when $x = 12.5$.

But x represents a certain number of people and must be a whole number. Hence, $x = 12.5$ cannot be the solution to this practical optimization problem. To find the optimal *integer* value of x, observe that R

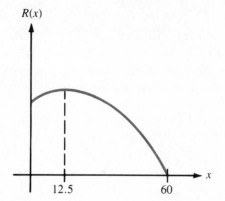

Figure 4.9 The revenue function $R(x) = (35 + x)(60 - x)$.

is increasing for $0 < x < 12.5$ and decreasing for $x > 12.5$, as shown in Figure 4.9.

It follows that the optimal integer value of x is either $x = 12$ or $x = 13$. Since

$$R(12) = 2,256 \quad \text{and} \quad R(13) = 2,256$$

you can conclude that the bus company's revenue will be greatest when the group contains either 12 or 13 people in excess of 35, that is, for groups of 47 or 48. The revenue in either case will be $2,256.

In the preceding example, the graph of revenue as a function of x was actually a collection of discrete points corresponding to the integer values of x as indicated in Figure 4.10a. Since calculus cannot be used to study such a function, we worked with the differentiable function $R(x) = (35 + x)(60 - x)$, which is defined for all values of x and whose graph (Figure 4.10b) "connects" the points in Figure 4.10a. After applying calculus to this continuous model, we obtained a mathematical solution that was not the solution of the discrete practical problem, but that did suggest where to look for the practical solution.

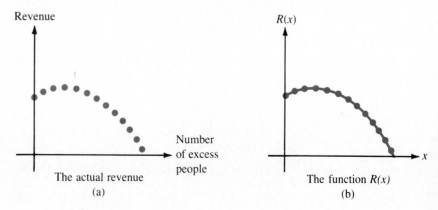

Figure 4.10 A discrete function and its continuous model.

The actual revenue
(a)

The function $R(x)$
(b)

Problems

Retail sales

1. A bookstore can obtain a certain gift book from the publisher at a cost of $3 per book. The bookstore has been offering the book at a price of $15 per copy and, at this price, has been selling 200 copies a month. The bookstore is planning to lower its price to stimulate sales and estimates that for each $1 reduction in the price, 20 more books will be sold each month. At what price should the bookstore sell the book to generate the greatest possible profit?

Manufacturing

2. A manufacturer has been selling lamps for $6 apiece, and, at this price, consumers have been buying 3,000 lamps per month. The manufacturer wishes to raise the price and estimates that for each $1 increase in the price, 1,000 fewer lamps will be sold each month. The manufacturer can produce the lamps at a cost of $4 per lamp. At what price should the manufacturer sell the lamps to generate the greatest possible profit?

Agricultural yield

3. A Florida citrus grower estimates that if 60 orange trees are planted, the average yield per tree will be 400 oranges. The average yield will decrease by 4 oranges per tree for each additional tree planted on the same acreage. How many trees should the grower plant to maximize the total yield?

Harvesting

4. Farmers can get $2 per bushel for their potatoes on July first, and after that, the price drops by 2 cents per bushel per day. On July first, a farmer has 80 bushels of potatoes in the field and estimates that the crop is increasing at the rate of 1 bushel per day. When should the farmer harvest the potatoes to maximize revenue?

Recycling

5. To raise money, a service club has been collecting used bottles that it plans to deliver to a local glass company for recycling. Since the project began 80 days ago, the club has collected 24,000 pounds of glass for which the glass company currently offers 1 cent per pound. However, because bottles are accumulating faster than they can be recycled, the company plans to reduce by 1 cent each day the price it will pay for 100 pounds of used glass. Assume that the club can continue to collect bottles at the same rate and that transportation costs make more than one trip to the glass company unfeasible. What is the most advantageous time for the club to conclude its project and deliver the bottles?

Fencing

6. There are 320 yards of fencing available to enclose a rectangular field. How should this fencing be used so that the enclosed area is as large as possible?

Geometry

7. Prove that of all rectangles with a given perimeter, the square has the largest area.

Fencing   8. A city recreation department plans to build a rectangular playground having an area of 3,600 square meters and surround it by a fence. How can this be done using the least amount of fencing?

Geometry   9. Prove that of all rectangles with a given area, the square has the smallest perimeter.

Construction cost   10. A closed box with a square base is to have a volume of 250 cubic meters. The material for the top and bottom of the box costs $2 per square meter, and the material for the sides costs $1 per square meter. Can the box be constructed for less than $300?

Construction cost   11. A carpenter has been asked to build an open box with a square base. The sides of the box will cost $3 per square meter, and the base will cost $4 per square meter. What are the dimensions of the box of greatest volume that can be constructed for $48?

Postal regulations   12. According to postal regulations, the girth plus length of parcels sent by fourth-class mail may not exceed 72 inches. What is the largest possible volume of a rectangular parcel with two square sides that can be sent by fourth-class mail?

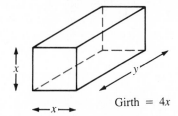

Girth = $4x$

Construction   13. An open box is to be made from a square piece of cardboard, 18 inches by 18 inches, by removing a small square from each corner and folding up the flaps to form the sides. What are the dimensions of the box of greatest volume that can be constructed in this way?

The distance between moving objects   14. A truck is 300 miles due east of a car and is traveling west at the constant speed of 30 miles per hour. Meanwhile, the car is going north at the constant speed of 60 miles per hour. At what time will the car and truck be closest to each other? (*Hint:* You will simplify the calculation if you minimize the *square* of the distance between the car and truck rather than the distance itself. Can you explain why this simplification is justified?)

Installation cost   15. A cable is to be run from a power plant on one side of a river 1,200 meters wide to a factory on the other side, 1,500 meters downstream. The cost of running the cable under the water is $25 per meter, while the cost over land is $20 per meter. What is the most economical route over which to run the cable?

Installation cost   16. Find the most economical route in Problem 15 if the power plant is 2,000 meters downstream from the factory.

Installation cost   17. For the summer, the company that is installing the cable in Example 4.3 has hired a temporary employee with a Ph.D. in mathematics. The mathematician, recalling a problem from first-year calculus, asserts that no matter how far downstream the factory is located (beyond 1,200 meters), it would be most economical to have the cable reach the opposite bank 1,200 meters downstream from the power plant. The supervisor, amused by the naïveté of the over-educated employee, replies, "Any fool can see that if the factory is further away, the cable should reach the opposite bank further downstream. It's just common sense." Who is right? And why?

Spy story   18. It is noon, and the hero of a popular spy story (the same fellow who escaped from the diamond smugglers in Chapter 1, Section 4, Problem 27) is driving a jeep through the sandy desert in the tiny principality of Alta Loma. He is 32 kilometers from the nearest point on a straight, paved road. Down the road 16 kilometers is a power plant in which a band of international terrorists has placed a time bomb set to explode at 12:50 P.M. The jeep can travel 48 kilometers per hour in the sand and 80 kilometers per hour on the paved road. If he arrives at the power plant in the shortest possible time, how long will our hero have to defuse the bomb? (*Hint:* The goal is to minimize time, which is distance divided by speed.)

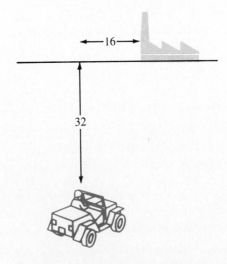

Poster design   19. A printer receives an order to produce a rectangular poster containing 25 square centimeters of print surrounded by margins of 2 centimeters on each side and 4 centimeters on the top and bottom. What are the dimensions of the smallest piece of paper that can be used to make the poster? (*Hint:* An unwise choice of variables will make the calculations unnecessarily complicated.)

Packaging   20. Use the fact that 12 fluid ounces is (approximately) 6.89π cubic inches to find the dimensions of the 12-ounce beer can that can be constructed using the least amount of metal. Compare these dimensions with those of one of the beer cans in your refrigerator. What do you think accounts for the difference?

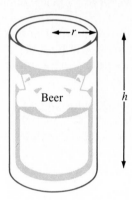

Packaging   21. A cylindrical can is to hold 4π cubic inches of frozen orange juice. The cost per square inch of constructing the metal top and bottom is twice the cost per square inch of constructing the cardboard side. What are the dimensions of the least expensive can?

Volume   22. What is the maximum possible volume of a cylindrical can with no top that can be made from 27π square inches of metal?

Volume   23. A cylindrical can (with top) is to be constructed using a fixed amount of metal. Use calculus to derive a simple relationship between the radius and height of the can having the greatest volume.

Construction cost   24. A cylindrical container with no top is to be constructed to hold a fixed volume of liquid. The cost of the material used for the bottom is 3 cents per square inch and the cost of the material used for the curved side is 2 cents per square inch. Use calculus to derive a simple relationship between the radius and height of the least expensive container.

Production cost   25. A plastics firm has received an order from the city recreation department to manufacture 8,000 special Styrofoam kickboards for its summer swimming program. The firm owns 10 machines, each of which can produce 30 kickboards an hour. The cost of setting up the machines to produce the kickboards is $20 per machine. Once the machines have been set up, the operation is fully automated and can be overseen by a single production supervisor earning $4.80 per hour.

(a) How many of the machines should be used to minimize the cost of production?

(b) How much will the supervisor earn during the production run if the optimal number of machines is used?

(c) How much will it cost to set up the optimal number of machines?

Production cost 26. A manufacturing firm receives an order for Q units of a certain commodity. Each of the firm's machines can produce n units per hour. The setup cost is s dollars per machine, and the operating cost is p dollars per hour.

(a) Derive a formula for the number of machines that should be used to keep total cost as low as possible.

(b) Prove that when the total cost is minimal, the cost of setting up the machines is equal to the cost of operating the machines.

Efficiency 27. An efficiency study of the morning shift (from 8:00 A.M. to 12:00 noon) at a certain factory indicates that an average worker who arrives on the job at 8:00 A.M. will have assembled $Q(t) = -t^3 + 6t^2 + 15t$ transistor radios t hours later.

(a) At what time during the morning is the worker performing most efficiently?

(b) At what time during the morning is the worker performing least efficiently?

Efficiency 28. An efficiency study of the morning shift (from 8:00 A.M. to 12:00 noon) at a factory indicates that an average worker who arrives on the job at 8:00 A.M. will have produced $Q(t) = -t^3 + \frac{9}{2}t^2 + 15t$ units t hours later.

(a) At what time during the morning is the worker performing most efficiently?

(b) At what time during the morning is the worker performing least efficiently?

The steepness of a curve 29. At what point does the tangent to the curve $y = 2x^3 - 3x^2 + 6x$ have the smallest slope? What is the slope of the tangent at this point?

The steepness of a curve 30. For what value of x in the interval $-1 \leq x \leq 4$ is the graph of the function $f(x) = 2x^2 - \frac{1}{3}x^3$ steepest? What is the slope of the tangent at this point?

Population growth 31. A 5-year projection of population trends suggests that t years from now, the population of a certain community will be $P(t) = -t^3 + 9t^2 + 48t + 50$ thousand.

(a) At what time during the 5-year period will the population be growing most rapidly?

(b) At what time during the 5-year period will the population be growing least rapidly?

Production cost 32. Each machine at a certain factory can produce 50 units per hour. The setup cost is $80 per machine, and the operating cost is $5 per hour. How many machines should be used to produce 8,000 units at the least possible cost? (Remember that the answer should be a whole number.)

Construction cost 33. It is estimated that the cost of constructing an office building that is n floors high is $C(n) = 2n^2 + 500n + 600$ thousand dollars. How many floors should the building have in order to minimize the average cost per floor? (Remember that your answer should be a whole number.)

Retail sales 34. A retailer has bought several cases of a certain imported wine. As the wine ages, its value initially increases, but eventually the wine will pass its prime and its value will decrease. Suppose that x years from now, the value of a case will be changing at the rate of $53 - 10x$ dollars per year. Suppose, in addition, that storage rates will remain fixed at $3 per case per year. When should the retailer sell the wine to obtain the greatest possible profit?

Worker efficiency 35. An efficiency study of the morning shift at a certain factory indicates that an average worker who arrives on the job at 8:00 A.M. will have assembled $f(x) = -x^3 + 6x^2 + 15x$ transistor radios x hours later. The study indicates further that after a 15-minute coffee break, the worker can assemble $g(x) = -\frac{1}{3}x^3 + x^2 + 23x$ radios in x hours.

Determine the time between 8:00 A.M. and noon at which a 15-minute coffee break should be scheduled so that the worker will assemble the maximum number of radios by lunchtime at 12:15 P.M. (*Hint:* If the coffee break begins x hours after 8:00 A.M., $4 - x$ hours will remain after the break.)

Transportation cost 36. For speeds between 40 and 65 miles per hour, a truck gets $\frac{480}{x}$ miles per gallon when driven at a constant speed of x miles per hour. Gasoline costs $1.12 per gallon, and the driver is paid $8.40 per hour. What is the most economical constant speed between 40 and 65 miles per hour at which to drive the truck?

5   Applications to Business and Economics

In this section, you will see three applications of calculus to business and economics. The first is an optimization problem from business in which inventory size is selected to minimize cost. The second involves an important relationship between marginal and average quantities in economics. And the third deals with the economic concept of elasticity of demand, which is closely related to the analysis of total revenue.

An Inventory Problem

For each shipment of raw materials, a manufacturer must pay an ordering fee to cover handling and transportation. When the raw materials arrive, they must be stored until needed and storage costs result. If each shipment of raw materials is large, few shipments will be needed and ordering costs will be low. Storage costs, however, will be high. On the other hand, if each shipment is small, ordering costs will be high because many shipments will be needed, but storage costs will be low. A manufacturer would like to determine the shipment size that will minimize total cost. The problem can be solved using calculus. Here is an example.

EXAMPLE 5.1   A bicycle manufacturer buys 6,000 tires a year from a distributor. The ordering fee is $20 per shipment, the storage cost is 96 cents per tire per year, and each tire costs 25 cents. Suppose that the tires are used at a constant rate throughout the year, and that each shipment arrives just as the preceding shipment has been used up. How many tires should the manufacturer order each time to minimize cost?

SOLUTION
The goal is to minimize the total cost, which can be written as

   Total cost = storage cost + ordering cost + purchase cost

Let x denote the number of tires in each shipment and $C(x)$ the corresponding total cost in dollars. Then,

   Ordering cost = (ordering cost per shipment)(number of shipments)

Since 6,000 tires are ordered during the year and each shipment contains x tires, the number of shipments is $\dfrac{6,000}{x}$ and so

$$\text{Ordering cost} = 20 \left(\frac{6,000}{x} \right) = \frac{120,000}{x}$$

Moreover,

   Purchase cost = (total number of tires ordered)(cost per tire)
               = 6,000(0.25) = 1,500

The storage cost is slightly more complicated. When a shipment arrives, all x tires are placed in storage and then withdrawn for use at a constant rate. The inventory decreases linearly until there are no tires left, at which time the next shipment arrives. The situation is illustrated in Figure 5.1a.

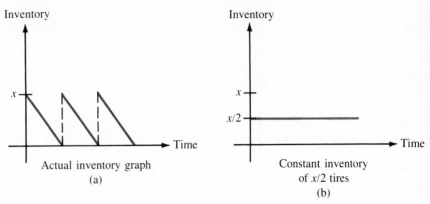

Figure 5.1 Inventory graphs.

The average number of tires in storage during the year is $\frac{x}{2}$, and the total yearly storage cost is the same as if $\frac{x}{2}$ tires were kept in storage for the entire year (Figure 5.1b). (This assertion, although reasonable, is not really obvious, and you have every right to be unconvinced. In Chapter 7, you will learn how to prove this fact mathematically using integral calculus.) It follows that

$$\text{Storage cost} = \frac{x}{2}(\text{cost of storing 1 tire for 1 year})$$

$$= \frac{x}{2}(0.96) = 0.48x$$

Putting it all together, the total cost is

$$C(x) = 0.48x + \frac{120{,}000}{x} + 1{,}500$$

The goal is to find the absolute minimum of $C(x)$ on the interval $0 < x \leq 6{,}000$. The derivative is

$$C'(x) = 0.48 - \frac{120{,}000}{x^2}$$

which is zero when

$$x^2 = \frac{120{,}000}{0.48} = 250{,}000 \qquad \text{or} \qquad x = \pm 500$$

Only the positive value $x = 500$ is in the relevant interval $0 < x \leq 6{,}000$. Since this is the only critical point in the interval, you can apply the second-derivative test for absolute extrema. In particular, the second derivative of the cost function is

$$C''(x) = \frac{240{,}000}{x^3}$$

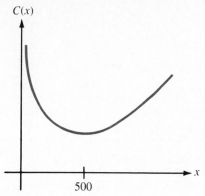

Figure 5.2 Total cost: $C(x) = 0.48x + \dfrac{120,000}{x} + 1,500$.

which is positive when $x > 0$. Hence the absolute minimum of the total cost $C(x)$ on the interval $0 < x \leq 6,000$ occurs when $x = 500$, that is, when the manufacturer orders the tires in lots of 500.

For reference, the graph of the total cost function is sketched in Figure 5.2.

Since the derivative of the (constant) purchase price of $1,500 in Example 5.1 was zero, this component of the total cost had no bearing on the optimization problem. In general, economists distinguish between **fixed costs** (such as the total purchase price) and **variable costs** (such as the storage and ordering costs). To minimize total cost, it is sufficient to minimize the sum of all the variable components of cost.

Average Cost and Marginal Cost

The purpose of this discussion is to derive an important relationship between average cost and marginal cost. This relationship generalizes to any pair of average and marginal quantities in economics. To begin, consider the following computational example.

EXAMPLE 5.2   Suppose the total cost in dollars of manufacturing q units of a certain commodity is $C(q) = 3q^2 + q + 48$.

(a) At what level of production is the average cost per unit the smallest?
(b) At what level of production is the average cost per unit equal to the marginal cost?
(c) On the same set of axes, graph the average cost and marginal cost functions.

SOLUTION

(a) The average cost per unit $A(q)$ is the total cost divided by the number of units produced. That is,

$$A(q) = \frac{C(q)}{q} = \frac{3q^2 + q + 48}{q} = 3q + 1 + \frac{48}{q}$$

The goal is to find the absolute minimum of $A(q)$ on the interval $q > 0$. The first derivative is

$$A'(q) = 3 - \frac{48}{q^2} = \frac{3q^2 - 48}{q^2} = \frac{3(q^2 - 16)}{q^2} = \frac{3(q + 4)(q - 4)}{q^2}$$

which is zero on the interval $q > 0$ only when $q = 4$. Since the second derivative $A''(q) = \frac{96}{q^3}$ is positive when $q > 0$, it follows from the second-derivative test that the average cost $A(q)$ is minimal on $q > 0$ when $q = 4$, that is, when 4 units are produced.

(b) The marginal cost is the derivative $C'(q) = 6q + 1$ of the total cost function and equals average cost when

$$C'(q) = A(q)$$

$$6q + 1 = 3q + 1 + \frac{48}{q}$$

$$3q = \frac{48}{q}$$

$$q^2 = 16 \quad \text{or} \quad q = 4$$

which is the same level of production as that in part (a) for which average cost is minimal.

(c) The marginal cost $C'(q) = 6q + 1$ is a linear function, and its graph is a straight line with slope 6 and vertical intercept 1.

To graph the average cost function $A(q) = 3q + 1 + \frac{48}{q}$, observe from part (a) that its derivative

$$A'(q) = \frac{3(q + 4)(q - 4)}{q^2}$$

is negative for $0 < q < 4$ and positive for $q > 4$. Hence, $A(q)$ is decreasing for $0 < q < 4$, increasing for $q > 4$, and has a relative minimum at $q = 4$.

The graphs of the marginal and average cost functions for $q > 0$ are sketched in Figure 5.3.

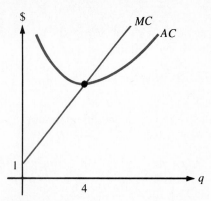

Figure 5.3 Average cost and marginal cost curves for Example 5.2.

Notice that the average cost in Figure 5.3 is decreasing when the marginal cost is less than the average cost. The average cost is increasing when the marginal cost is greater than the average cost. And the average cost has a first-order critical point (a relative minimum in this case) when the marginal cost is equal to the average cost.

This relationship between average cost and marginal cost is not accidental but holds in general. Here is a summary of the situation.

The Relationship Between Average and Marginal Cost

Suppose AC and MC denote the average cost and marginal cost, respectively. Then:

AC is decreasing when $MC < AC$.
AC is increasing when $MC > AC$.
AC has a first-order critical point (usually a relative minimum) when $MC = AC$.

Economic Explanation

Here is the informal explanation of this relationship that is often given in economics texts.

The marginal cost is (approximately) the cost of producing one additional unit. If the additional unit costs less to produce than the average cost of the existing units (i.e., if $MC < AC$), then this less expensive unit will cause the average cost per unit to decrease. On the other hand, if the additional unit costs more than the average cost of the existing units (i.e., if $MC > AC$), then this more expensive unit will cause the average cost per unit to increase. If the cost of the additional unit is equal to the average cost of the existing units (i.e., if $MC = AC$), then the average cost will neither increase nor decrease, that is, it will have a first-order critical point.

Mathematical Derivation

The relationship between average cost and marginal cost can be derived mathematically using calculus. If $C(q)$ is the total cost of manufacturing q units, then the average cost per unit is

$$A(q) = \frac{C(q)}{q}$$

By the quotient rule, the first derivative is

$$A'(q) = \frac{qC'(q) - C(q)(1)}{q^2}$$

which is zero when its numerator is zero, that is, when

$$qC'(q) - C(q) = 0$$
$$qC'(q) = C(q)$$
$$C'(q) = \frac{C(q)}{q}$$

or

$$MC = AC$$

That is, when $A(q)$ has a first-order critical point, marginal cost is equal to average cost.

If $MC < AC$, then

$$C'(q) < \frac{C(q)}{q} \qquad \text{or} \qquad qC'(q) < C(q)$$

and

$$A'(q) = \frac{qC'(q) - C(q)}{q^2} < \frac{C(q) - C(q)}{q^2} = 0$$

which implies that average cost $A(q)$ is decreasing.

Similarly, if $MC > AC$, then

$$C'(q) > \frac{C(q)}{q} \qquad \text{or} \qquad qC'(q) > C(q)$$

and

$$A'(q) = \frac{qC'(q) - C(q)}{q^2} > \frac{C(q) - C(q)}{q^2} = 0$$

which implies that average cost $A(q)$ is increasing.

Generalization

The relationship between average cost and marginal cost is actually a special case of a general relationship between any pair of average and marginal quantities in economics (such as average revenue and marginal revenue, for example). The only possible change is the nature of the first-order critical point that occurs when the average quantity equals the marginal quantity. (Average revenue, for example, usually has a relative maximum when average revenue equals marginal revenue.)

Here is a summary of the general situation. Its informal economic explanation and mathematical derivation are virtually identical to those for the special case of average and marginal cost.

General Relationship Between Average and Marginal Quantities

Suppose AQ and MQ are the average and marginal values, respectively, of some quantity Q. Then:

AQ is decreasing when $MQ < AQ$.
AQ is increasing when $MQ > AQ$.
AQ has a first-order critical point when $MQ = AQ$.

Elasticity of Demand

Consumer demand for a product is usually related to its price. In most cases, the demand decreases as the price increases. The sensitivity of demand to changes in price varies from one product to another. For some products, such as soap, flashlight batteries, or salt, small percentage changes in price have little effect on demand. For other products, such as airline tickets, designer furniture, or home loans, small percentage changes in price can have considerable effect on demand.

A convenient measure of the sensitivity of demand to changes in price is the percentage change in demand that is generated by a 1 percent increase in price. If p denotes the price, q the corresponding number of units demanded, and Δp a (small) change in price, the approximation formula for percentage change (from Chapter 2, Section 4) gives

$$\text{Percentage change in } q \simeq 100 \frac{(dq/dp)\,\Delta p}{q}.$$

If, in particular, the change in p is a 1 percent increase, then $\Delta p = 0.01p$ and

$$\text{Percentage change in } q \simeq 100 \frac{(dq/dp)(0.01p)}{q} = \frac{p}{q}\frac{dq}{dp}.$$

The expression on the right-hand side of this approximation is known in economics as the **elasticity of demand**.

Elasticity of Demand

> If q denotes the demand for a commodity and p its price, the elasticity of demand, denoted by the Greek letter η (eta), is given by the formula
>
> $$\eta = \frac{p}{q} \frac{dq}{dp}$$
>
> and has the following interpretation:
>
> $$\eta \simeq \frac{\text{percentage change in demand}}{\text{due to a 1 percent increase in price}}$$

Here is an example.

EXAMPLE 5.3   Suppose the demand q and price p for a certain commodity are related by the linear equation $q = 240 - 2p$ (for $0 \le p \le 120$).

(a) Express the elasticity of demand as a function of p.
(b) Calculate the elasticity of demand when the price is $p = 100$. Interpret your answer.
(c) Calculate the elasticity of demand when the price is $p = 50$. Interpret your answer.
(d) At what price is the elasticity of demand equal to -1? What is the economic significance of this price?

SOLUTION
(a) The elasticity of demand is

$$\eta = \frac{p}{q} \frac{dq}{dp} = \frac{p}{q}(-2) = -\frac{2p}{240 - 2p} = -\frac{p}{120 - p}$$

(b) When $p = 100$, the elasticity of demand is

$$\eta = -\frac{100}{120 - 100} = -5$$

That is, when the price is $p = 100$, a 1 percent increase in price will produce a decrease in demand of approximately 5 percent.

(c) When $p = 50$, the elasticity of demand is

$$\eta = -\frac{50}{120 - 50} \simeq -0.71$$

That is, when the price is $p = 50$, a 1 percent increase in price will produce a decrease in demand of approximately 0.71 percent.

(d) The elasticity of demand will be equal to -1 when

$$-1 = -\frac{p}{120 - p} \qquad 120 - p = p \qquad 2p = 120 \qquad \text{or} \qquad p = 60$$

At this price, a 1 percent increase in price will result in a decrease in demand of approximately the same percent.

Levels of Elasticity of Demand

In general, the elasticity of demand η is negative, since demand decreases as price increases. If $|\eta| > 1$, the percentage decrease in demand is greater than the percentage increase in price that caused it. This was the case in part (b) of the preceding example, in which $\eta = -5$, implying that an increase in price of 1 percent produces a decrease in demand of approximately 5 percent. At prices for which $|\eta| > 1$, the demand is relatively sensitive to changes in price, and economists say that demand is **elastic** with respect to price.

If $|\eta| < 1$, the percentage decrease in demand is less than the percentage increase in price that caused it. This was the case in part (c) of the preceding example, in which $\eta \simeq -0.71$, implying that an increase in price of 1 percent produces a decrease in demand of approximately only 0.71 percent. At prices for which $|\eta| < 1$, the demand is relatively insensitive to changes in price, and economists say that demand is **inelastic** with respect to price.

If $|\eta| = 1$, as in part (d) of the preceding example, the percentage changes in price and demand are (approximately) equal, and the demand is said to be of **unit elasticity.**

Levels of Elasticity

If $|\eta| > 1$, demand is said to be elastic with respect to price.

If $|\eta| < 1$, demand is said to be inelastic with respect to price.

If $|\eta| = 1$, demand is said to be of unit elasticity with respect to price.

Elasticity and the Total Revenue

The total revenue generated by the sale of a commodity is the price per unit times the number of units sold. That is,

$$R = pq$$

where R is the revenue, p the price per unit, and q the number of units sold (i.e., the demand).

The level of the elasticity of demand with respect to price gives useful information about the total revenue obtained from the sale of the product. In particular, if the demand is inelastic ($|\eta| < 1$), the total revenue increases as the price increases (even though demand drops). The idea is that, in this case, the relatively small percentage decrease in demand is offset by the larger percentage increase in price, and hence the revenue, which is price times demand, increases. If the demand is elastic ($|\eta| > 1$), the total revenue decreases as the price increases. In this case, the relatively large percentage decrease in demand is not offset by the smaller percentage increase in price.

Here is a summary of the situation.

Elasticity and Revenue

> If demand is inelastic ($|\eta| < 1$), total revenue increases as price increases.
>
> If demand is elastic ($|\eta| > 1$), total revenue decreases as price increases.

A graph showing the usual relationship between revenue and price is drawn in Figure 5.4. The graph is increasing when the demand is inelastic ($|\eta| < 1$), decreasing when the demand is elastic ($|\eta| > 1$), and has a maximum at the price for which the demand is of unit elasticity ($|\eta| = 1$).

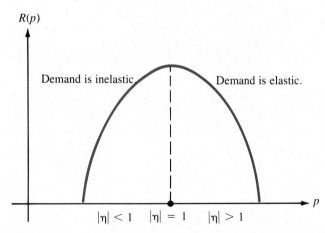

Figure 5.4 Revenue as a function of price.

Derivation of the Relationship Between Elasticity and Revenue

To see why the level of elasticity of demand determines the increase or decrease of total revenue, start with the revenue equation

$$R = pq$$

and differentiate both sides implicitly with respect to p to get

$$\frac{dR}{dp} = p\frac{dq}{dp} + q \qquad \text{by the product rule}$$

To get the elasticity $\eta = \dfrac{p}{q}\dfrac{dq}{dp}$ into the picture, simply multiply the expression on the right-hand side by $\dfrac{q}{q}$ as follows:

$$\frac{dR}{dp} = \frac{q}{q}\left(p\frac{dq}{dp} + q\right) = q\left(\frac{p}{q}\frac{dq}{dp} + 1\right) = q(\eta + 1)$$

If $|\eta| < 1$, then $\eta > -1$ (since η is negative) and

$$\frac{dR}{dp} > q(-1 + 1) = 0$$

That is, the derivative of R with respect to p is positive and so revenue is an increasing function of price.

If $|\eta| > 1$, then $\eta < -1$ (again, since η is negative) and

$$\frac{dR}{dp} < q(-1 + 1) = 0$$

which implies that revenue is a decreasing function of price.

The relationship between elasticity of demand and total revenue is illustrated in the next example.

EXAMPLE 5.4   Suppose the demand q and price p for a certain commodity are related by the equation $q = 300 - p^2$ (for $0 \le p \le \sqrt{300}$).

(a) Determine where the demand is elastic, inelastic, and of unit elasticity with respect to price.

(b) Use the results of part (a) to describe the behavior of the total revenue as a function of price.

(c) Find the total revenue function explicitly and use its first derivative to determine its intervals of increase and decrease and the price at which revenue is maximized.

SOLUTION

(a) The elasticity of demand is

$$\eta = \frac{p}{q}\frac{dq}{dp} = \frac{p}{300 - p^2}(-2p) = -\frac{2p^2}{300 - p^2}$$

The demand is of unit elasticity when $|\eta| = 1$, that is, when

$$\frac{2p^2}{300 - p^2} = 1 \qquad 2p^2 = 300 - p^2 \qquad 3p^2 = 300$$

$$p^2 = 100 \qquad \text{or} \qquad p = \pm 10$$

of which only $p = 10$ is in the relevant interval $0 \le p \le \sqrt{300}$.
    If $0 \le p < 10$,

$$|\eta| = \frac{2p^2}{300 - p^2} < \frac{2(10)^2}{300 - (10)^2} = 1$$

and hence the demand is inelastic.
    If $10 < p \le \sqrt{300}$,

$$|\eta| = \frac{2p^2}{300 - p^2} > \frac{2(10)^2}{300 - (10)^2} = 1$$

and hence the demand is elastic.
(b) The total revenue is an increasing function of p when demand is inelastic, that is, on the interval $0 \le p < 10$, and a decreasing function of p when demand is elastic, that is, on the interval $10 < p \le \sqrt{300}$. At the price $p = 10$ of unit elasticity, the revenue function has a relative maximum.
(c) The revenue function is

$$R = pq$$

or

$$R(p) = p(300 - p^2) = 300p - p^3$$

Its derivative is

$$R'(p) = 300 - 3p^2 = 3(10 - p)(10 + p)$$

which is zero when $p = \pm 10$, of which only $p = 10$ is in the relevant interval $0 \le p \le \sqrt{300}$.
    On the interval $0 \le p < 10$, $R'(p)$ is positive and so $R(p)$ is increasing. On the interval $10 < p \le \sqrt{300}$, $R'(p)$ is negative and so

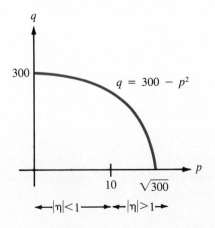

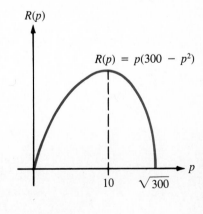

Figure 5.5 Demand and revenue curves for Example 5.4.

$R(p)$ is decreasing. At the critical point $p = 10$, $R(p)$ stops increasing and starts decreasing and hence has a relative maximum.

The relevant portions of the graphs of the demand and revenue functions are sketched in Figure 5.5.

A Technical Detail

The analysis of elasticity and revenue presented here differs slightly from that usually found in economics texts. For technical reasons, economists prefer to treat demand rather than price as the independent variable in the discussion of these concepts. The approach adopted in this section (in which price was the independent variable) was chosen because it is the more straightforward and natural for those not already accustomed to mathematical economics. However, so that you will have some familiarity and experience with the other approach, the development of this material from that point of view is outlined in the problems at this end of the section.

Problems

Inventory
1. An electronics firm uses 600 cases of transistors each year. The cost of storing one case for a year is 90 cents, and the ordering fee is $30 per shipment. How many cases should the firm order each time to keep total cost at a minimum? (Assume that the transistors are used at a constant rate throughout the year and that each shipment arrives just as the preceding shipment has been used up.)

Inventory
2. A local tavern expects to use 800 bottles of bourbon this year. The bourbon costs $4 per bottle, the ordering fee is $10 per shipment, and the cost of storing the bourbon is 40 cents per bottle per year. The bourbon is consumed at a constant rate throughout the year, and each shipment arrives just as the preceding shipment has been used up.
   (a) How many bottles should the tavern order in each shipment to minimize cost?
   (b) How often should the tavern order the bourbon?

Inventory
3. Through its franchised stations, an oil company gives out 16,000 road maps per year. The cost of setting up a press to print the maps is $100 for each production run. In addition, production costs are 6 cents per map and storage costs are 20 cents per map per year. The maps are distributed at a uniform rate throughout the year and are printed in equal batches timed so that each arrives just as the preceding batch has been used up. How many maps should the oil company print in each batch to minimize cost?

Inventory

4. A manufacturing firm receives raw materials in equal shipments arriving at regular intervals throughout the year. The cost of storing the raw materials is directly proportional to the size of each shipment, while the total yearly ordering cost is inversely proportional to the shipment size. Show that the total cost is lowest when the total storage cost and total ordering cost are equal.

Average cost

5. Suppose the total cost in dollars of manufacturing q units of a certain commodity is $C(q) = 3q^2 + 5q + 75$.
   (a) At what level of production is the average cost per unit the smallest?
   (b) At what level of production is the average cost per unit equal to the marginal cost?
   (c) On the same set of axes, graph the average cost and marginal cost functions for $q > 0$.

Average cost

6. Suppose the total cost in dollars of manufacturing q units of a certain commodity is $C(q) = q^3 + 5q + 162$.
   (a) At what level of production is the average cost per unit smallest?
   (b) At what level of production is the average cost per unit equal to the marginal cost?
   (c) On the same set of axes, graph the average cost and marginal cost functions for $q > 0$.

Average revenue

7. Suppose the total revenue in dollars from the sale of q units of a certain commodity is $R(q) = -2q^2 + 68q - 128$.
   (a) At what level of sales is the average revenue per unit equal to the marginal revenue?
   (b) Verify that the average revenue is increasing if the level of sales is less than the level in part (a) and decreasing if the level of sales is greater than the level in part (a).
   (c) On the same set of axes, graph the relevant portions of the average and marginal revenue functions.

Average productivity

8. The output Q at a certain factory is a function of the number L of worker-hours of labor that are used. Use calculus to prove that when the average output per worker-hour is greatest, the average output is equal to the marginal output per worker-hour. (*Hint:* The marginal output per worker-hour is the derivative of output Q with respect to labor L.) You may assume without proof that the first-order critical point of the average output function is actually the desired absolute maximum.

Elasticity

9. Suppose that the demand equation for a certain commodity is $q = 60 - 0.1p$ (for $0 \le p \le 600$).
   (a) Express the elasticity of demand as a function of p.
   (b) Calculate the elasticity of demand when the price is $p = 200$.

Interpret your answer.

(c) At what price is the elasticity of demand equal to -1?

Elasticity 10. Suppose that the demand equation for a certain commodity is $q = 200 - 2p^2$ (for $0 \le p \le 10$).

(a) Express the elasticity of demand as a function of p.

(b) Calculate the elasticity of demand when the price is $p = 6$. Interpret your answer.

(c) At what price is the elasticity of demand equal to -1?

Elasticity and revenue 11. Suppose that the demand equation for a certain commodity is $q = 500 - 2p$ (for $0 \le p \le 250$).

(a) Determine where the demand is elastic, inelastic, and of unit elasticity with respect to price.

(b) Use the results of part (a) to determine the intervals of increase and decrease of the revenue function and the price at which revenue is maximized.

(c) Find the total revenue function explicitly and use its first derivative to determine its intervals of increase and decrease and the price at which revenue is maximized.

(d) Graph the relevant portions of the demand and revenue functions.

Elasticity and revenue 12. Suppose that the demand equation for a certain commodity is $q = 120 - 0.1p^2$ (for $0 \le p \le \sqrt{1,200}$).

(a) Determine where the demand is elastic, inelastic, and of unit elasticity with respect to price.

(b) Use the results of part (a) to determine the intervals of increase and decrease of the revenue function and the price at which revenue is maximized.

(c) Find the total revenue function explicitly and use its first derivative to determine its intervals of increase and decrease and the price at which revenue is maximized.

(d) Graph the relevant portions of the demand and revenue functions.

Elasticity 13. Suppose the demand equation for a certain commodity is $q = \dfrac{a}{p^m}$, where a and m are positive constants. Show that the elasticity of demand is equal to $-m$ for all values of p. Interpret this result.

Elasticity and revenue 14. Suppose the demand equation for a certain commodity is linear, that is, $q = b - ap$ $\left(\text{for } 0 \le p \le \dfrac{b}{a} \right)$, where a and b are positive constants.

(a) Express the elasticity of demand as a function of p.

(b) Show that the demand is of unit elasticity at the midpoint $p = \dfrac{b}{2a}$ of the relevant interval $0 \le p \le \dfrac{b}{a}$.

(c) Find the intervals on which the demand is elastic and on which it is inelastic.

(d) Find the total revenue function explicitly and use its first derivative to determine its intervals of increase and decrease.

(e) Graph the relevant portions of the demand and revenue functions.

Elasticity and Revenue as Functions of Demand

For technical reasons, economists often treat elasticity and revenue as functions of demand q rather than price p. In this context, the elasticity of demand is defined as

$$\eta = \frac{p/q}{dp/dq}$$

which can be shown to be equivalent to the definition with which you are already familiar. (The proof of this equivalence involves properties of inverse functions, which are discussed in more advanced texts.)

In Problems 15 through 18, use this alternative definition of elasticity of demand to carry out the required calculations and derivations.

Elasticity 15. Suppose the demand q and price p for a certain commodity are related by the equation $p = 60 - 2q$ (for $0 \le q \le 30$).

(a) Express the elasticity of demand as a function of q.

(b) Calculate the elasticity of demand when $q = 10$. Interpret your answer.

(c) Substitute for q in the formula in part (a) to express the elasticity of demand as a function of p.

(d) Use our original definition of η to express the elasticity of demand as a function of p. [The answer should be the same as in part (c).]

Elasticity and revenue 16. Prove that the derivative of total revenue R with respect to demand q is

$$\frac{dR}{dq} = p\left(1 + \frac{1}{\eta}\right)$$

(*Hint:* As in the derivation given in the text, start with the equation $R = pq$, but this time differentiate with respect to q.)

Elasticity and revenue 17. Use the formula derived in Problem 16 to show that revenue R is an increasing function of demand q when $|\eta| > 1$ and a decreasing function of q when $|\eta| < 1$.

Elasticity and revenue 18. Suppose that the demand equation for a certain commodity is $p = 600 - 2q^2$ (for $0 \le q \le \sqrt{300}$).

(a) Express the elasticity of demand as a function of q and determine where $|\eta| = 1$, $|\eta| < 1$, and $|\eta| > 1$.

(b) Use the results of part (a) to determine the intervals of increase and decrease of revenue as a function of demand q. At what level of demand is the revenue greatest?

(c) Express total revenue explicitly as a function of q and use its first derivative to determine its intervals of increase and decrease and the demand at which revenue is maximized.

(d) Graph the relevant portions of the functions expressing price and revenue in terms of q.

Chapter Summary and Review Problems

Important Terms, Symbols, and Formulas

Relative maxima and minima

f is increasing: $f'(x) > 0$

f is decreasing: $f'(x) < 0$

First-order critical point: $f'(x) = 0$ or $f'(x)$ is undefined

Concave upward: Slope of tangent is increasing; $f''(x) > 0$

Concave downward: Slope of tangent is decreasing; $f''(x) < 0$

Inflection point: Concavity changes

Second-order critical point: $f''(x) = 0$ or $f''(x)$ is undefined

Second-derivative test for relative extrema: Suppose $f'(a) = 0$.

If $f''(a) > 0$, then f has a relative minimum at $x = a$

If $f''(a) < 0$, then f has a relative maximum at $x = a$

Absolute maxima and minima

Closed interval: $a \leq x \leq b$

Absolute extrema on closed intervals: At critical points or endpoints

Second-derivative test for absolute extrema: Suppose $f'(a) = 0$ and $x = a$ is the only critical point on the interval.

If $f''(a) > 0$, then $f(a)$ is the absolute minimum on the interval

If $f''(a) < 0$, then $f(a)$ is the absolute maximum on the interval

Average and marginal quantities

Elasticity of demand:

$$\eta = \frac{p}{q}\frac{dq}{dp} \simeq \frac{\text{percentage change in demand due to}}{\text{a 1 percent increase in price}}$$

Elastic demand, inelastic demand, unit elasticity

Elasticity and the revenue function

Review Problems

1. Determine where the given function is increasing and where it is decreasing, find its relative extrema, and sketch the graph.

   (a) $f(x) = -2x^3 + 3x^2 + 12x - 5$

   (b) $f(x) = 3x^5 - 20x^3$

   (c) $f(x) = \dfrac{x^2}{x + 1}$

   (d) $f(x) = 2x + \dfrac{8}{x} + 2$

2. Sketch the graph of a function that has all of the following properties:

   (a) $f'(x) > 0$ when $x < 0$ and when $x > 5$

   (b) $f'(x) < 0$ when $0 < x < 5$

   (c) $f''(x) > 0$ when $-6 < x < -3$ and when $x > 2$

   (d) $f''(x) < 0$ when $x < -6$ and when $-3 < x < 2$

3. Determine where the given function is increasing, decreasing, concave upward, and concave downward. Find the relative extrema and inflection points and draw the graph.

   (a) $f(x) = x^2 - 6x + 1$ (b) $f(x) = x^3 - 3x^2 + 2$

   (c) $f(x) = \dfrac{x^2 + 3}{x - 1}$ (d) $f(x) = \dfrac{x - 1}{(x + 1)^2}$

4. Use the second derivative test to find the relative maxima and minima of the given function.

   (a) $f(x) = -2x^3 + 3x^2 + 12x - 5$

   (b) $f(x) = \dfrac{x^2}{x + 1}$

   (c) $f(x) = 2x + \dfrac{8}{x} + 2$

5. Find the absolute maximum and absolute minimum (if any) of the given function on the specified interval.

   (a) $f(x) = -2x^3 + 3x^2 + 12x - 5;\ -3 \le x \le 3$

(b) $f(x) = -3x^4 + 8x^3 - 10; 0 \leq x \leq 3$

(c) $f(x) = \dfrac{x^2}{x + 1}; -\dfrac{1}{2} \leq x \leq 1$

(d) $f(x) = 2x + \dfrac{8}{x} + 2; x > 0$

6. It is estimated that between the hours of noon and 7:00 P.M., the speed of highway traffic flowing past a certain downtown exit is approximately $S(t) = t^3 - 9t^2 + 15t + 45$ miles per hour, where t is the number of hours past noon. At what time between noon and 7:00 P.M. is the traffic moving the fastest, and at what time between noon and 7:00 P.M. is it moving the slowest?

7. The rate at which a rumor spreads through a community is jointly proportional to the number of people who have heard the rumor and the number who have not. Show that the rumor is spreading most rapidly when half the people have heard it.

8. A retailer can obtain cameras from the manufacturer at a cost of $50 apiece. The retailer has been selling the cameras at a price of $80 apiece, and, at this price, consumers have been buying 40 cameras a month. The retailer is planning to lower the price to stimulate sales and estimates that for each $5 reduction in the price, 10 more cameras will be sold each month. At what price should the retailer sell the cameras to maximize profit?

9. You wish to use 300 meters of fencing to surround two identical adjacent rectangular plots as shown in the accompanying figure. How should you do this to make the combined area of the plots as large as possible?

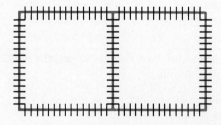

10. A cylindrical container with no top is to be constructed for a fixed amount of money. The cost of the material used for the bottom is 3 cents per square inch, and the cost of the material used for the curved side is 2 cents per square inch. Use calculus to derive a simple relationship between the radius and height of the container having the greatest volume.

11. You are standing on the bank of a river that is 1 mile wide and want to get to a town on the opposite bank, 1 mile upstream. You plan to row on a straight line to some point P on the opposite bank and then walk the remaining distance along the bank. To what point P should you row in order to reach the town in the shortest possible time if you can row 4 miles per hour and walk 5 miles per hour?

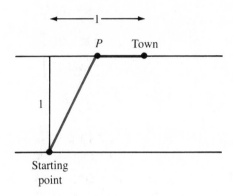

12. A postal clerk spends 4 hours each morning sorting mail. During that time, the clerk can sort approximately $f(t) = -t^3 + 7t^2 + 200t$ letters in t hours. At what time during this period is the clerk performing at peak efficiency?

13. A manufacturing firm has received an order to make 400,000 souvenir medals commemorating the 15th anniversary of the landing of Apollo 11 on the moon. The firm owns 20 machines, each of which can produce 200 medals per hour. The cost of setting up the machines to produce the medals is $80 per machine, and the total operating cost is $5.76 per hour. How many machines should be used to minimize the cost of producing the 400,000 medals?

14. A citrus grower estimates that if 60 lemon trees are planted in a grove, the average yield per tree will be 475 lemons. The average yield will decrease by 5 lemons per tree for each additional tree planted in the grove. How many trees should the grower plant to maximize the total yield? (Remember that the answer should be a whole number.)

15. Suppose the consumer demand for a certain commodity is $D(p) = mp + b$ units per month when the market price is p dollars per unit.
    (a) Assume that $m < 0$ and $b > 0$ and sketch this demand function. Explain in economic terms why the assumptions about the signs of m and b are reasonable.
    (b) Express consumers' total monthly expenditure for the commodity as a function of p and sketch the graph of this function.

(c) Use calculus to show that the market price at which consumer expenditure will be greatest is the value of p that is midway between the origin and the p intercept of the demand curve.

16. A manufacturer uses q units of a certain raw material each year. The cost of storing one unit for a year is s dollars, the ordering fee is b dollars per shipment, and the purchase cost of the raw material is p dollars per unit. Use calculus to derive a formula for the number of units that should be ordered in each shipment to minimize total cost. (As usual, assume that the raw material is used at a constant rate throughout the year and that each shipment arrives just as the preceding shipment has been used up.)

17. Suppose the total cost in dollars of manufacturing q units of a certain commodity is given by the quadratic function $C(q) = aq^2 + bq + c$, where a, b, and c are positive constants.
    (a) Find the level of production q at which average cost equals marginal cost.
    (b) Verify that the average cost is decreasing at levels of production less than the one found in part (a) and increasing at levels of production greater than the one found in part (a).

18. Suppose that the demand equation for a certain commodity is $q = 27 - 0.01p^2$ (for $0 \leq p \leq \sqrt{2,700}$).
    (a) Determine where the demand is elastic, inelastic, and of unit elasticity with respect to price.
    (b) Use the results of part (a) to determine the intervals of increase and decrease of the revenue function and the price at which revenue is maximized.
    (c) Find the total revenue function explicitly and use its first derivative to determine its intervals of increase and decrease and the price at which revenue is maximized.
    (d) Graph the relevant portions of the demand and revenue functions.

Chapter

Exponential and Logarithmic Functions

1 Exponential Functions

An **exponential function** is a function of the form $f(x) = a^x$, where a is a positive constant. In an exponential function, the independent variable x is the **exponent** of a positive constant known as the **base** of the function. Thus, an exponential function is fundamentally different from a power function $f(x) = x^n$ in which the base is the variable and the exponent is the constant.

Exponential functions play a central role in applied mathematics. They are used in demography to forecast population size, in finance to calculate the value of investments, in archaeology to date ancient artifacts, in psychology to study learning phenomena, in public health to analyze the spread of epidemics, and in industry to estimate the reliability of products. You will see some of these applications in Section 2.

The Algebra of Exponents

Working with exponential functions requires the use of exponential notation and the algebraic laws of exponents. For reference, the major facts are summarized here. A more complete review, including worked examples and practice problems, is in Appendix A at the back of the book. Be sure that you are comfortable with this material before you proceed with this chapter.

Definition of a^x for Rational Values of x (and $a > 0$)

Integer powers: If n is a positive integer,

$$a^n = a \cdot a \cdots a$$

where the product $a \cdot a \cdots a$ has n factors.

Fractional powers: If n and m are positive integers,

$$a^{n/m} = (\sqrt[m]{a})^n$$

where $\sqrt[m]{}$ denotes the positive mth root.

Negative powers: $a^{-x} = \dfrac{1}{a^x}$

Zero power: $a^0 = 1$

For example,

$$3^4 = 3 \cdot 3 \cdot 3 \cdot 3 = 81 \qquad 3^{-4} = \frac{1}{3^4} = \frac{1}{81}$$

$$4^{1/2} = \sqrt{4} = 2 \qquad 4^{3/2} = (\sqrt{4})^3 = 2^3 = 8$$

$$4^{-3/2} = \frac{1}{4^{3/2}} = \frac{1}{8} \qquad 27^{-2/3} = \frac{1}{(\sqrt[3]{27})^2} = \frac{1}{3^2} = \frac{1}{9}$$

The Laws of Exponents

The product law: $a^r a^s = a^{r+s}$

The quotient law: $\dfrac{a^r}{a^s} = a^{r-s}$

The power law: $(a^r)^s = a^{rs}$

For example,

$$(2^{-2})^3 = 2^{-6} = \frac{1}{2^6} = \frac{1}{64}$$

$$\frac{3^3}{3^{1/3}(3^{2/3})} = \frac{3^3}{3^{(1/3+2/3)}} = \frac{3^3}{3^1} = 3^2 = 9$$

$$2^{7/4}(8^{-1/4}) = 2^{7/4}(2^3)^{-1/4} = 2^{7/4}(2^{-3/4}) = 2^{(7/4-3/4)} = 2^1 = 2$$

The Graphs of Exponential Functions

The graphs of four exponential functions are shown in Figure 1.1.

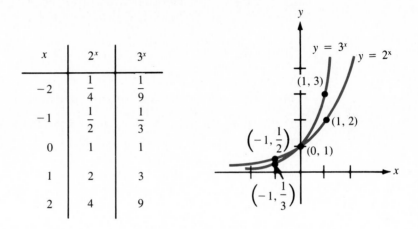

x	2^x	3^x
-2	$\frac{1}{4}$	$\frac{1}{9}$
-1	$\frac{1}{2}$	$\frac{1}{3}$
0	1	1
1	2	3
2	4	9

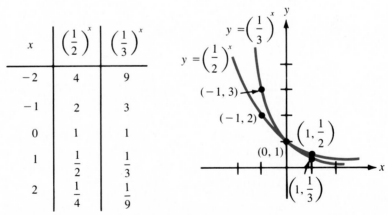

x	$\left(\frac{1}{2}\right)^x$	$\left(\frac{1}{3}\right)^x$
-2	4	9
-1	2	3
0	1	1
1	$\frac{1}{2}$	$\frac{1}{3}$
2	$\frac{1}{4}$	$\frac{1}{9}$

Figure 1.1 The graphs of four exponential functions.

One way to obtain a rough sketch of an exponential function very quickly is to find its y intercept and determine its behavior as x increases without bound and as x decreases without bound. The technique is illustrated in the following example.

EXAMPLE 1.1   Sketch the function $f(x) = a^x$ if $0 < a < 1$ and if $a > 1$.

SOLUTION
In both cases, the y intercept is $(0, 1)$ since $f(0) = a^0 = 1$. There are no x intercepts since, for positive a, a^x is always positive. To determine the behavior of the graph as x increases or decreases without bound, consider the two cases $0 < a < 1$ and $a > 1$ separately.

If $0 < a < 1$, the value of the product $a^n = a \cdot a \cdots a$ approaches zero as the number n of factors increases. For example,

$$\left(\frac{1}{10}\right)^n = \left(\frac{1}{10}\right)\left(\frac{1}{10}\right) \cdots \left(\frac{1}{10}\right) = \frac{1}{10^n}$$

which is very close to zero when n is large. This suggests that a^x approaches zero as x increases without bound. On the other hand, the value of the product $a^{-n} = \frac{1}{a} \cdot \frac{1}{a} \cdots \frac{1}{a}$ increases without bound (since $\frac{1}{a} > 1$) as the number n of factors increases. For example,

$$\left(\frac{1}{10}\right)^{-n} = 10 \cdot 10 \cdots 10 = 10^n$$

which is a very large number when n is large. This suggests that a^x increases without bound as x decreases without bound. A graph with these features is sketched in Figure 1.2a.

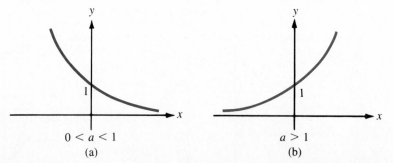

Figure 1.2 The graph of the exponential function $y = a^x$.

(a) $0 < a < 1$ (b) $a > 1$

If $a > 1$, a^x increases without bound as x increases without bound and a^x approaches zero as x decreases without bound. The corresponding graph is sketched in Figure 1.2b.

Irrational Exponents

The graphs in Figure 1.2 should not really have been drawn as unbroken curves since a^x has been defined only for *rational* values of x. However, it can be shown (using techniques beyond the scope of this book) that because there are "so many" rational numbers, there can be only *one* unbroken curve that passes through all the points (x, a^x) for which x is rational. That is, there exists a unique continuous function $f(x)$ that is defined for all real numbers x and that is equal to a^x when x is rational. When x is irrational, one *defines* a^x to be the value $f(x)$ of this function. Fortunately, in practical work you will rarely, if ever, have to deal explicitly with a^x for an irrational value of x.

The Number e

First impressions can be deceiving. At first glance, the expression $\left(1 + \dfrac{1}{n}\right)^n$ may look no more interesting than many other algebraic expressions. Yet the number it approaches as n increases without bound turns out to be one of the most important and useful numbers in mathematics. This special number will appear as the base in most of the exponential functions you will encounter.

To get a feel for what happens to the expression $\left(1 + \dfrac{1}{n}\right)^n$ as n increases without bound, use your calculator to complete the following table. (Round off your answers to three decimal places.)

n	1	2	5	10	100	1,000	10,000	100,000
$\left(1 + \dfrac{1}{n}\right)^n$	2.000	2.250	2.488	2.594	2.705			

It turns out that as n increases without bound, the expression $\left(1 + \dfrac{1}{n}\right)^n$ approaches an irrational number, traditionally denoted by the letter e, whose value is approximately 2.718. (A rigorous proof of this fact requires techniques beyond the scope of this book.)

Here is the definition of the number e, written compactly with the aid of limit notation. The symbol "$\lim_{n \to \infty}$" is read "the limit as n approaches infinity" and denotes the behavior of the subsequent expression as n increases without bound.

The Number e

$$e = \lim_{n \to \infty} \left(1 + \frac{1}{n}\right)^n \simeq 2.718$$

To illustrate how the number e might arise in practice, here is a brief discussion of compound interest.

Compound Interest

Suppose a sum of money is invested and the interest is compounded only once. If P is the initial investment (the principal) and r is the interest rate (expressed as a decimal), the balance B after the interest is added will be

$$B = P + Pr = P(1 + r) \qquad \text{dollars}$$

That is, to compute the balance at the end of an interest period, you multiply the balance at the beginning of the period by the expression $1 + r$, where r is the interest rate per period.

At most banks, interest is compounded more than once a year. The interest that is added to the account during one period will itself earn interest during the subsequent periods. If the annual interest rate is r and interest is compounded k times per year, the year is divided into k equal interest periods and the interest rate during each is $\frac{r}{k}$. To compute the balance at the end of any period, you multiply the balance at the beginning of that period by the expression $1 + \frac{r}{k}$. Hence, the balance at the end of the first period is

$$P\left(1 + \frac{r}{k}\right)$$

The balance at the end of the second period is

$$P\left(1 + \frac{r}{k}\right)\left(1 + \frac{r}{k}\right) = P\left(1 + \frac{r}{k}\right)^2$$

The balance at the end of the third period is

$$P\left(1 + \frac{r}{k}\right)^2\left(1 + \frac{r}{k}\right) = P\left(1 + \frac{r}{k}\right)^3$$

At the end of 1 year, the interest has been compounded k times, and the balance is

$$P\left(1 + \frac{r}{k}\right)^k$$

and at the end of t years, the interest has been compounded kt times, and the balance is given by the function

$$B(t) = P\left(1 + \frac{r}{k}\right)^{kt}$$

Compound Interest

If P dollars are invested at an annual interest rate r (expressed as a decimal) and interest is compounded k times per year, the balance $B(t)$ after t years will be

$$B(t) = P\left(1 + \frac{r}{k}\right)^{kt} \qquad \text{dollars}$$

Continuously Compounded Interest

As the frequency with which the interest is compounded increases, the corresponding balance $B(t)$ also increases. Hence a bank that compounds interest frequently may attract more customers than one that offers the same interest rate but that compounds interest less often. The question arises: What happens to the balance at the end of t years as the frequency with which the interest is compounded increases without bound? That is, what will the balance be at the end of t years if interest is compounded not quarterly, not monthly, not daily, but continuously? In mathematical terms: What happens to the expression $P\left(1 + \dfrac{r}{k}\right)^{kt}$ as k increases without bound? The answer turns out to involve the number e. Here is the argument.

To simplify the calculation, let $n = \dfrac{k}{r}$. Then, $k = nr$ and so

$$P\left(1 + \frac{r}{k}\right)^{kt} = P\left(1 + \frac{1}{n}\right)^{nrt} = P\left[\left(1 + \frac{1}{n}\right)^{n}\right]^{rt}$$

Since n increases without bound as k does, and since $\left(1 + \dfrac{1}{n}\right)^{n}$ approaches e as n increases without bound, it follows that the balance after t years is

$$B(t) = \lim_{k \to \infty} P\left(1 + \frac{r}{k}\right)^{kt} = \lim_{n \to \infty} P\left[\left(1 + \frac{1}{n}\right)^{n}\right]^{rt} = Pe^{rt}$$

Here is a summary of the situation.

Continuously Compounded Interest

> If P dollars are invested at an annual interest rate r (expressed as a decimal) and interest is compounded continuously, the balance $B(t)$ after t years will be
>
> $$B(t) = Pe^{rt} \quad \text{dollars}$$

The following example illustrates what happens to a bank balance as the interest is compounded with increasing frequency. The numerical computations were done on a calculator.

EXAMPLE 1.2 Suppose $1,000 is invested at an annual interest rate of 6 percent. Compute the balance after 10 years if the interest is compounded

(a) quarterly   (b) monthly   (c) daily   (d) continuously

SOLUTION

(a) To compute the balance after 10 years if the interest is compounded quarterly, use the formula $B(t) = P\left(1 + \dfrac{r}{k}\right)^{kt}$, with $t = 10$, $P = 1,000$, $r = 0.06$, and $k = 4$ to get

$$B(10) = 1,000\left(1 + \frac{0.06}{4}\right)^{40} \simeq \$1,814.02$$

(b) This time, take $t = 10$, $P = 1,000$, $r = 0.06$, and $k = 12$ to get

$$B(10) = 1,000\left(1 + \frac{0.06}{12}\right)^{120} \simeq \$1,819.40$$

(c) Take $t = 10$, $P = 1,000$, $r = 0.06$, and $k = 365$ to get

$$B(10) = 1,000\left(1 + \frac{0.06}{365}\right)^{3,650} \simeq \$1,822.03$$

(d) For continuously compounded interest use the formula $B(t) = Pe^{rt}$, with $t = 10$, $P = 1,000$, and $r = 0.06$ to get

$$B(10) = 1,000e^{0.6} \simeq \$1,822.12$$

This value, $1,822.12, is an upper bound for the possible balance. No matter how often interest is compounded, $1,000 invested at an annual interest rate of 6 percent cannot grow to more than $1,822.12 in 10 years.

Graphs of Functions Involving Powers of e

Since $e > 1$, the graph of the function $y = e^x$ (Figure 1.3a) resembles the graph of the general exponential function $y = a^x$ for $a > 1$, which you saw in Figure 1.2b. Since $\dfrac{1}{e} < 1$, the graph of the function $y = e^{-x} = \left(\dfrac{1}{e}\right)^x$ (Figure 1.3b) resembles the graph of the general exponential function $y = a^x$ for $0 < x < 1$, which you saw in Figure 1.2a.

Notice that as x increases without bound, e^x increases without bound while e^{-x} approaches zero. As x decreases without bound, e^x approaches zero while e^{-x} increases without bound. As you will see, these basic properties of the functions e^x and e^{-x} will help you analyze and graph more complicated functions involving powers of e. Here is a summary of these properties, written compactly with the aid of limit notation.

Limits at Infinity for e^x and e^{-x}

$$\lim_{x \to \infty} e^x = \infty \quad \text{and} \quad \lim_{x \to -\infty} e^x = 0$$

$$\lim_{x \to \infty} e^{-x} = 0 \quad \text{and} \quad \lim_{x \to -\infty} e^{-x} = \infty$$

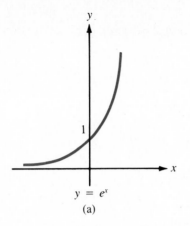

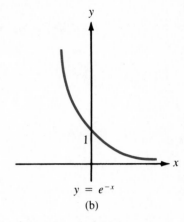

Figure 1.3 The graphs of the functions $y = e^x$ and $y = e^{-x}$.

$y = e^x$
(a)

$y = e^{-x}$
(b)

By now, you are probably having no trouble interpreting the limit notation. For example, the expression $\lim\limits_{x \to -\infty} e^{-x} = \infty$ is read "the limit of e^{-x} as x approaches minus infinity is infinity" and expresses the fact that as x decreases without bound, e^{-x} increases without bound.

More complicated functions involving powers of e arise frequently in applications. You can obtain a rough sketch of such a function using the following procedure.

How to Sketch Functions Involving Powers of e

Step 1. Find the y intercept (by setting x equal to zero).

Step 2. Find $\lim\limits_{x \to \infty} f(x)$ and $\lim\limits_{x \to -\infty} f(x)$.

Step 3. Sketch the graph using these clues.

Here are two examples.

EXAMPLE 1.3   Sketch the graph of the function $f(x) = 4 - 2e^{-x}$.

SOLUTION
Since $f(0) = 4 - 2e^0 = 4 - 2 = 2$, the y intercept is $(0, 2)$.

As x increases without bound, e^{-x} approaches zero and hence $f(x) = 4 - 2e^{-x}$ approaches $4 - 0 = 4$. That is,

$$\lim_{x \to \infty} f(x) = 4$$

As x decreases without bound, e^{-x} increases without bound. Hence, $-2e^{-x}$ *decreases* without bound (because of the minus sign) and so $f(x) = 4 - 2e^{-x}$ decreases without bound. That is,

$$\lim_{x \to -\infty} f(x) = -\infty$$

A graph with these properties is sketched in Figure 1.4.

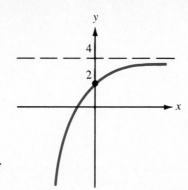

Figure 1.4 The graph of $y = 4 - 2e^{-x}$.

EXAMPLE 1.4   Sketch the graph of the function $f(x) = \dfrac{3}{1 + 2e^{-5x}}$.

SOLUTION

Since $f(0) = \dfrac{3}{1 + 2e^0} = \dfrac{3}{1 + 2} = 1$, the y intercept is $(0, 1)$.

As x increases without bound, e^{-5x} approaches zero. Hence the denominator $1 + 2e^{-5x}$ approaches 1 and $f(x) = \dfrac{3}{1 + 2e^{-5x}}$ approaches $\dfrac{3}{1} = 3$. That is,

$$\lim_{x \to \infty} f(x) = 3$$

As x decreases without bound, e^{-5x} increases without bound. Hence the denominator $1 + 2e^{-5x}$ increases without bound and $f(x) = \dfrac{3}{1 + 2e^{-5x}}$ approaches zero. That is,

$$\lim_{x \to -\infty} f(x) = 0$$

A graph with these properties is sketched in Figure 1.5.

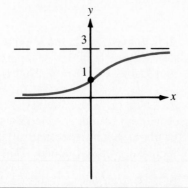

Figure 1.5 The graph of $y = \dfrac{3}{1 + 2e^{-5x}}$.

An Application of the Power Law of Exponents

To conclude this introductory section, here is an example illustrating a simple but useful computational technique based on the power law for exponents. You will use this technique several times in Section 2.

EXAMPLE 1.5   Find $f(6)$ if $f(x) = e^{kx}$ and $f(2) = 5$.

SOLUTION
You do not have to know the value of k or of e to solve this problem! The fact that $f(2) = 5$ tells you that

$$e^{2k} = 5$$

and using the power law for exponents, you can rewrite the expression for $f(6)$ in terms of this quantity to get

$$f(6) = e^{6k} = (e^{2k})^3 = 5^3 = 125$$

Problems

1. Learn how to use your calculator to find powers of e. In particular, find e^2, e^{-2}, $e^{0.05}$, $e^{-0.05}$, e^0, e, $\sqrt{e}$, and $\dfrac{1}{\sqrt{e}}$. (Round off your answers to three decimal places.)

2. Sketch the curves $y = 3^x$ and $y = 4^x$ on the same set of axes.

3. Sketch the curves $y = \left(\dfrac{1}{3}\right)^x$ and $y = \left(\dfrac{1}{4}\right)^x$ on the same set of axes.

In Problems 4 through 15, sketch the given function.

4. $f(x) = e^x$ 5. $f(x) = e^{-x}$

6. $f(x) = 2 + e^x$ 7. $f(x) = 3 + e^{-x}$

8. $f(x) = 2 - 3e^x$ 9. $f(x) = 3 - 2e^x$

10. $f(x) = 5 - 3e^{-x}$ 11. $f(x) = 3 - 5e^{-x}$

12. $f(x) = \dfrac{2}{1 + 3e^{-2x}}$ 13. $f(x) = \dfrac{2}{1 + 3e^{2x}}$

14. $f(x) = 1 - \dfrac{3}{1 + 2e^{-3x}}$ 15. $f(x) = 1 - \dfrac{6}{2 + e^{3x}}$

16. Find $f(2)$ if $f(x) = e^{kx}$ and $f(1) = 20$.

17. Find $f(9)$ if $f(x) = e^{kx}$ and $f(3) = 2$.

18. Find $f(8)$ if $f(x) = Ae^{kx}$, $f(0) = 20$, and $f(2) = 40$.

19. Find $f(4)$ if $f(x) = 50 - Ae^{-kx}$, $f(0) = 20$, and $f(2) = 30$.

20. Find $f(2)$ if $f(x) = 50 - Ae^{kx}$, $f(0) = 30$, and $f(4) = 5$.

Compound interest   21. Suppose \$1,000 is invested at an annual interest rate of 7 percent. Compute the balance after 10 years if the interest is compounded
   (a) annually   (b) quarterly
   (c) monthly   (b) continuously

Compound interest   22. A sum of money is invested at a certain fixed interest rate, and the interest is compounded continuously. After 10 years, the money has doubled. How will the balance at the end of 20 years compare with the initial investment?

Compound interest   23. (a) Solve the equation $B = Pe^{rt}$ for P.
   (b) Use the result from part (a) to determine how much money should be invested today at an annual interest rate of 6 percent compounded continuously so that 10 years from now it will be worth \$10,000.

Effective interest rate   24. When a bank offers an annual interest rate of $100r$ percent and compounds the interest more than once a year, the total interest earned during a year is greater than $100r$ percent of the balance at the beginning of that year. The actual percentage by which the balance grows during a year is sometimes called the **effective interest rate,** while the advertised rate of $100r$ percent is called the **nominal interest rate.** Find the effective interest rate if the nominal rate is 6 percent and interest is compounded
   (a) quarterly   (b) continuously

25. Program a computer or use a calculator to evaluate $\left(1 + \dfrac{1}{n}\right)^n$ for $n = 1{,}000, 2{,}000, \ldots, 50{,}000$.

26. Program a computer or use a calculator to evaluate $\left(1 + \dfrac{1}{n}\right)^n$ for $n = -1{,}000, -2{,}000, \ldots, -50{,}000$. On the basis of these calculations, what can you conjecture about the behavior of $\left(1 + \dfrac{1}{n}\right)^n$ as n *decreases* without bound?

2   Exponential Models

Functions involving powers of e play a central role in applied mathematics. Here is a sampling of practical situations from the social, managerial, and natural sciences that can be described mathematically in terms of such functions.

Exponential Growth

A quantity $Q(t)$ that increases according to a law of the form $Q(t) = Q_0 e^{kt}$, where Q_0 and k are positive constants, is said to experience **exponential growth.** For example, if interest is compounded continuously, the resulting bank balance $B(t) = Pe^{rt}$ grows exponentially. Also, in the absence of environmental constraints, population increases exponentially. As you will see later in this book, quantities that grow exponentially are characterized by the fact that their rate of growth is proportional to their size.

Exponential Growth

> $Q(t)$ grows exponentially if
>
> $$Q(t) = Q_0 e^{kt}$$
>
> where k is a positive constant and Q_0 is the initial value $Q(0)$.

To sketch the function $Q(t) = Q_0 e^{kt}$, observe that $Q(t)$ is always positive, that $Q(0) = Q_0$, that $Q(t)$ increases without bound as t increases without bound, and that $Q(t)$ approaches zero as t decreases without bound. The graph is sketched in Figure 2.1.

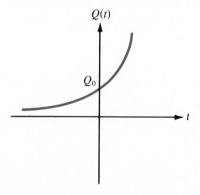

Figure 2.1 Exponential growth: $Q(t) = Q_0 e^{kt}$.

Here is an example from biology.

EXAMPLE 2.1   Biologists have determined that under ideal conditions, the number of bacteria in a culture grows exponentially. Suppose that 2,000 bacteria are initially present in a certain culture and that 6,000 are present 20 minutes later. How many bacteria will be present at the end of 1 hour?

SOLUTION

Let $Q(t)$ denote the number of bacteria present after t minutes. Since the number of bacteria grows exponentially and since 2,000 bacteria were initially present, you know that Q is a function of the form

$$Q(t) = 2,000 e^{kt}$$

Since 6,000 bacteria are present after 20 minutes, it follows that

$$6,000 = 2,000e^{20k} \quad \text{or} \quad e^{20k} = 3$$

To find the number of bacteria present at the end of 1 hour (60 minutes), compute $Q(60)$ using the power law for exponents as follows:

$$Q(60) = 2,000e^{60k} = 2,000(e^{20k})^3 = 2,000(3)^3 = 54,000$$

That is, 54,000 bacteria will be present at the end of 1 hour.

Exponential Decay

A quantity $Q(t)$ that decreases according to a law of the form $Q(t) = Q_0e^{-kt}$, where Q_0 and k are positive constants, is said to experience **exponential decay** or, equivalently, to **decrease exponentially.** Radioactive substances decay exponentially. Sales of many products decrease exponentially when advertising is discontinued. Quantities that decrease exponentially are characterized by the fact that their rate of decrease is proportional to their size.

Exponential Decay

> $Q(t)$ decreases (or decays) exponentially if
>
> $$Q(t) = Q_0e^{-kt}$$
>
> where k is a positive constant and Q_0 is the initial value $Q(0)$.

A graph of the function $Q(t) = Q_0e^{-kt}$ is sketched in Figure 2.2.

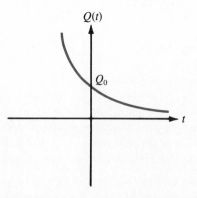

Figure 2.2 Exponential decay: $Q(t) = Q_0e^{-kt}$.

EXAMPLE 2.2   A certain industrial machine depreciates so that its value after t years is given by a function of the form $Q(t) = Q_0e^{-0.04t}$. After 20 years, the machine is worth $8,986.58. What was its original value?

SOLUTION

The goal is to find Q_0. Since $Q(20) = 8,986.58$,

$$Q_0 e^{-0.8} = 8,986.58$$

Multiply both sides of this equation by $e^{0.8}$ to get

$$Q_0 = 8,986.58 e^{0.8} \simeq \$20,000$$

Learning Curves

The graph of a function of the form $Q(t) = B - Ae^{-kt}$, where B, A, and k are positive constants, is sometimes called a **learning curve**. The name arose when psychologists discovered that functions of this form often describe the relationship between the efficiency with which an individual performs a task and the amount of training or experience the individual has had.

To sketch the function $Q(t) = B - Ae^{-kt}$, observe that the vertical intercept is $Q(0) = B - A$. As t increases without bound, e^{-kt} approaches zero and so

$$\lim_{t \to \infty} Q(t) = B - 0 = B$$

As t decreases without bound, e^{-kt} increases without bound, and so $Q(t) = B - Ae^{-kt}$ *decreases* without bound (because of the minus sign preceding the A). That is

$$\lim_{t \to -\infty} Q(t) = -\infty$$

A graph with these features is sketched in Figure 2.3. The behavior of the graph as t increases without bound reflects the fact that eventually an individual will approach peak efficiency, and additional training will have little effect on performance.

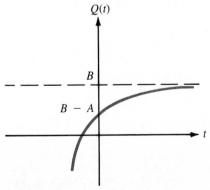

Figure 2.3 A learning curve: $Q(t) = B - Ae^{-kt}$.

EXAMPLE 2.3 The rate at which a postal clerk can sort mail is a function of the clerk's experience. Suppose the postmaster of a large city estimates that after t months on the job, the average clerk can sort $Q(t) = 700 - 400e^{-0.5t}$ letters per hour.

(a) How many letters can a new employee sort per hour?
(b) How many letters can a clerk with 6 months' experience sort per hour?
(c) Approximately how many letters will the average clerk ultimately be able to sort per hour?

SOLUTION

(a) The number of letters a new employee can sort per hour is

$$Q(0) = 700 - 400 = 300$$

(b) After 6 months, the average clerk can sort

$$Q(6) = 700 - 400e^{-0.5(6)} = 700 - 400e^{-3} \approx 680 \qquad \text{letters per hour}$$

(c) As t increases without bound, $Q(t)$ approaches 700. Hence, the average clerk will ultimately be able to sort approximately 700 letters per hour.

The graph of the function $Q(t)$ is sketched in Figure 2.4.

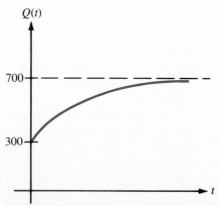

Figure 2.4 Worker efficiency: $Q(t) = 700 - 400e^{-0.5t}$.

Logistic Curves

The graph of a function of the form $Q(t) = \dfrac{B}{1 + Ae^{-Bkt}}$, where B, A, and k are positive constants, is an S-shaped or **sigmoidal curve.** The term **logistic curve** is also used to refer to such a graph.

To sketch the logistic function $Q(t) = \dfrac{B}{1 + Ae^{-Bkt}}$, notice that the vertical intercept is

$$Q(0) = \frac{B}{1 + Ae^0} = \frac{B}{1 + A}$$

As t increases without bound, e^{-Bkt} approaches zero (since B and k are positive). Hence,

$$\lim_{t \to \infty} Q(t) = \lim_{t \to \infty} \frac{B}{1 + Ae^{-Bkt}} = \frac{B}{1 + 0} = B$$

As t decreases without bound, e^{-Bkt} increases without bound (because of the minus sign in the exponent). Hence the denominator $1 + Ae^{-Bkt}$ increases without bound and

$$\lim_{t \to -\infty} Q(t) = \lim_{t \to -\infty} \frac{B}{1 + Ae^{-Bkt}} = 0$$

A graph with these properties is sketched in Figure 2.5.

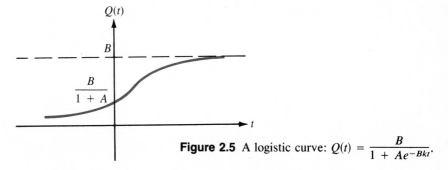

Figure 2.5 A logistic curve: $Q(t) = \dfrac{B}{1 + Ae^{-Bkt}}$.

Logistic curves are rather accurate models of population growth when environmental factors impose an upper bound on the possible size of the population. They also describe the spread of epidemics and rumors in a community. Here is a typical example.

EXAMPLE 2.4 Public health records indicate that t weeks after the outbreak of a certain form of influenza, approximately $Q(t) = \dfrac{20}{1 + 19e^{-1.2t}}$ thousand people had caught the disease.

(a) How many people had the disease when it first broke out?
(b) How many had caught the disease by the end of the 2nd week?
(c) If the trend continues, approximately how many people in all will contract the disease?

SOLUTION
(a) Since

$$Q(0) = \frac{20}{1 + 19} = 1$$

it follows that 1,000 people had the disease initially.
(b) Since

$$Q(2) = \frac{20}{1 + 19e^{-1.2(2)}} \simeq 7.343$$

it follows that approximately 7,343 people had caught the disease by
the end of the 2nd week.
(c) Since $Q(t)$ approaches 20 as t increases without bound, it follows that
approximately 20,000 people will eventually contract the disease.

For reference, the graph of $Q(t)$ is sketched in Figure 2.6.

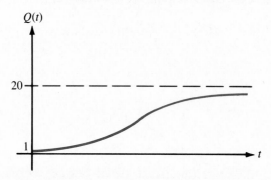

Figure 2.6 The spread of an epidemic: $Q(t) = \dfrac{20}{1 + 19e^{-1.2t}}.$

Problems

Population growth

1. It is projected that t years from now, the population of a certain
country will be $P(t) = 50e^{0.02t}$ million.
(a) What is the current population?
(b) What will the population be 30 years from now?

Compound interest

2. How much money should be invested today at an annual interest rate
of 7 percent compounded continuously, so that 20 years from now it
will be worth $20,000? [Recall that the balance after t years is $B(t) =
Pe^{rt}$, where r is the interest rate expressed as a decimal and P is the
initial deposit.]

Population growth

3. It is estimated that the population of a certain country grows exponentially. If the population was 60 million in 1980 and 90 million in
1985, what will the population be in 1995?

Growth of bacteria 4. The following data were compiled by a researcher during the first 10 minutes of an experiment designed to study the growth of bacteria:

Number of minutes	0	10
Number of bacteria	5,000	8,000

Assuming that the number of bacteria grows exponentially, how many bacteria will be present after 30 minutes?

Gross national product 5. The gross national product (GNP) of a certain country was 100 billion dollars in 1975 and 180 billion dollars in 1985. Assuming that the GNP is growing exponentially, what will the GNP be in 1995?

Retail sales 6. The total number of hamburgers sold by a national fast-food chain is growing exponentially. If 4 billion had been sold by 1980 and 12 billion had been sold by 1985, how many will have been sold by 1990?

Population density 7. The population density x miles from the center of a certain city is $D(x) = 12e^{-0.07x}$ thousand people per square mile.
(a) What is the population density at the center of the city?
(b) What is the population density 10 miles from the center of the city?

Radioactive decay 8. The amount of a sample of a radioactive substance remaining after t years is given by a function of the form $Q(t) = Q_0 e^{-0.0001t}$. At the end of 5,000 years, 2,000 grams of the substance remain. How many grams were present initially?

Radioactive decay 9. A radioactive substance decays exponentially. If 500 grams of the substance were present initially and 400 grams are present 50 years later, how many grams will be present after 200 years?

Product reliability 10. A statistical study indicates that the fraction of the electric toasters manufactured by a certain company that are still in working condition after t years of use is approximately $f(t) = e^{-0.2t}$.
(a) What fraction of the toasters can be expected to work for at least 3 years?
(b) What fraction of the toasters can be expected to fail during the 3rd year of use?
(c) What fraction of the toasters can be expected to fail before 1 year of use?

Product reliability 11. A manufacturer of toys has found that the fraction of its plastic battery-operated toy oil tankers that sink in fewer than t days is approximately $f(t) = 1 - e^{-0.03t}$.
(a) Sketch this reliability function. What happens to the graph as t increases without bound?
(b) What fraction of the tankers can be expected to float for at least 10 days?

(c) What fraction of the tankers can be expected to sink between the 15th and 20th days?

Sales 12. Once the initial publicity surrounding the release of a new book is over, sales of the hardcover edition tend to decrease exponentially. At the time publicity was discontinued, a certain book was experiencing sales of 25,000 copies per month. One month later, sales of the book had dropped to 10,000 copies per month. What will the sales be after 1 more month?

Recall from memory 13. Psychologists believe that when a person is asked to recall a set of facts, the number of facts recalled after t minutes is given by a function of the form $Q(t) = A(1 - e^{-kt})$, where k is a positive constant and A is the total number of relevant facts in the person's memory.
(a) Sketch the graph of $Q(t)$.
(b) What happens to the graph as t increases without bound? Explain this behavior in practical terms.

Advertising 14. When professors select texts for their courses, they usually choose from among the books already on their shelves. For this reason, most publishers send complimentary copies of new texts to professors teaching related courses. The mathematics editor at a major publishing house estimates that if x thousand complimentary copies are distributed, the first-year sales of a certain new mathematics text will be approximately $f(x) = 20 - 15e^{-0.2x}$ thousand copies.
(a) Sketch this sales function.
(b) How many copies can the editor expect to sell in the first year if no complimentary copies are sent out?
(c) How many copies can the editor expect to sell in the first year if 10,000 complimentary copies are sent out?
(d) If the editor's estimate is correct, what is the most optimistic projection for the first-year sales of the text?

Depreciation 15. When a certain industrial machine is t years old, its resale value will be $V(t) = 4,800e^{-t/5} + 400$ dollars.
(a) Sketch the graph of $V(t)$. What happens to the value of the machine as t increases without bound?
(b) How much was the machine worth when it was new?
(c) How much will the machine be worth after 10 years?

Efficiency 16. The daily output of a worker who has been on the job for t weeks is given by a function of the form $Q(t) = 40 - Ae^{-kt}$. Initially the worker could produce 20 units a day, and after 1 week the worker can produce 30 units a day. How many units will the worker produce per day after 3 weeks?

Newton's law of heating   17. A cool drink is removed from a refrigerator on a hot summer day and placed in a room whose temperature is 30° Celsius. According to a law of physics, the temperature of the drink t minutes later is given by a function of the form $f(t) = 30 - Ae^{-kt}$. If the temperature of the drink was 10° Celsius when it left the refrigerator and 15° Celsius after 20 minutes, what will the temperature of the drink be after 40 minutes?

The spread of an epidemic   18. Public health records indicate that t weeks after the outbreak of a certain form of influenza, approximately $f(t) = \dfrac{2}{1 + 3e^{-0.8t}}$ thousand people had caught the disease.
(a) Sketch the graph of $f(t)$.
(b) How many people had the disease initially?
(c) How many had caught the disease by the end of 3 weeks?
(d) If the trend continues, approximately how many people in all will contract the disease?

Population growth   19. It is estimated that t years from now, the population of a certain country will be $P(t) = \dfrac{20}{2 + 3e^{-0.06t}}$ million.
(a) Sketch the graph of $P(t)$.
(b) What is the current population?
(c) What will the population be 50 years from now?
(d) What will happen to the population in the long run?

The spread of an epidemic   20. An epidemic spreads through a community so that t weeks after its outbreak, the number of people who have been infected is given by a function of the form $f(t) = \dfrac{B}{1 + Ce^{-kt}}$, where B is the number of residents in the community who are susceptible to the disease. If $\frac{1}{5}$ of the susceptible residents were infected initially and $\frac{1}{2}$ had been infected by the end of the 4th week, what fraction of the susceptible residents will have been infected by the end of the 8th week?

The spread of a rumor   21. A traffic accident was witnessed by $\frac{1}{10}$ of the residents of a small town. The number of residents who had heard about the accident t hours later is given by a function of the form $f(t) = \dfrac{B}{1 + Ce^{-kt}}$, where B is the population of the town. If $\frac{1}{4}$ of the residents had heard about the accident after 2 hours, how long did it take for $\frac{1}{2}$ of the residents to hear the news?

3   The Natural Logarithm

In many practical problems, a number a is known and the goal is to find the corresponding number b such that $a = e^b$. This number b is called the **natural logarithm** of a and is denoted by the symbol ln a. The letter l in the symbol ln stands for "logarithm" and the letter n for "natural." (The word "ln" is virtually unpronounceable. It can be read as "el en," "log," or "Lynn.")

The Natural Logarithm

> Corresponding to each positive number a there is a unique power b such that $a = e^b$. This power b is called the natural logarithm of a and is denoted by ln a. Thus,
>
> $$b = \ln a \qquad \text{if and only if} \qquad a = e^b$$

The situation is illustrated in Figure 3.1.

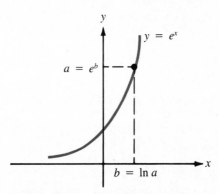

Figure 3.1 $b = \ln a$ if and only if $a = e^b$.

EXAMPLE 3.1   Find:

(a) ln e (b) ln 1   (c) ln $\sqrt{e}$ (d) ln 2

SOLUTION
(a) According to the definition, ln e is the unique number b such that $e = e^b$. Clearly this number is $b = 1$. Hence, ln $e = 1$.
(b) Ln 1 is the unique number b such that $1 = e^b$. Since $e^0 = 1$, it follows that ln $1 = 0$.
(c) Ln $\sqrt{e} = \ln e^{1/2}$ is the unique number b such that $e^{1/2} = e^b$. Clearly this number is $b = \dfrac{1}{2}$. Hence, ln $\sqrt{e} = \dfrac{1}{2}$.

(d) Ln 2 is the unique number b such that $2 = e^b$. The value of this number is not obvious, and you will have to use your calculator (or Table II at the back of the book) to find that $\ln 2 \simeq 0.69315$.

The Relationship Between e^x and ln x

The next example establishes two important identities which show that logarithmic and exponential functions have a certain "neutralizing" effect on each other.

EXAMPLE 3.2   Simplify the following expressions:

(a) $e^{\ln x}$ (for $x > 0$)   (b) $\ln e^x$

SOLUTION
(a) According to the definition, $\ln x$ is the unique number b for which $x = e^b$. Hence, $e^{\ln x} = e^b = x$.
(b) Similarly, $\ln e^x$ is the unique number b for which $e^x = e^b$. Clearly this number b is x itself. Hence, $\ln e^x = x$.

The two identities derived in Example 3.2 show that the composite functions $\ln e^x$ and $e^{\ln x}$ leave the variable x unchanged. In general, two functions f and g for which $f[g(x)] = x$ and $g[f(x)] = x$ are said to be **inverses** of one another. Thus the exponential function $y = e^x$ and the logarithmic function $y = \ln x$ are inverses.

The Inverse Relationship Between e^x and ln x

$$e^{\ln x} = x \quad (\text{if } x > 0) \quad \text{and} \quad \ln e^x = x$$

The next example illustrates how you can use the inverse relationship between e^x and $\ln x$ to solve equations.

EXAMPLE 3.3   Solve each of the following equations for x:

(a) $3 = e^{20x}$ (b) $2 \ln x = 1$

SOLUTION
(a) Take the natural logarithm of each side of the equation to get

$$\ln 3 = \ln e^{20x} \quad \text{or} \quad \ln 3 = 20x$$

Solve for x (using a calculator or the natural logarithm table at the back of the book to find ln 3).

$$x = \frac{\ln 3}{20} \simeq \frac{1.0986}{20} \simeq 0.0549$$

(b) First isolate ln x on the left-hand side of the equation by dividing both sides by 2.

$$\ln x = \frac{1}{2}$$

Then apply the exponential function to both sides of the equation to get

$$e^{\ln x} = e^{1/2} \qquad \text{or} \qquad x = e^{1/2} = \sqrt{e}$$

The Graph of ln x

There is an easy way to obtain the graph of the logarithmic function $y = \ln x$ from that of the exponential function $y = e^x$. The method is based on the geometric fact that the point (b, a) is the reflection across the line $y = x$ of the point (a, b), as illustrated in Figure 3.2.

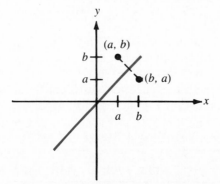

Figure 3.2 Reflection of points across the line $y = x$.

The idea is that since $y = \ln x$ means $x = e^y$, the graph of $y = \ln x$ is the graph of $y = e^x$ with x and y reversed. More precisely, suppose that (a, b) is a point on the curve $y = \ln x$. Then $b = \ln a$, or, equivalently, $a = e^b$. Hence the reflected point (b, a) can be written as (b, e^b), which is a point on the curve $y = e^x$. Conversely, if (a, b) is on the curve $y = e^x$, it follows that $b = e^a$ and so $a = \ln b$. Hence the reflected point is $(b, a) = (b, \ln b)$, which lies on the curve $y = \ln x$.

The graph of the exponential function $y = e^x$, the line $y = x$, and the graph of the logarithmic function $y = \ln x$ are sketched in Figure 3.3. Notice that ln x is defined only for positive values of x, that $\ln 1 = 0$, that ln x increases without bound as x increases without bound, and that ln x decreases without bound as x approaches zero from the right.

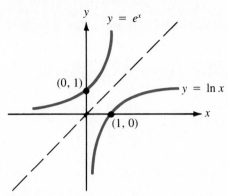

Figure 3.3 The graphs of $y = e^x$ and $y = \ln x$.

Properties of the Natural Logarithm

The laws of exponents can be used to derive the following important properties of logarithms.

Properties of Logarithms

> The logarithm of a product: $\ln uv = \ln u + \ln v$
>
> The logarithm of a quotient: $\ln \dfrac{u}{v} = \ln u - \ln v$
>
> The logarithm of a power: $\ln u^v = v \ln u$

The first of these properties states that logarithms transform multiplication into the simpler operation of addition. The derivation of this property will be given in Example 3.5. The second property states that logarithms transform division into subtraction, and the third states that logarithms reduce exponentiation to multiplication. The derivations of these two properties are left as problems for you to do.

The next example illustrates the use of these algebraic properties of logarithms.

EXAMPLE 3.4   (a) Find $\ln \sqrt{ab}$ if $\ln a = 3$ and $\ln b = 7$.

(b) Show that $\ln \dfrac{1}{x} = -\ln x$.

(c) Find x if $2^x = e^3$.

SOLUTION

(a) $\ln \sqrt{ab} = \ln (ab)^{1/2} = \frac{1}{2} \ln ab = \frac{1}{2}(\ln a + \ln b) = \frac{1}{2}(3 + 7) = 5$

(b) $\ln \frac{1}{x} = \ln 1 - \ln x = 0 - \ln x = -\ln x$

(c) Take the natural logarithm of each side of the equation $2^x = e^3$ and solve for x to get

$$x \ln 2 = 3 \qquad \text{or} \qquad x = \frac{3}{\ln 2} \simeq 4.33$$

The formula for the logarithm of a product is derived in the next example.

EXAMPLE 3.5   Derive the formula $\ln uv = \ln u + \ln v$.

SOLUTION
The fact that e^x and $\ln x$ are inverses implies that

$$e^{\ln uv} = uv \qquad e^{\ln u} = u \qquad e^{\ln v} = v$$

Use these identities together with the product law for exponents to get

$$e^{\ln uv} = uv = e^{\ln u}e^{\ln v} = e^{\ln u + \ln v}$$

Compare the powers of e at each end of this string of equalities to conclude that

$$\ln uv = \ln u + \ln v$$

Logarithms with Other Bases

You may be wondering how natural logarithms are related to the logarithms you studied in high school. The following discussion will show you the connection.

The graph of the exponential function $y = a^x$ (shown in Figure 3.4 for $a > 1$) suggests that for each positive number y there corresponds a unique number x such that $y = a^x$. This power x is called the **logarithm of y to the base a** and is denoted by $\log_a y$. Thus,

$$x = \log_a y \qquad \text{if and only if} \qquad y = a^x$$

The logarithms you used in high school algebra to simplify numerical calculations were logarithms to the base 10. For numerical work, 10 is a particularly convenient base for logarithms because the standard decimal representation of numbers is based on powers of 10. Natural logarithms

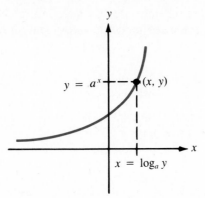

Figure 3.4 The graph of the function $y = a^x$ for $a > 1$.

are logarithms to the base e. Because of the importance of the special exponential function e^x, natural logarithms arise frequently in applied work.

Doubling Time

In the next example, you will see how to use logarithms to determine how much time it takes a quantity that grows exponentially to double in size.

EXAMPLE 3.6   The world's population is growing at the rate of about 2 percent per year. As you will see in Chapter 8, this implies that if the percentage rate of growth remains constant, the population t years from now will be given by an exponential function of the form $P(t) = P_0 e^{0.02t}$, where P_0 is the current population. If this model of population growth is correct, how long will it take for the world's population to double?

SOLUTION
The goal is to find the value of t for which $P(t) = 2P_0$, that is, for which

$$2P_0 = P_0 e^{0.02t}$$

To do this, divide each side of the equation by P_0 to get

$$2 = e^{0.02t}$$

and then take the logarithm of each side and solve for t to get

$$\ln 2 = 0.02t \quad \text{or} \quad t = \frac{\ln 2}{0.02} \approx 34.66$$

That is, the world's population will double in about $34\frac{2}{3}$ years.

Notice that the constant P_0 representing the current population was eliminated from the equation at an early stage and that the final answer is therefore independent of this quantity.

Exponential Curve Fitting

In the next example, you will see how to use logarithms to fit an exponential curve to a set of data.

EXAMPLE 3.7   The population density x miles from the center of a city is given by a function of the form $Q(x) = Ae^{-kx}$. Find this function if the population density at the center of the city is 15 thousand people per square mile and the density 10 miles from the center is 9 thousand people per square mile.

SOLUTION
For simplicity, express the density in units of 1,000 people per square mile. The fact that $Q(0) = 15$ tells you that $A = 15$. The fact that $Q(10) = 9$ tells you that

$$9 = 15e^{-10k} \quad \text{or} \quad \frac{3}{5} = e^{-10k}$$

Taking the logarithm of each side of this equation you get

$$\ln \frac{3}{5} = -10k \quad \text{or} \quad k = -\frac{\ln 3/5}{10} \approx 0.051$$

Hence the exponential function for the population density is (approximately) $Q(x) = 15e^{-0.051x}$.

Carbon Dating

Carbon 14 (^{14}C) is a radioactive isotope of carbon that is widely used to date ancient fossils and artifacts. Here is an outline of the technique.

The carbon dioxide in the air contains ^{14}C as well as carbon 12 (^{12}C), a nonradioactive isotope. Scientists have found that the ratio of ^{14}C to ^{12}C in the air has remained approximately constant throughout history. Living plants absorb carbon dioxide from the air, and so the ratio of ^{14}C to ^{12}C in a living plant is the same as that in the air itself. When a plant dies, the absorption of carbon dioxide ceases. The ^{12}C already in the plant remains while the ^{14}C decays, and the ratio of ^{14}C to ^{12}C decreases exponentially. The ratio of ^{14}C to ^{12}C in a fossil t years after it was alive is approximately $R(t) = R_0e^{-kt}$, where $k = \dfrac{\ln 2}{5,730}$ and R_0 is the ratio of ^{14}C to ^{12}C in the atmosphere. By comparing $R(t)$ with R_0, archaeologists can estimate the age of the fossil.

EXAMPLE 3.8 An archaeologist has found a fossil in which the ratio of ^{14}C to ^{12}C is $\frac{1}{5}$ the ratio found in the atmosphere. Approximately how old is the fossil?

SOLUTION

The age of the fossil is the value of t for which $R(t) = \frac{1}{5}R_0$, that is, for which

$$\frac{1}{5}R_0 = R_0 e^{-kt}$$

Dividing by R_0 and taking logarithms, you get

$$\ln \frac{1}{5} = -kt \quad \text{or} \quad t = -\frac{\ln 1/5}{k} = \frac{\ln 5}{k}$$

Since $k = \frac{\ln 2}{5{,}730}$, it follows that

$$t = \frac{5{,}730 \ln 5}{\ln 2} \simeq 13{,}304.65$$

That is, the fossil is approximately 13,305 years old.

Problems

1. Learn how to use your calculator to find natural logarithms. In particular, find $\ln 1$, $\ln 2$, $\ln e$, $\ln 5$, $\ln \frac{1}{5}$, and $\ln e^2$. What happens if you try to find $\ln 0$ or $\ln -2$? Why?

In Problems 2 through 7, evaluate the given expression without using tables or a calculator.

2. $\ln e^3$

3. $\ln \sqrt{e}$

4. $e^{\ln 5}$

5. $e^{2 \ln 3}$

6. $e^{3 \ln 2 - 2 \ln 5}$

7. $\ln \dfrac{e^3 \sqrt{e}}{e^{1/3}}$

In Problems 8 through 19, solve the given equation for x.

8. $2 = e^{0.06x}$

9. $\frac{1}{2}Q_0 = Q_0 e^{-1.2x}$

10. $3 = 2 + 5e^{-4x}$

11. $-2 \ln x = b$

12. $-\ln x = \dfrac{t}{50} + C$

13. $5 = 3 \ln x - \dfrac{1}{2} \ln x$

14. $\ln x = \dfrac{1}{3}(\ln 16 + 2 \ln 2)$

15. $\ln x = 2(\ln 3 - \ln 5)$

16. $3^x = e^2$

17. $a^k = e^{kx}$

18. $a^{x+1} = b$

19. $x^{\ln x} = e$

20. Find $\ln \dfrac{1}{\sqrt{ab^3}}$ if $\ln a = 2$ and $\ln b = 3$.

21. Find $\dfrac{1}{a} \ln \left(\dfrac{\sqrt{b}}{c} \right)^a$ if $\ln b = 6$ and $\ln c = -2$.

Compound interest

22. How quickly will money double if it is invested at an annual interest rate of 6 percent compounded continuously?

Compound interest

23. Money deposited in a certain bank doubles every 13 years. The bank compounds interest continuously. What annual interest rate does the bank offer?

Compound interest

24. How quickly will money double if it is invested at an annual interest rate of 7 percent and interest is compounded
(a) continuously   (b) annually   (c) quarterly

Population growth

25. Based on the estimate that there are 10 billion acres of arable land on the earth and that each acre can produce enough food to feed 4 people, some demographers believe that the earth can support a population of no more than 40 billion people. The population of the earth was approximately 3 billion in 1960 and 4 billion in 1975. If the population of the earth were growing exponentially, when would it reach the theoretical limit of 40 billion?

The **half-life** of a radioactive substance is the time it takes for 50 percent of a sample of the substance to decay. Problems 26 through 29 deal with this concept.

Radioactive decay

26. The amount of a certain radioactive substance remaining after t years is given by a function of the form $Q(t) = Q_0 e^{-0.003t}$. Find the half-life of the substance.

Radioactive decay

27. Radium decays exponentially. Its half-life is 1,690 years. How long will it take for a 50-gram sample of radium to be reduced to 5 grams?

Radioactive decay

28. A radioactive substance decays exponentially. Show that the amount $Q(t)$ of the substance remaining after t years is $Q(t) = Q_0 e^{-(\ln 2/\lambda)t}$, where Q_0 is the amount present initially and λ is the half-life of the substance.

Radioactive decay 29. A radioactive substance decays exponentially. Show that the amount $Q(t)$ of the substance remaining after t years is $Q(t) = Q_0\left(\frac{1}{2}\right)^{t/\lambda}$, where Q_0 is the amount present initially and λ is the half-life of the substance. (*Hint:* Use the result of Problem 28.)

Advertising 30. The mathematics editor at a major publishing house estimates that if x thousand complimentary copies are distributed to instructors, the first-year sales of a new mathematics text will be approximately $f(x) = 20 - 15e^{-0.2x}$ thousand copies. According to this estimate, approximately how many complimentary copies should the editor send out to generate first-year sales of 12,000 copies?

Growth of bacteria 31. A medical student studying the growth of bacteria in a certain culture has compiled the following data:

Number of minutes	0	20
Number of bacteria	6,000	9,000

Use these data to find an exponential function of the form $Q(t) = Q_0 e^{kt}$ expressing the number of bacteria in the culture as a function of time.

Gross national product 32. An economist has compiled the following data on the gross national product (GNP) of a certain country:

Year	1975	1985
GNP (in billions)	100	180

Use these data to predict the GNP in 1995 if the GNP is growing
(a) linearly   (b) exponentially

Worker efficiency 33. An efficiency expert hired by a manufacturing firm has compiled the following data relating workers' output to their experience:

Experience (months)	0	6
Output (units per hour)	300	410

The expert believes that the output Q is related to experience t by a function of the form $Q(t) = 500 - Ae^{-kt}$. Find the function of this form that fits the data.

Population growth 34. According to a logistic model based on the assumption that the earth can support no more than 40 billion people, the world's population (in billions) t years after 1960 is given by a function of the form $P(t) = \dfrac{40}{1 + Ce^{-kt}}$, where C and k are positive constants. Find the function of this form that is consistent with the fact that the world's population was approximately 3 billion in 1960 and 4 billion in 1975.

Carbon dating 35. An archaeologist has found a fossil in which the ratio of ^{14}C to ^{12}C is $\dfrac{1}{3}$ the ratio found in the atmosphere. Approximately how old is the fossil?

Carbon dating 36. How old is a fossil in which the ratio of ^{14}C to ^{12}C is $\dfrac{1}{2}$ the ratio found in the atmosphere?

37. Use one of the laws of exponents to prove that $\ln \dfrac{u}{v} = \ln u - \ln v$.

38. Use one of the laws of exponents to prove that $\ln u^v = v \ln u$.

39. Express $\log_a x$ in terms of the natural logarithm.

4 Differentiation of Logarithmic and Exponential Functions

Both the logarithmic function $y = \ln x$ and the exponential function $y = e^x$ have simple derivatives.

Logarithmic Functions

The derivative of $\ln x$ is simply $\dfrac{1}{x}$.

The Derivative of ln x

$$\frac{d}{dx}(\ln x) = \frac{1}{x}$$

You will see an outline of the proof of this formula a little later in this section. First, here are some examples illustrating its use.

EXAMPLE 4.1 Differentiate the function $f(x) = x \ln x$.

SOLUTION
Combine the product rule with the formula for the derivative of $\ln x$ to get

$$f'(x) = x\left(\frac{1}{x}\right) + \ln x = 1 + \ln x$$

EXAMPLE 4.2 Differentiate the function $f(x) = \ln (2x^3 + 1)$.

SOLUTION
Think of the function $f(x) = \ln (2x^3 + 1)$ as the composite function $f(x) = g[h(x)]$, where

$$g(u) = \ln u \quad \text{and} \quad h(x) = 2x^3 + 1$$

Then,

$$g'(u) = \frac{1}{u} \quad \text{and} \quad h'(x) = 6x^2$$

and, by the chain rule,

$$f'(x) = g'[h(x)]h'(x) = \frac{1}{2x^3 + 1}(6x^2) = \frac{6x^2}{2x^3 + 1}$$

The preceding example illustrates the following general rule for differentiating the logarithm of a differentiable function.

The Chain Rule for Logarithmic Functions

$$\frac{d}{dx}[\ln h(x)] = \frac{h'(x)}{h(x)}$$

That is, to differentiate $\ln h$, simply divide the derivative of h by h itself.

Here is another example.

EXAMPLE 4.3 Differentiate the function $f(x) = \ln (x^2 + 1)^3$.

SOLUTION
As a preliminary step, simplify $f(x)$ using a property of logarithms to get

$$f(x) = 3 \ln (x^2 + 1)$$

Now apply the chain rule for logarithms and conclude that

$$f'(x) = 3\frac{2x}{x^2 + 1} = \frac{6x}{x^2 + 1}$$

For practice, convince yourself that the final answer would have been the same if you had not made the initial simplification of $f(x)$.

The Proof that $\frac{d}{dx}(\ln x) = \frac{1}{x}$

The proof that $\frac{d}{dx}(\ln x) = \frac{1}{x}$ is based on the fact that $\left(1 + \frac{1}{n}\right)^n$ approaches e as n increases (or decreases) without bound. To derive the formula for the derivative of $f(x) = \ln x$, form the difference quotient and rewrite it using the properties of logarithms as follows:

$$\frac{f(x + \Delta x) - f(x)}{\Delta x} = \frac{\ln(x + \Delta x) - \ln x}{\Delta x}$$

$$= \frac{1}{\Delta x} \ln\left(\frac{x + \Delta x}{x}\right)$$

$$= \ln\left(\frac{x + \Delta x}{x}\right)^{1/\Delta x}$$

$$= \ln\left(1 + \frac{\Delta x}{x}\right)^{1/\Delta x}$$

To find the derivative of $\ln x$, let Δx approach zero in the simplified difference quotient. This will be easier to do if you first let $n = \frac{x}{\Delta x}$. Then,

$$\frac{\Delta x}{x} = \frac{1}{n} \qquad \text{and} \qquad \frac{1}{\Delta x} = \frac{n}{x}$$

and so

$$\frac{f(x + \Delta x) - f(x)}{\Delta x} = \ln\left(1 + \frac{1}{n}\right)^{n/x} = \ln\left[\left(1 + \frac{1}{n}\right)^n\right]^{1/x}$$

As Δx approaches zero, $n = \frac{x}{\Delta x}$ increases or decreases without bound, depending on the sign of Δx. Since

$$\left(1 + \frac{1}{n}\right)^n \to e$$

as n increases or decreases without bound, it follows that

$$\ln\left[\left(1 + \frac{1}{n}\right)^n\right]^{1/x} \to \ln e^{1/x} = \frac{1}{x}$$

as n increases or decreases without bound. Hence,

$$\frac{d}{dx}(\ln x) = \frac{1}{x}$$

and the proof is complete.

Exponential Functions

To find the formula for the derivative of e^x, differentiate both sides of the equation

$$\ln e^x = x$$

with respect to x, using the chain rule for logarithms to differentiate $\ln e^x$. You get

$$\frac{d/dx(e^x)}{e^x} = 1 \quad \text{or} \quad \frac{d}{dx}(e^x) = e^x$$

That is, e^x is its own derivative!

The Derivative of e^x

$$\frac{d}{dx}(e^x) = e^x$$

Combining the formula for the derivative of e^x with the chain rule gives the following formula for the derivative of $e^{h(x)}$, where h is a differentiable function of x.

The Chain Rule for Exponential Functions

$$\frac{d}{dx}[e^{h(x)}] = h'(x)e^{h(x)}$$

That is, to compute the derivative of $e^{h(x)}$, simply multiply $e^{h(x)}$ by the derivative of the exponent $h(x)$.

Here are two examples that illustrate the use of these rules.

EXAMPLE 4.4   Differentiate the function $f(x) = e^{x^2+1}$.

SOLUTION
By the chain rule,

$$f'(x) = \left[\frac{d}{dx}(x^2 + 1)\right]e^{x^2+1} = 2xe^{x^2+1}$$

EXAMPLE 4.5   Differentiate the function $f(x) = xe^{2x}$.

SOLUTION
By the product rule,

$$f'(x) = x\frac{d}{dx}(e^{2x}) + e^{2x}\frac{d}{dx}(x) = 2xe^{2x} + e^{2x} = (2x + 1)e^{2x}$$

Implicit Differentiation

In the next example, the chain rule for exponential functions is combined with the technique of implicit differentiation from Chapter 2, Section 6.

EXAMPLE 4.6 Find $\frac{dy}{dx}$ if $e^{xy} = 3xy^2$.

SOLUTION
Differentiate both sides of the equation with respect to x (using the product rule twice) to get

$$\left[\frac{d}{dx}(xy)\right]e^{xy} = 3x\left(2y\frac{dy}{dx}\right) + 3y^2$$

$$\left(x\frac{dy}{dx} + y\right)e^{xy} = 6xy\frac{dy}{dx} + 3y^2$$

Now solve for $\frac{dy}{dx}$ to get

$$xe^{xy}\frac{dy}{dx} + ye^{xy} = 6xy\frac{dy}{dx} + 3y^2$$

$$xe^{xy}\frac{dy}{dx} - 6xy\frac{dy}{dx} = 3y^2 - ye^{xy}$$

$$(xe^{xy} - 6xy)\frac{dy}{dx} = 3y^2 - ye^{xy}$$

or

$$\frac{dy}{dx} = \frac{3y^2 - ye^{xy}}{xe^{xy} - 6xy} = \frac{y(3y - e^{xy})}{x(e^{xy} - 6y)}$$

To simplify the answer further, you may substitute $e^{xy} = 3xy^2$ (from the original equation) to get

$$\frac{dy}{dx} = \frac{y(3y - 3xy^2)}{x(3xy^2 - 6y)} = \frac{3y^2(1 - xy)}{3xy(xy - 2)} = \frac{y(1 - xy)}{x(xy - 2)}$$

Exponential Growth

In Section 2 you saw that a quantity $Q(t)$ that increases according to a law of the form $Q(t) = Q_0e^{kt}$, where Q_0 and k are positive constants, is said to experience exponential growth. The next example shows that if a quantity grows exponentially, its rate of growth is proportional to its size and its percentage rate of growth is constant. In Problems 34 and 35 you will derive the analogous results for exponential decay. The converse of these facts will be established in Chapter 8.

EXAMPLE 4.7   Suppose that $Q(t)$ grows exponentially.

(a) Show that the rate of change of Q with respect to t is proportional to its size.

(b) Show that the percentage rate of change of Q with respect to t is constant.

SOLUTION

If $Q(t)$ grows exponentially, it is given by a function of the form $Q(t) = Q_0 e^{kt}$, where Q_0 and k are positive constants.

(a) The rate of change of Q with respect to t is the derivative

$$Q'(t) = kQ_0 e^{kt} = kQ(t)$$

This says that the rate of change, $Q'(t)$, is proportional to $Q(t)$ itself and that the constant k that appears in the exponent of $Q(t)$ is the constant of proportionality.

(b) The percentage rate of change of Q with respect to t is

$$100\,\frac{Q'(t)}{Q(t)} = 100\frac{kQ(t)}{Q(t)} = 100k$$

This says that the constant k which appears in the exponent of the function $Q(t) = Q_0 e^{kt}$ is the percentage rate of change of Q with respect to t (expressed as a decimal). Note that this is consistent with what you already know about compound interest; namely that if interest is compounded continuously, the balance after t years is $B(t) = Pe^{rt}$, where r is the interest rate expressed as a decimal.

Exponential Growth

> If $Q(t)$ grows exponentially according to the law $Q(t) = Q_0 e^{kt}$, then
>
> 1. The rate of change of Q with respect to t is proportional to its size. In particular,
>
>   Rate of change $= Q'(t) = kQ(t)$
>
> 2. The percentage rate of change of Q with respect to t is constant. In particular,
>
>   Percentage rate of change $= 100\dfrac{Q'(t)}{Q(t)} = 100k$

Optimization

Now that you know how to differentiate exponential functions, you can use calculus to find maximum and minimum values of functions involving powers of e. Here is an example.

EXAMPLE 4.8

The consumer demand for a certain commodity is $D(p) = 5{,}000e^{-0.02p}$ units per month when the market price is p dollars per unit. Determine the market price that will result in the greatest consumer expenditure.

SOLUTION

The consumer expenditure $E(p)$ for the commodity is the price per unit times the number of units sold. That is,

$$E(p) = pD(p) = 5{,}000pe^{-0.02p}$$

Since only nonnegative values of p are meaningful in this context, the goal is to find the absolute maximum of $E(p)$ on the interval $p \geq 0$.

The derivative of E is

$$E'(p) = 5{,}000(-0.02pe^{-0.02p} + e^{-0.02p}) = 5{,}000(1 - 0.02p)e^{-0.02p}$$

Since $e^{-0.02p}$ is never zero, $E'(p) = 0$ if and only if

$$1 - 0.02p = 0 \qquad \text{or} \qquad p = \frac{1}{0.02} = 50$$

The easiest way to verify that $p = 50$ actually gives the absolute maximum of $E(p)$ on the interval $p \geq 0$ is to check the sign of $E'(p)$ for $0 < p < 50$ and for $p > 50$. Since $5{,}000e^{-0.02p}$ is always positive, it follows that $E'(p)$ is positive if $0 < p < 50$ and negative if $p > 50$. Hence, $E(p)$ is increasing for $0 < p < 50$, decreasing for $p > 50$, and has an absolute maximum when $p = 50$, as shown in Figure 4.1. Thus, consumer expenditure will be greatest when the market price is \$50 per unit.

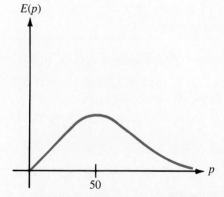

Figure 4.1 Consumer expenditure: $E(p) = 5{,}000pe^{-0.02p}$.

Curve Sketching

In the next example, calculus is used to obtain a detailed sketch of the function $f(x) = \dfrac{1}{\sqrt{2\pi}} e^{-x^2/2}$. This function is known as the **standard normal probability density function** and is one of the most important functions in probability and statistics. Its graph is the famous "bell-shaped curve" that is used by social scientists to describe the distributions of many quantities including people's heights and IQ scores.

EXAMPLE 4.9   Determine where the function

$$f(x) = \frac{1}{\sqrt{2\pi}} e^{-x^2/2}$$

is increasing, decreasing, concave upward, and concave downward. Find the relative extrema and inflection points and draw the graph.

SOLUTION
The first derivative is

$$f'(x) = \frac{-x}{\sqrt{2\pi}} e^{-x^2/2}$$

Since $e^{-x^2/2}$ is always positive, $f'(x)$ is zero if and only if $x = 0$. Hence the corresponding point

$$\left(0, \frac{1}{\sqrt{2\pi}}\right) \simeq (0, 0.40)$$

is the only first-order critical point.

By the product rule, the second derivative is

$$f''(x) = \frac{x^2}{\sqrt{2\pi}} e^{-x^2/2} - \frac{1}{\sqrt{2\pi}} e^{-x^2/2} = \frac{1}{\sqrt{2\pi}} (x^2 - 1) e^{-x^2/2}$$

which is zero if $x = \pm 1$. Hence the corresponding points

$$\left(1, \frac{e^{-1/2}}{\sqrt{2\pi}}\right) \simeq (1, 0.24) \qquad \text{and} \qquad \left(-1, \frac{e^{-1/2}}{\sqrt{2\pi}}\right) \simeq (-1, 0.24)$$

are the second-order critical points.

Plot the critical points and then check the signs of the first and second derivatives on each of the intervals defined by the x coordinates of these points.

Interval	f'(x)	f"(x)	Increase or Decrease of f	Concavity of f	Shape of f
$x < -1$	+	+	Increasing	C-up	
$-1 < x < 0$	+	−	Increasing	C-down	
$0 < x < 1$	−	−	Decreasing	C-down	
$x > 1$	−	+	Decreasing	C-up	

Complete the graph as shown in Figure 4.2 by connecting the critical points with a curve of appropriate shape on each interval. Notice that the graph has no x intercepts since $e^{-x^2/2}$ is always positive and that the graph approaches the x axis as $|x|$ increases without bound since $e^{-x^2/2}$ approaches zero.

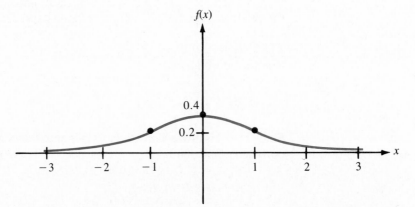

Figure 4.2 The standard normal density function:

$$f(x) = \frac{1}{\sqrt{2\pi}}\, e^{-x^2/2}.$$

Logarithmic Differentiation

Sometimes you can simplify the work involved in differentiating a function if you first take its logarithm. This technique, called **logarithmic differentiation,** is illustrated in the following example.

EXAMPLE 4.10 Differentiate the function $f(x) = \dfrac{\sqrt[3]{x+1}}{x-1}$.

SOLUTION
You could do this problem using the quotient rule, but the resulting computation would be somewhat tedious. (Try it.)

A more efficient approach is to work with the natural logarithm of f:

$$\ln f(x) = \frac{1}{3} \ln (x + 1) - \ln (x - 1)$$

(Notice that by introducing the logarithm, you eliminate both the quotient and the cube root.)

Now use the chain rule for logarithms to differentiate both sides of this equation to get

$$\frac{f'(x)}{f(x)} = \frac{1}{3(x + 1)} - \frac{1}{x - 1} = \frac{(x - 1) - 3(x + 1)}{3(x + 1)(x - 1)} = \frac{-2(x + 2)}{3(x + 1)(x - 1)}$$

and solve for $f'(x)$ by multiplying both sides of this equation by $f(x) = \dfrac{\sqrt[3]{x + 1}}{x - 1}$ to get

$$f'(x) = \frac{-2(x + 2)}{3(x + 1)(x - 1)} \frac{\sqrt[3]{x + 1}}{x - 1} = \frac{-2(x + 2)}{3(x + 1)^{2/3}(x - 1)^2}$$

The Proof of the Power Rule

In the next example, logarithmic differentiation is used to prove the power rule from Chapter 2, Section 2.

EXAMPLE 4.11 Use logarithmic differentiation to prove the power rule:

$$\frac{d}{dx}(x^n) = nx^{n-1}$$

SOLUTION
Let $f(x) = x^n$. Then,

$$\ln f(x) = n \ln x$$

and differentiation gives

$$\frac{f'(x)}{f(x)} = \frac{n}{x}$$

or

$$f'(x) = \frac{n}{x}f(x) = \frac{n}{x}(x^n) = nx^{n-1}$$

and the power rule is proved.

Problems

In Problems 1 through 26, differentiate the given function.

1. $f(x) = e^{5x}$

2. $f(x) = 3e^{4x+1}$

3. $f(x) = e^{x^2 + 2x - 1}$

4. $f(x) = e^{1/x}$

5. $f(x) = 30 + 10e^{-0.05x}$

6. $f(x) = x^2 e^x$

7. $f(x) = (x^2 + 3x + 5)e^{6x}$

8. $f(x) = xe^{-x^2}$

9. $f(x) = \dfrac{x}{e^x}$

10. $f(x) = \dfrac{x^2}{e^x}$

11. $f(x) = (1 - 3e^x)^2$

12. $f(x) = \sqrt{1 + e^x}$

13. $f(x) = e^{\sqrt{3x}}$

14. $f(x) = e^{-1/(2x)}$

15. $f(x) = \ln x^3$

16. $f(x) = \ln 2x$

17. $f(x) = \ln (x^2 + 5x - 2)$

18. $f(x) = \ln \sqrt{x^2 + 1}$

19. $f(x) = x^2 \ln x$

20. $f(x) = x \ln \sqrt{x}$

21. $f(x) = \dfrac{\ln x}{x}$

22. $f(x) = \dfrac{x}{\ln x}$

23. $f(x) = \ln\left(\dfrac{x + 1}{x - 1}\right)$

24. $f(x) = e^x \ln x$

25. $f(x) = \ln e^{2x}$

26. $f(x) = e^{\ln x + 2 \ln 3x}$

In Problems 27 through 30, find $\dfrac{dy}{dx}$ by implicit differentiation. Simplify your answers.

27. $e^{xy} = xy^3$

28. $e^{x/y} = 3y^5$

29. $\ln \dfrac{y}{x} = x^2 y^3$

30. $\ln xy^2 = \dfrac{x}{3y}$

Population growth 31. It is projected that t years from now, the population of a certain country will be $P(t) = 50e^{0.02t}$ million.
 (a) At what rate will the population be changing with respect to time 10 years from now?
 (b) At what percentage rate will the population be changing with respect to time t years from now? Does this percentage rate depend on t or is it constant?

Compound interest 32. Money is deposited in a bank offering interest at an annual rate of 6 percent compounded continuously. Find the percentage rate of change of the balance with respect to time.

Depreciation 33. A certain industrial machine depreciates so that its value after t years is $Q(t) = 20,000e^{-0.4t}$ dollars.

(a) At what rate is the value of the machine changing with respect to time after 5 years?

(b) At what percentage rate is the value of the machine changing with respect to time after t years? Does this percentage rate depend on t or is it constant?

Exponential decay

34. Show that if a quantity decreases exponentially, its rate of decrease is proportional to its size.

Exponential decay

35. Show that if a quantity decreases exponentially, its percentage rate of change is constant.

The spread of an epidemic

36. Public health records indicate that t weeks after the outbreak of a certain form of influenza, approximately $Q(t) = \dfrac{80}{4 + 76e^{-1.2t}}$ thousand people had caught the disease. At what rate was the disease spreading at the end of the 2nd week?

Population growth

37. According to a logistic model based on the assumption that the earth can support no more than 40 billion people, the world's population (in billions) t years after 1960 will be approximately $P(t) = \dfrac{40}{1 + 12e^{-0.08t}}$.

(a) If this model is correct, at what rate will the world's population be increasing with respect to time in 1995?

(b) If this model is correct, at what percentage rate will the world's population be increasing with respect to time in 1995?

Newton's law of heating

38. A cool drink is removed from a refrigerator on a hot summer day and placed in a room whose temperature is 30° Celsius. According to a law of physics, the temperature of the drink t minutes later is given by a function of the form $f(t) = 30 - Ae^{-kt}$. Show that the rate of change of the temperature of the drink with respect to time is proportional to the difference between the temperature of the room and that of the drink.

Marginal analysis

39. The mathematics editor at a major publishing house estimates that if x thousand complimentary copies are distributed to professors, the first-year sales of a certain new text will be $f(x) = 20 - 15e^{-0.2x}$ thousand copies. Currently, the editor is planning to distribute 10,000 complimentary copies.

(a) Use marginal analysis to estimate the increase in first-year sales that will result if 1,000 additional complimentary copies are distributed.

(b) Calculate the actual increase in first-year sales that will result from the distribution of the additional 1,000 complimentary copies. Is the estimate in part (a) a good one?

Consumer expenditure

40. The consumer demand for a certain commodity is $D(p) = 3,000e^{-0.01p}$ units per month when the market price is p dollars per unit. Express consumers' total monthly expenditure for the commodity as a function

of p and determine the market price that will result in the greatest consumer expenditure.

Profit maximization 41. A manufacturer can produce radios at a cost of $5 apiece and estimates that if they are sold for x dollars apiece, consumers will buy approximately $1,000e^{-0.1x}$ radios per week. At what price should the manufacturer sell the radios to maximize profit?

In Problems 42 through 51, determine where the given function is increasing, decreasing, concave upward, and concave downward. Find the relative extrema and inflection points and draw the graph.

42. $f(x) = xe^x$

43. $f(x) = xe^{-x}$

44. $f(x) = xe^{2-x}$

45. $f(x) = e^{-x^2}$

46. $f(x) = x^2e^{-x}$

47. $f(x) = e^x + e^{-x}$

48. $f(x) = \dfrac{6}{1 + e^{-x}}$

49. $f(x) = x - \ln x$ (for $x > 0$)

50. $f(x) = (\ln x)^2$ (for $x > 0$)

51. $f(x) = \dfrac{\ln x}{x}$ (for $x > 0$)

In Problems 52 through 60, use logarithmic differentiation to find $f'(x)$.

52. $f(x) = \sqrt{\dfrac{x + 1}{x - 1}}$

53. $f(x) = \dfrac{(x + 2)^5}{(3x - 5)^6}$

54. $f(x) = (x + 1)^3(x - 6)^2(2x + 1)$ 55. $f(x) = 2^x$

56. $f(x) = a^x$ (for $a > 0$)

57. $f(x) = x^x$ (for $x > 0$)

58. $f(x) = (x^x)(e^x)$ (for $x > 0$)

59. $f(x) = (2x)^{\ln x}$ (for $x > 0$)

60. $f(x) = xe^{3x}$ (for $x > 0$)

Population growth 61. It is estimated that t years from now the population of a certain country will be $P(t) = \dfrac{160}{1 + 8e^{-0.01t}}$ million. When will the population be growing most rapidly?

The spread of an epidemic 62. An epidemic spreads through a community so that t weeks after its outbreak, the number of residents who have been infected is given by a function of the form $f(t) = \dfrac{A}{1 + Ce^{-kt}}$, where A is the total number of susceptible residents. Show that the epidemic is spreading most rapidly when half of the susceptible residents have been infected.

Percentage rate of change 63. Show that the percentage rate of change of f with respect to x is $100\dfrac{d}{dx}[\ln f(x)]$.

In Problems 64 and 65, use the formula from Problem 63 to find the specified percentage rate of change.

Exponential growth 64. A quantity grows exponentially according to the law $Q(t) = Q_0 e^{kt}$. Find the percentage rate of change of Q with respect to t.

Population growth 65. It is projected that x years from now the population of a certain town will be approximately $P(x) = 5,000\sqrt{x^2 + 4x + 19}$. At what percentage rate will the population be changing with respect to time 3 years from now?

5 Compound Interest

In this section, you will learn more about the concept of compound interest, which was introduced in Section 1. To begin the discussion, here is a summary of what you should already know.

When money is invested it (usually) earns **interest.** The amount of money invested is called the **principal.** Interest that is computed on the principal alone is called **simple interest,** and the balance after t years is given by the following formula.

Simple Interest

If P dollars are invested at an annual simple interest rate r (expressed as a decimal), the balance $B(t)$ after t years will be

$$B(t) = P(1 + rt) \quad \text{dollars}$$

Interest that is computed on the principal plus the previous interest is called **compound interest.** If the annual interest rate is r and interest is compounded k times per year, the year is divided into k equal **interest periods** and the interest rate during each is $\frac{r}{k}$. The balance at the end of t years is given by the following formula.

Compound Interest

If P dollars are invested at an annual interest rate r and interest is compounded k times per year, the balance $B(t)$ after t years will be

$$B(t) = P\left(1 + \frac{r}{k}\right)^{kt} \quad \text{dollars}$$

If the number of times interest is compounded per year is allowed to increase without bound, the interest is said to be **compounded continuously.** In this case, the balance at the end of t years is given by the following formula.

Continuously Compounded Interest

If P dollars are invested at an annual interest rate r and interest is compounded continuously, the balance $B(t)$ after t years will be

$$B(t) = Pe^{rt} \quad \text{dollars}$$

The use of these formulas is illustrated in the next example.

EXAMPLE 5.1 Suppose $1,000 is invested at an annual interest rate of 8 percent. Compute the balance after 10 years in each of the following cases:

(a) The interest is simple interest.
(b) The interest is compounded quarterly.
(c) The interest is compounded continuously.

SOLUTION
(a) Use the formula $B(t) = P(1 + rt)$ with $P = 1,000$, $r = 0.08$, and $t = 10$ to get

$$B(10) = 1,000(1.8) = \$1,800$$

(b) Use the formula $B(t) = P\left(1 + \dfrac{r}{k}\right)^{kt}$ with $P = 1,000$, $r = 0.08$, $k = 4$, and $t = 10$ to get

$$B(10) = 1,000(1.02)^{40} \simeq \$2,208.04$$

(c) Use the formula $B(t) = Pe^{rt}$ with $P = 1,000$, $r = 0.08$, and $t = 10$ to get

$$B(10) = 1,000e^{0.8} \simeq \$2,225.54$$

Doubling Time

Investors often want to know how long it will take for a given investment to grow to a particular size. In the next example, you will see how to use the compound interest formulas to calculate the time it takes for money to double.

EXAMPLE 5.2 How quickly will money double if it is invested at an annual interest rate of 8 percent and interest is compounded

(a) quarterly   (b) continuously

SOLUTION

(a) The balance after t years is

$$B(t) = P\left(1 + \frac{r}{k}\right)^{kt} = P\left(1 + \frac{0.08}{4}\right)^{4t} = P(1.02)^{4t}$$

The goal is to find the value of t for which $B(t) = 2P$, that is, for which

$$2P = P(1.02)^{4t} \quad \text{or} \quad 2 = (1.02)^{4t}$$

Take the natural logarithm of each side and solve for t to get

$$\ln 2 = 4t \ln 1.02$$

or

$$t = \frac{\ln 2}{4 \ln 1.02} \simeq 8.75 \text{ years}$$

That is, at 8 percent compounded quarterly, money will double in approximately 8.75 years.

(b) This time, use the formula

$$B(t) = Pe^{rt} = Pe^{0.08t}$$

and solve the equation

$$2P = Pe^{0.08t}$$

to get

$$2 = e^{0.08t} \quad \ln 2 = 0.08t \quad \text{or} \quad t = \frac{\ln 2}{0.08} \simeq 8.66 \text{ years}$$

That is, at 8 percent compounded continuously, money will double in approximately 8.66 years.

Effective Interest Rate

Money deposited in a bank offering interest at an annual rate of 6 percent will increase in value by more than 6 percent in a year if the interest is compounded more than once. This is because interest compounded during one period will itself earn interest during subsequent periods. The actual percentage by which an investment grows during a year is called the **effective interest rate,** while the corresponding annual compound interest rate is called the **nominal interest rate.** In other words, the effective interest rate is the *simple* interest rate that is equivalent to the nominal compound interest rate. In the next example, you will see how to derive a formula for the effective interest rate.

EXAMPLE 5.3   Find a formula for the effective interest rate if the nominal rate is r and interest is compounded k times per year.

SOLUTION
If P dollars are invested at a nominal rate r and interest is compounded k times per year, the balance at the end of 1 year will be

$$B = P\left(1 + \frac{r}{k}\right)^k$$

On the other hand, if x is the simple interest rate, the corresponding balance at the end of 1 year will be

$$B = P(1 + x)$$

The goal is to find the value of x for which these two expressions for the balance are equal, that is,

$$P\left(1 + \frac{r}{k}\right)^k = P(1 + x)$$

$$\left(1 + \frac{r}{k}\right)^k = 1 + x$$

or

$$x = \left(1 + \frac{r}{k}\right)^k - 1$$

Using a similar argument, you can derive a formula for the effective interest rate when interest is compounded continuously. This derivation is left as a problem for you to do. For reference, here is a summary of the situation.

Effective Interest Rate

If interest is compounded k times per year at an annual interest rate r, then

$$\text{Effective interest rate} = \left(1 + \frac{r}{k}\right)^k - 1$$

If interest is compounded continuously at an annual interest rate r, then

$$\text{Effective interest rate} = e^r - 1$$

EXAMPLE 5.4   One bank offers interest at an annual rate of 6.1 percent compounded quarterly, and a competing bank offers interest at an annual rate of 6 percent compounded continuously. Compare the effective interest rates at the two banks.

SOLUTION

The effective interest rate at the first bank is

$$\left(1 + \frac{0.061}{4}\right)^4 - 1 \simeq 0.0624$$

or 6.24 percent, and the effective interest rate at the second bank is only

$$e^{0.06} - 1 \simeq 0.0618$$

or 6.18 percent.

Present Value

The **present value of B dollars payable t years from now** is the amount P that should be invested today so that it will be worth B dollars at the end of t years. To derive a formula for the present value, simply solve the compound interest formula for the principal P. For example, if interest is compounded k times per year, the balance after t years is

$$B = P\left(1 + \frac{r}{k}\right)^{kt}$$

and the present value of B dollars payable t years from now is

$$P = B\left(1 + \frac{r}{k}\right)^{-kt}$$

If interest is compounded continuously, the balance is

$$B = Pe^{rt}$$

and the present value is

$$P = Be^{-rt}$$

Present Value of Future Money

If interest is compounded k times per year at an annual interest rate r, the present value of B dollars payable t years from now is

$$P = B\left(1 + \frac{r}{k}\right)^{-kt} \quad \text{dollars}$$

If interest is compounded continuously at an annual interest rate r, the present value of B dollars payable t years from now is

$$P = Be^{-rt} \quad \text{dollars}$$

The concept of present value is illustrated in the next example.

EXAMPLE 5.5   How much should you invest now at an annual interest rate of 8 percent so that your balance 20 years from now will be $10,000 if the interest is compounded

(a) quarterly   (b) continuously

SOLUTION
(a) The amount you should invest is the present value of $10,000 payable 20 years from now. Using the formula $P = B\left(1 + \dfrac{r}{k}\right)^{-kt}$ with $B = 10,000$, $r = 0.08$, $k = 4$, and $t = 20$, you find that this amount is

$$P = 10,000(1.02)^{-80} \approx \$2,051.10$$

(b) Using the present-value formula $P = Be^{-rt}$ with $B = 10,000$, $r = 0.08$, and $t = 20$, you find that the amount you should invest is

$$P = 10,000e^{-1.6} \approx \$2,018.97$$

The Amount of an Annuity

An **annuity** is a sequence of equal payments or deposits made at regular intervals over a specified period of time. Once a deposit is made, it earns interest at a fixed rate until the expiration or **term** of the annuity. The total amount of money (deposits plus interest) accumulated in this way is known as the **amount** of the annuity. The next example illustrates how to use the compound interest formula to calculate the amount of an annuity.

EXAMPLE 5.6   Each year on January first, you deposit $1,000 in an account that pays interest at an annual rate of 8 percent compounded continuously. How much will you have in the account at the end of 3 years (just before you make your 4th deposit)?

SOLUTION
The three deposits of $1,000 and the value of each (at 8 percent compounded continuously) at the end of the 3-year term are shown in Figure 5.1. The amount of the annuity at the end of the term is the sum

$$\text{Amount} = 1,000e^{0.08(3)} + 1,000e^{0.08(2)} + 1,000e^{0.08(1)}$$
$$= 1,000(e^{0.24} + e^{0.16} + e^{0.08}) \approx \$3,528.05$$

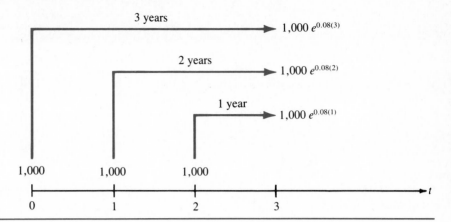

Figure 5.1 The annuity for Example 5.6.

The Present Value of an Annuity

The **present value of an annuity** that consists of a periodic sequence of payments over a specified term is the amount of money that must be deposited today to permit periodic withdrawals generating the same sequence of payments over the specified term, after which nothing will be left in the account. The concept and its calculation are illustrated in the next example.

EXAMPLE 5.7 A bank pays interest at the annual rate of 8 percent compounded quarterly. How much should you deposit today so that you will be able to withdraw $500 at the end of each of the next 3 years, after which nothing will be left in the account? (That is, what is the present value of an annuity that pays $500 at the end of each year for a term of 3 years if the annual interest rate is 8 percent compounded quarterly?)

SOLUTION

The present value of the annuity is the sum of the present values of the 3 withdrawals. Since $r = 0.08$ and $k = 4$, the appropriate present-value formula is

$$P(t) = B\left(1 + \frac{r}{k}\right)^{-kt} = B\left(1 + \frac{0.08}{4}\right)^{-4t} = B(1.02)^{-4t}$$

The withdrawals and corresponding present values are shown in Figure 5.2. The present value of the annuity is the sum

$$\text{Present value} = 500(1.02)^{-4(1)} + 500(1.02)^{-4(2)} + 500(1.02)^{-4(3)}$$

$$= 500[(1.02)^{-4} + (1.02)^{-8} + (1.02)^{-12}] \approx \$1{,}282.91$$

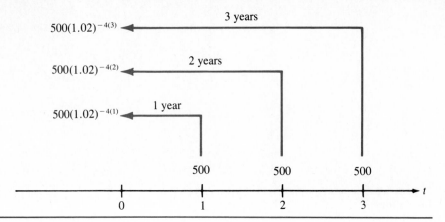

Figure 5.2 The annuity for Example 5.7.

Optimal Holding Time

Suppose you own an asset whose value increases with time. The longer you hold the asset, the more it will be worth. However, if there comes a time after which money invested at the prevailing interest rate grows more quickly than the value of the asset, you will be better off selling the asset and investing the proceeds. You determine the most advantageous time to sell the asset and invest the proceeds by comparing the prevailing interest rate with the percentage rate of growth of the value of the asset. The technique is illustrated in the next example.

EXAMPLE 5.8 Suppose you own a parcel of land whose market price t years from now will be $V(t) = 20,000e^{\sqrt{t}}$ dollars. If the prevailing interest rate remains constant at 7 percent compounded continuously, when will it be most advantageous to sell the land?

SOLUTION
The percentage rate of change of the market price of the land (expressed in decimal form) is

$$\frac{V'(t)}{V(t)} = \frac{10,000t^{-1/2}e^{\sqrt{t}}}{20,000e^{\sqrt{t}}} = \frac{1}{2\sqrt{t}}$$

This will be equal to the prevailing interest rate when

$$\frac{1}{2\sqrt{t}} = 0.07 \qquad \text{or} \qquad t = \left[\frac{1}{2(0.07)}\right]^2 \approx 51.02$$

Moreover,

$$\frac{1}{2\sqrt{t}} > 0.07 \qquad \text{when } 0 < t < 51.02$$

and

$$\frac{1}{2\sqrt{t}} < 0.07 \qquad \text{when } t > 51.02$$

which implies that the percentage rate of growth of the value of the land is greater than the prevailing interest rate when $0 < t < 51.02$ and less than the prevailing interest rate when $t > 51.02$. It follows that you should sell the land 51.02 years from now and invest the proceeds at the prevailing interest rate of 7 percent.

___

Economists have a different way of solving problems like the one in Example 5.8. They determine the optimal time to sell an asset by finding the time at which the *present value* of the selling price of the asset is greatest. That is, they maximize today's dollar equivalent of the selling price. Here is Example 5.8 solved as an economist would do it. Notice that the final answer is exactly the same as the one we obtained before.

___

EXAMPLE 5.9 Suppose you own a parcel of land whose market price t years from now will be $V(t) = 20{,}000e^{\sqrt{t}}$ dollars. If the prevailing interest rate remains constant at 7 percent compounded continuously, when will the present value of the market price of the land be greatest?

SOLUTION
In t years, the market price of the land will be $V(t) = 20{,}000e^{\sqrt{t}}$. The present value of this price is

$$P(t) = V(t)e^{-0.07t} = 20{,}000e^{\sqrt{t}}e^{-0.07t} = 20{,}000e^{\sqrt{t}-0.07t}$$

The goal is to maximize $P(t)$ for $t \geq 0$. The derivative of P is

$$P'(t) = 20{,}000e^{\sqrt{t}-0.07t}\left(\frac{1}{2\sqrt{t}} - 0.07\right)$$

which is zero if and only if

$$\frac{1}{2\sqrt{t}} - 0.07 = 0 \qquad \text{or} \qquad t = \left[\frac{1}{2(0.07)}\right]^2 \approx 51.02$$

(and which is undefined when $t = 0$).

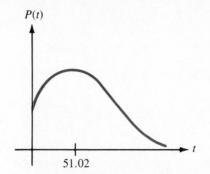

$P(t)$

Figure 5.3 Present value:
$P(t) = 20,000e^{\sqrt{t}-0.07t}$.

51.02

   Since $P'(t)$ is positive if $0 < t < 51.02$ and negative if $t > 51.02$, it follows that the graph of P is increasing for $0 < t < 51.02$ and decreasing for $t > 51.02$ as shown in Figure 5.3 and that $t = 51.02$ is the time at which the present value will be greatest.

Problems

1. Suppose \$5,000 is invested at an annual interest rate of 5 percent. Compute the balance after 20 years in each of the following cases:
   (a) The interest is simple interest.
   (b) The interest is compounded semiannually.
   (c) The interest is compounded continuously.

2. Suppose \$5,000 is invested at an annual interest rate of 6 percent. Compute the balance after 20 years in each of the following cases:
   (a) The interest is simple interest.
   (b) The interest is compounded semiannually.
   (c) The interest is compounded continuously.

3. How quickly will money double if it is invested at an annual interest rate of 12 percent and interest is compounded
   (a) quarterly   (b) continuously

4. How quickly will money double if it is invested at an annual simple interest rate of 8 percent?

5. Derive a formula for the amount of time it takes for money to double if the annual interest rate is r and interest is compounded k times per year.

6. Derive a formula for the amount of time it takes for money to double if the annual interest rate is r and interest is compounded continuously.

7. How quickly will money triple if it is invested at an annual interest rate of 6 percent compounded semiannually?

8. How quickly will money triple if it is invested at an annual interest rate of 6 percent compounded continuously?

9. Derive a formula for the amount of time it takes for money to triple if the annual interest rate is r and interest is compounded k times per year.

10. Derive a formula for the amount of time it takes for money to triple if the annual interest rate is r and interest is compounded continuously.

11. Determine how quickly $1,000 will grow to $2,500 if the annual interest rate is 6 percent and interest is compounded
    (a) quarterly   (b) continuously

12. Determine how quickly $600 will grow to $1,000 if the annual interest rate is 10 percent and interest is compounded
    (a) semiannually   (b) continuously

13. Find the effective interest rate if the nominal interest rate is 6 percent per year and interest is compounded
    (a) quarterly   (b) continuously

14. Which investment has the greater effective interest rate: 8.2 percent per year compounded quarterly or 8.1 percent per year compounded continuously?

15. Which investment has the greater effective interest rate: 10.25 percent per year compounded semiannually or 10.20 percent per year compounded continuously?

16. Derive the formula for the effective interest rate if the nominal rate is r and interest is compounded continuously.

17. A bank compounds interest quarterly. What (nominal) interest rate does it offer if $1,000 grows to $2,203.76 in 8 years?

18. A bank compounds interest continuously. What (nominal) interest rate does it offer if $1,000 grows to $2,054.44 in 12 years?

19. A certain bank offers an interest rate of 6 percent per year compounded annually. A competing bank compounds its interest continuously. What (nominal) interest rate should the competing bank offer so that the effective interest rates of the two banks will be equal?

20. How much should you invest now at an annual interest rate of 6 percent so that your balance 10 years from now will be $5,000 if interest is compounded
    (a) quarterly   (b) continuously

21. How much should you invest now at an annual interest rate of 8 percent so that your balance 15 years from now will be $20,000 if interest is compounded
    (a) quarterly   (b) continuously

22. Find the present value of $8,000 payable 10 years from now if the annual interest rate is 6.25 percent and interest is compounded
    (a) semiannually   (b) continuously

The amount of an annuity

23. At the end of each year you deposit $2,000 in an account that pays interest at an annual rate of 10 percent compounded continuously. How much will you have in your account immediately after you make your 3rd deposit?

The amount of an annuity

24. At the end of each interest period you deposit $100 in an account that pays interest at an annual rate of 8 percent compounded quarterly. How much will you have in the account just after you make your 4th deposit?

The amount of an annuity

25. At the end of each interest period you deposit $50 in an account that pays interest at an annual rate of 6 percent compounded semiannually. How much will you have in the account just after you make your 4th deposit?

The present value of an annuity

26. A bank offers interest at an annual rate of 5 percent compounded continuously. How much should you deposit today so that you will be able to make withdrawals of $2,000 at the end of each of the next 3 years, after which nothing will be left in the account?

The present value of an annuity

27. How much should you invest now (at the beginning of an interest period) at an annual interest rate of 6 percent compounded annually to enable you to make withdrawals of $500 at the end of each of the next 4 years, after which nothing will be left in the account?

The present value of an annuity

28. Find the present value of an annuity that makes 4 annual payments of $1,000 (starting today) if the prevailing annual interest rate is 9 percent compounded continuously.

Optimal holding time

29. Suppose you own a parcel of land whose value t years from now will be $V(t) = 8,000e^{\sqrt{t}}$ dollars. If the prevailing interest rate remains constant at 6 percent per year compounded continuously, when should you sell the land to maximize its present value?

Optimal holding time

30. Suppose your family owns a rare book whose value t years from now will be $V(t) = 200e^{\sqrt{2t}}$ dollars. If the prevailing interest rate remains constant at 6 percent per year compounded continuously, when will it be most advantageous for your family to sell the book and invest the proceeds?

Optimal holding time

31. Suppose you own a stamp collection that is currently worth $1,200 and whose value increases linearly at the rate of $200 per year. If the prevailing interest rate remains constant at 8 percent per year compounded continuously, when will it be most advantageous for you to sell the collection and invest the proceeds?

Chapter Summary and Review Problems

Important Terms, Symbols, and Formulas

Exponential function: $f(x) = a^x$

Laws of exponents: $a^r a^s = a^{r+s}$ $\dfrac{a^r}{a^s} = a^{r-s}$ $(a^r)^s = a^{rs}$

The number e: $e = \lim\limits_{n \to \infty} \left(1 + \dfrac{1}{n}\right)^n \approx 2.718$

Exponential growth: $Q(t) = Q_0 e^{kt}$

$$Q'(t) = kQ(t)$$

$$\dfrac{\text{Percentage}}{\text{rate of change}} = 100k$$

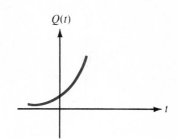

Exponential decay: $Q(t) = Q_0 e^{-kt}$

$$Q'(t) = -kQ(t)$$

$$\dfrac{\text{Percentage}}{\text{rate of change}} = -100k$$

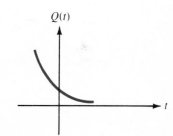

Learning curve: $Q(t) = B - Ae^{-kt}$

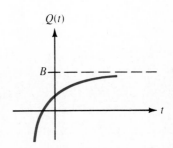

Logistic curve: $Q(t) = \dfrac{B}{1 + Ae^{-Bkt}}$

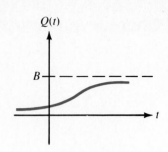

The natural logarithm: $y = \ln x$ if and only if $x = e^y$

Inverse relationship: $e^{\ln x} = x$ and $\ln e^x = x$

Properties of logarithms:

$$\ln uv = \ln u + \ln v \qquad \ln \frac{u}{v} = \ln u - \ln v \qquad \ln u^v = v \ln u$$

Differentiation formulas:

$$\frac{d}{dx}(\ln x) = \frac{1}{x} \qquad \text{and} \qquad \frac{d}{dx}[\ln h(x)] = \frac{h'(x)}{h(x)}$$

$$\frac{d}{dx}(e^x) = e^x \qquad \text{and} \qquad \frac{d}{dx}[e^{h(x)}] = h'(x)e^{h(x)}$$

Logarithmic differentiation

Simple interest formula: $B(t) = P(1 + rt)$

Interest compounded k times per year:

Balance $= B(t) = P\left(1 + \dfrac{r}{k}\right)^{kt}$

Effective rate $= \left(1 + \dfrac{r}{k}\right)^{k} - 1$

Present value $= B\left(1 + \dfrac{r}{k}\right)^{-kt}$

Continuously compounded interest:

Balance $= B(t) = Pe^{rt}$

Effective rate $= e^r - 1$

Present value $= Be^{-rt}$

The amount of an annuity; the present value of an annuity

Optimal holding time

Review Problems

1. Sketch the following functions:

   (a) $f(x) = 5e^{-x}$

   (b) $f(x) = 5 - 2e^{-x}$

   (c) $f(x) = 1 - \dfrac{6}{2 + e^{-3x}}$

2. Find $f(9)$ if $f(x) = 30 + Ae^{-kx}$, $f(0) = 50$, and $f(3) = 40$.

3. The value of a certain industrial machine is decreasing exponentially. If the machine was originally worth \$50,000 and was worth \$20,000 when it was 5 years old, how much will it be worth when it is 10 years old?

4. It is estimated that if x thousand dollars are spent on advertising, approximately $Q(x) = 50 - 40e^{-0.1x}$ thousand units of a certain commodity will be sold.

   (a) Sketch the relevant portion of this sales function.

   (b) How many units will be sold if no money is spent on advertising?

   (c) How many units will be sold if \$8,000 is spent on advertising?

   (d) How much should be spent on advertising to generate sales of 35,000 units?

   (e) According to this model, what is the most optimistic sales projection?

5. The daily output of a worker who has been on the job for t weeks is $Q(t) = 120 - Ae^{-kt}$ units. Initially the worker could produce 30 units per day, and after 8 weeks the worker could produce 80 units per day. How many units could the worker produce per day after 4 weeks?

6. It is estimated that t years from now the population of a certain country will be $P(t) = \dfrac{30}{1 + 2e^{-0.05t}}$ million.

   (a) Sketch the graph of $P(t)$.

   (b) What is the current population?

   (c) What will the population be 20 years from now?

   (d) What will happen to the population in the long run?

7. Evaluate the following expressions without using tables or a calculator.

   (a) $\ln e^5$ (b) $e^{\ln 2}$ (c) $e^{3\ln 4 - \ln 2}$

8. Solve for x.

   (a) $8 = 2e^{0.04x}$ (b) $5 = 1 + 4e^{-6x}$

   (c) $4 \ln x = 8$ (d) $5^x = e^3$

9. The number of bacteria in a certain culture grows exponentially. If 5,000 bacteria were intially present and 8,000 were present 10 minutes later, how long will it take for the number of bacteria to double?

10. Differentiate the following functions.

   (a) $f(x) = 2e^{3x+5}$ (b) $f(x) = x^2 e^{-x}$

   (c) $f(x) = \ln \sqrt{x^2 + 4x + 1}$ (d) $f(x) = x \ln x^2$

   (e) $f(x) = \dfrac{x}{\ln 2x}$

11. Use implicit differentiation to find $\dfrac{dy}{dx}$. Simplify your answers.

   (a) $e^{y/x} = 3xy^2$ (b) $\ln \dfrac{x}{y} = x^3 y^2$

12. An environmental study of a certain suburban community suggests that t years from now, the average level of carbon monoxide in the air will be $Q(t) = 4e^{0.03t}$ parts per million.
   (a) At what rate will the carbon monoxide level be changing with respect to time 2 years from now?
   (b) At what percentage rate will the carbon monoxide level be changing with respect to time t years from now? Does this percentage rate of change depend on t or is it constant?

13. A manufacturer can produce cameras at a cost of $40 apiece and estimates that if they are sold for p dollars apiece, consumers will buy approximately $D(p) = 800e^{-0.01p}$ cameras per week. At what price should the manufacturer sell the cameras to maximize profit?

14. Determine where the given function is increasing, decreasing, concave upward, and concave downward. Find the relative extrema and inflection points and draw the graph.

   (a) $f(x) = xe^{-2x}$ (b) $f(x) = e^x - e^{-x}$

   (c) $f(x) = \dfrac{4}{1 + e^{-x}}$ (d) $f(x) = \ln (x^2 + 1)$

15. Find $f'(x)$ by logarithmic differentiation.

   (a) $f(x) = \sqrt{(x^2 + 1)(x^2 + 2)}$

   (b) $f(x) = x^{x^2}$ (for $x > 0$)

   (c) $f(x) = (x^x)(2^x)$ (for $x > 0$)

16. How quickly will $2,000 grow to $5,000 when invested at an annual interest rate of 8 percent if interest is compounded
   (a) quarterly   (b) continuously

17. Which investment has the greater effective interest rate: 8.25 percent per year compounded quarterly or 8.20 percent per year compounded continuously?

18. How much should you invest now at an annual interest rate of 6.25 percent so that your balance 10 years from now will be $2,000 if interest is compounded
    (a) semiannually   (b) continuously

19. Find the present value of an annuity that makes 4 semiannual payments of $2,000 (starting today) if the prevailing annual interest rate is 10 percent compounded continuously.

20. Suppose you own a coin collection whose value t years from now will be $V(t) = 2,000e^{\sqrt{t}}$ dollars. If the prevailing interest rate remains constant at 7 percent per year compounded continuously, when will it be most advantageous for you to sell the collection and invest the proceeds?

Chapter

5 Antidifferentiation

1   Antiderivatives

In many problems, the derivative of a function is known and the goal is to find the function itself. For example, a sociologist who knows the rate at which the population is growing may wish to use this information to predict future population levels; a physicist who knows the speed of a moving body may wish to calculate the future position of the body; an economist who knows the rate of inflation may wish to estimate future prices.

The process of obtaining a function from its derivative is called **antidifferentiation** or **integration.**

Antiderivative

A function $F(x)$ for which

$$F'(x) = f(x)$$

for every x in the domain of f is said to be an antiderivative (or indefinite integral) of f.

276

Later in this section you will learn techniques you can use to find antiderivatives. Once you have found what you believe to be an antiderivative of a function, you can always check your answer by differentiating. You should get the original function back. Here is an example.

EXAMPLE 1.1   Verify that $F(x) = \frac{1}{3}x^3 + 5x + 2$ is an antiderivative of $f(x) = x^2 + 5$.

SOLUTION

$F(x)$ is an antiderivative of $f(x)$ if and only if $F'(x) = f(x)$. Differentiate F and you will find that

$$F'(x) = x^2 + 5 = f(x)$$

as required.

The General Antiderivative of a Function

A function has more than one antiderivative. For example, one antiderivative of the function $f(x) = 3x^2$ is $F(x) = x^3$, since

$$F'(x) = 3x^2 = f(x)$$

But so is $G(x) = x^3 + 12$, since the derivative of the constant 12 is zero and

$$G'(x) = 3x^2 = f(x)$$

In general, if F is one antiderivative of f, any function obtained by adding a constant to F is also an antiderivative of f. In fact, it turns out that all the antiderivatives of f can be obtained by adding constants to any one particular antiderivative of f.

The Antiderivatives of a Function

If F and G are antiderivatives of f, then there is a constant C such that

$$G(x) = F(x) + C$$

There is a simple geometric explanation for the fact that any two antiderivatives of the same function must differ by a constant. If F is an antiderivative of f, then $F'(x) = f(x)$. This says that for each value of x, $f(x)$ is the slope of the tangent to the graph of $F(x)$. If G is another antiderivative of f, the slope of its tangent is also $f(x)$. Hence the graph of G is "parallel" to the graph of F and can be obtained by translating the graph of F vertically. That is, there is some constant C for which

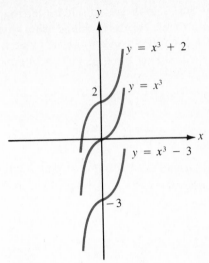

Figure 1.1 Some antiderivatives of $3x^2$.

$G(x) = F(x) + C$. The situation is illustrated in Figure 1.1, which shows several antiderivatives of the function $f(x) = 3x^2$.

Integral Notation

It is customary to write

$$\int f(x) \, dx = F(x) + C$$

to express the fact that every antiderivative of $f(x)$ is of the form $F(x) + C$. For example, you can express the fact that every antiderivative of $3x^2$ is of the form $x^3 + C$ by writing

$$\int 3x^2 \, dx = x^3 + C$$

The symbol $\int$ is called an **integral sign** and indicates that you are to find the most general form of the antiderivative of the function following it. The integral sign resembles an elongated "s." The "s" stands for "sum." In Chapter 7 you will see a surprising connection between antiderivatives and sums, which is so important it is known as the **fundamental theorem of calculus.**

In the expression $\int f(x) \, dx = F(x) + C$, the function $f(x)$ that is to be integrated is called the **integrand.** The (unspecified) constant C that is added to $F(x)$ to give the most general form of the antiderivative is known as the **constant of integration.**

The symbol dx that appears after the integrand may seem mysterious. Its role is to indicate that x is the variable with respect to which the integration is to be performed. Analogous notation is used if the function is expressed in terms of a variable other than x. For example, $\int 3t^2\, dt = t^3 + C$. In the expression $\int 3px^2\, dx$, the dx tells you that x rather than p is the variable. Thus, $\int 3px^2\, dx = px^3 + C$. (Do you see how to evaluate $\int 3px^2\, dp$?)

Here is a restatement of the definition of the indefinite integral (or antiderivative) in integral notation.

The Indefinite Integral

$$\int f(x)\, dx = F(x) + C$$

if and only if

$$F'(x) = f(x)$$

for every x in the domain of f.

Integration Rules

Integration is the reverse of differentiation. Hence many rules for integration can be obtained by stating the corresponding rules for differentiation in reverse.

The Power Rule for Integrals

According to the power rule for derivatives, $\dfrac{d}{dx}(x^n) = nx^{n-1}$. That is, to differentiate a power function, reduce its power by 1 and multiply by the original power. Stated in reverse, this rule says that to integrate a power function, *raise* its power by 1 and *divide* by the *new* power. Here is a more precise statement of the rule.

The Power Rule for Integrals

For $n \neq -1$,

$$\int x^n\, dx = \frac{1}{n + 1}x^{n+1} + C$$

That is, to integrate x^n (for $n \neq -1$), increase the power of x by 1 and divide by the new power.

To verify this rule, simply observe that

$$\frac{d}{dx}\left(\frac{1}{n+1}x^{n+1}\right) = \frac{n+1}{n+1}x^n = x^n$$

The rule holds for all values of n except, of course, for $n = -1$, in which case $\frac{1}{n+1}$ is undefined. This special case will be discussed shortly. In the meantime, see if you can think of an antiderivative of x^{-1} on your own.

The rule for integrating power functions is illustrated in the next example.

EXAMPLE 1.2   Find the following integrals:

(a) $\displaystyle\int x^{3/5}\, dx$ (b) $\displaystyle\int 1\, dx$ (c) $\displaystyle\int \frac{1}{\sqrt{x}}\, dx$

SOLUTION
(a) Increase the power of x by 1 and then divide by the new power to get

$$\int x^{3/5}\, dx = \frac{5}{8}x^{8/5} + C$$

(b) Since $1 = x^0$, it follows that

$$\int 1\, dx = x + C$$

(c) Begin by rewriting $\frac{1}{\sqrt{x}}$ as $x^{-1/2}$. Then,

$$\int \frac{1}{\sqrt{x}}\, dx = \int x^{-1/2}\, dx = 2x^{1/2} + C = 2\sqrt{x} + C$$

The Integral of $\frac{1}{x}$

Were you able to discover an antiderivative of x^{-1}? That is, were you able to think of a function whose derivative is $\frac{1}{x}$? The natural logarithm $\ln x$ is such a function, and so it appears that $\int \frac{1}{x}\, dx = \ln x + C$. Actually, this is true only when x is positive, since $\ln x$ is not defined for negative values of x. When x is negative, it turns out that the function $\ln |x|$ is an antiderivative of $\frac{1}{x}$. To see this, observe that if x is negative, $|x| = -x$ and so

$$\frac{d}{dx}(\ln |x|) = \frac{d}{dx}[\ln (-x)] = \left(\frac{1}{-x}\right)(-1) = \frac{1}{x}$$

Since $|x| = x$ when x is positive, the situation can be summarized using a single formula as follows.

The Integral of $\dfrac{1}{x}$

$$\int \frac{1}{x}\, dx = \ln|x| + C$$

The Integral of e^x

Integration of the exponential function e^x is trivial since e^x is its own derivative.

The Integral of e^x

$$\int e^x\, dx = e^x + C$$

The Constant Multiple Rule and Sum Rule

It is easy to rewrite the constant multiple rule and the sum rule for differentiation as rules for integration.

The Constant Multiple Rule for Integrals

For any constant c,

$$\int cf(x)\, dx = c\int f(x)\, dx$$

That is, the integral of a constant times a function is equal to the constant times the integral of the function.

The Sum Rule for Integrals

$$\int [f(x) + g(x)]\, dx = \int f(x)\, dx + \int g(x)\, dx$$

That is, the integral of a sum is the sum of the individual integrals.

Here is an example illustrating the use of these rules.

EXAMPLE 1.3 Find $\displaystyle\int \left(3e^x + \frac{2}{x} - \frac{1}{2}x^2\right) dx.$

SOLUTION

$$\int \left(3e^x + \frac{2}{x} - \frac{1}{2}x^2 \right) dx = 3\int e^x \, dx + 2\int \frac{1}{x} \, dx - \frac{1}{2}\int x^2 \, dx$$

$$= 3e^x + 2 \ln |x| - \frac{1}{6}x^3 + C$$

Notice that in place of separate constants of integration for each of the three antiderivatives computed in Example 1.3, a single constant C was added at the end.

Integration of Products and Quotients

You may have noticed that no general rules have been given for the integration of products and quotients. This is because there are no general rules. Occasionally, you will be able to rewrite a product or a quotient in a form in which you can integrate it using the techniques you have learned in this section. Here is an example.

EXAMPLE 1.4   Find $\displaystyle\int \frac{3x^5 + 2x - 5}{x^3} \, dx$.

SOLUTION
Perform the indicated division to get

$$\frac{3x^5 + 2x - 5}{x^3} = 3x^2 + \frac{2}{x^2} - \frac{5}{x^3} = 3x^2 + 2x^{-2} - 5x^{-3}$$

and then integrate term by term to get

$$\int \frac{3x^5 + 2x - 5}{x^3} \, dx = \int (3x^2 + 2x^{-2} - 5x^{-3}) \, dx$$

$$= x^3 - 2x^{-1} + \frac{5}{2}x^{-2} + C = x^3 - \frac{2}{x} + \frac{5}{2x^2} + C$$

Practical Applications

Here are two problems in which the rate of change of a quantity is known and the goal is to find an expression for the quantity itself. Since the rate of change is the derivative of the quantity, you find the expression for the quantity itself by antidifferentiation.

EXAMPLE 1.5   It is estimated that x months from now the population of a certain town will be changing at the rate of $2 + 6\sqrt{x}$ people per month. The current population is 5,000. What will the population be 9 months from now?

SOLUTION

Let $P(x)$ denote the population of the town x months from now. Then the rate of change of the population with respect to time is the derivative

$$\frac{dP}{dx} = 2 + 6\sqrt{x}$$

It follows that the population function $P(x)$ is an antiderivative of $2 + 6\sqrt{x}$. That is,

$$P(x) = \int \frac{dP}{dx}\, dx = \int (2 + 6\sqrt{x})\, dx = 2x + 4x^{3/2} + C$$

for some constant C. To determine C, use the fact that at present (when $x = 0$) the population is 5,000. That is,

$$5{,}000 = 2(0) + 4(0)^{3/2} + C \qquad \text{or} \qquad C = 5{,}000$$

Hence,

$$P(x) = 2x + 4x^{3/2} + 5{,}000$$

and the population 9 months from now will be

$$P(9) = 2(9) + 4(27) + 5{,}000 = 5{,}126$$

EXAMPLE 1.6   A manufacturer has found that marginal cost is $3q^2 - 60q + 400$ dollars per unit when q units have been produced. The total cost of producing the first 2 units is $900. What is the total cost of producing the first 5 units?

SOLUTION

Recall that the marginal cost is the derivative of the total cost function $C(q)$. Thus,

$$C'(q) = 3q^2 - 60q + 400$$

and so $C(q)$ must be the antiderivative

$$C(q) = \int C'(q)\, dq = \int (3q^2 - 60q + 400)\, dq = q^3 - 30q^2 + 400q + K$$

for some constant K. (The letter K was used for the constant to avoid confusion with the cost function C.)

The value of K is determined by the fact that $C(2) = 900$. In particular,

$$900 = (2)^3 - 30(2)^2 + 400(2) + K \qquad \text{or} \qquad K = 212$$

Hence,

$$C(q) = q^3 - 30q^2 + 400q + 212$$

and the cost of producing the first 5 units is

$$C(5) = (5)^3 - 30(5)^2 + 400(5) + 212 = \$1,587$$

Geometric Application

In the next example you will see how to use integration to find the equation of a curve whose slope is known.

EXAMPLE 1.7 Find the function $f(x)$ whose tangent has slope $3x^2 + 1$ for each value of x and whose graph passes through the point $(2, 6)$.

SOLUTION

The slope of the tangent is the derivative of f. Thus

$$f'(x) = 3x^2 + 1$$

and so $f(x)$ is the antiderivative

$$f(x) = \int f'(x)\, dx = \int (3x^2 + 1)\, dx = x^3 + x + C$$

To find C, use the fact that the graph of f passes through $(2, 6)$. That is, substitute $x = 2$ and $f(2) = 6$ into the equation for $f(x)$ and solve for C to get

$$6 = (2)^3 + 2 + C \qquad \text{or} \qquad C = -4$$

That is, the desired function is $f(x) = x^3 + x - 4$.

Problems

In Problems 1 through 20, find the indicated integral. Check your answers by differentiation.

1. $\displaystyle\int x^5\, dx$

2. $\displaystyle\int x^{3/4}\, dx$

3. $\displaystyle\int \frac{1}{x^2}\, dx$

4. $\displaystyle\int \sqrt{x}\, dx$

5. $\displaystyle\int 5\, dx$

6. $\displaystyle\int 3e^x\, dx$

7. $\displaystyle\int (3x^2 - 5x + 2)\, dx$

8. $\displaystyle\int (x^{1/2} - 3x^{2/3} + 6)\, dx$

9. $\displaystyle\int \left(3\sqrt{x} - \frac{2}{x^3} + \frac{1}{x}\right) dx$

10. $\displaystyle\int \left(\frac{1}{2x} - \frac{2}{x^2} + \frac{3}{\sqrt{x}}\right) dx$

11. $\int \left(2e^x + \dfrac{6}{x} + \ln 2 \right) dx$

12. $\int \left(\dfrac{1}{3x} - \dfrac{3}{2x^2} + e^2 + \dfrac{\sqrt{x}}{2} \right) dx$

13. $\int \left(\sqrt{x^3} - \dfrac{1}{2\sqrt{x}} + \sqrt{2} \right) dx$

14. $\int \dfrac{x + 1}{x^3} dx$

15. $\int \dfrac{x^2 + 2x + 1}{x^2} dx$

16. $\int \dfrac{x^2 + 3x - 2}{\sqrt{x}} dx$

17. $\int \left(\dfrac{2x^3 - 5x}{3x} \right)^2 dx$

18. $\int x^3 \left(2x + \dfrac{1}{x} \right) dx$

19. $\int \sqrt{x}(x^2 - 1) dx$

20. $\int x(2x + 1)^2 dx$

Population growth
21. It is estimated that t months from now the population of a certain town will be changing at the rate of $4 + 5t^{2/3}$ people per month. If the current population is 10,000, what will the population be 8 months from now?

Retail prices
22. In a certain section of the country, the price of large Grade A eggs is currently $1.60 per dozen. Studies indicate that x weeks from now, the price will be changing at the rate of $0.2 + 0.003x^2$ cents per week. How much will eggs cost 10 weeks from now?

Depreciation
23. The resale value of a certain industrial machine decreases at a rate that changes with time. When the machine is t years old, the rate at which its value is changing is $220(t - 10)$ dollars per year. If the machine was bought new for $12,000, how much will it be worth 10 years later?

The speed of a moving object
24. An object is moving so that its speed after t minutes is $3 + 2t + 6t^2$ meters per minute. How far does the object travel during the 2nd minute?

Marginal cost
25. A manufacturer has found that marginal cost is $6q + 1$ dollars per unit when q units have been produced. The total cost (including overhead) of producing the 1st unit is $130. What is the total cost of producing the first 10 units?

Marginal profit
26. A manufacturer estimates marginal revenue to be $100q^{-1/2}$ dollars per unit when the level of production is q units. The corresponding marginal cost has been found to be $0.4q$ dollars per unit. Suppose the manufacturer's profit is $520 when the level of production is 16 units. What is the manufacturer's profit when the level of production is 25 units?

Marginal profit
27. The marginal profit (the derivative of profit) of a certain company is $100 - 2q$ dollars per unit when q units are produced. If the company's profit is $700 when 10 units are produced, what is the company's maximum possible profit?

Air pollution   28. An environmental study of a certain community suggests that t years from now the level of carbon monoxide in the air will be changing at the rate of $0.1t + 0.1$ parts per million per year. If the current level of carbon monoxide in the air is 3.4 parts per million, what will the level be 3 years from now?

29. Find the function whose tangent has slope $4x + 1$ for each value of x and whose graph passes through the point $(1, 2)$.

30. Find the function whose tangent has slope $3x^2 + 6x - 2$ for each value of x and whose graph passes through the point $(0, 6)$.

31. Find the function whose tangent has slope $x^3 - \dfrac{2}{x^2} + 2$ for each value of x and whose graph passes through the point $(1, 3)$.

32. Find a function whose graph has a relative minimum when $x = 1$ and a relative maximum when $x = 4$.

In Problems 33 through 38, find the indicated integral. In each case, start with an educated guess, check your answer by differentiation, and then modify your guess if necessary.

33. $\displaystyle\int e^{3x}\, dx$ (*Hint:* Try $e^{3x} + C$.)

34. $\displaystyle\int 3e^{5x}\, dx$

35. $\displaystyle\int (2x + 3)^5\, dx$ $\left[\textit{Hint:} \text{ Try } \dfrac{1}{6}(2x + 3)^6 + C.\right]$

36. $\displaystyle\int 5(3x + 1)^4\, dx$

37. $\displaystyle\int \dfrac{1}{2x + 1}\, dx$

38. $\displaystyle\int \dfrac{2}{3x + 2}\, dx$

2 Integration by Substitution

This section deals with the integral version of the chain rule. To refresh your memory, here is a typical application of the chain rule for differentiation.

According to the chain rule, the derivative of the function $(x^2 + 3x + 5)^9$ is

$$\frac{d}{dx}[(x^2 + 3x + 5)^9] = 9(x^2 + 3x + 5)^8(2x + 3)$$

Notice the form of this derivative. In particular, notice that the derivative is a product and that one of its factors, $2x + 3$, is the derivative of an expression, $x^2 + 3x + 5$, which occurs in the other factor. More precisely, the derivative is a product of the form

$$g(u)\frac{du}{dx}$$

where, in this case, $g(u) = 9u^8$ and $u = x^2 + 3x + 5$.

You can integrate many products of the form $g(u)\frac{du}{dx}$ by applying the chain rule in reverse. Specifically, if G is an antiderivative of g, then

$$\int g(u)\frac{du}{dx}\, dx = G(u) + C$$

since, by the chain rule,

$$\frac{d}{dx}[G(u)] = G'(u)\frac{du}{dx} = g(u)\frac{du}{dx}$$

Here is a summary of the procedure.

Integral Form of the Chain Rule

$$\int g(u)\frac{du}{dx}\, dx = \int g(u)\, du$$

That is, to integrate a product of the form $g(u)\frac{du}{dx}$ in which one of the factors $\frac{du}{dx}$ is the derivative of an expression u, which appears in the other factor:

1. Find the antiderivative $\int g(u)\, du$ of the factor $g(u)$ with respect to u.

2. Replace u in the answer by its expression in terms of x.

Here are two examples.

EXAMPLE 2.1 Find $\int 9(x^2 + 3x + 5)^8(2x + 3)\, dx.$

SOLUTION

The integrand $9(x^2 + 3x + 5)^8(2x + 3)$ is a product in which one of the factors $2x + 3$ is the derivative of an expression $x^2 + 3x + 5$, which appears in the other factor. In particular,

$$9(x^2 + 3x + 5)^8(2x + 3) = g(u)\frac{du}{dx}$$

where

$$g(u) = 9u^8 \quad \text{and} \quad u = x^2 + 3x + 5$$

Hence, by the integral form of the chain rule,

$$\int 9(x^2 + 3x + 5)^8(2x + 3)\, dx = \int g(u)\, du = \int 9u^8\, du = u^9 + C$$

$$= (x^2 + 3x + 5)^9 + C$$

The product to be integrated in the next example is not exactly of the form $g(u)\dfrac{du}{dx}$. However, it is a constant multiple of such a product and you can integrate it by combining the constant multiple rule with the integral form of the chain rule.

EXAMPLE 2.2   Find $\displaystyle\int x^3 e^{x^4+2}\, dx$.

SOLUTION
First use the constant multiple rule to rewrite the integral as

$$\int x^3 e^{x^4+2}\, dx = \int \frac{1}{4}(4x^3 e^{x^4+2})\, dx = \frac{1}{4}\int 4x^3 e^{x^4+2}\, dx$$

so that the new integrand $4x^3 e^{x^4+2}$ is a product in which one of the factors $4x^3$ is the derivative of an expression $x^4 + 2$, which appears in the other factor. In particular,

$$4x^3 e^{x^4+2} = g(u)\frac{du}{dx}$$

where

$$g(u) = e^u \quad \text{and} \quad u = x^4 + 2$$

Hence, by the integral form of the chain rule,

$$\int x^3 e^{x^4+2}\, dx = \frac{1}{4}\int 4x^3 e^{x^4+2}\, dx = \frac{1}{4}\int g(u)\, du$$

$$= \frac{1}{4}\int e^u\, du = \frac{1}{4}e^u + C$$

$$= \frac{1}{4}e^{x^4+2} + C$$

Change of Variables

The integral form of the chain rule may be thought of as a technique for simplifying an integral by changing the variable of integration. In particular, you start with an integral $\int g(u)\dfrac{du}{dx}\,dx$ in which the variable of integration is x and transform it into the simpler integral $\int g(u)\,du$ in which the variable of integration is u. In this transformation, the expression $\dfrac{du}{dx}\,dx$ in the original integral is replaced in the simplified integral by the symbol du. You can remember this relationship between $\dfrac{du}{dx}\,dx$ and du by pretending that $\dfrac{du}{dx}$ is a quotient and writing

$$\frac{du}{dx}\,dx \;=\; du$$

These observations suggest the following general integration technique called **integration by substitution** in which the variable u is formally substituted for an appropriate expression in x, and the original integral is transformed into a simpler one in which the variable of integration is u.

Integration by Substitution

Step 1. Introduce the letter u to stand for some expression in x that is chosen with the simplification of the integral as the goal. (If the integrand is a product or quotient of two terms in which one term is a constant multiple of the derivative of an expression that appears in the other, then this expression is probably a good choice for u.)

Step 2. Rewrite the integral in terms of u. To rewrite dx, compute $\dfrac{du}{dx}$ and solve algebraically as if the symbol $\dfrac{du}{dx}$ were a quotient.

Step 3. Evaluate the resulting integral and then replace u by its expression in terms of x in the answer.

The method of integration by substitution is illustrated in the next example using the integral from Example 2.1.

EXAMPLE 2.3   Find $\displaystyle\int 9(x^2 + 3x + 5)^8(2x + 3)\,dx$.

SOLUTION

The integrand is a product in which one of the factors $2x + 3$ is the derivative of an expression $x^2 + 3x + 5$, which appears in the other factor. This suggests that you let $u = x^2 + 3x + 5$. Then

$$\frac{du}{dx} = 2x + 3 \qquad \text{and so} \qquad du = (2x + 3)\, dx$$

Substituting $u = x^2 + 3x + 5$ and $du = (2x + 3)\, dx$, you get

$$\int 9(x^2 + 3x + 5)^8(2x + 3)\, dx = \int 9u^8\, du = u^9 + C$$

$$= (x^2 + 3x + 5)^9 + C$$

Here are some additional examples illustrating the method of integration by substitution.

EXAMPLE 2.4 Find $\int \dfrac{3x}{x^2 - 1}\, dx$.

SOLUTION
Observe that

$$\frac{d}{dx}(x^2 - 1) = 2x = \frac{2}{3}(3x)$$

Thus the integrand is a quotient in which one term $3x$ is a constant multiple of the derivative of an expression $x^2 - 1$, which appears in the other term. This suggests that you let $u = x^2 - 1$. Then,

$$\frac{du}{dx} = 2x \qquad du = 2x\, dx \qquad \text{or} \qquad \frac{3}{2}du = 3x\, dx$$

Substituting $u = x^2 - 1$ and $\frac{3}{2}du = 3x\, dx$, you get

$$\int \frac{3x}{x^2 - 1}\, dx = \int \frac{1}{u}\left(\frac{3}{2}\right) du = \frac{3}{2} \int \frac{1}{u}\, du$$

$$= \frac{3}{2} \ln |u| + C$$

$$= \frac{3}{2} \ln |x^2 - 1| + C$$

EXAMPLE 2.5 Find $\int \dfrac{3x + 6}{\sqrt{2x^2 + 8x + 3}}\, dx$.

SOLUTION
Observe that

$$\frac{d}{dx}(2x^2 + 8x + 3) = 4x + 8 = 4(x + 2) = \frac{4}{3}(3x + 6)$$

Thus, the integrand is a quotient in which one term $3x + 6$ is a constant multiple of the derivative of an expression $2x^2 + 8x + 3$, which appears in the other term. This suggests that you let $u = 2x^2 + 8x + 3$. Then,

$$\frac{du}{dx} = 4x + 8$$

$$du = (4x + 8)\ dx = 4(x + 2)\ dx = \frac{4}{3}(3x + 6)\ dx$$

or

$$\frac{3}{4}\ du = (3x + 6)\ dx$$

Substituting $u = 2x^2 + 8x + 3$ and $\frac{3}{4}du = (3x + 6)\ dx$, you get

$$\int \frac{3x + 6}{\sqrt{2x^2 + 8x + 3}}\ dx = \int \frac{3}{4}\frac{1}{\sqrt{u}}\ du = \frac{3}{4}\int u^{-1/2}\ du$$

$$= \frac{3}{4}(2u^{1/2}) + C = \frac{3}{2}\sqrt{2x^2 + 8x + 3} + C$$

EXAMPLE 2.6   Find $\displaystyle\int \frac{(\ln x)^2}{x}\ dx$.

SOLUTION
Observe that

$$\frac{d}{dx}(\ln x) = \frac{1}{x}$$

Thus, the integrand

$$\frac{(\ln x)^2}{x} = (\ln x)^2\left(\frac{1}{x}\right)$$

is a product in which one factor $\dfrac{1}{x}$ is the derivative of an expression $\ln x$, which appears in the other factor. This suggests that you let $u = \ln x$. Then,

$$\frac{du}{dx} = \frac{1}{x} \quad \text{or} \quad du = \frac{1}{x}\,dx$$

Substituting $u = \ln x$ and $du = \dfrac{1}{x}\,dx$, you get

$$\int \frac{(\ln x)^2}{x}\,dx = \int u^2\,du = \frac{1}{3}u^3 + C = \frac{1}{3}(\ln x)^3 + C$$

The next example is designed to show you the versatility of the formal method of substitution. It deals with an integral that does not seem to be of the form $\displaystyle\int g(u)\frac{du}{dx}\,dx$ but that nevertheless can be simplified significantly by a clever change of variables.

EXAMPLE 2.7 Find $\displaystyle\int \frac{x}{x+1}\,dx$.

SOLUTION

There seems to be no easy way to integrate this quotient as it stands. But watch what happens if you make the substitution $u = x + 1$. Then $du = dx$ and $x = u - 1$ and so

$$\int \frac{x}{x+1}\,dx = \int \frac{u-1}{u}\,du = \int 1\,du - \int \frac{1}{u}\,du$$

$$= u - \ln |u| + C = x + 1 - \ln |x+1| + C$$

A Comparison of the Two Techniques

The integral form of the chain rule, which was used in Examples 2.1 and 2.2, is attractive because it involves nothing more than a familiar rule for differentiation applied in reverse. With practice in using this method you should be able to find integrals like those in Examples 2.1 through 2.6 by inspection, without writing down any intermediate steps.

Nevertheless, many people prefer to use the method of formal substitution. They like the fact that it involves straightforward manipulation of symbols, and they appreciate the convenience of the notation. This method of substitution is also somewhat more versatile, as you saw in Example 2.7.

For many of the integrals you will encounter, both methods work well and you should feel free to use the one with which you are most comfortable.

Problems

In Problems 1 through 26, find the indicated integral and check your answer by differentiation.

1. $\displaystyle\int (2x + 6)^5\, dx$

2. $\displaystyle\int e^{5x}\, dx$

3. $\displaystyle\int \sqrt{4x - 1}\, dx$

4. $\displaystyle\int \frac{1}{3x + 5}\, dx$

5. $\displaystyle\int e^{1-x}\, dx$

6. $\displaystyle\int [(x - 1)^5 + 3(x - 1)^2 + 5]\, dx$

7. $\displaystyle\int xe^{x^2}\, dx$

8. $\displaystyle\int 2xe^{x^2-1}\, dx$

9. $\displaystyle\int x(x^2 + 1)^5\, dx$

10. $\displaystyle\int 3x\sqrt{x^2 + 8}\, dx$

11. $\displaystyle\int x^2(x^3 + 1)^{3/4}\, dx$

12. $\displaystyle\int x^5 e^{1-x^6}\, dx$

13. $\displaystyle\int \frac{2x^4}{x^5 + 1}\, dx$

14. $\displaystyle\int \frac{x^2}{(x^3 + 5)^2}\, dx$

15. $\displaystyle\int (x + 1)(x^2 + 2x + 5)^{12}\, dx$

16. $\displaystyle\int (3x^2 - 1)e^{x^3-x}\, dx$

17. $\displaystyle\int \frac{3x^4 + 12x^3 + 6}{x^5 + 5x^4 + 10x + 12}\, dx$

18. $\displaystyle\int \frac{10x^3 - 5x}{\sqrt{x^4 - x^2 + 6}}\, dx$

19. $\displaystyle\int \frac{3x - 3}{(x^2 - 2x + 6)^2}\, dx$

20. $\displaystyle\int \frac{6x - 3}{4x^2 - 4x + 1}\, dx$

21. $\displaystyle\int \frac{\ln 5x}{x}\, dx$

22. $\displaystyle\int \frac{1}{x \ln x}\, dx$

23. $\displaystyle\int \frac{1}{x(\ln x)^2}\, dx$

24. $\displaystyle\int \frac{\ln x^2}{x}\, dx$

25. $\displaystyle\int \frac{2x \ln (x^2 + 1)}{x^2 + 1}\, dx$

26. $\displaystyle\int \frac{e^{\sqrt{x}}}{\sqrt{x}}\, dx$

In Problems 27 through 33, use an appropriate change of variables to find the indicated integral.

27. $\displaystyle\int \frac{x}{x - 1}\, dx$

28. $\displaystyle\int x\sqrt{x + 1}\, dx$

29. $\displaystyle\int \frac{x}{(x - 5)^6}\, dx$

30. $\displaystyle\int (x + 1)(x - 2)^9\, dx$

31. $\displaystyle\int \frac{x + 3}{(x - 4)^2}\, dx$

32. $\displaystyle\int \frac{x}{2x + 1}\, dx$

33. $\displaystyle\int (2x + 3)\sqrt{2x - 1}\, dx$

34. Find the function whose tangent has slope $x\sqrt{x^2 + 5}$ for each value of x and whose graph passes through the point (2, 10).

35. Find the function whose tangent has slope $\dfrac{2x}{1 - 3x^2}$ for each value of x and whose graph passes through the point (0, 5).

36. A tree has been transplanted and after x years is growing at the rate of $1 + \dfrac{1}{(x + 1)^2}$ meters per year. After 2 years it has reached a height of 5 meters. How tall was it when it was transplanted?

Depreciation 37. The resale value of a certain industrial machine decreases at a rate that changes with time. When the machine is t years old, the rate at which its value is changing is $-960e^{-t/5}$ dollars per year. If the machine was bought new for $5,000, how much will it be worth 10 years later?

Population growth 38. It is projected that t years from now the population of a certain country will be changing at the rate of $e^{0.02t}$ million per year. If the current population is 50 million, what will the population be 10 years from now?

Land value 39. It is estimated that x years from now the value of an acre of farmland will be increasing at the rate of $\dfrac{0.4x^3}{\sqrt{0.2x^4 + 8,000}}$ dollars per year. If the land is currently worth $500 per acre, how much will it be worth in 10 years?

3 Integration by Parts

In this section you will see a technique you can use to integrate certain products $f(x)g(x)$, in which one of the factors, say $g(x)$, can be easily integrated and the other, $f(x)$, can be simplified by differentiation. The technique is called **integration by parts,** and, as you will see, it is a restatement of the product rule for differentiation. Here is the formal statement of the technique.

Integration by Parts

$$\int f(x)g(x)\ dx = f(x)G(x) - \int f'(x)G(x)\ dx$$

where G is an antiderivative of g.

Why Integration by Parts Works

Integration by parts is simply a restatement of what happens when the product rule is used to differentiate $f(x)G(x)$, where G is an antiderivative of g. In particular,

$$\frac{d}{dx}[f(x)G(x)] = f'(x)G(x) + f(x)G'(x) = f'(x)G(x) + f(x)g(x)$$

Expressed in terms of integrals, this says

$$f(x)G(x) = \int f'(x)G(x)\ dx + \int f(x)g(x)\ dx$$

or

$$\int f(x)g(x)\ dx = f(x)G(x) - \int f'(x)G(x)\ dx$$

which is precisely the formula for integration by parts.

How and When to Use Integration by Parts

Integration by parts is a technique for integrating products $f(x)g(x)$, in which one of the factors, say $g(x)$, can be easily integrated and the other, $f(x)$, becomes simpler when differentiated. To evaluate such an integral, $\int f(x)g(x)\ dx$, using integration by parts, first integrate g and multiply the result by f to get

$$f(x)G(x)$$

where G is an antiderivative of g. Then multiply the antiderivative G by the derivative of f and subtract the integral of this product from the result of the first step to get

$$f(x)G(x) - \int f'(x)G(x)\ dx$$

This expression will be equal to the original integral $\int f(x)g(x)\ dx$, and, if you are lucky, the new integral $\int f'(x)G(x)\ dx$ will be easier to find than the original one.

Here is an informal, step-by-step summary of the procedure.

How to Use Integration by Parts to Integrate a Product

Step 1. Select one of the factors of the product as the one to be integrated and the other as the one to be differentiated. The factor selected for integration should be easy to integrate, and the factor selected for differentiation should become simpler when differentiated.

Step 2. Integrate the designated factor and multiply it by the other factor.

Step 3. Differentiate the designated factor, multiply it by the integrated factor from Step 2, and subtract the integral of this product from the result of Step 2.

Step 4. Complete the procedure by finding the new integral that was formed in Step 3.

Here are some examples illustrating the procedure. In each example, $g(x)$ is used to denote the factor that is to be integrated and $f(x)$ to denote the factor that is to be differentiated. As reminders, the letters I (for integrate) and D (for differentiate) are placed above the appropriate factors in the integrand.

With practice, you will become familiar with the pattern and should find that you can do integration by parts without the intermediate step of writing down the functions $g(x)$, $f(x)$, $G(x)$, and $f'(x)$.

EXAMPLE 3.1   Find $\int xe^{2x}\, dx$.

SOLUTION

In this case, both factors x and e^{2x} are easy to integrate. Both are also easy to differentiate, but the process of differentiation simplifies x while it makes e^{2x} slightly more complicated. This suggests that you should try integration by parts with

$$g(x) = e^{2x} \quad \text{and} \quad f(x) = x$$

Then,

$$G(x) = \frac{1}{2}e^{2x} \quad \text{and} \quad f'(x) = 1$$

and so

$$\int \overset{\text{D I}}{xe^{2x}}\, dx = \left(\frac{1}{2}e^{2x}\right)(x) - \int\left(\frac{1}{2}e^{2x}\right)(1)\, dx$$

$$= \frac{1}{2}xe^{2x} - \frac{1}{2}\int e^{2x}\, dx$$

$$= \frac{1}{2}xe^{2x} - \frac{1}{4}e^{2x} + C = \frac{1}{2}\left(x - \frac{1}{2}\right)e^{2x} + C$$

EXAMPLE 3.2   Find $\int x\sqrt{x + 5}\ dx$.

SOLUTION

Again, both factors in the product are easy to integrate and differentiate. However, the factor x is simplified by differentiation, whereas the derivative of $\sqrt{x + 5}$ is even more complicated than $\sqrt{x + 5}$ itself. This suggests that you should try integration by parts with

$$g(x) = \sqrt{x + 5} \qquad \text{and} \qquad f(x) = x$$

Then,

$$G(x) = \frac{2}{3}(x + 5)^{3/2} \qquad \text{and} \qquad f'(x) = 1$$

and so

$$\int \overset{D \quad I}{x\sqrt{x + 5}}\ dx = \frac{2}{3}x(x + 5)^{3/2} - \frac{2}{3}\int (x + 5)^{3/2}\ dx$$

$$= \frac{2}{3}x(x + 5)^{3/2} - \frac{4}{15}(x + 5)^{5/2} + C$$

EXAMPLE 3.3   Find $\int x \ln x\ dx$.

SOLUTION

In this case, the factor x is easy to integrate, while the factor $\ln x$ is simplified by differentiation. This suggests that you try integration by parts with

$$g(x) = x \qquad \text{and} \qquad f(x) = \ln x$$

Then,

$$G(x) = \frac{1}{2}x^2 \qquad \text{and} \qquad f'(x) = \frac{1}{x}$$

and so

$$\int \overset{I \quad D}{x \ln x}\ dx = \frac{1}{2}x^2 \ln x - \frac{1}{2}\int x^2\left(\frac{1}{x}\right) dx = \frac{1}{2}x^2 \ln x - \frac{1}{2}\int x\ dx$$

$$= \frac{1}{2}x^2 \ln x - \frac{1}{4}x^2 + C = \frac{1}{2}x^2\left(\ln x - \frac{1}{2}\right) + C$$

The Integral of ln x

In the next example, you will see how to use integration by parts to integrate the natural logarithm ln x.

EXAMPLE 3.4 Find $\int \ln x \, dx$.

SOLUTION

The trick is to write ln x as the product $1(\ln x)$, in which the factor 1 is easy to integrate and the factor ln x is simplified by differentiation. This suggests that you use integration by parts with

$$g(x) = 1 \quad \text{and} \quad f(x) = \ln x$$

Then,

$$G(x) = x \quad \text{and} \quad f'(x) = \frac{1}{x}$$

and so

$$\int \ln x \, dx = \int \overset{I}{1} \overset{D}{(\ln x)} \, dx = x \ln x - \int x\left(\frac{1}{x}\right) dx = x \ln x - \int 1 \, dx$$

$$= x \ln x - x + C = x(\ln x - 1) + C$$

Repeated Applications of Integration by Parts

Sometimes integration by parts leads to a new integral that also must be integrated by parts. This situation is illustrated in the next example.

EXAMPLE 3.5 Find $\int x^2 e^x \, dx$.

SOLUTION

Since the factor e^x is easy to integrate and the factor x^2 is simplified by differentiation, try integration by parts with

$$g(x) = e^x \quad \text{and} \quad f(x) = x^2$$

Then,

$$G(x) = e^x \quad \text{and} \quad f'(x) = 2x$$

and so

$$\int \overset{D}{x^2} \overset{I}{e^x} \, dx = x^2 e^x - 2 \int x e^x \, dx$$

To find $\int x e^x \, dx$, you have to integrate by parts again, this time with

$$g(x) = e^x \quad \text{and} \quad f(x) = x$$

Then,

$$G(x) = e^x \quad \text{and} \quad f'(x) = 1$$

and so

$$\int x^2 e^x \, dx = x^2 e^x - 2 \int \overset{\text{D I}}{x \, e^x} \, dx$$

$$= x^2 e^x - 2\left(x e^x - \int e^x \, dx\right)$$

$$= x^2 e^x - 2(x e^x - e^x) + C = (x^2 - 2x + 2)e^x + C$$

Problems

In Problems 1 through 25, use integration by parts to find the given integral.

1. $\int x e^{-x} \, dx$

2. $\int x e^{x/2} \, dx$

3. $\int x e^{-x/5} \, dx$

4. $\int x e^{0.1x} \, dx$

5. $\int (1 - x)e^x \, dx$

6. $\int (3 - 2x)e^{-x} \, dx$

7. $\int x \ln 2x \, dx$

8. $\int x \ln x^2 \, dx$

9. $\int x\sqrt{x - 6} \, dx$

10. $\int x\sqrt{1 - x} \, dx$

11. $\int x(x + 1)^8 \, dx$

12. $\int (x + 1)(x + 2)^6 \, dx$

13. $\int \frac{x}{\sqrt{x + 2}} \, dx$

14. $\int \frac{x}{\sqrt{2x + 1}} \, dx$

15. $\int x^2 e^{-x} \, dx$

16. $\int x^2 e^{3x} \, dx$

17. $\int x^3 e^x \, dx$

18. $\int x^3 e^{2x} \, dx$

19. $\int x^2 \ln x \, dx$

20. $\int x(\ln x)^2 \, dx$

21. $\int \dfrac{\ln x}{x^2}\, dx$
22. $\int \dfrac{\ln x}{x^3}\, dx$

23. $\int x^3 e^{x^2}\, dx$ [*Hint:* Rewrite the integrand as $x^2(xe^{x^2})$.]

24. $\int x^3(x^2 - 1)^{10}\, dx$
25. $\int x^7(x^4 + 5)^8\, dx$

Distance   26. After t seconds, an object is moving at the speed of $te^{-t/2}$ meters per second. Express the distance the object travels as a function of time.

Efficiency   27. After t hours on the job, a factory worker can produce $100te^{-0.5t}$ units per hour. How many units does the worker produce during the first 3 hours?

Fund-raising   28. After t weeks, contributions in response to a local fund-raising campaign were coming in at the rate of $2{,}000te^{-0.2t}$ dollars per week. How much money was raised during the first 5 weeks?

Reduction formula   29. (a) Use integration by parts to derive the formula

$$\int x^n e^{ax}\, dx = \frac{1}{a}x^n e^{ax} - \frac{n}{a}\int x^{n-1} e^{ax}\, dx$$

(b) Use the formula in part (a) to find $\int x^3 e^{5x}\, dx$.

4 The Use of Integral Tables

Most of the integrals you will encounter in the social, managerial, and life sciences can be evaluated using the techniques you have learned so far in this chapter. From time to time, however, an integral will turn up that cannot be handled by these techniques. For such occasions, it is helpful to know how to use a **table of integrals.**

A table of integrals is a list of integration formulas. Extensive tables listing several hundred formulas can be found in most mathematics handbooks, and condensed versions appear in many calculus texts. Here is a tiny sampling of the formulas that appear in a table of integrals.

A Small Table of Integrals

1. $\displaystyle\int \frac{dx}{p^2 - x^2} = \frac{1}{2p}\ln\left|\frac{p + x}{p - x}\right|$

2. $\displaystyle\int \frac{dx}{x(ax + b)} = \frac{1}{b}\ln\left|\frac{x}{ax + b}\right|$

3. $\int \dfrac{dx}{\sqrt{x^2 \pm p^2}} = \ln\left|x + \sqrt{x^2 \pm p^2}\right|$

4. $\int x^n e^{ax}\, dx = \dfrac{1}{a}x^n e^{ax} - \dfrac{n}{a}\int x^{n-1}e^{ax}\, dx$

In these formulas, the letters a, b, p, and n denote constants. The term p^2 in the first and third formulas may be any *positive* constant (since any positive number is the square of its square root). The compact fractional notation $\dfrac{dx}{p^2 - x^2}$ in the first formula is an abbreviation for $\dfrac{1}{p^2 - x^2}\, dx$. Similar notation occurs in some of the other formulas. Also, to keep the formulas simple, the constant C is omitted from each of the integrals in the table. You add the C.

For convenience, most tables of integrals are divided into sections. Integrals containing similar expressions are grouped together in the same section. For example, the first formula would be found in a section entitled "Expressions Containing $p^2 - x^2$," the second in a section called "Expressions Containing $ax + b$," and the fourth in the section "Expressions Containing Exponential and Logarithmic Functions."

The use of these integration formulas is illustrated in the following examples.

EXAMPLE 4.1 Find $\int \dfrac{1}{x(3x - 6)}\, dx$.

SOLUTION
Apply the second formula with $a = 3$ and $b = -6$ to get

$$\int \frac{1}{x(3x - 6)}\, dx = -\frac{1}{6}\ln\left|\frac{x}{3x - 6}\right| + C$$

EXAMPLE 4.2 Find $\int \dfrac{1}{6 - 3x^2}\, dx$.

SOLUTION
If the coefficient of x^2 were 1 instead of 3, you could use the first formula. This suggests that you first rewrite the integrand as

$$\frac{1}{6 - 3x^2} = \frac{1}{3}\left(\frac{1}{2 - x^2}\right)$$

and then apply the first formula with $p = \sqrt{2}$ to get

$$\int \frac{1}{6 - 3x^2} \, dx = \frac{1}{3} \int \frac{1}{2 - x^2} \, dx$$

$$= \frac{1}{3}\left(\frac{1}{2\sqrt{2}}\right) \ln \left|\frac{\sqrt{2} + x}{\sqrt{2} - x}\right| + C$$

$$= \frac{\sqrt{2}}{12} \ln \left|\frac{\sqrt{2} + x}{\sqrt{2} - x}\right| + C$$

EXAMPLE 4.3 Find $\int \frac{1}{3x^2 - 6} \, dx$.

SOLUTION
Since

$$\frac{1}{3x^2 - 6} = -\frac{1}{6 - 3x^2}$$

you can apply the first formula as in Example 4.2 to get

$$\int \frac{1}{3x^2 - 6} \, dx = -\int \frac{1}{6 - 3x^2} \, dx = -\frac{\sqrt{2}}{12} \ln \left|\frac{\sqrt{2} + x}{\sqrt{2} - x}\right| + C$$

EXAMPLE 4.4 Find $\int \frac{1}{x^2 + 2x} \, dx$.

SOLUTION
Factor the integrand as

$$\frac{1}{x^2 + 2x} = \frac{1}{x(x + 2)}$$

and apply the second formula with $a = 1$ and $b = 2$ to get

$$\int \frac{1}{x^2 + 2x} \, dx = \int \frac{1}{x(x + 2)} \, dx = \frac{1}{2} \ln \left|\frac{x}{x + 2}\right| + C$$

EXAMPLE 4.5 Find $\int \frac{1}{3x^2 + 6} \, dx$.

SOLUTION
It is natural to try to match this integral to the one in the first formula by writing

$$\int \frac{1}{3x^2 + 6} \, dx = -\frac{1}{3} \int \frac{1}{-2 - x^2} \, dx$$

However, since -2 is negative, it cannot be written as the square p^2 of any real number p, and so the formula does not apply.

There is a formula for integrals of the form $\int \dfrac{1}{x^2 + p^2} \, dx$ that can be used in this case. You can find it in a table of integrals under a heading like "Expressions Containing $x^2 \pm p^2$." However, the antiderivative will be written in terms of inverse trigonometric functions and cannot be expressed in more elementary terms.

EXAMPLE 4.6   Find $\int \dfrac{1}{\sqrt{4x^2 - 9}} \, dx$.

SOLUTION
To put this integral in the form of the third formula, rewrite the integrand as

$$\frac{1}{\sqrt{4x^2 - 9}} = \frac{1}{\sqrt{4(x^2 - 9/4)}} = \frac{1}{2\sqrt{x^2 - 9/4}}$$

Then apply the formula with $p^2 = \dfrac{9}{4}$, using minus signs in place of the symbol $\pm$, to get

$$\int \frac{1}{\sqrt{4x^2 - 9}} \, dx = \frac{1}{2} \int \frac{1}{\sqrt{x^2 - 9/4}} \, dx = \frac{1}{2} \ln \left| x + \sqrt{x^2 - 9/4} \right| + C$$

Reduction Formulas

The fourth formula expresses an integral in terms of a simpler one of the same type. If the formula is subsequently applied to the new integral, further simplification may occur. Successive applications of the formula usually lead to an integral that can be found by elementary methods. A formula of this type is called a **reduction formula.** The use of reduction formulas is illustrated in the next example.

EXAMPLE 4.7   Find $\int x^2 e^{5x} \, dx$.

SOLUTION
Apply the fourth formula with $n = 2$ and $a = 5$ to get

$$\int x^2 e^{5x} \, dx = \frac{1}{5} x^2 e^{5x} - \frac{2}{5} \int x e^{5x} \, dx$$

Now apply the fourth formula again, this time with $n = 1$ and $a = 5$, to get

$$\int x e^{5x} \, dx = \frac{1}{5} x e^{5x} - \frac{1}{5} \int e^{5x} \, dx = \frac{1}{5} x e^{5x} - \frac{1}{25} e^{5x} + C$$

Combine these results to conclude that

$$\int x^2 e^{5x}\, dx = \frac{1}{5}x^2 e^{5x} - \frac{2}{5}\left(\frac{1}{5}x e^{5x} - \frac{1}{25}e^{5x}\right) + C$$

$$= \left(\frac{1}{5}x^2 - \frac{2}{25}x + \frac{2}{125}\right)e^{5x} + C$$

A Word of Advice

No special formula was really needed to find the integral in Example 4.7. You could have found this integral quite easily using integration by parts. Indeed, if the formula had not been so conveniently displayed at the beginning of this section, you would have been much better off integrating by parts directly than hunting through a table of integrals for the appropriate formula. Try not to succumb to the temptation to rely excessively on tables when computing integrals. Most of the integrals you will encounter can be found quite easily without the aid of formulas. Moreover, before you can use a formula, you must find it, and this can be time-consuming. In general, it is good strategy to use a table of integrals only as a last resort.

Problems

In Problems 1 through 10, use one of the integration formulas listed in this section to find the given integral.

1. $\displaystyle\int \frac{1}{x(2x-3)}\, dx$

2. $\displaystyle\int \frac{3}{4x(x-5)}\, dx$

3. $\displaystyle\int \frac{1}{\sqrt{x^2+25}}\, dx$

4. $\displaystyle\int \frac{1}{\sqrt{9x^2-4}}\, dx$

5. $\displaystyle\int \frac{1}{4-x^2}\, dx$

6. $\displaystyle\int \frac{1}{3x^2-9}\, dx$

7. $\displaystyle\int \frac{1}{3x^2+2x}\, dx$

8. $\displaystyle\int \frac{4}{x^2-x}\, dx$

9. $\displaystyle\int x^2 e^{3x}\, dx$

10. $\displaystyle\int x^3 e^{-x}\, dx$

Locate a table of integrals and use it to find the integrals in Problems 11 through 18.

11. $\displaystyle\int \frac{x}{2-x^2}\, dx$

12. $\displaystyle\int \frac{x+3}{\sqrt{2x+4}}\, dx$

13. $\displaystyle\int (\ln 2x)^2\, dx$

14. $\displaystyle\int (x^2+1)^{3/2}\, dx$

15. $\int \dfrac{1}{3x\sqrt{2x + 5}}\, dx$

16. $\int \dfrac{x}{\sqrt{4 - x^2}}\, dx$

17. $\int \dfrac{1}{2 - 3e^{-x}}\, dx$

18. $\int \dfrac{1}{\sqrt{3x^2 - 6x + 2}}\, dx$

19. One table of integrals lists the formula

$$\int \frac{dx}{\sqrt{x^2 \pm p^2}} = \ln \left| \frac{x + \sqrt{x^2 \pm p^2}}{p} \right|$$

while another table lists

$$\int \frac{dx}{\sqrt{x^2 \pm p^2}} = \ln \left| x + \sqrt{x^2 \pm p^2} \right|$$

Can you reconcile this apparent contradiction?

20. The following two formulas appear in a table of integrals:

$$\int \frac{dx}{p^2 - x^2} = \frac{1}{2p} \ln \left| \frac{p + x}{p - x} \right|$$

and

$$\int \frac{dx}{a + bx^2} = \frac{1}{2\sqrt{-ab}} \ln \left| \frac{a + x\sqrt{-ab}}{a - x\sqrt{-ab}} \right| \qquad (\text{for } -ab \geq 0)$$

(a) Use the second formula to derive the first.

(b) Apply both formulas to the integral $\int \dfrac{1}{9 - 4x^2}\, dx$. Which do you find easier to use in this problem?

Chapter Summary and Review Problems

Important Terms, Symbols, and Formulas

Antiderivative; indefinite integral:

$$\int f(x)\, dx = F(x) + C \text{ if and only if } F'(x) = f(x)$$

Power rule: $\displaystyle\int x^n\, dx = \dfrac{1}{n + 1} x^{n+1} + C \qquad (\text{for } n \neq -1)$

The integral of $\dfrac{1}{x}$: $\displaystyle\int \dfrac{1}{x}\, dx = \ln |x| + C$

Constant multiple rule: $\displaystyle\int cf(x)\, dx = c \int f(x)\, dx$

Sum rule: $\displaystyle\int [f(x) + g(x)]\, dx = \int f(x)\, dx + \int g(x)\, dx$

The integral of e^x: $\displaystyle\int e^x \, dx = e^x + C$

Integration by substitution:

$$\int g(u) \frac{du}{dx} \, dx = G(u) + C \qquad \text{where } G \text{ is an antiderivative of } g$$

Integration by parts:

$$\int f(x)g(x) \, dx = f(x)G(x) - \int f'(x)G(x) \, dx \qquad \text{where } G \text{ is an anti-}$$
derivative of g

Table of integrals

Reduction formula

Review Problems

In Problems 1 through 20, find the indicated integral.

1. $\displaystyle\int \left(x^5 - 3x^2 + \frac{1}{x^2} \right) dx$

2. $\displaystyle\int \left(x^{2/3} - \frac{1}{x} + 5 + \sqrt{x} \right) dx$

3. $\displaystyle\int \sqrt{3x + 1} \, dx$

4. $\displaystyle\int (3x + 1)\sqrt{3x^2 + 2x + 5} \, dx$

5. $\displaystyle\int (x + 2)(x^2 + 4x + 2)^5 \, dx$

6. $\displaystyle\int \frac{x + 2}{x^2 + 4x + 2} \, dx$

7. $\displaystyle\int \frac{3x + 6}{(2x^2 + 8x + 3)^2} \, dx$

8. $\displaystyle\int (x - 5)^{12} \, dx$

9. $\displaystyle\int x(x - 5)^{12} \, dx$

10. $\displaystyle\int 5e^{3x} \, dx$

11. $\displaystyle\int 5xe^{3x} \, dx$

12. $\displaystyle\int xe^{-x/2} \, dx$

13. $\displaystyle\int x^5 e^{x^3} \, dx$

14. $\displaystyle\int (2x + 1)e^{0.1x} \, dx$

15. $\displaystyle\int x \ln 3x \, dx$

16. $\displaystyle\int \ln 3x \, dx$

17. $\displaystyle\int \frac{\ln 3x}{x} \, dx$

18. $\displaystyle\int \frac{\ln 3x}{x^2} \, dx$

19. $\displaystyle\int x^3 (x^2 + 1)^8 \, dx$

20. $\displaystyle\int 2x \ln (x^2 + 1) \, dx$

21. Find the equation of the curve whose tangent has slope $x(x^2 + 1)^3$ for each value of x and that passes through the point $(1, 5)$.

22. It is estimated that x weeks from now, the number of commuters using a new subway line will be increasing at the rate of $18x^2 + 500$ per week. Currently, 8,000 commuters use the subway. How many will be using it 5 weeks from now?

23. Statistics compiled by the local department of corrections indicate that x years from now the number of inmates in county prisons will be increasing at the rate of $280e^{0.2x}$ per year. Currently 2,000 inmates are housed in county prisons. How many inmates should the county expect 10 years from now?

24. A manufacturer estimates marginal revenue to be $200q^{-1/2}$ dollars per unit when the level of production is q units. The corresponding marginal cost has been found to be $0.4q$ dollars per unit. If the manufacturer's profit is $2,000 when the level of production is 25 units, what is the profit when the level of production is 36 units?

In Problems 25 through 27, use one of the integration formulas listed in Section 4 to find the given integral.

25. $\displaystyle\int \frac{5}{8 - 2x^2}\, dx$

26. $\displaystyle\int \frac{2}{\sqrt{9x^2 + 16}}\, dx$

27. $\displaystyle\int x^2 e^{-x/2}\, dx$

Chapter

 Integration:
Basic Concepts

1   The Definite Integral

Suppose you know the rate $f(x) = \dfrac{dF}{dx}$ at which a certain quantity F is changing and wish to find the amount by which the quantity F will change between $x = a$ and $x = b$. You would first find F by antidifferentiation and then compute the difference

$$\begin{array}{l}\text{Change in } F \text{ between} \\ \quad x = a \text{ and } x = b \end{array} = F(b) - F(a)$$

The numerical result of such a computation is called a **definite integral** of the function f and is denoted by the symbol $\displaystyle\int_{a}^{b} f(x)\ dx$.

The Definite Integral

> The definite integral of f from a to b is the difference
>
> $$\int_a^b f(x)\,dx = F(b) - F(a)$$
>
> where F is an antiderivative of f. That is, the definite integral is the net change in the antiderivative between $x = a$ and $x = b$.

The symbol $\int_a^b f(x)\,dx$ is read "the (definite) integral of f from a to b." The numbers a and b are called **limits of integration.** In computations involving definite integrals, it is often convenient to use the symbol $F(x)\Big|_a^b$ to stand for the difference $F(b) - F(a)$.

Applications

Here are three practical problems involving net change whose solutions are definite integrals.

EXAMPLE 1.1 A study indicates that x months from now the population of a certain town will be increasing at the rate of $2 + 6\sqrt{x}$ people per month. By how much will the population of the town increase during the next 4 months?

SOLUTION
Let $P(x)$ denote the population of the town x months from now. Then, the rate of change of the population with respect to time is $\dfrac{dP}{dx} = 2 + 6\sqrt{x}$, and the amount by which the population will increase during the next 4 months is the definite integral

$$P(4) - P(0) = \int_0^4 (2 + 6\sqrt{x})\,dx = (2x + 4x^{3/2} + C)\Big|_0^4$$

$$= [2(4) + 4(4)^{3/2} + C] - [2(0) + 4(0)^{3/2} + C]$$

$$= (40 + C) - (0 + C) = 40 \text{ people}$$

Notice what happens to the constant C in the evaluation of a definite integral. It appears in the expressions for both $F(b)$ and $F(a)$ and is eventually eliminated by the subtraction. You may therefore omit the constant C altogether when evaluating definite integrals.

EXAMPLE 1.2 At a certain factory, the marginal cost is $3(q - 4)^2$ dollars per unit when the level of production is q units. By how much will the total manufacturing cost increase if the level of production is raised from 6 units to 10 units?

SOLUTION

Let $C(q)$ denote the total cost of producing q units. Then the marginal cost is the derivative $\dfrac{dC}{dq} = 3(q - 4)^2$, and the increase in cost if production is raised from 6 units to 10 units is the definite integral

$$C(10) - C(6) = \int_6^{10} 3(q - 4)^2 \, dq = (q - 4)^3 \Big|_6^{10}$$

$$= (10 - 4)^3 - (6 - 4)^3 = 216 - 8 = \$208$$

EXAMPLE 1.3 In a certain community, the demand for gasoline is increasing exponentially at the rate of 5 percent per year. If the current demand is 4 million gallons per year, how much gasoline will be consumed in the community during the next 3 years?

SOLUTION

Let $Q(t)$ denote the total consumption (in million-gallon units) of gasoline in the community over the next t years. Then, the demand (million gallons per year) is the rate of change $\dfrac{dQ}{dt}$ of the total consumption with respect to time. The fact that this demand is increasing exponentially at the rate of 5 percent per year and is currently equal to 4 million gallons per year implies that

$$\frac{dQ}{dt} = 4e^{0.05t} \qquad \text{million gallons per year}$$

Hence, the total consumption during the next 3 years is the definite integral

$$Q(3) - Q(0) = \int_0^3 4e^{0.05t} \, dt = 80e^{0.05t} \Big|_0^3$$

$$= 80(e^{0.15} - 1) \simeq 12.95 \text{ million gallons}$$

The Evaluation of Definite Integrals by Substitution

In the next example, you will see how the method of substitution can be used to evaluate a definite integral.

EXAMPLE 1.4   Evaluate $\int_0^1 8x(x^2 + 1)^3 \, dx$.

SOLUTION
The integrand is a product in which one factor $8x$ is a constant multiple of the derivative of an expression $x^2 + 1$, which appears in the other factor. This suggests that you let $u = x^2 + 1$. Then $du = 2x \, dx$, and so

$$\int 8x(x^2 + 1)^3 \, dx = \int 4u^3 \, du = u^4$$

The limits of integration, 0 and 1, refer to the variable x and not to u. You may, therefore, proceed in one of two ways. Either you can rewrite the antiderivative in terms of x, or you can find the values of u that correspond to $x = 0$ and $x = 1$.

If you choose the first alternative, you find that

$$\int 8x(x^2 + 1)^3 \, dx = u^4 = (x^2 + 1)^4$$

and so

$$\int_0^1 8x(x^2 + 1)^3 \, dx = (x^2 + 1)^4 \Big|_0^1 = 16 - 1 = 15$$

If you choose the second alternative, you use the fact that $u = x^2 + 1$ to conclude that $u = 1$ when $x = 0$, and $u = 2$ when $x = 1$. Hence,

$$\int_0^1 8x(x^2 + 1)^3 \, dx = u^4 \Big|_1^2 = 16 - 1 = 15$$

Probably the most efficient approach is to adopt the second alternative and write the solution compactly as follows:

$$\int_0^1 8x(x^2 + 1)^3 \, dx = \int_1^2 4u^3 \, du = u^4 \Big|_1^2 = 16 - 1 = 15$$

Here is a summary of the method of substitution for definite integrals.

Integration by Substitution

$$\int_a^b g[u(x)] \frac{du}{dx} \, dx = \int_{u(a)}^{u(b)} g(u) \, du$$

Here is one more example.

EXAMPLE 1.5   Evaluate $\displaystyle\int_1^e \frac{\ln x}{x}\, dx$.

SOLUTION

The integrand can be written as the product $(\ln x)\left(\dfrac{1}{x}\right)$ in which one of the factors $\dfrac{1}{x}$ is the derivative of the other $\ln x$. This suggests that you let $u = \ln x$. Then, $du = \dfrac{1}{x}\, dx$, $u(1) = 0$, and $u(e) = 1$. Hence,

$$\int_1^e \frac{\ln x}{x}\, dx = \int_0^1 u\, du = \frac{1}{2}u^2\Big|_0^1 = \frac{1}{2} - 0 = \frac{1}{2}$$

The Evaluation of Definite Integrals by Parts

The formula for integration by parts can be rephrased for definite integrals as follows.

Integration by Parts

$$\int_a^b f(x)g(x)\, dx = f(x)G(x)\Big|_a^b - \int_a^b f'(x)G(x)\, dx$$

where G is an antiderivative of g.

Here are two examples.

EXAMPLE 1.6   Evaluate $\displaystyle\int_0^{\ln 2} xe^x\, dx$.

SOLUTION

Both factors x and e^x are easy to integrate, while the factor x is simplified by differentiation. This suggests that you try integration by parts with

$$g(x) = e^x \quad \text{and} \quad f(x) = x$$

Then,

$$G(x) = e^x \quad \text{and} \quad f'(x) = 1$$

and so

$$\int_0^{\ln 2} xe^x\, dx = xe^x\Big|_0^{\ln 2} - \int_0^{\ln 2} e^x\, dx = (xe^x - e^x)\Big|_0^{\ln 2}$$

$$= (2\ln 2 - 2) - (0 - 1) = 2\ln 2 - 1$$

EXAMPLE 1.7   Evaluate $\displaystyle\int_1^e x \ln x \, dx$.

SOLUTION

Since the factor x is easy to integrate and the factor $\ln x$ is simplified by differentiation, try integration by parts with

$$g(x) = x \quad \text{and} \quad f(x) = \ln x$$

Then,

$$G(x) = \frac{1}{2}x^2 \quad \text{and} \quad f'(x) = \frac{1}{x}$$

and so

$$\int_1^e x \ln x \, dx = \frac{1}{2}x^2 \ln x \Big|_1^e - \frac{1}{2}\int_1^e x^2\left(\frac{1}{x}\right) dx$$

$$= \frac{1}{2}x^2 \ln x \Big|_1^e - \frac{1}{2}\int_1^e x \, dx$$

$$= \left(\frac{1}{2}x^2 \ln x - \frac{1}{4}x^2\right)\Big|_1^e$$

$$= \left(\frac{1}{2}e^2 \ln e - \frac{1}{4}e^2\right) - \left[\frac{1}{2}(1)^2 \ln 1 - \frac{1}{4}(1)^2\right]$$

$$= \frac{1}{2}e^2 - \frac{1}{4}e^2 + \frac{1}{4} = \frac{1}{4}(e^2 + 1)$$

Definite Integrals and Sums

In this text, we *define* the definite integral to be the net change in the antiderivative and then *prove* (in Chapter 7) that this net change is equal to the "limiting value" of a certain sum. Some calculus texts take the opposite approach. In particular, they *define* the definite integral as the limiting value of a sum and then *prove* that it can be computed as the net change in the antiderivative. In either case, the important result is the connection between antiderivatives and sums. This relationship is known as the **fundamental theorem of calculus** and will be derived and applied in Chapter 7 of this text.

Problems

In Problems 1 through 24, evaluate the given definite integral.

1. $\displaystyle\int_0^1 (x^4 - 3x^3 + 1) \, dx$

2. $\displaystyle\int_{-1}^0 (3x^5 - 3x^2 + 2x - 1) \, dx$

3. $\displaystyle\int_2^5 (2 + 2t + 3t^2) \, dt$

4. $\displaystyle\int_1^9 \left(\sqrt{t} - \frac{1}{\sqrt{t}}\right) dt$

5. $\int_1^3 \left(1 + \frac{1}{x} + \frac{1}{x^2}\right) dx$

6. $\int_{\ln 1/2}^{\ln 2} (e^t - e^{-t}) \, dt$

7. $\int_{-3}^{-1} \frac{t+1}{t^3} \, dt$

8. $\int_0^6 x^2(x-1) \, dx$

9. $\int_1^2 (2x-4)^5 \, dx$

10. $\int_{-3}^0 (2x+6)^4 \, dx$

11. $\int_0^4 \frac{1}{\sqrt{6t+1}} \, dt$

12. $\int_1^2 \frac{x^2}{(x^3+1)^2} \, dx$

13. $\int_0^1 (t^3 + t) \sqrt{t^4 + 2t^2 + 1} \, dt$

14. $\int_0^1 \frac{6x}{x^2+1} \, dx$

15. $\int_2^{e+1} \frac{x}{x-1} \, dx$

16. $\int_1^2 (t+1)(t-2)^9 \, dt$

17. $\int_1^{e^2} \ln t \, dt$

18. $\int_{1/2}^{e/2} t \ln 2t \, dt$

19. $\int_{-2}^2 xe^{-x} \, dx$

20. $\int_0^1 x^2 e^{2x} \, dx$

21. $\int_1^{e^2} \frac{(\ln x)^2}{x} \, dx$

22. $\int_e^{e^2} \frac{1}{x \ln x} \, dx$

23. $\int_0^{10} (20 + t)e^{-0.1t} \, dt$

24. $\int_0^5 te^{-(5-t)/20} \, dt$

Population growth 25. A study indicates that x months from now the population of a certain town will be increasing at the rate of $5 + 3x^{2/3}$ people per month. By how much will the population of the town increase over the next 8 months?

Distance and speed 26. An object is moving so that its speed after t minutes is $5 + 2t + 3t^2$ meters per minute. How far does the object travel during the 2nd minute?

Depreciation 27. The resale value of a certain industrial machine decreases over a 10-year period at a rate that changes with time. When the machine is x years old, the rate at which its value is changing is $220(x - 10)$ dollars per year. By how much does the machine depreciate during the 2nd year?

Admission to events 28. The promoters of a county fair estimate that t hours after the gates open at 9:00 A.M. visitors will be entering the fair at the rate of $-4(t + 2)^3 + 54(t + 2)^2$ people per hour. How many people will enter the fair between 10:00 A.M. and noon?

Marginal cost 29. At a certain factory, the marginal cost is $6(q - 5)^2$ dollars per unit when the level of production is q units. By how much will the total manufacturing cost increase if the level of production is raised from 10 to 13 units?

Oil production 30. A certain oil well that yields 400 barrels of crude oil a month will run dry in 2 years. The price of crude oil is currently $18 per barrel and is expected to rise at a constant rate of 3 cents per barrel per month. If the oil is sold as soon as it is extracted from the ground, what will the total future revenue from the well be?

Farming 31. It is estimated that t days from now a farmer's crop will be increasing at the rate of $0.3t^2 + 0.6t + 1$ bushels per day. By how much will the value of the crop increase during the next 5 days if the market price remains fixed at $3 per bushel?

Energy consumption 32. It is estimated that the demand for oil is increasing exponentially at the rate of 10 percent per year. If the demand for oil is currently 30 billion barrels per year, how much oil will be consumed during the next 10 years?

Sales revenue 33. It is estimated that the demand for a manufacturer's product is increasing exponentially at the rate of 2 percent per year. If the current demand is 5,000 units per year and if the price remains fixed at $400 per unit, how much revenue will the manufacturer receive from the sale of the product over the next 2 years?

Efficiency 34. After t hours on the job, a factory worker can produce $100te^{-0.5t}$ units per hour. How many units does a worker who arrives on the job at 8:00 A.M. produce between 10:00 A.M. and noon?

35. (a) Show that $\displaystyle\int_a^b f(x)\, dx + \int_b^c f(x)\, dx = \int_a^c f(x)\, dx.$

(b) Use the formula in part (a) to evaluate $\displaystyle\int_{-1}^1 |x|\, dx.$ (*Hint:* Evaluate $\displaystyle\int_{-1}^0 |x|\, dx$ and $\displaystyle\int_0^1 |x|\, dx$ and combine the results.)

(c) Evaluate $\displaystyle\int_0^4 (1 + |x - 3|)^2\, dx.$

Even and odd functions 36. (a) Show that if F is an antiderivative of f, then

$$\int_a^b f(-x)\, dx = -F(-b) + F(-a)$$

(*Hint:* Use the method of substitution with $u = -x$.)
(b) A function f is said to be **even** if $f(-x) = f(x)$. [For example, $f(x) = x^2$ is even.] Use Problem 35 and part (a) to show that if f is even, then

$$\int_{-a}^a f(x)\, dx = 2 \int_0^a f(x)\, dx$$

(c) Use part (b) to evaluate $\int_{-1}^{1} |x| \, dx$ and $\int_{-2}^{2} x^2 \, dx$.

(d) A function f is said to be **odd** if $f(-x) = -f(x)$. Use Problem 35 and part (a) to show that if f is odd, then

$$\int_{-a}^{a} f(x) \, dx = 0$$

(e) Evaluate $\int_{-12}^{12} x^3 \, dx$.

2   Area and Integration

There is a surprising connection between definite integrals and the geometric concept of area. If $f(x)$ is continuous and nonnegative on the interval $a \le x \le b$ and R is the region under the graph of f between $x = a$ and $x = b$ as shown in Figure 2.1, then the area of R is equal to the definite integral $\int_{a}^{b} f(x) \, dx$.

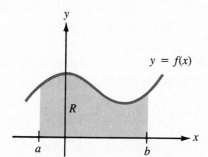

Figure 2.1 The area under the curve $y = f(x)$.

The Area Under a Curve

If $f(x)$ is continuous and nonnegative on the interval $a \le x \le b$ and R is the region bounded by the graph of f, the vertical lines $x = a$ and $x = b$, and the x axis, then

$$\text{Area of } R = \int_{a}^{b} f(x) \, dx$$

The use of this formula is illustrated in the next example for a region whose area you already know.

EXAMPLE 2.1   Find the area of the region bounded by the lines $y = 2x$ and $x = 2$, and the x axis.

SOLUTION

The region in question is the triangle in Figure 2.2, and its area is

$$\text{Area} = \frac{1}{2}(\text{base})(\text{height}) = \frac{1}{2}(2)(4) = 4$$

To compute this area using calculus, apply the integral formula with $f(x) = 2x$. Take $b = 2$ since the region is bounded on the right by the line $x = 2$, and take $a = 0$ since, on the left, the boundary consists of the single point $(0, 0)$, which is part of the vertical line $x = 0$. You will find, as expected, that

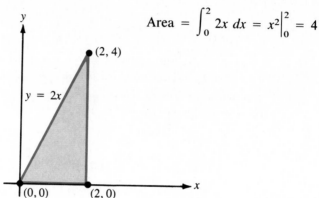

$$\text{Area} = \int_0^2 2x \, dx = x^2 \Big|_0^2 = 4$$

Figure 2.2 The region under $y = 2x$ from $x = 0$ to $x = 2$.

Here is another example.

EXAMPLE 2.2   Find the area of the region bounded by the curve $y = -x^2 + 4x - 3$ and the x axis.

SOLUTION

From the factored form of the polynomial

$$y = -x^2 + 4x - 3 = -(x - 3)(x - 1)$$

you see that the x intercepts of the curve are $(1, 0)$ and $(3, 0)$. From the corresponding graph (Figure 2.3) you see that the region in question is below the curve $y = -x^2 + 4x - 3$ and extends from $x = 1$ to $x = 3$. Hence,

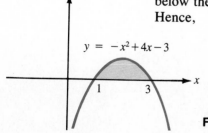

$$\text{Area} = \int_1^3 (-x^2 + 4x - 3) \, dx = \left(-\frac{1}{3}x^3 + 2x^2 - 3x\right)\Big|_1^3$$

$$= (-9 + 18 - 9) - \left(-\frac{1}{3} + 2 - 3\right) = \frac{4}{3}$$

Figure 2.3 The region bounded by $y = -x^2 + 4x - 3$ and the x axis.

Why the Integral Formula for Area Works

To see why the integral formula for area works, suppose $f(x)$ is continuous and nonnegative on the interval $a \leq x \leq b$. For any value of x in this interval, let $A(x)$ denote the area of the region under the graph of f between a and x as shown in Figure 2.4. Note that $A(x)$ is a function of x on the interval $a \leq x \leq b$.

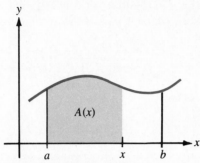

Figure 2.4 The area $A(x)$.

Your goal is to show that $A(b) = \displaystyle\int_a^b f(x)\, dx$. The key step is to establish that the derivative $A'(x)$ of the area function is equal to $f(x)$. To do this, consider the difference quotient

$$\frac{A(x + \Delta x) - A(x)}{\Delta x}$$

The expression $A(x + \Delta x) - A(x)$ in the numerator is just the area under the curve between x and $x + \Delta x$. If Δx is small, this area is approximately the same as the area of the rectangle whose height is $f(x)$ and whose width is Δx, as indicated in Figure 2.5. That is,

$$A(x + \Delta x) - A(x) \simeq f(x)\, \Delta x$$

or, equivalently,

$$\frac{A(x + \Delta x) - A(x)}{\Delta x} \simeq f(x)$$

As Δx approaches zero, the error in the approximation approaches zero and it follows that

$$\lim_{\Delta x \to 0} \frac{A(x + \Delta x) - A(x)}{\Delta x} = f(x)$$

But, by the definition of the derivative,

$$\lim_{\Delta x \to 0} \frac{A(x + \Delta x) - A(x)}{\Delta x} = A'(x)$$

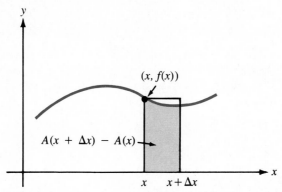

Figure 2.5 The area $A(x + \Delta x) - A(x)$ and an approximating rectangle.

Hence,

$$A'(x) = f(x)$$

and so

$$\int_a^b f(x)\ dx = \int_a^b A'(x)\ dx = A(b) - A(a)$$

But $A(a)$ is the area under the curve between $x = a$ and $x = a$, which is clearly zero. Hence,

$$\int_a^b f(x)\ dx = A(b)$$

and the area formula is proved.

Areas of More Complex Regions

The region in the next example is not bounded above by a single curve. However, it can be broken into two regions that are, and the area of each can be computed using the integral formula.

EXAMPLE 2.3   Find the area of the region R in the first quadrant that lies under the curve $y = \dfrac{1}{x}$ and is bounded by this curve and the lines $y = x$, $y = 0$, and $x = 2$.

SOLUTION
First sketch the region as shown in Figure 2.6.
    Observe that to the left of $x = 1$, R is bounded above by the line $y = x$, while to the right of $x = 1$, it is bounded above by the curve $y = \dfrac{1}{x}$.

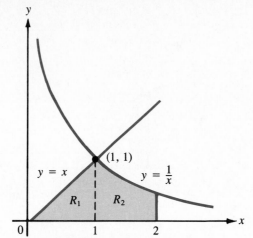

Figure 2.6 The region in the first quadrant bounded by $y = \dfrac{1}{x}$, $y = x$, $y = 0$, and $x = 2$.

This suggests that you break R into two subregions, R_1 and R_2, as shown in Figure 2.6 and apply the integral formula for area to each subregion separately. In particular,

$$\text{Area of } R_1 = \int_0^1 x \, dx = \frac{1}{2}x^2 \Big|_0^1 = \frac{1}{2}$$

and

$$\text{Area of } R_2 = \int_1^2 \frac{1}{x} \, dx = \ln |x| \Big|_1^2 = \ln 2 - \ln 1 = \ln 2$$

The area of R is the sum of these two areas. That is,

$$\text{Area of } R = \text{area of } R_1 + \text{area of } R_2 = \frac{1}{2} + \ln 2 \approx 1.19$$

Probability Density Functions

One of the most important applications in the social, managerial, and life sciences of the integral formula for area is the computation of probabilities. Here, by way of a very brief introduction, is a simplified outline of a typical situation. A more detailed discussion of the topic will be given in Section 4 of the next chapter.

The life span x of an electronic component selected at random from a manufacturer's stock is a quantity that cannot be predicted with certainty. The **probability** that the life span x lies in some interval $a \le x \le b$ is denoted by $P(a \le x \le b)$ and is defined to be the fraction of all the components manufactured by the company that can be expected to have life spans in this range. A probability of the form $P(a \le x \le b)$ can be interpreted as the area under a certain curve and evaluated as a definite integral. The curve is the graph of a function $f(x)$ known as the **probability density function** for x. More precisely, a probability density function for the variable x is a positive continuous function $f(x)$ with the property that $P(a \leq x \leq b)$ is the area under its graph between $x = a$ and $x = b$. The situation is illustrated in Figure 2.7.

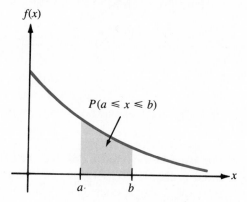

Figure 2.7 A probability density function.

Probability density functions are constructed by statisticians from experimental data and theoretical considerations using techniques explained in most probability and statistics texts. The purpose of the discussion in this section is to show you how to use integral calculus to compute probabilities once the appropriate probability density function is known. Here is an example.

EXAMPLE 2.4   Suppose that the probability density function for the life span of electronic components produced by a certain company is $f(x) = 0.02e^{-0.02x}$, where x denotes the life span (in months) of a randomly selected component.

(a) What is the probability that the life span of a component selected at random will be between 20 and 30 months?
(b) What is the probability that the life span of a component selected at random will be less than or equal to 20 months?
(c) What is the probability that the life span of a component selected at random will be greater than 20 months?

SOLUTION
(a) The desired probability $P(20 \leq x \leq 30)$ is the area (Figure 2.8a) under the graph of the density function between $x = 20$ and $x = 30$. Using the integral formula to compute this area, you get

$$P(20 \leq x \leq 30) = \int_{20}^{30} 0.02e^{-0.02x}\, dx = -e^{-0.02x}\Big|_{20}^{30}$$

$$\approx -0.5488 + 0.6703 \approx 0.1215$$

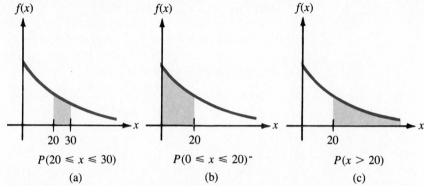

Figure 2.8 Areas under the probability density function $f(x) = 0.02e^{-0.02x}$.

That is, approximately 12.15 percent of the components manufactured by the company will have life spans of between 20 and 30 months.

(b) The desired probability is $P(0 \leq x \leq 20)$, which is the area (Figure 2.8b) under the density function between $x = 0$ and $x = 20$. That is,

$$P(0 \leq x \leq 20) = \int_0^{20} 0.02e^{-0.02x}\, dx = -e^{-0.02x}\Big|_0^{20}$$

$$\simeq -0.6703 + 1 \simeq 0.3297$$

That is, roughly $\frac{1}{3}$ of the components will fail during the first 20 months.

(c) The fraction of components whose life span is greater than 20 months is 1 (the fraction representing 100 percent of all the components) minus the fraction whose life span is less than or equal to 20 months. Hence,

$$P(x > 20) = 1 - P(0 \leq x \leq 20) \simeq 1 - 0.3297 \simeq 0.6703$$

In geometric terms, this is the area (Figure 2.8c) under the density function to the right of $x = 20$.

The Area Between Two Curves

In some practical problems you may have to compute the area between two curves. Suppose $f(x)$ and $g(x)$ are nonnegative functions and that $f(x) \geq g(x)$ on the interval $a \leq x \leq b$ as shown in Figure 2.9a.

To find the area of the region R between the curves from $x = a$ to $x = b$, you simply subtract the area under the lower curve $y = g(x)$ (Figure 2.9c) from the area under the upper curve $y = f(x)$ (Figure 2.9b). That is,

$$\text{Area of } R = \int_a^b f(x)\, dx - \int_a^b g(x)\, dx = \int_a^b [f(x) - g(x)]\, dx$$

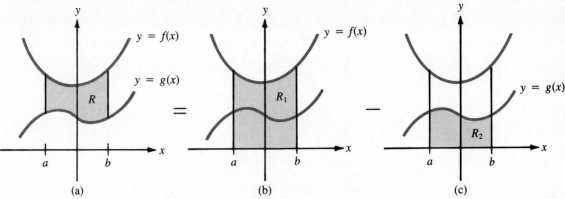

Figure 2.9 Area of R = area of R_1 − area of R_2.

It can be shown that this formula remains valid, even if the functions f and g are not assumed to be nonnegative.

The Area Between Two Curves

> If $f(x)$ and $g(x)$ are continuous on the interval $a \le x \le b$ with $f(x) \ge g(x)$, and if R is the region bounded by the graphs of f and g and the vertical lines $x = a$ and $x = b$, then
>
> $$\text{Area of } R = \int_a^b [f(x) - g(x)]\, dx$$

Here are two examples.

EXAMPLE 2.5   Find the area of the region bounded by the curves $y = x^2 + 1$ and $y = 2x - 2$ between $x = -1$ and $x = 2$.

SOLUTION

So that you can visualize the situation, begin by sketching the region as shown in Figure 2.10. Then apply the integral formula with $f(x) = x^2 + 1$, $g(x) = 2x - 2$, $a = -1$, and $b = 2$ to get

$$\text{Area} = \int_{-1}^{2} [(x^2 + 1) - (2x - 2)]\, dx = \int_{-1}^{2} (x^2 - 2x + 3)\, dx$$

$$= \left(\frac{1}{3}x^3 - x^2 + 3x \right)\Big|_{-1}^{2} = \frac{14}{3} - \left(-\frac{13}{3} \right) = 9$$

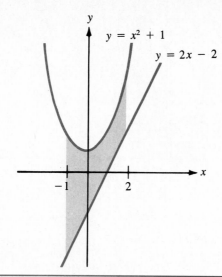

Figure 2.10 The region bounded by $y = x^2 + 1$ and $y = 2x - 2$ between $x = -1$ and $x = 2$.

EXAMPLE 2.6   Find the area of the region bounded by the curves $y = x^3$ and $y = x^2$.

SOLUTION

Sketch the curves as shown in Figure 2.11.

Find the points of intersection by solving the equations of the two curves simultaneously to get

$$x^3 = x^2 \qquad x^3 - x^2 = 0 \qquad x^2(x - 1) = 0$$

or

$$x = 0 \quad \text{and} \quad x = 1$$

The corresponding points $(0, 0)$ and $(1, 1)$ are the points of intersection.

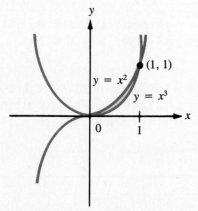

Figure 2.11 The region bounded by $y = x^3$ and $y = x^2$.

Notice that for $0 \le x \le 1$, the graph of $y = x^2$ lies above that of $y = x^3$ (since the square of a fraction between 0 and 1 is greater than its cube). Hence the region in question is bounded above by the curve $y = x^2$ and below by the curve $y = x^3$ and extends from $x = 0$ to $x = 1$. Hence,

$$\text{Area} = \int_0^1 (x^2 - x^3) \, dx = \left(\frac{1}{3}x^3 - \frac{1}{4}x^4 \right)\Big|_0^1 = \frac{1}{12}$$

Problems

In Problems 1 through 14, use calculus to find the area of the region R.

1. R is the triangle bounded by the line $y = 4 - 3x$ and the coordinate axes.

2. R is the triangle with vertices $(-4, 0)$, $(2, 0)$, and $(2, 6)$.

3. R is the rectangle with vertices $(1, 0)$, $(-2, 0)$, $(-2, 5)$, and $(1, 5)$.

4. R is the trapezoid bounded by the lines $y = x + 6$ and $x = 2$ and the coordinate axes.

5. R is the region bounded by the curve $y = \sqrt{x}$, the lines $x = 4$ and $x = 9$, and the x axis.

6. R is the region bounded by the curve $y = 4x^3$, the line $x = 2$, and the x axis.

7. R is the region bounded by the curve $y = 1 - x^2$ and the x axis.

8. R is the region bounded by the curve $y = -x^2 - 6x - 5$ and the x axis.

9. R is the region bounded by the curve $y = e^x$, the lines $x = 0$ and $x = \ln \frac{1}{2}$, and the x axis.

10. R is the region in the first quadrant bounded by the curve $y = 4 - x^2$ and the lines $y = 3x$ and $y = 0$.

11. R is the region bounded by the curve $y = \sqrt{x}$ and the lines $y = 2 - x$ and $y = 0$.

12. R is the region in the first quadrant that lies under the curve $y = \frac{16}{x}$ and that is bounded by this curve and the lines $y = x$, $y = 0$, and $x = 8$.

13. R is the region that lies below the curve $y = x^2 + 4$ and is bounded by this curve, the line $y = -x + 10$, and the coordinate axes.

14. R is the region bounded by the curve $y = x^2 - 2x$ and the x axis. (*Hint*: Reflect the region across the x axis and integrate the corresponding function.)

Product reliability

15. The probability density function for the life span of the light bulbs manufactured by a certain company is $f(x) = 0.01e^{-0.01x}$, where x denotes the life span (in hours) of a randomly selected bulb.
    (a) What is the probability that the life span of a randomly selected bulb will be between 50 and 60 hours?
    (b) What is the probability that the life span of a randomly selected bulb will be less than or equal to 60 hours?
    (c) What is the probability that the life span of a randomly selected bulb will be greater than 60 hours?

Duration of telephone calls

16. The probability density function for the duration of telephone calls in a certain city is $f(x) = 0.5e^{-0.5x}$, where x denotes the duration (in minutes) of a randomly selected call.
    (a) What percentage of the calls can be expected to last between 2 and 3 minutes?
    (b) What percentage of the calls can be expected to last 2 minutes or less?
    (c) What percentage of the calls can be expected to last more than 2 minutes?

Airplane arrivals

17. The probability density function for the time interval between the arrivals of successive planes at a certain airport is $f(x) = 0.2e^{-0.2x}$, where x is the time (in minutes) between the arrivals of a randomly selected pair of successive planes.
    (a) What is the probability that two successive planes selected at random will arrive within 5 minutes of one another?
    (b) What is the probability that two successive planes selected at random will arrive more than 6 minutes apart?

In Problems 18 through 28, find the area of the region R.

18. R is the region bounded by the curves $y = x^2 + 3$ and $y = 1 - x^2$ between $x = -2$ and $x = 1$.

19. R is the region bounded by the curves $y = x^2 + 5$ and $y = -x^2$, the line $x = 3$, and the y axis.

20. R is the region bounded by the curve $y = e^x$ and the lines $y = 1$ and $x = 1$.

21. R is the region bounded by the curve $y = x^2$ and the line $y = x$.

22. R is the region bounded by the curve $y = x^2$ and the line $y = 4$.

23. R is the region bounded by the curves $y = \sqrt{x}$ and $y = x^2$.

24. (a) R is the region to the right of the y axis that is bounded above by the curve $y = 4 - x^2$ and below by the line $y = 3$.
    (b) R is the region to the right of the y axis that lies below the line $y = 3$ and is bounded by the curve $y = 4 - x^2$, the line $y = 3$, and the coordinate axes.

25. R is the region bounded by the curves $y = x^3 - 6x^2$ and $y = -x^2$.

26. R is the region bounded by the line $y = x$ and the curve $y = x^3$.

27. R is the region bounded by the curve $y = \dfrac{1}{x^2}$ and the lines $y = x$ and $y = \dfrac{x}{8}$.

28. R is the region in the first quadrant bounded by the curve $y = x^2 + 2$ and the lines $y = 11 - 8x$ and $y = 11$.

3   Applications to Business and Economics

In this section you will see four examples from business and economics in which quantities can be expressed as definite integrals and represented geometrically as areas between or under curves. The first two examples deal with the general concept of **net gain,** while the last two involve consumers' willingness to spend and a related concept known as **consumers' surplus.**

Net Excess Profit

Suppose that x years from now, two investment plans will be generating profit at the rates of $R_1(x)$ and $R_2(x)$ dollars per year, respectively, and that for the next N years the rate $R_2(x)$ will be greater than the rate $R_1(x)$, as illustrated in Figure 3.1.

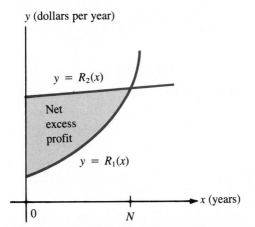

Figure 3.1 Net excess profit as the area between two curves.

The functions $R_2(x)$ and $R_1(x)$ represent the rates at which profit is generated by the second and first plans, respectively. Hence, their difference, $R_2(x) - R_1(x)$, represents the rate at which the profit generated by the second plan exceeds that generated by the first plan, and the net *excess* profit generated by the second plan during the next N years is the definite integral of this rate of change from $x = 0$ to $x = N$. That is,

$$\text{Net excess profit} = \int_0^N [R_2(x) - R_1(x)]\, dx$$

which can be interpreted geometrically as the area between the curves $y = R_2(x)$ and $y = R_1(x)$ from $x = 0$ to $x = N$. Here is an example.

EXAMPLE 3.1 Suppose that x years from now, one investment plan will be generating profit at the rate of $R_1(x) = 50 + x^2$ dollars per year, while a second plan will be generating profit at the rate of $R_2(x) = 200 + 5x$ dollars per year.

(a) For how many years will the second plan be the more profitable one?
(b) Compute your net excess profit if you invest in the second plan instead of the first for the period of time in part (a).
(c) Interpret the net excess profit from part (b) as the area between two curves.

SOLUTION
To help you visualize the situation, begin by sketching the curves $y = R_1(x)$ and $y = R_2(x)$ as shown in Figure 3.2.

(a) As the graph indicates, the rate $R_2(x)$ at which the second plan generates profit is initially greater than the rate $R_1(x)$ at which the first plan generates profit. The second plan will be the more profitable one until $R_1(x) = R_2(x)$, that is, until

$$50 + x^2 = 200 + 5x \qquad x^2 - 5x - 150 = 0$$

$$(x - 15)(x + 10) = 0 \qquad \text{or} \qquad x = 15 \text{ years}$$

(b) For $0 \le x \le 15$, the rate at which the profit generated by the second plan exceeds that of the first plan is $R_2(x) - R_1(x)$ dollars per year. Hence the net excess profit generated over the 15-year period by the second plan is the definite integral

$$\int_0^{15} [R_2(x) - R_1(x)]\, dx = \int_0^{15} [(200 + 5x) - (50 + x^2)]\, dx$$

$$= \int_0^{15} (150 + 5x - x^2)\, dx$$

$$= \left(150x + \frac{5}{2}x^2 - \frac{1}{3}x^3 \right)\Big|_0^{15} = \$1{,}687.50$$

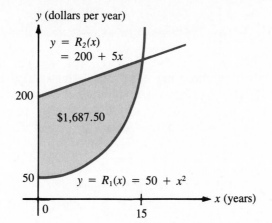

y (dollars per year)

$y = R_2(x)$
$= 200 + 5x$

200

$1,687.50

50

$y = R_1(x) = 50 + x^2$

x (years)

0   15

Figure 3.2 Net excess profit for Example 3.1.

(c) In geometric terms, the definite integral giving the net excess profit in part (b) is the area of the shaded region in Figure 3.2 between the curves $y = R_2(x)$ and $y = R_1(x)$ from $x = 0$ to $x = 15$.

Net Earnings from Industrial Equipment

The net earnings generated by an industrial machine over a period of time is the difference between the total revenue generated by the machine and the total cost of operating and servicing the machine. In the next example, the net earnings of a machine is calculated as a definite integral and interpreted as the area between two curves.

EXAMPLE 3.2   Suppose that when it is x years old, an industrial machine generates revenue at the rate of $R(x) = 5,000 - 20x^2$ dollars per year and results in costs that accumulate at the rate of $C(x) = 2,000 + 10x^2$ dollars per year.

(a) For how many years is the use of the machine profitable?

(b) What are the net earnings generated by the machine during the period of time in part (a)?

(c) Interpret the net earnings in part (b) as the area between two curves.

SOLUTION

To help you visualize the situation, begin by sketching the curves $y = R(x)$ and $y = C(x)$ as shown in Figure 3.3.

(a) Use of the machine will be profitable as long as the rate at which revenue is generated is greater than the rate at which costs accumulate, that is, until $R(x) = C(x)$ or

$$5,000 - 20x^2 = 2,000 + 10x^2 \qquad 30x^2 = 3,000 \qquad x^2 = 100$$

or

$$x = 10 \text{ years}$$

(b) The functions $R(x)$ and $C(x)$ represent the rates of change of total revenue and total cost, respectively, and hence their difference, $R(x) - C(x)$, represents the rate of change of the net earnings generated by the machine. It follows that the net earnings for the next 10 years is the definite integral

$$\int_0^{10} [R(x) - C(x)] \, dx = \int_0^{10} [(5,000 - 20x^2) - (2,000 + 10x^2)] \, dx$$

$$= \int_0^{10} (3,000 - 30x^2) \, dx$$

$$= (3,000x - 10x^3) \Big|_0^{10} = \$20,000$$

(c) In geometric terms, the definite integral representing the net earnings is the area of the shaded region in Figure 3.3 between the curves $y = R(x)$ and $y = C(x)$ from $x = 0$ to $x = 10$.

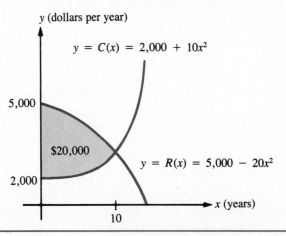

Figure 3.3 The net earnings from an industrial machine.

The Consumers' Demand Curve and Willingness to Spend

In studying consumer behavior, economists often assume that the price a consumer or group of consumers is willing to pay to buy an additional unit of a commodity is a function of the number of units of the commodity that the consumer or group has already bought.

For example, a young couple on a limited budget might be willing to spend up to $500 to own one television set. For the convenience of having two sets (to eliminate the conflict between *Monday Night Football* and the network news, for example), the couple might be willing to spend an additional \$300 to buy a second set. Since there would be very little use for more than two sets, the couple might be willing to spend no more than \$50 for a third set.

A function $p = D(q)$ giving the price per unit that consumers are willing to pay to get the qth unit of a commodity is known in economics as the **consumers' demand function** for the commodity. As in the preceding illustration, the consumers' demand function is usually a decreasing function of q. That is, the price that consumers are willing to pay to get one additional unit usually decreases as the number of units already bought increases. A typical consumers' demand function is sketched in Figure 3.4a.

The consumers' demand function $p = D(q)$ giving the price per unit that consumers are willing to pay when the level of consumption is q units can be thought of as the rate of change with respect to q of the *total* amount consumers are willing to spend for q units. [In economic terminology, $D(q)$ is the **marginal willingness to spend.**] If $A(q)$ is the total amount (in dollars) that consumers are willing to spend to get q units of the commodity [and if $A(q)$ is differentiable], then $D(q) = \dfrac{dA}{dq}$. Hence, the total amount that consumers are willing to spend to get q_0 units of the commodity is the definite integral

$$A(q_0) = A(q_0) - A(0) = \int_0^{q_0} \frac{dA}{dq}\, dq = \int_0^{q_0} D(q)\, dq$$

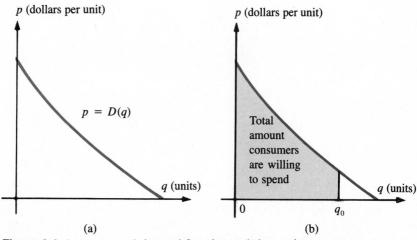

Figure 3.4 A consumers' demand function and the total amount consumers are willing to spend.

In geometric terms, this total willingness to spend is the area of the region in Figure 3.4b under the demand curve $p = D(q)$ between $q = 0$ and $q = q_0$.

EXAMPLE 3.3 Suppose the consumers' demand function for a certain commodity is $D(q) = 4(25 - q^2)$ dollars per unit.

(a) Find the total amount of money consumers are willing to spend to get 3 units of the commodity.
(b) Sketch the demand curve and interpret the answer to part (a) as an area.

SOLUTION

(a) Since the demand function $D(q) = 4(25 - q^2)$, measured in dollars per unit, is the rate of change with respect to q of consumers' willingness to spend, the total amount that consumers are willing to spend to get 3 units of the commodity is the definite integral

$$\int_0^3 D(q)\, dq = 4 \int_0^3 (25 - q^2)\, dq = 4\left(25q - \frac{1}{3}q^3\right)\Big|_0^3 = \$264$$

(b) The consumers' demand curve is sketched in Figure 3.5. In geometric terms, the total amount, \$264, that consumers are willing to spend to get 3 units of the commodity is the area under the demand curve from $q = 0$ to $q = 3$.

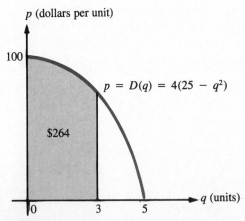

Figure 3.5 The consumers' demand curve $p = 4(25 - q^2)$ and consumers' willingness to spend for 3 units.

Consumers' Surplus

In a competitive economy, the total amount that consumers actually spend on a commodity is usually less than the total amount they would have been willing to spend. The difference between the two amounts can be thought of as a savings realized by consumers and is known in economics as the **consumers' surplus.** That is,

$$\text{Consumers' surplus} = \begin{pmatrix} \text{total amount consumers} \\ \text{would be willing to spend} \end{pmatrix}$$
$$- \begin{pmatrix} \text{actual consumer} \\ \text{expenditure} \end{pmatrix}$$

Market conditions determine the price per unit at which a commodity is sold. Once the price, say p_0, is known, the demand equation $p = D(q)$ determines the number of units q_0 that consumers will buy. The actual consumer expenditure for q_0 units of the commodity at the price of p_0 dollars per unit is $p_0 q_0$ dollars. The consumers' surplus is calculated by subtracting this amount from the total amount consumers would have been willing to spend to get q_0 units of the commodity.

To get a better feel for the concept of consumer surplus, consider once again the example of the couple that was willing to spend $500 for its first television set, $300 for a second set, and $50 for a third set. Suppose the market price for television sets is $300 per set. Then the couple would buy only two sets and would spend a total of $2 \times \$300 = \$600$. This is less than the $\$500 + \$300 = \$800$ that the couple would have been willing to spend to get the two sets. The savings of $\$800 - \$600 = \$200$ is the couple's consumer surplus.

Consumers' surplus has a simple geometric interpretation, which is illustrated in Figure 3.6. The symbols p_0 and q_0 denote the market price and corresponding demand, respectively. Figure 3.6a shows the region under the demand curve from $q = 0$ to $q = q_0$. Its area, as we have seen, represents the total amount that consumers are willing to spend to get q_0 units of the commodity. The rectangle in Figure 3.6b has an area of $p_0 q_0$ and hence represents the actual consumer expenditure for q_0 units at

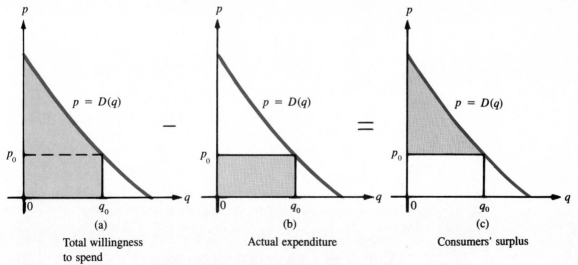

(a)
Total willingness
to spend

(b)
Actual expenditure

(c)
Consumers' surplus

Figure 3.6 Geometric interpretation of consumers' surplus.

p_0 dollars per unit. The difference between these two areas (Figure 3.6c) represents the consumers' surplus. That is, consumers' surplus is the area of the region between the demand curve $p = D(q)$ and the horizontal line $p = p_0$ and hence is equal to the definite integral

$$\int_0^{q_0} [D(q) - p_0] \, dq = \int_0^{q_0} D(q) \, dq - \int_0^{q_0} p_0 \, dq$$

$$= \int_0^{q_0} D(q) \, dq - p_0 q \Big|_0^{q_0}$$

$$= \int_0^{q_0} D(q) \, dq - p_0 q_0$$

Here is a summary of the situation.

Consumers' Surplus

> If q_0 units of a commodity are sold at a price of p_0 dollars per unit, and if $p = D(q)$ is the consumers' demand function for the commodity, then
>
> $$\text{Consumers' surplus} = \begin{pmatrix} \text{total amount consumers are} \\ \text{willing to spend for } q_0 \text{ units} \end{pmatrix}$$
>
> $$- \begin{pmatrix} \text{actual consumer expenditure} \\ \text{for } q_0 \text{ units} \end{pmatrix}$$
>
> $$= \int_0^{q_0} D(q) \, dq - p_0 q_0$$

EXAMPLE 3.4 Suppose the consumers' demand function for a certain commodity is $D(q) = 4(25 - q^2)$ dollars per unit.

(a) Find the consumers' surplus if the commodity is sold for $64 per unit.
(b) Sketch the demand curve and interpret the consumers' surplus as an area.

SOLUTION
(a) First find the number of units that will be bought by solving the demand equation $p = D(q)$ for q when $p = \$64$ to get

$$64 = 4(25 - q^2) \qquad 16 = 25 - q^2 \qquad q^2 = 9 \qquad \text{or} \qquad q = 3$$

That is, 3 units will be bought when the price is $64 per unit. The corresponding consumers' surplus is

$$\int_0^3 D(q)\, dq - 64(3) = 4 \int_0^3 (25 - q^2)\, dq - 192$$

$$= 4\left(25q - \frac{1}{3}q^3\right)\Big|_0^3 - 192 = 264 - 192 = \$72$$

(b) The consumers' demand curve is sketched in Figure 3.7. The consumers' surplus from part (a) is equal to the area of the region between the demand curve and the horizontal line $p = 64$.

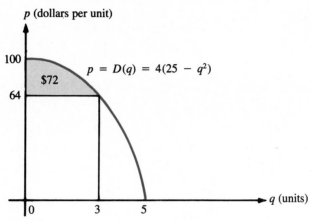

p (dollars per unit)

100

\$72

64

$p = D(q) = 4(25 - q^2)$

0   3   5   *q* (units)

Figure 3.7 The consumers' demand curve $p = 4(25 - q^2)$ and corresponding consumers' surplus when the price is \$64 per unit.

Problems

Investment
1. Suppose that x years from now, one investment plan will be generating profit at the rate of $R_1(x) = 100 + x^2$ dollars per year, while a second plan will be generating profit at the rate of $R_2(x) = 220 + 2x$ dollars per year.
   (a) For how many years will the second plan be the more profitable one?
   (b) How much excess profit will you earn if you invest in the second plan instead of the first for the period of time in part (a)?
   (c) Interpret the excess profit from part (b) as the area between two curves.

Investment
2. Suppose that x years from now, one investment plan will be generating profit at the rate of $R_1(x) = 60e^{0.12x}$ dollars per year, while a second plan will be generating profit at the rate of $R_2(x) = 160e^{0.08x}$ dollars per year.
   (a) For how many years will the second plan be the more profitable one?

(b) How much excess profit will you earn if you invest in the second plan instead of the first for the period of time in part (a)?

(c) Interpret the excess profit from part (b) as the area between two curves.

Industrial equipment 3. Suppose that when it is x years old, an industrial machine generates revenue at the rate of $R(x) = 6{,}025 - 10x^2$ dollars per year and results in costs that accumulate at the rate of $C(x) = 4{,}000 + 15x^2$ dollars per year.

(a) For how many years is the use of the machine profitable?

(b) What are the net earnings generated by the machine during the period of time in part (a)?

(c) Interpret the net earnings in part (b) as the area between two curves.

Efficiency 4. After t hours on the job, one factory worker is producing $Q_1(t) = 60 - 2(t - 1)^2$ units per hour while a second worker is producing $Q_2(t) = 50 - 5t$ units per hour.

(a) If both arrive on the job at 8:00 A.M., how many more units will the first worker have produced by noon than the second worker?

(b) Interpret the answer in part (a) as the area between two curves.

Fund-raising 5. It is estimated that t weeks from now, contributions in response to a fund-raising campaign will be coming in at the rate of $R(t) = 5{,}000e^{-0.2t}$ dollars per week, while campaign expenses are expected to accumulate at the constant rate of $676 per week.

(a) For how many weeks will the campaign be profitable?

(b) What net earnings will be generated by the campaign during the period of time in part (a)?

(c) Interpret the net earnings in part (b) as the area between two curves.

Consumers' willingness to spend For the consumers' demand functions $D(q)$ in Problems 6 through 9

(a) Find the total amount of money consumers are willing to spend to get q_0 units of the commodity.

(b) Sketch the demand curve and interpret the consumers' willingness to spend in part (a) as an area.

6. $D(q) = 2(64 - q^2)$ dollars per unit; $q_0 = 6$ units

7. $D(q) = \dfrac{300}{(0.1q + 1)^2}$ dollars per unit; $q_0 = 5$ units

8. $D(q) = \dfrac{400}{0.5q + 2}$ dollars per unit; $q_0 = 12$ units

9. $D(q) = 40e^{-0.05q}$ dollars per unit; $q_0 = 10$ units

Consumers' surplus For the consumers' demand functions $D(q)$ in Problems 10 through 13

(a) Find the consumers' surplus if the market price of the commodity is p_0 dollars per unit.

(b) Sketch the demand curve and interpret the consumers' surplus as an area.

10. $D(q) = 2(64 - q^2)$ dollars per unit; $p_0 = \$110$ per unit

11. $D(q) = \dfrac{300}{(0.1q + 1)^2}$ dollars per unit; $p_0 = \$12$ per unit

12. $D(q) = \dfrac{400}{0.5q + 2}$ dollars per unit; $p_0 = \$20$ per unit

13. $D(q) = 40e^{-0.05q}$ dollars per unit; $p_0 = \$11.46$ per unit

Chapter Summary and Review Problems

Important Terms, Symbols, and Formulas

Definite integral: $\displaystyle\int_a^b f(x)\,dx = F(b) - F(a)$

Integration by substitution:

$$\int_a^b g[u(x)]\frac{du}{dx}\,dx = \int_{u(a)}^{u(b)} g(u)\,du$$

Integration by parts:

$$\int_a^b f(x)g(x)\,dx = f'(x)G(x)\Big|_a^b - \int_a^b f'(x)G(x)\,dx$$

where G is an antiderivative of g

Area under a curve:

Area of $R = \displaystyle\int_a^b f(x)\,dx$

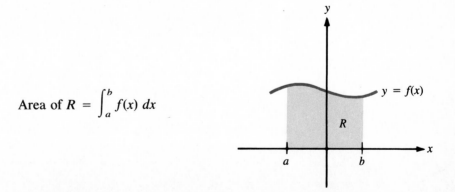

Probability density function: $P(a \le x \le b) = \displaystyle\int_a^b f(x)\,dx$

Area between two curves:

$$\text{Area of } R = \int_a^b [f(x) - g(x)] \, dx$$

Net gain

Demand function: $p = D(q)$

Consumers' willingness to spend

Consumers' surplus

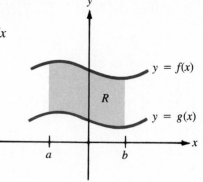

Review Problems

In Problems 1 through 10, evaluate the given definite integral.

1. $\displaystyle\int_0^1 (5x^4 - 8x^3 + 1) \, dx$

2. $\displaystyle\int_1^4 (\sqrt{x} + x^{-3/2}) \, dx$

3. $\displaystyle\int_{-1}^2 30(5x - 2)^2 \, dx$

4. $\displaystyle\int_0^1 2xe^{x^2-1} dx$

5. $\displaystyle\int_0^1 (x - 3)(x^2 - 6x + 2)^3 \, dx$

6. $\displaystyle\int_{-1}^1 \frac{3x + 6}{(x^2 + 4x + 5)^2} \, dx$

7. $\displaystyle\int_{-1}^1 xe^x \, dx$

8. $\displaystyle\int_e^{e^2} \frac{1}{x \, (\ln x)^2} \, dx$

9. $\displaystyle\int_1^e x^2 \ln x \, dx$

10. $\displaystyle\int_0^{10} (2x + 1)e^{0.2x} \, dx$

11. A study indicates that x months from now the population of a certain town will be increasing at the rate of $10 + 2\sqrt{x}$ people per month. By how much will the population of the town increase over the next 9 months?

12. It is estimated that t days from now a farmer's crop will be increasing at the rate of $0.3t^2 + 0.6t + 1$ bushels per day. By how much will the value of the crop increase during the next 6 days if the market price remains fixed at $2 per bushel?

13. It is estimated that the demand for oil is increasing exponentially at the rate of 10 percent per year. If the demand for oil is currently 40 billion barrels per year, how much oil will be consumed during the next 5 years?

In Problems 14 through 19, find the area of the region R.

14. R is the region bounded by the curve $y = 3x^2 + 2$, the lines $x = -1$ and $x = 3$, and the x axis.

15. R is the region bounded by the curve $y = \dfrac{1}{x^2}$, the x axis, and the vertical lines $x = 1$ and $x = 4$.

16. R is the region bounded by the curve $y = 2 + x - x^2$ and the x axis.

17. R is the region bounded by the curves $y = \dfrac{8}{x}$ and $y = \sqrt{x}$ and the lines $y = 0$ and $x = 8$.

18. R is the region bounded by the curve $y = x^4$ and the line $y = x$.

19. R is the region in the first quadrant bounded by the line $y = 7x$ and the curves $y = 8 - x^2$ and $y = x^2$.

20. The probability density function for the duration of telephone calls in a certain city is $f(x) = 0.4e^{-0.4x}$, where x denotes the duration (in minutes) of a randomly selected call.
    (a) What percentage of the calls can be expected to last between 1 and 2 minutes?
    (b) What percentage of the calls can be expected to last 2 minutes or less?
    (c) What percentage of the calls can be expected to last more than 2 minutes?

21. When it is x years old, a certain industrial machine generates revenue at the rate of $R(x) = 4{,}575 - 5x^2$ dollars per year and results in costs that accumulate at the rate of $C(x) = 1{,}200 + 10x^2$ dollars per year.
    (a) For how many years is the use of the machine profitable?
    (b) What are the net earnings generated by the machine during the period of time in part (a)?
    (c) Interpret the net earnings in part (b) as the area between two curves.

22. Suppose the consumers' demand function for a certain commodity is $D(q) = 30 - 2q - q^2$ dollars per unit.
    (a) Find the number of units that will be bought if the market price is $15 per unit.
    (b) Compute the consumers' willingness to spend to get the number of units in part (a).
    (c) Compute the consumers' surplus when the market price is $15 per unit.
    (d) Sketch the demand curve and interpret the consumers' willingness to spend and the consumer's surplus as areas.

Chapter

7 Integration: Further Topics

1 The Definite Integral as the Limit of a Sum

In the preceding chapter, we defined the definite integral of a function to be a certain net change in its antiderivative. In particular,

$$\int_a^b f(x)\ dx = F(b) - F(a)$$

where F is an antiderivative of f. In this section we shall introduce an important relationship between antiderivatives and limits of sums known as the **fundamental theorem of calculus.** By combining this theorem with our definition of the definite integral, we shall be able to interpret certain limits of sums arising in practical problems as definite integrals, which we can then evaluate by antidifferentiation.

The discussion begins with a geometric derivation of the theorem based on the interpretation of definite integrals as areas.

Geometric Derivation of the Fundamental Theorem of Calculus

Suppose that $f(x)$ is nonnegative and continuous on the interval $a \leq x \leq b$. You can approximate the area under the graph of f between $x = a$ and $x = b$ as follows: First divide the interval $a \leq x \leq b$ into n equal subintervals of width Δx and let x_j denote the beginning of the jth subinterval. Then draw n rectangles such that the base of the jth rectangle is the jth subinterval and the height of the jth rectangle is $f(x_j)$. The situation is illustrated in Figure 1.1.

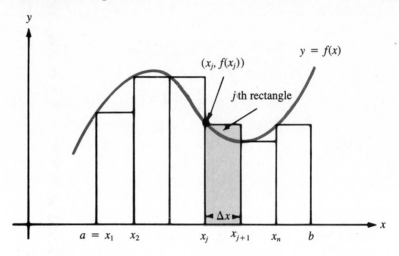

Figure 1.1 An approximation by rectangles of the area under a curve.

The area of the jth rectangle is $f(x_j) \, \Delta x$ and is an approximation to the area under the curve from $x = x_j$ to $x = x_{j+1}$. The sum of the areas of all n rectangles is

$$f(x_1) \, \Delta x + f(x_2) \, \Delta x + \cdots + f(x_n) \, \Delta x$$

This sum is an approximation to the total area under the curve from $x = a$ to $x = b$ and hence an approximation to the corresponding definite integral

$$\int_a^b f(x) \, dx$$

That is,

$$f(x_1) \, \Delta x + f(x_2) \, \Delta x + \cdots + f(x_n) \, \Delta x \simeq \int_a^b f(x) \, dx$$

As Figure 1.2 suggests, the sum of the areas of the rectangles approaches the actual area under the curve as the number of rectangles increases without bound. That is, as n increases without bound,

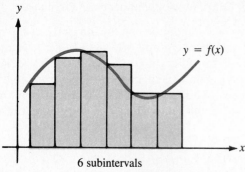

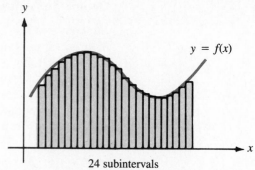

Figure 1.2 The approximation improves as the number of subintervals increases.

$$f(x_1)\,\Delta x + f(x_2)\,\Delta x + \cdots + f(x_n)\,\Delta x \rightarrow \int_a^b f(x)\,dx = F(b) - F(a)$$

where F is an antiderivative of f.

This is the relationship between sums and integrals that is known as the fundamental theorem of calculus. Although this argument establishes it only for nonnegative functions, it actually holds for any function that is continuous on the interval $a \le x \le b$.

Summation Notation

You can write the relationship between definite integrals and sums more compactly if you use the following **summation notation.** (The use of summation notation is discussed in more detail in the algebra review in Section A of the appendix.) To describe the sum

$$f(x_1)\,\Delta x + f(x_2)\,\Delta x + \cdots + f(x_n)\,\Delta x$$

specify the general term $f(x_j)\,\Delta x$ and use the symbol

$$\sum_{j=1}^{n} f(x_j)\,\Delta x$$

to indicate that n terms of this form are to be added together, starting with the term in which $j = 1$ and ending with the term in which $j = n$. Thus,

$$\sum_{j=1}^{n} f(x_j)\,\Delta x = f(x_1)\,\Delta x + f(x_2)\,\Delta x + \cdots + f(x_n)\,\Delta x$$

and the fundamental theorem of calculus can be stated compactly as follows.

The Fundamental Theorem of Calculus

Suppose f is continuous on the interval $a \le x \le b$, which is divided into n equal subintervals of length Δx by $x_1, x_2, \ldots, x_n$, where x_j is the left-hand endpoint of the jth subinterval. Then,

$$\lim_{n \to \infty} \sum_{j=1}^{n} f(x_j) \, \Delta x = \int_a^b f(x) \, dx = F(b) - F(a)$$

where F is any antiderivative of f.

Generalizations

Actually this is a somewhat restricted version of a more general characterization of definite integrals. The relationship between definite integrals and limits of sums is still valid if $f(x_j)$ in the jth term of the sum is replaced by $f(x_j')$, where x_j' is *any* point whatsoever in the jth subinterval. Moreover, the n subintervals need not have equal width, as long as the width of the largest eventually approaches zero as n increases. For most applications, however, the restricted characterization is sufficient, and you will have no need to use the more general result.

Applications

The steps involved in applying the fundamental theorem of calculus are outlined below.

How to Apply the Fundamental Theorem of Calculus

Step 1. Approximate the quantity in question by a sum of the form

$$\sum_{j=1}^{n} f(x_j) \, \Delta x$$

Step 2. Observe that the approximation approaches the true value of the quantity as n increases without bound.

Step 3. Replace the limit

$$\lim_{n \to \infty} \sum_{j=1}^{n} f(x_j) \, \Delta x$$

with the corresponding definite integral

$$\int_a^b f(x) \, dx$$

Step 4. Find the quantity in question by evaluating the definite integral by antidifferentiation.

To illustrate the procedure, here are two examples from economics. Actually, each of these could be solved rather easily using the methods of

Section 1 of the preceding chapter, and they have been chosen as illustrative examples in this chapter because of their relative simplicity. There are, however, many important applied problems that cannot be solved easily without the characterization of the definite integral as the limit of a sum. Some of these will be presented in the next section.

Total Revenue

In the first example, the relationship between integrals and limits of sums is used to compute total revenue.

EXAMPLE 1.1   A certain oil well that yields 300 barrels of crude oil a month will run dry in 3 years. It is estimated that t months from now the price of crude oil will be $P(t) = 18 + 0.3\sqrt{t}$ dollars per barrel. If the oil is sold as soon as it is extracted from the ground, what will the total future revenue from the well be?

SOLUTION
To approximate the total revenue during the 36-month period, divide the interval $0 \leq t \leq 36$ into n equal subintervals of length Δt months and let t_j denote the beginning of the jth subinterval as shown in Figure 1.3.

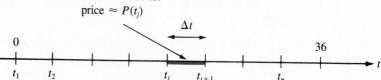

Figure 1.3 The division into subintervals of a 36-month period.

During the jth subinterval,

$$\text{Number of barrels produced} = (\text{barrels per month})(\text{number of months})$$

$$= 300\Delta t$$

Moreover, if Δt is small, the price of crude oil throughout the jth subinterval will be approximately $P(t_j)$ dollars per barrel (the price that was in effect at the beginning of the subinterval). Hence,

$$\begin{array}{l} \text{Revenue from} \\ \text{jth subinterval} \end{array} = \left(\begin{array}{c} \text{number of} \\ \text{barrels produced} \end{array}\right)\left(\begin{array}{c} \text{price per} \\ \text{barrel} \end{array}\right)$$

$$\simeq (300\Delta t)P(t_j) = 300P(t_j)\,\Delta t$$

The total revenue over the 36-month period is the sum of the revenues from each of the n subintervals. Thus,

$$\text{Total revenue} \simeq \sum_{j=1}^{n} 300P(t_j)\,\Delta t$$

As n increases, the length Δt of the subintervals decreases and the approximation improves. In fact,

$$\text{Total revenue} = \lim_{n \to \infty} \sum_{j=1}^{n} 300P(t_j)\ \Delta t$$

But according to the characterization of the definite integral as the limit of a sum,

$$\lim_{n \to \infty} \sum_{j=1}^{n} 300P(t_j)\ \Delta t = \int_{0}^{36} 300P(t)\ dt$$

Hence,

$$\text{Total revenue} = \int_{0}^{36} 300P(t)\ dt = 300 \int_{0}^{36} (18 + 0.3\sqrt{t}\,)\ dt$$

$$= 300(18t + 0.2t^{3/2})\Big|_{0}^{36} = \$207{,}360$$

Inventory Storage Costs

In the next example, you will see how to calculate the total cost resulting from the storage of unused inventory.

EXAMPLE 1.2   A retailer receives a shipment of 10,000 kilograms of rice that will be used up over a 5-month period at the constant rate of 2,000 kilograms per month. If storage costs are 1 cent per kilogram per month, how much will the retailer pay in storage costs over the next 5 months?

SOLUTION
Let $Q(t)$ denote the number of kilograms of rice in storage after t months. Then $Q(t) = 10{,}000 - 2{,}000t$.

To approximate the total storage cost over the 5-month period, divide the interval $0 \le t \le 5$ into n equal subintervals of length Δt months and let t_j denote the beginning of the jth subinterval as shown in Figure 1.4.

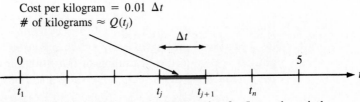

Figure 1.4 The division into subintervals of a 5-month period.

During the jth subinterval,

$$\text{Cost per kilogram} = \binom{\text{cost per kilogram}}{\text{per month}}\binom{\text{number}}{\text{of months}} = 0.01\Delta t$$

and

$$\text{Number of kilograms in storage} \simeq Q(t_j)$$

Hence,

$$\binom{\text{Storage cost during}}{\text{jth subinterval}} = \binom{\text{cost per}}{\text{kilogram}}\binom{\text{number of}}{\text{kilograms}} \simeq 0.01Q(t_j)\,\Delta t$$

and so

$$\text{Total storage cost} \simeq \sum_{j=1}^{n} 0.01Q(t_j)\,\Delta t$$

The approximation improves as n increases without bound, and it follows from the characterization of the definite integral as the limit of a sum that

$$\text{Total storage cost} = \lim_{n \to \infty} \sum_{j=1}^{n} 0.01Q(t_j)\,\Delta t$$

$$= \int_0^5 0.01Q(t)\,dt$$

$$= \int_0^5 0.01(10{,}000 - 2{,}000t)\,dt$$

$$= \int_0^5 (100 - 20t)\,dt = (100t - 10t^2)\Big|_0^5 = \$250$$

Observe that this total storage cost is the same as the cost of storing 5,000 kilograms (that is, one-half of the original shipment) for the entire 5 months.

The Volume of a Solid of Revolution

The next application is a geometric one in which the characterization of the definite integral as the limit of a sum is used to find the volume of a **solid of revolution** formed by revolving a region R in the xy plane about the x axis.

The technique is to express the volume of the solid as the limit of a sum of the volumes of approximating disks. In particular, suppose that S is the solid (Figure 1.5b) formed by rotating about the x axis the region R (Figure 1.5a) under the curve $y = f(x)$ from $x = a$ to $x = b$.

Divide the interval $a \leq x \leq b$ into n equal subintervals of length Δx, and let x_j denote the beginning of the jth subinterval. Then approximate the region R by the n rectangles shown in Figure 1.6a and the solid S by

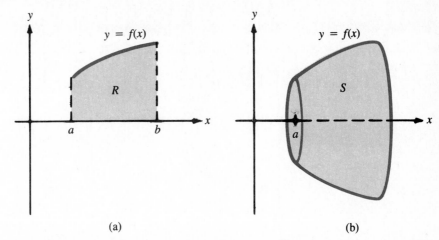

Figure 1.5 The region under the curve $y = f(x)$ from $x = a$ to $x = b$ and the corresponding solid of revolution.

(a)   (b)

the corresponding n cylindrical disks (Figure 1.6b) formed by revolving these rectangles about the x axis.

The radius r_j of the jth disk (Figure 1.7b) is the height $f(x_j)$ of the jth rectangle (Figure 1.7a), and so

Volume of jth disk = (area of circular cross section)(width)

$$= \pi r_j^2(\text{width}) = \pi[f(x_j)]^2 \, \Delta x$$

The total volume of S is approximately the sum of the volumes of the n disks. That is,

$$\text{Volume of } S \simeq \sum_{j=1}^{n} \pi[f(x_j)]^2 \, \Delta x$$

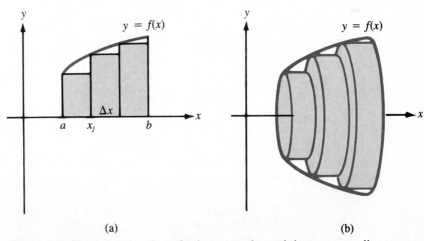

(a)   (b)

Figure 1.6 The approximation of R by rectangles and the corresponding approximation of S by disks.

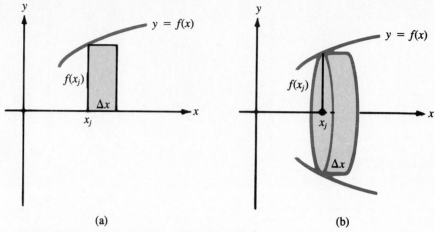

Figure 1.7 The jth rectangle and the jth disk.

The approximation improves as n increases without bound and

$$\text{Volume of } S = \lim_{n \to \infty} \sum_{j=1}^{n} \pi[f(x_j)]^2 \, \Delta x = \pi \int_a^b [f(x)]^2 \, dx$$

Volume Formula

> If $f(x)$ is nonnegative and R is the region under the graph of f from $x = a$ to $x = b$ and S is the solid formed by revolving R about the x axis, then
>
> $$\text{Volume of } S = \pi \int_a^b [f(x)]^2 \, dx$$

Here is an example.

EXAMPLE 1.3 Find the volume of the solid S formed by revolving the region under the curve $y = x^2 + 1$ from $x = 0$ to $x = 2$ about the x axis.

SOLUTION
The region, the solid of revolution, and the jth disk are shown in Figure 1.8.

The radius of the jth disk is $f(x_j) = x_j^2 + 1$. Hence,

$$\text{Volume of } j\text{th disk} = \pi[f(x_j)]^2 \, \Delta x = \pi(x_j^2 + 1)^2 \, \Delta x$$

and

$$\text{Volume of } S = \lim_{n \to \infty} \sum_{j=1}^{n} \pi(x_j^2 + 1)^2 \, \Delta x$$

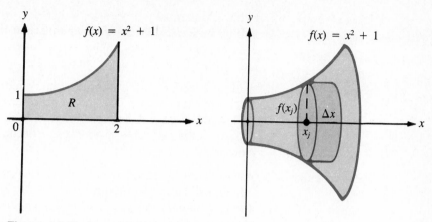

Figure 1.8 Region, solid of revolution, and jth disk for Example 1.3.

$$= \pi \int_0^2 (x^2 + 1)^2 \, dx$$

$$= \pi \int_0^2 (x^4 + 2x^2 + 1) \, dx$$

$$= \pi \left(\frac{1}{5} x^5 + \frac{2}{3} x^3 + x \right) \Big|_0^2 = \frac{206}{15} \pi \approx 43.14$$

Problems

Use the characterization of the definite integral as the limit of a sum to solve the following problems.

Distance and speed

1. An object is moving so that its speed after t minutes is $S(t) = 1 + 4t + 3t^2$ meters per minute. How far does the object travel during the 3rd minute?

Growth

2. A tree has been transplanted and after x years is growing at the rate of $f(x) = 0.5 + \dfrac{1}{(x + 1)^2}$ meters per year. By how much does the tree grow during the 2nd year?

Land values

3. It is estimated that t years from now the value of a certain parcel of land will be increasing at the rate of $r(t)$ dollars per year. Find an expression for the amount by which the value of the land will increase during the next 5 years.

Admission to events

4. The promoters of a county fair estimate that t hours after the gates open at 9:00 A.M., visitors will be entering the fair at the rate of $r(t)$ people per hour. Find an expression for the number of people who will enter the fair between 11:00 A.M. and 1:00 P.M.

Sales revenue
5. A bicycle manufacturer expects that x months from now consumers will be buying 5,000 bicycles per month at the price of $P(x) = 80 + 3\sqrt{x}$ dollars per bicycle. What is the total revenue the manufacturer can expect from the sale of the bicycles over the next 16 months?

Sales revenue
6. A bicycle manufacturer expects that x months from now consumers will be buying $f(x) = 5,000 + 60\sqrt{x}$ bicycles per month at the price of $P(x) = 80 + 3\sqrt{x}$ dollars per bicycle. What is the total revenue the manufacturer can expect from the sale of the bicycles over the next 16 months?

Sales revenue
7. A manufacturer expects that x months from now consumers will be buying $n(x)$ lamps per month at the price of $p(x)$ dollars per lamp. Find an expression for the total revenue the manufacturer can expect from the sale of the lamps over the next 12 months.

Oil production
8. Suppose that t months from now an oil well will be producing crude oil at the rate of $r(t)$ barrels per month and that the price of crude oil will be $p(t)$ dollars per barrel. Assuming that the oil is sold as soon as it is extracted from the ground, find an expression for the total revenue from the oil well over the next 2 years.

Farming
9. It is estimated that t days from now a farmer's crop will be increasing at the rate of $0.3t^2 + 0.6t + 1$ bushels per day. By how much will the value of the crop increase during the next 5 days if the market price remains fixed at \$3 per bushel?

Water pollution
10. It is estimated that t years from now the population of a certain lakeside community will be changing at the rate of $0.6t^2 + 0.2t + 0.5$ thousand people per year. Environmentalists have found that the level of pollution in the lake increases at the rate of approximately 5 units per 1,000 people. By how much will the pollution in the lake increase during the next 2 years?

Storage cost
11. A retailer receives a shipment of 12,000 pounds of soybeans that will be used at a constant rate of 300 pounds per week. If the cost of storing the soybeans is 0.2 cent per pound per week, how much will the retailer have to pay in storage costs over the next 40 weeks?

Storage cost
12. A manufacturer receives N units of a certain raw material that are initially placed in storage and then withdrawn and used at a constant rate until the supply is exhausted 1 year later. Suppose storage costs remain fixed at p dollars per unit per year.
   (a) Find an expression for the total storage cost the manufacturer will pay during the year.
   (b) Show that the total storage cost in part (a) is the same as the cost of storing $\dfrac{N}{2}$ units for the entire year.

Volume of a solid of revolution

In Problems 13 through 20, find the volume of the solid of revolution formed by revolving the region R about the x axis.

13. R is the region under the line $y = 3x + 1$ from $x = 0$ to $x = 1$.

14. R is the region under the curve $y = \sqrt{x}$ from $x = 1$ to $x = 4$.

15. R is the region under the curve $y = x^2 + 2$ from $x = -1$ to $x = 3$.

16. R is the region under the curve $y = 4 - x^2$ from $x = -2$ to $x = 2$.

17. R is the region under the curve $y = \sqrt{4 - x^2}$ from $x = -2$ to $x = 2$.

18. R is the region under the curve $y = \dfrac{1}{x}$ from $x = 1$ to $x = 10$.

19. R is the region under the curve $y = \dfrac{1}{\sqrt{x}}$ from $x = 1$ to $x = e^2$.

20. R is the region under the curve $y = e^{-0.1x}$ from $x = 0$ to $x = 10$.

Volume of a sphere

21. Use integral calculus to prove that the volume of a sphere of radius r is $\frac{4}{3}\pi r^3$. (*Hint:* Think of the sphere as the solid formed by revolving the region under the semicircle $y = \sqrt{r^2 - x^2}$ in the accompanying figure from $x = -r$ to $x = r$ about the x axis.)

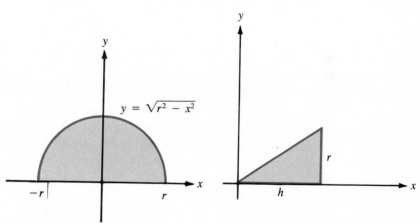

Figure for Problem 21.   Figure for Problem 22.

Volume of a cone

22. Use integral calculus to prove that the volume of a right circular cone of height h and radius r is $\frac{1}{3}\pi r^2 h$. (*Hint:* Think of the cone as the solid formed by revolving the triangle shown in the accompanying figure about the x axis.)

2 Further Applications of the Definite Integral

In this section you will see some further applications involving the characterization of the definite integral as the limit of a sum. The first of these is an integral formula for the average value of a function over an interval. The others come from finance, the social sciences, and biology.

The Average Value of a Function

In many practical situations one is interested in the **average value** of a continuous function over an interval, such as the average level of air pollution over a 24-hour period, the average speed of a truck during a 3-hour trip, the average productivity of a worker during a production run, and the average blood pressure of a patient during an operation. Here is a simple formula involving a definite integral that you can use to compute averages of this type.

The Average Value of a Function

The average value of a continuous function $f(x)$ over the interval $a \le x \le b$ is given by the formula

$$\text{Average value} = \frac{1}{b-a} \int_a^b f(x)\ dx$$

Derivation of the Average-Value Formula

To see why the average-value formula is valid, divide the interval $a \le x \le b$ into n equal subintervals of length Δx and let x_j denote the beginning of the jth subinterval. The numerical average of the corresponding n function values $f(x_1), f(x_2), \ldots, f(x_n)$ is

$$\frac{f(x_1) + f(x_2) + \cdots + f(x_n)}{n} = \frac{1}{n} \sum_{j=1}^{n} f(x_j)$$

As n increases without bound, this numerical average becomes increasingly sensitive to fluctuations in f and therefore approximates with increasing accuracy the average value of f over the entire interval $a \le x \le b$. That is,

$$\begin{array}{c} \text{Average value of } f \\ \text{over } a \le x \le b \end{array} = \lim_{n \to \infty} \frac{1}{n} \sum_{j=1}^{n} f(x_j)$$

To write this limit of a sum as a definite integral, observe that if the interval $a \leq x \leq b$ is divided into n equal subintervals of length Δx, then $\Delta x = \dfrac{b - a}{n}$. Hence, $\dfrac{1}{n} = \dfrac{\Delta x}{b - a}$ and so

$$\frac{1}{n} \sum_{j=1}^{n} f(x_j) = \frac{\Delta x}{b - a} \sum_{j=1}^{n} f(x_j) = \frac{1}{b - a} \sum_{j=1}^{n} f(x_j) \, \Delta x$$

It follows from the characterization of the definite integral as the limit of a sum that

$$\begin{array}{rl} \text{Average value of } f \\ \text{over } a \leq x \leq b \end{array} = \lim_{n \to \infty} \frac{1}{n} \sum_{j=1}^{n} f(x_j)$$

$$= \lim_{n \to \infty} \frac{1}{b - a} \sum_{j=1}^{n} f(x_j) \, \Delta x$$

$$= \frac{1}{b - a} \int_{a}^{b} f(x) \, dx$$

and the average-value formula is proved.

Application of the Average-Value Formula

The use of the integral formula for average value is illustrated in the next example.

EXAMPLE 2.1   For several weeks, the highway department has been recording the speed of freeway traffic flowing past a certain downtown exit. The data suggest that between the hours of 1:00 and 6:00 P.M. on a normal weekday, the speed of the traffic at the exit is approximately $S(t) = t^3 - 10.5t^2 + 30t + 20$ miles per hour, where t is the number of hours past noon. Compute the average speed of the traffic between the hours of 1:00 and 6:00 P.M.

SOLUTION
The goal is to find the average value of $S(t)$ on the interval $1 \leq t \leq 6$. From the integral formula,

$$\text{Average speed} = \frac{1}{6 - 1} \int_{1}^{6} (t^3 - 10.5t^2 + 30t + 20) \, dt$$

$$= \frac{1}{5} \left(\frac{1}{4} t^4 - \frac{10.5}{3} t^3 + 15t^2 + 20t \right) \bigg|_{1}^{6}$$

$$= \frac{1}{5} (228 - 31.75) = 39.25 \text{ miles per hour}$$

Geometric Interpretation of Average Value

The integral formula for average value has an interesting geometric interpretation. To see this, multiply both sides of the equation by $b - a$ to get

$$(b - a)(\text{average value}) = \int_a^b f(x)\,dx$$

If $f(x)$ is nonnegative, the integral on the right-hand side of this equation is equal to the area under the graph of f from $x = a$ to $x = b$. The product on the left-hand side is the area of a rectangle whose width is $b - a$ and whose height is the average value of f over the interval $a \leq x \leq b$. It follows that the average value of $f(x)$ over the interval $a \leq x \leq b$ is equal to the height of the rectangle whose base is this interval and whose area is the same as the area under the graph of f from $x = a$ to $x = b$. The situation is illustrated in Figure 2.1.

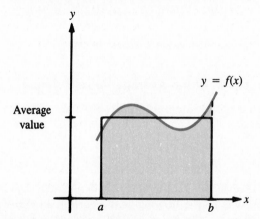

Figure 2.1 Geometric interpretation of average value: Area of rectangle = area under curve.

Further Applications

In the remainder of this section you will see four additional applications of the characterization of the definite integral as the limit of a sum. Although the applications are diverse, the same strategy is used for each. Here is a summary of that strategy.

Applications Involving Limits of Sums

The following steps often lead to the numerical value of a quantity associated with an interval $a \leq x \leq b$.

Step 1. Divide the interval $a \leq x \leq b$ into n equal subintervals of length Δx and let x_j denote the left-hand endpoint of the jth subinterval.

Step 2. Approximate the contribution to the total quantity that comes from the jth subinterval.

Step 3. Add the approximations from Step 2 to approximate the total quantity.

Step 4. Take the limit as n increases without bound to pass from the approximation to the exact value of the quantity.

Step 5. Use the fact that

$$\lim_{n \to \infty} \sum_{j=1}^{n} f(x_j) \, \Delta x = \int_a^b f(x) \, dx$$

to transform the limit of the sum in Step 4 to a definite integral.

Step 6. Evaluate the definite integral using antidifferentiation.

The Amount of an Income Stream

Suppose that a stream of income is transferred continuously into an account in which it earns interest over a specified period of time (or **term**). The **amount** (or **future value**) of the income stream is the total amount (money transferred into the account plus interest) that is accumulated in this way during the specified term.

The calculation of the amount of an income stream is illustrated in the next example. The strategy is to approximate the continuous income stream by a sequence of discrete deposits (i.e., by an **annuity**). The amount of the approximating annuity is a certain sum whose limit (a definite integral) is the amount of the income stream. (The amount of an annuity was calculated in Example 5.6 of Chapter 4, and, before proceeding, you may wish to reread that example.)

EXAMPLE 2.2 Money is transferred continuously into an account at the constant rate of \$1,200 per year. The account earns interest at the annual rate of 8 percent compounded continuously. How much will be in the account at the end of 2 years?

SOLUTION

Recall from Chapter 4 that P dollars invested at 8 percent compounded continuously will be worth $Pe^{0.08t}$ dollars t years later.

To approximate the future value of the income stream, divide the 2-year time interval $0 \leq t \leq 2$ into n equal subintervals of length Δt years and let t_j denote the beginning of the jth subinterval. Then, during the jth subinterval (of length Δt years),

Money deposited = (dollars per year)(number of years) = $1{,}200 \Delta t$

If all this money were deposited at the beginning of the subinterval (at time t_j), it would remain in the account for $2 - t_j$ years and hence would grow to $(1{,}200 \, \Delta t)e^{0.08(2 - t_j)}$ dollars. Thus,

$$\begin{matrix} \text{Future value of money deposited} \\ \text{during } j\text{th subinterval} \end{matrix} \approx 1{,}200e^{0.08(2-t_j)}\,\Delta t$$

The situation is illustrated in Figure 2.2.

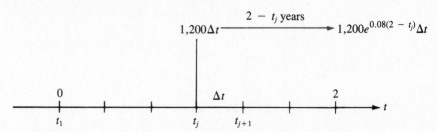

Figure 2.2 The (approximate) future value of the money deposited during the jth subinterval.

The future value of the entire income stream is the sum of the future values of the money deposited during each of the n subintervals. Hence,

$$\text{Future value of income stream} \approx \sum_{j=1}^{n} 1{,}200e^{0.08(2-t_j)}\,\Delta t$$

(Note that this is only an approximation because it is based on the assumption that all $1{,}200\Delta t$ dollars are deposited at time t_j rather than continuously throughout the jth subinterval.)

As n increases without bound, the length of each subinterval approaches zero and the approximation approaches the true future value of the income stream. Hence,

$$\begin{aligned}
\text{Future value of income stream} &= \lim_{n \to \infty} \sum_{j=1}^{n} 1{,}200e^{0.08(2-t_j)}\,\Delta t \\
&= \int_{0}^{2} 1{,}200e^{0.08(2-t)}\,dt \\
&= 1{,}200e^{0.16}\int_{0}^{2} e^{-0.08t}\,dt \\
&= -\frac{1{,}200}{0.08}e^{0.16}\left(e^{-0.08t}\right)\Big|_{0}^{2} \\
&= -15{,}000e^{0.16}\left(e^{-0.16} - 1\right) \\
&= -15{,}000 + 15{,}000e^{0.16} \\
&= -15{,}000(1 - e^{0.16}) \\
&\approx \$2{,}602.66
\end{aligned}$$

The Present Value of an Income Stream

The **present value** of an investment scheme or business venture that generates income continuously at a certain rate over a specified period of time is the amount of money that must be deposited today at the prevailing interest rate to generate the same income stream over the same term.

The calculation of the present value of a continuous income stream is illustrated in the next example. Once again, the strategy is to approximate the continuous income stream by a sequence of discrete payments (i.e., by an annuity). The present value of the approximating annuity is a certain sum whose limit (a definite integral) is the present value of the income stream. (The present value of an annuity was calculated in Example 5.7 of Chapter 4, and, before proceeding, you may wish to reread that example.)

EXAMPLE 2.3   The management of a national chain of ice-cream parlors is selling a 5-year franchise to operate its newest outlet in Madison, Wisconsin. Past experience in similar localities suggests that t years from now the franchise will be generating profit at the rate of $f(t) = 14{,}000 + 490t$ dollars per year. If the prevailing annual interest rate remains fixed during the next 5 years at 7 percent compounded continuously, what is the present value of the franchise?

SOLUTION

Recall from Chapter 4 that if the prevailing annual interest rate is 7 percent compounded continuously, the present value of B dollars payable t years from now is $Be^{-0.07t}$.

To approximate the present value of the franchise, divide the 5-year time interval $0 \le t \le 5$ into n equal subintervals of length Δt years and let t_j denote the beginning of the jth subinterval. Then,

$$\begin{array}{l}\text{Profit from the}\\ \text{jth interval}\end{array} = \text{(dollars per year)(number of years)} \simeq f(t_j)\,\Delta t$$

Hence

$$\begin{array}{l}\text{Present value of the profit}\\ \text{from the jth subinterval}\end{array} \simeq f(t_j)e^{-0.07t_j}\,\Delta t$$

and

$$\text{Present value of franchise} \simeq \sum_{j=1}^{n} f(t_j)e^{-0.07t_j}\,\Delta t$$

The situation is illustrated in Figure 2.3.

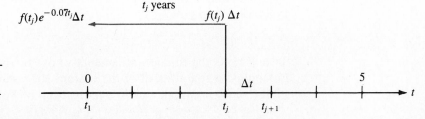

Figure 2.3 The (approximate) present value of the profit generated during the jth subinterval.

The approximation approaches the true present value as n increases without bound. Thus,

$$
\begin{aligned}
\text{Present value} \atop \text{of franchise} &= \lim_{n\to\infty} \sum_{j=1}^{n} f(t_j)e^{-0.07t_j}\,\Delta t \\[2mm]
&= \int_0^5 f(t)e^{-0.07t}\,dt \\[2mm]
&= \int_0^5 (14{,}000 + 490t)e^{-0.07t}\,dt \\[2mm]
&= -\frac{1}{0.07}(14{,}000 + 490t)e^{-0.07t}\Big|_0^5 \\[2mm]
&\quad + 7{,}000\int_0^5 e^{-0.07t}\,dt \qquad \text{integration by parts} \\[2mm]
&= \big[-1{,}000(200 + 7t)e^{-0.07t} - 100{,}000e^{-0.07t}\big]\Big|_0^5 \\[2mm]
&= -1{,}000(300 + 7t)e^{-0.07t}\Big|_0^5 \\[2mm]
&= -1{,}000[335e^{-0.35} - 300] \\[2mm]
&\simeq \$63{,}929.49
\end{aligned}
$$

Survival and Renewal Functions

In the next example, a **survival function** gives the fraction of individuals in a group or population that can be expected to remain in the group for any specified period of time. A **renewal function** giving the rate at which new members arrive is also known, and the goal is to predict the size of the group at some future time. Problems of this type arise in many fields, including sociology, demography, and ecology.

EXAMPLE 2.4 A new county mental health clinic has just opened. Statistics compiled at similar facilities suggest that the fraction of patients who will still be receiving treatment at the clinic t months after their initial visit is given by the function $f(t) = e^{-t/20}$. The clinic initially accepts 300 people for treatment and plans to accept new patients at the rate of 10 per month. Approximately how many people will be receiving treatment at the clinic 15 months from now?

SOLUTION

Since $f(15)$ is the fraction of patients whose treatment continues at least 15 months, it follows that of the current 300 patients, only $300f(15)$ will still be receiving treatment 15 months from now.

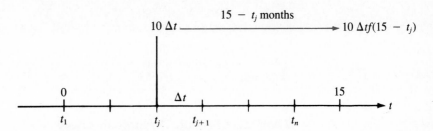

Figure 2.4 New members arriving during the *j*th subinterval.

To approximate the number of *new* patients who will be receiving treatment 15 months from now, divide the 15-month time interval $0 \le t \le 15$ into *n* equal subintervals of length Δt months and let t_j denote the beginning of the *j*th subinterval. Since new patients are accepted at the rate of 10 per month, the number of new patients accepted during the *j*th subinterval is $10\Delta t$. Fifteen months from now, approximately $15 - t_j$ months will have elapsed since these $10\Delta t$ new patients had their initial visits, and so approximately $(10\Delta t)f(15 - t_j)$ of them will still be receiving treatment at that time. (See Figure 2.4.) It follows that the total number of new patients still receiving treatment 15 months from now can be approximated by the sum

$$\sum_{j=1}^{n} 10f(15 - t_j)\, \Delta t$$

Adding this to the number of current patients who will still be receiving treatment in 15 months, you get

$$P \simeq 300f(15) + \sum_{j=1}^{n} 10f(15 - t_j)\, \Delta t$$

where *P* is the total number of patients who will be receiving treatment 15 months from now.

As *n* increases without bound, the approximation improves and approaches the true value of *P*. That is,

$$P = 300f(15) + \lim_{n \to \infty} \sum_{j=1}^{n} 10f(15 - t_j)\, \Delta t$$

$$= 300f(15) + \int_{0}^{15} 10f(15 - t)\, dt$$

Since $f(t) = e^{-t/20}$, it follows that $f(15) = e^{-3/4}$ and $f(15 - t) = e^{-(15-t)/20} = e^{-3/4}e^{t/20}$. Hence,

$$P = 300e^{-3/4} + 10e^{-3/4} \int_{0}^{15} e^{t/20}\, dt$$

$$= 300e^{-3/4} + \left(200e^{-3/4}e^{t/20}\Big|_{0}^{15}\right)$$

$$= 300e^{-3/4} + 200(1 - e^{-3/4})$$

$$= 100e^{-3/4} + 200$$

$$\approx 247.24$$

That is, 15 months from now, the clinic will be treating approximately 247 patients.

The Flow of Blood Through an Artery

Biologists have found that the speed of blood in an artery is a function of the distance of the blood from the artery's central axis. According to Poiseuille's law, the speed (in centimeters per second) of blood that is r centimeters from the central axis of the artery is $S(r) = k(R^2 - r^2)$, where R is the radius of the artery and k is a constant. In the next example, you will see how to use this information to compute the rate (in cubic centimeters per second) at which blood passes through the artery.

EXAMPLE 2.5   Find an expression for the rate (in cubic centimeters per second) at which blood flows through an artery of radius R if the speed of the blood r centimeters from the central axis is $S(r) = k(R^2 - r^2)$, where k is a constant.

SOLUTION

To approximate the volume of blood that flows through a cross section of the artery per second, divide the interval $0 \leq r \leq R$ into n equal subintervals of width Δr centimeters and let r_j denote the beginning of the jth subinterval. These subintervals determine n concentric rings as illustrated in Figure 2.5.

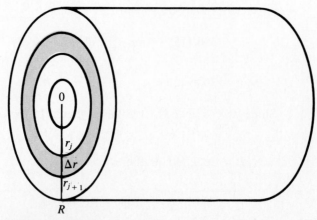

Figure 2.5 The division of the cross section into concentric rings.

If Δr is small, the area of the jth ring is approximately the area of a rectangle whose length is the circumference of the (inner) boundary of the ring and whose width is Δr. That is,

Area of jth ring $\simeq 2\pi r_j \, \Delta r$

If you multiply the area of the jth ring (square centimeters) by the speed (centimeters per second) of the blood flowing through this ring, you get the rate (cubic centimeters per second) at which blood flows through the jth ring. Since the speed of blood flowing through the jth ring is approximately $S(r_j)$ centimeters per second, it follows that

$$\begin{pmatrix} \text{Rate of flow} \\ \text{through } j\text{th ring} \end{pmatrix} \simeq \begin{pmatrix} \text{area of} \\ j\text{th ring} \end{pmatrix} \begin{pmatrix} \text{speed of blood} \\ \text{through } j\text{th ring} \end{pmatrix}$$

$$\simeq (2\pi r_j \, \Delta r)S(r_j)$$

$$= (2\pi r_j \, \Delta r)[k(R^2 - r_j^2)]$$

$$= 2\pi k(R^2 r_j - r_j^3) \, \Delta r$$

The rate of flow of blood through the entire cross section is the sum of n such terms, one for each of the n concentric rings. That is,

$$\text{Rate of flow} \simeq \sum_{j=1}^{n} 2\pi k(R^2 r_j - r_j^3) \, \Delta r$$

As n increases without bound, this approximation approaches the true value of the rate of flow. That is,

$$\text{Rate of flow} = \lim_{n \to \infty} \sum_{j=1}^{n} 2\pi k(R^2 r_j - r_j^3) \, \Delta r$$

$$= \int_0^R 2\pi k(R^2 r - r^3) \, dr$$

$$= 2\pi k \left(\frac{R^2}{2} r^2 - \frac{1}{4} r^4 \right) \Big|_0^R$$

$$= \frac{\pi k R^4}{2} \text{ cubic centimeters per second}$$

Problems

In Problems 1 through 4, find the average value of the given function over the specified interval. In each case, sketch the graph of the function along with the rectangle whose base is the given interval and whose height is the average value of the function.

1. $f(x) = x; \, 0 \le x \le 4$

2. $f(x) = 2x - x^2; \, 0 \le x \le 2$

3. $f(x) = (x + 2)^2; \, -4 \le x \le 0$

4. $f(x) = \dfrac{1}{x}; \, 1 \le x \le 2$

Temperature
5. Records indicate that t hours past midnight, the temperature at the local airport was $f(t) = -0.3t^2 + 4t + 10$ degrees Celsius. What was the average temperature at the airport between 9:00 A.M. and noon?

Food prices
6. Records indicate that t months after the beginning of the year, the price of ground beef in local supermarkets was $P(t) = 0.09t^2 - 0.2t + 1.6$ dollars per pound. What was the average price of ground beef during the first 3 months of the year?

Efficiency
7. After t months on the job, a postal clerk can sort mail at the rate of $Q(t) = 700 - 400e^{-0.5t}$ letters per hour. What is the average rate at which the clerk sorts mail during the first 3 months on the job?

Bacterial growth
8. The number of bacteria present in a certain culture after t minutes of an experiment was $Q(t) = 2,000e^{0.05t}$. What was the average number of bacteria present during the first 5 minutes of the experiment?

Speed and distance
9. A car is driven so that after t hours its speed is $S(t)$ miles per hour.
   (a) Write down a definite integral that gives the average speed of the car during the first N hours.
   (b) Write down a definite integral that gives the total distance the car travels during the first N hours.
   (c) Discuss the relationship between the integrals in parts (a) and (b).

Inventory
10. An inventory of 60,000 kilograms of a certain commodity is used at a constant rate and is exhausted after 1 year. What is the average inventory for the year?

The amount of an income stream
11. Money is transferred continuously into an account at the constant rate of $2,400 per year. The account earns interest at the annual rate of 6 percent compounded continuously. How much will be in the account at the end of 5 years?

The amount of an income stream
12. Money is transferred continuously into an account at the constant rate of $1,000 per year. The account earns interest at the annual rate of 10 percent compounded continuously. How much will be in the account at the end of 10 years?

The amount of an income stream
13. Money is transferred continuously into an account at the constant rate of $8,000 per year. The account earns interest at the annual rate of 8 percent compounded continuously. How much will be in the account at the end of 3 years?

The amount of an income stream
14. Money is transferred continuously into an account at the constant rate of $6,000 per year. The account earns interest at the annual rate of 6 percent compounded continuously. How much will be in the account at the end of 10 years?

The present value of an investment
15. An investment scheme will generate income continuously at the constant rate of $2,400 per year for 5 years. If the prevailing annual

interest rate remains fixed at 6 percent compounded continuously, what is the present value of the investment scheme?

The present value of an investment

16. An investment scheme will generate income continuously at the constant rate of $6,000 per year for 6 years. If the prevailing annual interest rate remains fixed at 8 percent compounded continuously, what is the present value of the investment scheme?

The present value of an investment

17. An investment scheme will generate income continuously at the constant rate of $1,200 per year for 5 years. If the prevailing annual interest rate remains fixed at 12 percent compounded continuously, what is the present value of the investment scheme?

The present value of a franchise

18. The management of a national chain of fast-food outlets is selling a 10-year franchise in Cleveland, Ohio. Past experience in similar localities suggests that t years from now the franchise will be generating profit at the rate of $f(t) = 10,000 + 500t$ dollars per year. If the prevailing annual interest rate remains fixed at 10 percent compounded continuously, what is the present value of the franchise?

Spy story

19. Having been left partially disabled after a head-on collision with a camel (see Chapter 8, Section 1, Problem 38), the hero of a popular spy story has been retired from the Secret Service. As compensation for many long years of dedicated public service, the government has offered the spy a choice between a 10-year pension of 5,000 pounds sterling per year or a flat sum of 35,000 pounds sterling to be paid immediately. Assuming that an annual interest rate of 10 percent compounded continuously will be available at banks throughout this period, decide which offer the spy should accept. (*Hint*: Compare the flat sum of 35,000 pounds with the present value of the pension. Assume that the pension is paid continuously.)

Present value

20. A certain investment scheme generates income continuously over a period of N years. After t years, the scheme will be generating income at the rate of $f(t)$ dollars per year. Derive an expression for the present value of this investment scheme if the prevailing annual interest rate remains fixed at $100r$ percent compounded continuously.

Computer dating

21. The operators of a new computer dating service estimate that the fraction of people who will retain their memberships in the service for at least t months is given by the function $f(t) = e^{-t/10}$. There are 8,000 charter members, and the operators expect to attract 200 new members per month. How many members will the service have 10 months from now?

Association membership

22. A national consumers' association has compiled statistics suggesting that the fraction of its members who are still active t months after joining is given by the function $f(t) = e^{-0.2t}$. A new local chapter has

200 charter members and expects to attract new members at the rate of 10 per month. How many members can the chapter expect to have at the end of 8 months?

Group membership 23. Let $f(t)$ denote the fraction of the membership of a certain group that will remain in the group for at least t years. Suppose that the group has just been formed with an initial membership of P_0 and that t years from now new members will be added to the group at the rate of $r(t)$ per year. Find an expression for the size of the group N years from now.

Poiseuille's law 24. Calculate the rate (in cubic centimeters per second) at which blood flows through an artery of radius 0.1 centimeter if the speed of the blood r centimeters from the central axis is $8 - 800r^2$ centimeters per second.

Fluid flow 25. Find an expression for the rate (in cubic centimeters per second) at which a fluid flows through a cylindrical pipe of radius R if the speed of the fluid r centimeters from the central axis of the pipe is $S(r)$ centimeters per second.

The area of a disk 26. Use the characterization of the definite integral as the limit of a sum to show that the area of a circular disk of radius R is πR^2. (*Hint*: Divide the disk into n concentric rings as in Example 2.5.)

Population density 27. The population density r miles from the center of a certain city is $D(r) = 5,000e^{-0.1r}$ people per square mile. How many people live within 3 miles of the center of the city? (*Hint*: Divide a circular disk of radius 3 into concentric rings.)

Population density 28. The population density t miles from the center of a certain city is $D(r) = 25,000e^{-0.05r}$ people per square mile. How many people live between 1 and 2 miles from the center of the city?

Nuclear waste 29. A certain nuclear power plant produces radioactive waste in the form of strontium 90 at the constant rate of 500 pounds per year. The waste decays exponentially with a half-life of 28 years. How much of this radioactive waste from the nuclear plant will be present after 140 years?

3 Improper Integrals

This section extends the concept of the definite integral to integrals of the form

$$\int_a^\infty f(x)\ dx$$

in which the upper limit of integration is not a finite number. Such integrals are known as **improper integrals** and arise in a variety of practical situations.

Geometric Interpretation

If f is nonnegative, the improper integral $\int_a^\infty f(x)\,dx$ may be interpreted as the area of the region under the graph of f to the right of $x = a$ (Figure 3.1a). Although this region has infinite extent, its area may be either finite or infinite, depending on how quickly $f(x)$ approaches zero as x increases.

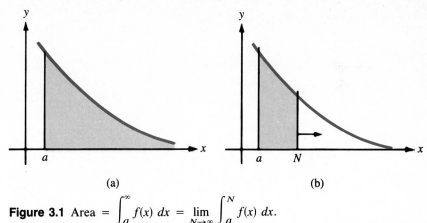

(a)   (b)

Figure 3.1 Area $= \displaystyle\int_a^\infty f(x)\,dx = \lim_{N\to\infty}\int_a^N f(x)\,dx.$

A reasonable strategy for finding the area of such a region is to first use a definite integral to compute the area from $x = a$ to some finite number $x = N$, and then to let N increase without bound in the resulting expression. That is,

$$\text{Total area} = \lim_{N\to\infty}(\text{area from } a \text{ to } N) = \lim_{N\to\infty}\int_a^N f(x)\,dx$$

This strategy is illustrated in Figure 3.1b and motivates the following definition of the improper integral.

The Improper Integral

$$\int_a^\infty f(x)\,dx = \lim_{N\to\infty}\int_a^N f(x)\,dx$$

If the limit defining the improper integral is a finite number, the integral is said to **converge**. Otherwise the integral is said to **diverge**. Here are some examples.

EXAMPLE 3.1   Evaluate $\int_1^\infty \frac{1}{x^2}\,dx$.

SOLUTION
First compute the integral from 1 to N and then let N approach infinity. Arrange your work compactly as follows:

$$\int_1^\infty \frac{1}{x^2}\,dx = \lim_{N\to\infty} \int_1^N \frac{1}{x^2}\,dx = \lim_{N\to\infty} \left(-\frac{1}{x}\Big|_1^N\right) = \lim_{N\to\infty}\left(-\frac{1}{N}+1\right) = 1$$

EXAMPLE 3.2   Evaluate $\int_1^\infty \frac{1}{x}\,dx$.

SOLUTION

$$\int_1^\infty \frac{1}{x}\,dx = \lim_{N\to\infty} \int_1^N \frac{1}{x}\,dx = \lim_{N\to\infty}\left(\ln|x|\Big|_1^N\right) = \lim_{N\to\infty} \ln N = \infty$$

Notice that the improper integral of the function $f(x) = \frac{1}{x^2}$ in Example 3.1 converges while that of the function $f(x) = \frac{1}{x}$ in Example 3.2 diverges. In geometric terms, this says that the area to the right of $x = 1$ under the curve $y = \frac{1}{x^2}$ is finite while the corresponding area under the curve $y = \frac{1}{x}$ is infinite. The reason for the difference is that, as x increases, $\frac{1}{x^2}$ approaches zero more quickly than does $\frac{1}{x}$. (See Figure 3.2.)

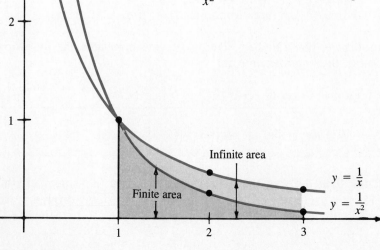

Infinite area

Finite area

$y = \frac{1}{x}$

$y = \frac{1}{x^2}$

Figure 3.2 The areas under $y = \frac{1}{x}$ and $y = \frac{1}{x^2}$.

A Useful Limit for Improper Integrals

The evaluation of improper integrals arising from practical problems often involves limits of the form

$$\lim_{N \to \infty} N^p e^{-kN} \quad \text{(for } k > 0)$$

If p and N are positive,

$$\lim_{N \to \infty} N^p = \infty \quad \text{while} \quad \lim_{N \to \infty} e^{-kN} = 0$$

That is, the factor N^p tends to drive the product to infinity, while the factor e^{-kN} tends to drive the product to zero. Without further analysis it is not clear which factor will dominate. Using techniques covered in more advanced texts, it can be shown that e^{-kN} approaches zero "so much more quickly" than N^p approaches infinity that their product approaches zero. Here, for reference, is a summary of this result.

A Useful Limit for Improper Integrals

> For any power p and positive number k,
> $$\lim_{N \to \infty} N^p e^{-kN} = 0$$

The next example involves a limit of this type and also requires the use of integration by parts.

EXAMPLE 3.3 Evaluate $\displaystyle\int_0^\infty xe^{-2x}\, dx$.

SOLUTION

$$\int_0^\infty xe^{-2x}\, dx = \lim_{N \to \infty} \int_0^N xe^{-2x}\, dx$$

$$= \lim_{N \to \infty} \left(-\frac{1}{2}xe^{-2x}\Big|_0^N + \frac{1}{2}\int_0^N e^{-2x}\, dx \right) \qquad \text{integration by parts}$$

$$= \lim_{N \to \infty} \left(-\frac{1}{2}xe^{-2x} - \frac{1}{4}e^{-2x} \right)\Big|_0^N$$

$$= \lim_{N \to \infty} \left(-\frac{1}{2}Ne^{-2N} - \frac{1}{4}e^{-2N} + 0 + \frac{1}{4} \right)$$

$$= \frac{1}{4} \qquad \text{since } Ne^{-2N} \to 0 \text{ and } e^{-2N} \to 0$$

Applications of the Improper Integral

The following applications of the improper integral generalize applications of the definite integral that you saw in the preceding section. In each, the strategy is to use the characterization of the definite integral as the limit of a sum to construct an appropriate definite integral and then to let the upper limit of integration increase without bound. As you read these examples, you may want to refer back to the corresponding examples in Section 2.

Present Value

In Example 2.3 of the preceding section you saw that the present value of an investment that generates income over a finite period of time is given by a definite integral. The present value of an investment that generates income in perpetuity (i.e., forever) is given by an improper integral. Here is an example.

EXAMPLE 3.4 A donor wishes to make a gift to a private college from which the college will draw $7,000 per year in perpetuity to support the operation of its computer center. Assuming that the prevailing annual interest rate will remain fixed at 14 percent compounded continuously, how much should the donor give the college? That is, what is the present value of the endowment?

SOLUTION
To find the present value of a gift that generates $7,000 per year for N years, divide the N-year time interval $0 \leq t \leq N$ into n subintervals of length Δt years and let t_j denote the beginning of the jth subinterval (Figure 3.3). Then,

$$\begin{array}{c}\text{Amount generated during} \\ \text{jth subinterval}\end{array} \simeq 7{,}000\,\Delta t$$

and

$$\begin{array}{c}\text{Present value of} \\ \text{amount generated during} \\ \text{jth subinterval}\end{array} \simeq 7{,}000e^{-0.14t_j}\,\Delta t$$

Hence,

$$\begin{array}{c}\text{Present value of} \\ \text{N-year gift}\end{array} = \lim_{n \to \infty} \sum_{j=1}^{n} 7{,}000e^{-0.14t_j}\,\Delta t$$

$$= \int_{0}^{N} 7{,}000e^{-0.14t}\,dt$$

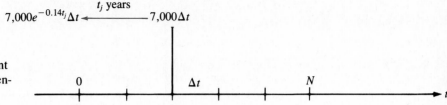

Figure 3.3 The present value of the money generated during the jth subinterval.

To find the present value of the total gift, take the limit of this integral as N approaches infinity. That is,

$$\begin{aligned}\text{Present value}\atop\text{of total gift} &= \lim_{N\to\infty} \int_0^N 7{,}000e^{-0.14t}\, dt \\[2mm]
&= \lim_{N\to\infty}\left(-50{,}000e^{-0.14t}\Big|_0^N\right) \\[2mm]
&= \lim_{N\to\infty} -50{,}000(e^{-0.14N} - 1) \\[2mm]
&= \$50{,}000 \qquad \text{since } e^{-0.14N} \to 0\end{aligned}$$

Nuclear Waste

The next example is similar in structure to the problem involving survival and renewal functions in Example 2.4 of the preceding section.

EXAMPLE 3.5 It is estimated that t years from now, a certain nuclear power plant will be producing radioactive waste at the rate of $f(t) = 400t$ pounds per year. What will happen to the accumulation of radioactive waste from the plant in the long run?

SOLUTION
To find the amount of radioactive waste present after N years, divide the N-year time interval $0 \le t \le N$ into n equal subintervals of length Δt years and let t_j denote the beginning of the jth subinterval (Figure 3.4). Then,

$$\text{Amount of waste produced}\atop\text{during }j\text{th subinterval} \simeq 400t_j\, \Delta t$$

Since the waste decays exponentially at the rate of 2 percent per year and since there are $N - t_j$ years between the times $t = t_j$ and $t = N$, it follows that

$$\begin{matrix}\text{Amount of waste produced}\\ \text{during }j\text{th subinterval}\\ \text{still present at time }N\end{matrix} \simeq 400t_j e^{-0.02(N - t_j)}\, \Delta t$$

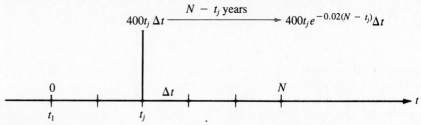

Figure 3.4 Radioactive waste generated during the jth subinterval.

Thus,

$$\begin{array}{c}\text{Amount of waste}\\\text{present after } N \text{ years}\end{array} = \lim_{n\to\infty} \sum_{j=1}^{n} 400t_j e^{-0.02(N-t_j)}\,\Delta t$$

$$= \int_0^N 400te^{-0.02(N-t)}\,dt$$

$$= 400e^{-0.02N} \int_0^N te^{0.02t}\,dt$$

The amount of radioactive waste present in the long run is the limit of this expression as N approaches infinity. That is,

$$\begin{array}{c}\text{Waste present}\\\text{in long run}\end{array} = \lim_{N\to\infty} 400e^{-0.02N} \int_0^N te^{0.02t}\,dt$$

$$= \lim_{N\to\infty} 400e^{-0.02N}(50te^{0.02t} - 2{,}500e^{0.02t})\Big|_0^N$$
$$\qquad\qquad\qquad\qquad\qquad\qquad \text{integration by parts}$$

$$= \lim_{N\to\infty} 400e^{-0.02N}(50Ne^{0.02N} - 2{,}500e^{0.02N} + 2{,}500)$$

$$= \lim_{N\to\infty} 400(50N - 2{,}500 + 2{,}500e^{-0.02N})$$

$$= \infty \quad \text{since } 50N \to \infty \text{ and } 2{,}500e^{-0.02N} \to 0$$

That is, in the long run, the accumulation of radioactive waste from the plant will increase without bound.

Other Types of Improper Integrals

In the next section you will see applications of improper integrals to probability. In the course of the discussion, you will encounter improper integrals of the form $\int_{-\infty}^{\infty} f(x)\,dx$. Here, for future reference, is the definition of such an integral.

Improper Integrals from $-\infty$ to ∞

$$\int_{-\infty}^{\infty} f(x)\ dx = \lim_{N \to \infty} \int_{-N}^{0} f(x)\ dx + \lim_{N \to \infty} \int_{0}^{N} f(x)\ dx$$

If both limits are finite, the improper integral is said to converge to their sum. Otherwise, the improper integral is said to diverge.

If f is nonnegative, the improper integral $\int_{-\infty}^{\infty} f(x)\ dx$ may be interpreted as the total area under the graph of f (Figure 3.5).

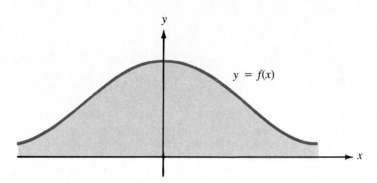

Figure 3.5

Area $= \displaystyle\int_{-\infty}^{\infty} f(x)\ dx.$

Problems

In Problems 1 through 24, evaluate the given improper integral.

1. $\displaystyle\int_{1}^{\infty} \frac{1}{x^3}\ dx$

2. $\displaystyle\int_{1}^{\infty} x^{-3/2}\ dx$

3. $\displaystyle\int_{1}^{\infty} \frac{1}{\sqrt{x}}\ dx$

4. $\displaystyle\int_{1}^{\infty} x^{-2/3}\ dx$

5. $\displaystyle\int_{3}^{\infty} \frac{1}{2x-1}\ dx$

6. $\displaystyle\int_{3}^{\infty} \frac{1}{\sqrt[3]{2x-1}}\ dx$

7. $\displaystyle\int_{3}^{\infty} \frac{1}{(2x-1)^2}\ dx$

8. $\displaystyle\int_{0}^{\infty} e^{-x}\ dx$

9. $\displaystyle\int_{0}^{\infty} 5e^{-2x}\ dx$

10. $\displaystyle\int_{1}^{\infty} e^{1-x}\ dx$

11. $\displaystyle\int_{1}^{\infty} \frac{x^2}{(x^3+2)^2}\ dx$

12. $\displaystyle\int_{1}^{\infty} \frac{x^2}{x^3+2}\ dx$

13. $\displaystyle\int_{1}^{\infty} \frac{x^2}{\sqrt{x^3+2}}\ dx$

14. $\displaystyle\int_{0}^{\infty} xe^{-x^2}\ dx$

15. $\displaystyle\int_1^\infty \frac{e^{-\sqrt{x}}}{\sqrt{x}}\, dx$

16. $\displaystyle\int_0^\infty xe^{-x}\, dx$

17. $\displaystyle\int_0^\infty 2xe^{-3x}\, dx$

18. $\displaystyle\int_0^\infty xe^{1-x}\, dx$

19. $\displaystyle\int_0^\infty 5xe^{10-x}\, dx$

20. $\displaystyle\int_1^\infty \frac{\ln x}{x}\, dx$

21. $\displaystyle\int_2^\infty \frac{1}{x \ln x}\, dx$

22. $\displaystyle\int_2^\infty \frac{1}{x\sqrt{\ln x}}\, dx$

23. $\displaystyle\int_0^\infty x^2 e^{-x}\, dx$

24. $\displaystyle\int_0^\infty x^3 e^{-x^2}\, dx$

Present value of an investment 25. An investment will generate \$2,400 per year in perpetuity. If the money is dispensed continuously throughout the year and if the prevailing annual interest rate remains fixed at 12 percent compounded continuously, what is the present value of the investment?

Present value 26. An investment will generate income continuously at the constant rate of Q dollars per year in perpetuity. Assuming a fixed annual interest rate of r compounded continuously, use an improper integral to show that the present value of the investment is $\dfrac{Q}{r}$ dollars.

Present value of rental property 27. It is estimated that t years from now an apartment complex will be generating profit for its owner at the rate of $f(t) = 10,000 + 500t$ dollars per year. If the profit is generated in perpetuity and the prevailing annual interest rate remains fixed at 10 percent compounded continuously, what is the present value of the apartment complex?

Present value of a franchise 28. The management of a national chain of fast-food outlets is selling a permanent franchise in Seattle, Washington. Past experience in similar localities suggests that t years from now, the franchise will be generating profit at the rate of $f(t) = 12,000 + 900t$ dollars per year. If the prevailing annual interest rate remains fixed at 10 percent compounded continuously, what is the present value of the franchise?

Present value 29. In t years, an investment will be generating $f(t) = A + Bt$ dollars per year, where A and B are constants. If the income is generated in perpetuity and the prevailing annual interest rate of r compounded continuously does not change, show that the present value of this investment is $\dfrac{A}{r} + \dfrac{B}{r^2}$.

Nuclear waste 30. A certain nuclear power plant produces radioactive waste at the rate of 600 pounds per year. The waste decays exponentially at the rate of

2 percent per year. How much radioactive waste from the plant will be present in the long run?

Health care 31. The fraction of patients who will still be receiving treatment at a certain mental health clinic t months after their initial visit is $f(t) = e^{-t/20}$. If the clinic accepts new patients at the rate of 10 per month, approximately how many patients will be receiving treatment at the clinic in the long run?

Population growth 32. Demographic studies conducted in a certain city indicate that the fraction of the residents that will remain in the city for at least t years is $f(t) = e^{-t/20}$. The current population of the city is 200,000, and it is estimated that new residents will be arriving at the rate of 100 people per year. If this estimate is correct, what will happen to the population of the city in the long run?

Medicine 33. A hospital patient receives intravenously 5 units of a certain drug per hour. The drug is eliminated exponentially, so that the fraction that remains in the patient's body for t hours is $f(t) = e^{-t/10}$. If the treatment is continued indefinitely, approximately how many units of the drug will be in the patient's body in the long run?

4 Probability Density Functions

Some of the most important applications of integration to the social, managerial, and life sciences are in the areas of probability and statistics. In Chapter 6, Section 2, you were introduced to the technique of integrating probability density functions to compute probabilities. The purpose of this section is to explore in more detail the relationship between integration and probability. Improper integrals will play an important role in the discussion.

Random Variables

The life span of a light bulb selected at random from a manufacturer's stock is a quantity that cannot be predicted with certainty. In the terminology of probability and statistics, the process of selecting the bulb at random is called a **random experiment,** and the life span of the bulb is said to be a **random variable.** In general, a random variable is a number associated with the outcome of a random experiment.

Discrete and Continuous Random Variables

A random variable that can take on only integer values is said to be **discrete.** The face value of a randomly selected playing card and the number of times heads comes up in three tosses of a coin are discrete random variables. So is the IQ of a randomly selected university student, since IQs are measured in whole numbers.

A random variable that can take on any value in some interval is said to be **continuous.** Some continuous random variables are the time a randomly selected motorist spends waiting at a traffic light, the time interval between the arrivals of randomly selected successive planes at an airport, and the time it takes a randomly selected subject to learn a particular task. Integral calculus is used in the study of continuous random variables.

Probability

The **probability** of an event that can result from a random experiment is a number between 0 and 1 that specifies the likelihood of the event. In particular, the probability is the fraction of the time the event can be expected to occur if the experiment is repeated a large number of times. For example, the probability that an evenly balanced tossed coin will come up heads is $\frac{1}{2}$ since this event can be expected approximately $\frac{1}{2}$ of the time if the coin is tossed repeatedly. In a group containing 13 men and 10 women, the probability is $\frac{10}{23}$ that a person selected at random is a woman. The probability of an event that is certain to occur is 1, while the probability of an event that cannot possibly occur is zero. For example, if you roll an ordinary die, the probability is 1 that you will get a number between 1 and 6, inclusive, while the probability is zero that you will get a 7.

Events Described in Terms of Random Variables

Consider again the random experiment in which a light bulb is selected at random from a manufacturer's stock. A possible event resulting from this experiment is that the life span of the selected bulb is between 20 and 35 hours. If x is the random variable denoting the life span of a randomly selected bulb, this event can be described by the inequality $20 \leq x \leq 35$ and its probability denoted by the symbol $P(20 \leq x \leq 35)$. Similarly, the probability that the bulb will burn for at least 50 hours is denoted by $P(x \geq 50)$ or $P(50 \leq x < \infty)$.

Probability Density Functions

A **probability density function** for a continuous random variable x is a nonnegative function f with the property that $P(a \leq x \leq b)$ is the area under the graph of f from $x = a$ to $x = b$. A possible probability density

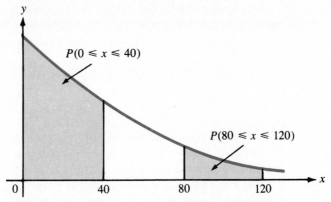

Figure 4.1 Probability density function for the life span of a light bulb.

function for the life span of a light bulb is sketched in Figure 4.1. Its shape reflects the fact that most bulbs burn out relatively quickly. For example, the probability that a bulb will fail within the first 40 hours is represented by the area under the curve between $x = 0$ and $x = 40$. This is much greater than the area under the curve between $x = 80$ and $x = 120$, which represents the probability that the bulb will fail between the 80th and 120th hour of use.

The basic property of probability density functions can be restated in terms of the integrals you would use to compute the appropriate areas.

Probability Density Functions

> A probability density function for a continuous random variable x is a nonnegative function f such that
>
> $$P(a \leq x \leq b) = \int_a^b f(x) \, dx$$

The values of a and b in this formula need not be finite. If either is infinite, the corresponding probability is given by an improper integral. For example, the probability that x is greater than or equal to a is

$$P(x \geq a) = P(a \leq x < \infty) = \int_a^\infty f(x) \, dx$$

The total area under the graph of a probability density function must be equal to 1. This is because the total area represents the probability that x is between $-\infty$ and ∞, which is an event that is certain to occur. This observation can be restated in terms of improper integrals as follows:

A Property of Probability Density Functions

If f is a probability density function for a continuous random variable x, then

$$\int_{-\infty}^{\infty} f(x)\ dx = 1$$

How to determine the appropriate probability density function for a particular random variable is a central problem in probability. It involves techniques beyond the scope of this book, which can be found in most probability and statistics texts. The purpose of the discussion in the remainder of this section is to show you some of the probability density functions that have proved to be useful, to illustrate their use, and to introduce some important properties of probability density functions in general.

Uniform Density Functions

A **uniform density function** (Figure 4.2) is constant over a bounded interval $A \leq x \leq B$ and zero outside the interval. A random variable that has a uniform density function is said to be **uniformly distributed.** Roughly speaking, a uniformly distributed random variable is one for which all the values in some bounded interval are "equally likely." More precisely, a continuous random variable is uniformly distributed if the probability that its value will be in a particular subinterval of the bounded interval is equal to the probability that it will be in any other subinterval that has the same length. An example of a uniformly distributed random variable is the waiting time of a motorist at a traffic light that remains red for, say, 40 seconds at a time. This random variable has a uniform distribution because all waiting times between 0 and 40 seconds are equally likely.

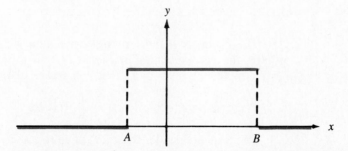

Figure 4.2 A uniform density function.

If k is the constant value of a uniform density function $f(x)$ on the interval $A \leq x \leq B$, the value of k is determined by the requirement that the total area under the graph of f must be equal to 1. In particular,

$$1 = \int_{-\infty}^{\infty} f(x) \, dx = \int_{A}^{B} f(x) \, dx \qquad \text{since } f(x) = 0 \text{ off the interval } A \leq x \leq B$$

$$= \int_{A}^{B} k \, dx = kx \Big|_{A}^{B} = k(B - A)$$

and so

$$k = \frac{1}{B - A}$$

This calculation leads to the following formula for a uniform density function.

Uniform Density Function

$$f(x) = \begin{cases} \dfrac{1}{B - A} & \text{if } A \leq x \leq B \\ 0 & \text{otherwise} \end{cases}$$

Here is a typical application involving a uniform density function.

EXAMPLE 4.1   A certain traffic light remains red for 40 seconds at a time. You arrive (at random) at the light and find it red. Use an appropriate uniform density function to find the probability that you will have to wait at least 15 seconds for the light to turn green.

SOLUTION
Let x denote the time (in seconds) that you must wait. Since all waiting times between 0 and 40 are "equally likely," x is uniformly distributed over the interval $0 \leq x \leq 40$. The corresponding uniform density function is

$$f(x) = \begin{cases} \dfrac{1}{40} & \text{if } 0 \leq x \leq 40 \\ 0 & \text{otherwise} \end{cases}$$

and the desired probability is

$$P(15 \leq x \leq 40) = \int_{15}^{40} \frac{1}{40} \, dx = \frac{x}{40} \Big|_{15}^{40} = \frac{40 - 15}{40} = \frac{5}{8}$$

Exponential Density Functions

An **exponential density function** is a function $f(x)$ that is zero for $x < 0$ and that decreases exponentially for $x \geq 0$. That is, for $x \geq 0$,

$$f(x) = Ae^{-kx}$$

where A and k are positive constants.

The value of A is determined by the requirement that the total area under the graph of f be equal to 1. Thus,

$$1 = \int_{-\infty}^{\infty} f(x)\, dx = \int_{0}^{\infty} Ae^{-kx}\, dx = \lim_{N \to \infty} \int_{0}^{N} Ae^{-kx}\, dx$$

$$= \lim_{N \to \infty} \left(-\frac{A}{k} e^{-kx} \Big|_{0}^{N} \right) = \lim_{N \to \infty} \left(-\frac{A}{k} e^{-kN} + \frac{A}{k} \right) = \frac{A}{k}$$

and so

$$A = k$$

This calculation leads to the following general formula for an exponential density function. The corresponding graph is shown in Figure 4.3.

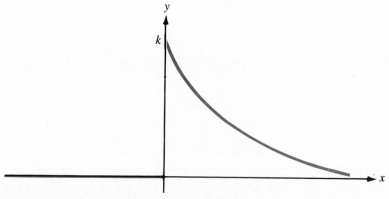

Figure 4.3 An exponential density function.

Exponential Density Function

$$f(x) = \begin{cases} ke^{-kx} & \text{if } x \geq 0 \\ 0 & \text{if } x < 0 \end{cases}$$

A random variable that has an exponential density function is said to be **exponentially distributed**. As you can see from the graph in Figure 4.3, the value of an exponentially distributed random variable is much more likely to be small than large. Such random variables include the life span of electronic components, the duration of telephone calls, and the interval between the arrivals of successive planes at an airport. The use of exponential density functions was illustrated in Example 2.4 of Chapter 6. Here is another example.

EXAMPLE 4.2 The probability density function for the duration of telephone calls in a certain city is

$$f(x) = \begin{cases} 0.5e^{-0.5x} & \text{if } x \geq 0 \\ 0 & \text{if } x < 0 \end{cases}$$

where x denotes the duration (in minutes) of a randomly selected call.

(a) Find the probability that a randomly selected call will last between 2 and 3 minutes.
(b) Find the probability that a randomly selected call will last at least 2 minutes.

SOLUTION

(a) $$P(2 \leq x \leq 3) = \int_2^3 0.5e^{-0.5x} \, dx = -e^{-0.5x}\Big|_2^3$$

$$= -e^{-1.5} + e^{-1} \approx 0.1447$$

(b) There are two ways to compute this probability. The first method is to evaluate an improper integral.

$$P(x \geq 2) = P(2 \leq x < \infty) = \int_2^\infty 0.5e^{-0.5x} \, dx$$

$$= \lim_{N \to \infty} \int_2^N 0.5e^{-0.5x} \, dx = \lim_{N \to \infty} \left(-e^{-0.5x}\Big|_2^N \right)$$

$$= \lim_{N \to \infty} (-e^{-0.5N} + e^{-1}) = e^{-1} \approx 0.3679$$

The second method is to compute 1 minus the probability that x is less than 2. That is,

$$P(x \geq 2) = 1 - \int_0^2 0.5e^{-0.5x} \, dx = 1 - \left(-e^{-0.5x}\Big|_0^2 \right)$$

$$= 1 - (-e^{-1} + 1) = e^{-1} \approx 0.3679$$

The Expected Value of a Random Variable

An important characteristic of a random variable is its "average" value. Familiar averages of random variables include the average highway mileage for a particular car model, the average waiting time at the check-in counter of a certain airline, and the average life span of workers in a hazardous profession. The average value of a random variable x is called its **expected value** or **mean** and is denoted by the symbol $E(x)$. For continuous random variables, you can calculate the expected value from the probability density function using the following formula.

Expected Value

If x is a continuous random variable with probability density function f, the expected value (or mean) of x is

$$E(x) = \int_{-\infty}^{\infty} xf(x)\, dx$$

Why the Expected Value Is the Average

To see why this integral gives the average value of a continuous random variable, first consider the simpler case in which x is a discrete random variable that takes on the values $x_1, x_2, \ldots, x_n$. If each of these values occurs with equal frequency, the probability of each is $\dfrac{1}{n}$ and the average value of x is

$$\frac{x_1 + x_2 + \cdots + x_n}{n} = x_1\!\left(\frac{1}{n}\right) + x_2\!\left(\frac{1}{n}\right) + \cdots + x_n\!\left(\frac{1}{n}\right)$$

More generally, if the values $x_1, x_2, \ldots, x_n$ occur with probabilities $p_1, p_2, \ldots, p_n$, respectively, the average value of x is the weighted sum

$$x_1 p_1 + x_2 p_2 + \cdots + x_n p_n = \sum_{j=1}^{n} x_j p_j$$

Now consider the continuous case. For simplicity, restrict your attention to a bounded interval $A \leq x \leq B$. Imagine that this interval is divided into n subintervals of width Δx, and let x_j denote the beginning of the jth subinterval. Then,

$$p_j = \begin{array}{c}\text{probability that } x \text{ is}\\ \text{in } j\text{th subinterval}\end{array} = \begin{array}{c}\text{area under graph}\\ \text{of } f \text{ from } x_j \text{ to } x_{j+1}\end{array} \approx f(x_j)\,\Delta x$$

where $f(x_j)\,\Delta x$ is the area of an approximating rectangle (Figure 4.4).

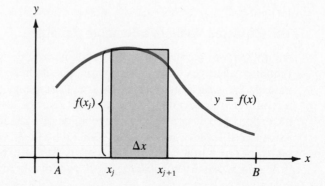

Figure 4.4 The jth subinterval with an approximating rectangle.

To approximate the average value of x over the interval $A \leq x \leq B$, treat x as if it were a discrete random variable that takes on the values x_1, $x_2, \ldots, x_n$ with probabilities $p_1, p_2, \ldots, p_n$, respectively. Then,

$$\text{Average value of } x \text{ on } A \leq x \leq B \simeq \sum_{j=1}^{n} x_j p_j \simeq \sum_{j=1}^{n} x_j f(x_j) \, \Delta x$$

The actual average value of x on the interval $A \leq x \leq B$ is the limit of this approximating sum as n increases without bound. That is,

$$\text{Average value of } x \text{ on } A \leq x \leq B = \lim_{n \to \infty} \sum_{j=1}^{n} x_j f(x_j) \, \Delta x = \int_{A}^{B} x f(x) \, dx$$

An extension of this argument shows that

$$\text{Average value of } x \text{ on } -\infty < x < \infty = \int_{-\infty}^{\infty} x f(x) \, dx$$

which is the formula for the expected value of a continuous random variable.

Geometric Interpretation of the Expected Value

To get a better feel for the expected value of a continuous random variable x, think of the probability density function for x as describing the distribution of mass on a beam lying on the x axis. Then the expected value of x is the point at which the beam will balance. If the graph of the density function is symmetric, the expected value is the point of symmetry, as illustrated in Figure 4.5.

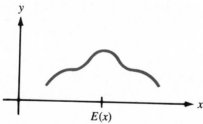

Figure 4.5 The expected value of a symmetric random variable.

Calculation of the Expected Value

The use of the integral formula to compute the expected value of a continuous random variable is illustrated in the next two examples.

EXAMPLE 4.3 Find the expected value of the uniformly distributed random variable from Example 4.1 with density function

$$f(x) = \begin{cases} \dfrac{1}{40} & \text{if } 0 \le x \le 40 \\ 0 & \text{otherwise} \end{cases}$$

SOLUTION

$$E(x) = \int_{-\infty}^{\infty} xf(x) \, dx = \int_0^{40} \frac{x}{40} \, dx = \frac{x^2}{80}\Big|_0^{40} = \frac{1{,}600}{80} = 20$$

In the context of Example 4.1, this says that the average waiting time at the red light is 20 seconds, a conclusion that should come as no surprise since the random variable is uniformly distributed between 0 and 40.

EXAMPLE 4.4   Find the expected value of the exponentially distributed random variable from Example 4.2 with density function

$$f(x) = \begin{cases} 0.5e^{-0.5x} & \text{if } x \ge 0 \\ 0 & \text{if } x < 0 \end{cases}$$

SOLUTION

$$E(x) = \int_{-\infty}^{\infty} xf(x) \, dx = \int_0^{\infty} 0.5xe^{-0.5x} \, dx$$

$$= \lim_{N \to \infty} \int_0^N 0.5xe^{-0.5x} \, dx$$

$$= \lim_{N \to \infty} \left(-xe^{-0.5x}\Big|_0^N + \int_0^N e^{-0.5x} \, dx \right) \qquad \text{integration by parts}$$

$$= \lim_{N \to \infty} (-xe^{-0.5x} - 2e^{-0.5x})\Big|_0^N$$

$$= \lim_{N \to \infty} (-Ne^{-0.5N} - 2e^{-0.5N} + 2)$$

$$= 2$$

That is, the average duration of telephone calls in the city in Example 4.2 is 2 minutes.

The Variance of a Random Variable

The expected value or mean of a random variable tells you the center of its distribution. Another concept that is useful in describing the distribution of a random variable is the **variance**, which tells you how spread out the distribution is. That is, the variance measures the tendency of the values of a random variable to cluster about their mean. Here is the definition.

Variance

> If x is a continuous random variable with probability density function f, the variance of x is
>
> $$\text{Var}(x) = \int_{-\infty}^{\infty} [x - E(x)]^2 f(x)\, dx$$

The definition of the variance is not as mysterious as it may seem at first glance. Notice that the formula for the variance is the same as that for the expected value, except that x has been replaced by the expression $[x - E(x)]^2$, which represents the square of the deviation of x from its mean $E(x)$. The variance, therefore, is simply the expected value or average of the squared deviations of the values of x from the mean. If the values of x tend to cluster about the mean as in Figure 4.6a, most of the deviations from the mean will be small and the variance, which is the average of the squares of these deviations, will also be small. On the other hand, if the values of the random variable are widely scattered as in Figure 4.6b, there will be many large deviations from the mean and the variance will be large.

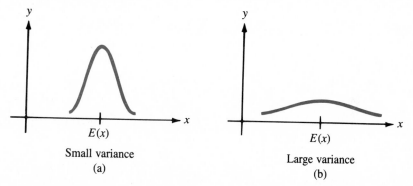

Figure 4.6 The variance as a measure of the spread of a distribution.

Small variance
(a)

Large variance
(b)

Calculation of the Variance

As you just saw, the formula

$$\text{Var}(x) = \int_{-\infty}^{\infty} [x - E(x)]^2 f(x)\, dx$$

defining the variance is fairly easily interpreted as a measure of the spread of the distribution of x. However, for all but the simplest density functions, this formula is cumbersome to use for the actual calculation of the variance. This is because the term $x - E(x)$ must be squared and multiplied by $f(x)$ before the integration is performed. Here is an equivalent formula for the variance, which is easier to use for computational purposes.

Variance Formula

$$\text{Var}(x) = \int_{-\infty}^{\infty} x^2 f(x) \, dx - [E(x)]^2$$

The derivation of this formula is straightforward. It involves expanding the integrand in the definition of the variance and rearranging the resulting terms until the new formula emerges. As you read the steps, keep in mind that the expected value $E(x)$ is a constant and can be brought outside integrals by the constant multiple rule.

$$\text{Var}(x) = \int_{-\infty}^{\infty} [x - E(x)]^2 f(x) \, dx \quad \text{definition of variance}$$

$$= \int_{-\infty}^{\infty} \{x^2 - 2xE(x) + [E(x)]^2\} f(x) \, dx \quad \text{integrand expanded}$$

$$= \int_{-\infty}^{\infty} x^2 f(x) \, dx - 2E(x) \int_{-\infty}^{\infty} xf(x) \, dx + [E(x)]^2 \int_{-\infty}^{\infty} f(x) \, dx$$

$$\text{sum rule and constant multiple rule}$$

$$= \int_{-\infty}^{\infty} x^2 f(x) \, dx - 2E(x) \cdot E(x) + [E(x)]^2 (1)$$

$$\text{definition of } E(x) \text{ and the fact that } \int_{-\infty}^{\infty} f(x) \, dx = 1$$

$$= \int_{-\infty}^{\infty} x^2 f(x) \, dx - [E(x)]^2$$

The use of this formula to compute the variance is illustrated in the next two examples.

EXAMPLE 4.5   Find the variance of the uniformly distributed random variable from Example 4.1 with density function

$$f(x) = \begin{cases} \dfrac{1}{40} & \text{if } 0 \leq x \leq 40 \\ 0 & \text{otherwise} \end{cases}$$

SOLUTION
The first step is to compute $E(x)$. This was done in Example 4.3, where you found that $E(x) = 20$. Using this value in the variance formula, you get

$$\text{Var}(x) = \int_{-\infty}^{\infty} x^2 f(x) \, dx - [E(x)]^2$$

$$= \int_0^{40} \frac{x^2}{40} \, dx - 400 = \frac{x^3}{120}\Big|_0^{40} - 400$$

$$= \frac{64{,}000}{120} - 400 = \frac{16{,}000}{120} = \frac{400}{3}$$

EXAMPLE 4.6 Find the variance of the exponentially distributed random variable from Example 4.2 with density function

$$f(x) = \begin{cases} 0.5e^{-0.5x} & \text{if } x \geq 0 \\ 0 & \text{if } x < 0 \end{cases}$$

SOLUTION

Using the value $E(x) = 2$ obtained in Example 4.4 and integrating by parts twice, you get

$$\mathrm{Var}(x) = \int_{-\infty}^{\infty} x^2 f(x) \, dx - [E(x)]^2$$

$$= \int_0^{\infty} 0.5x^2 e^{-0.5x} \, dx - 4$$

$$= \lim_{N \to \infty} \int_0^N 0.5x^2 e^{-0.5x} \, dx - 4$$

$$= \lim_{N \to \infty} \left(-x^2 e^{-0.5x}\Big|_0^N + 2 \int_0^N xe^{-0.5x} \, dx \right) - 4$$

$$= \lim_{N \to \infty} \left[(-x^2 e^{-0.5x} - 4xe^{-0.5x})\Big|_0^N + 4 \int_0^N e^{-0.5x} \, dx \right] - 4$$

$$= \lim_{N \to \infty} (-x^2 - 4x - 8)e^{-0.5x}\Big|_0^N - 4$$

$$= \lim_{N \to \infty} (-N^2 - 4N - 8)e^{-0.5N} + 8 - 4$$

$$= 4$$

Problems

In Problems 1 through 8, integrate the given probability density function to find the indicated probabilities.

1.
$$f(x) = \begin{cases} \dfrac{1}{3} & \text{if } 2 \leq x \leq 5 \\ 0 & \text{otherwise} \end{cases}$$

(a) $P(2 \leq x \leq 5)$ (b) $P(3 \leq x \leq 4)$ (c) $P(x \geq 4)$

2. $$f(x) = \begin{cases} \dfrac{x}{2} & \text{if } 0 \le x \le 2 \\ 0 & \text{otherwise} \end{cases}$$

    (a) $P(0 \le x \le 2)$ (b) $P(1 \le x \le 2)$ (c) $P(x \le 1)$

3. $$f(x) = \begin{cases} \dfrac{1}{8}(4 - x) & \text{if } 0 \le x \le 4 \\ 0 & \text{otherwise} \end{cases}$$

    (a) $P(0 \le x \le 4)$ (b) $P(2 \le x \le 3)$ (c) $P(x \ge 1)$

4. $$f(x) = \begin{cases} \dfrac{3}{32}(4x - x^2) & \text{if } 0 \le x \le 4 \\ 0 & \text{otherwise} \end{cases}$$

    (a) $P(0 \le x \le 4)$ (b) $P(1 \le x \le 2)$ (c) $P(x \le 1)$

5. $$f(x) = \begin{cases} \dfrac{3}{x^4} & \text{if } 1 \le x < \infty \\ 0 & \text{if } x < 1 \end{cases}$$

    (a) $P(1 \le x < \infty)$ (b) $P(1 \le x \le 2)$ (c) $P(x \ge 2)$

6. $$f(x) = \begin{cases} \dfrac{1}{10}e^{-x/10} & \text{if } x \ge 0 \\ 0 & \text{if } x < 0 \end{cases}$$

    (a) $P(0 \le x < \infty)$ (b) $P(x \le 2)$ (c) $P(x \ge 5)$

7. $$f(x) = \begin{cases} 2xe^{-x^2} & \text{if } x \ge 0 \\ 0 & \text{if } x < 0 \end{cases}$$

    (a) $P(x \ge 0)$ (b) $P(1 \le x \le 2)$ (c) $P(x \le 2)$

8. $$f(x) = \begin{cases} \dfrac{1}{4}xe^{-x/2} & \text{if } x \ge 0 \\ 0 & \text{if } x < 0 \end{cases}$$

    (a) $P(0 \le x < \infty)$ (b) $P(2 \le x \le 4)$ (c) $P(x \ge 6)$

Traffic flow   9. A certain traffic light remains red for 45 seconds at a time. You arrive (at random) at the light and find it red. Use an appropriate uniform density function to find the probability that the light will turn green within 15 seconds.

Commuting
10. During the morning rush hour, commuter trains from Long Island to Manhattan run every 20 minutes. You arrive (at random) at the station during the rush hour and find no train at the platform. Assuming that the trains are running on schedule, use an appropriate uniform density function to find the probability that you will have to wait at least 8 minutes for your train.

Movie theaters
11. A 2-hour movie runs continuously at a local theater. You leave for the theater without first checking the show times. Use an appropriate uniform density function to find the probability that you will arrive at the theater within 10 minutes (before or after) of the start of the film.

Experimental psychology
12. Suppose x is the length of time (in minutes) that it takes a laboratory rat to traverse a certain maze. If x is exponentially distributed with density function

$$f(x) = \begin{cases} \frac{1}{3}e^{-x/3} & \text{if } x \geq 0 \\ 0 & \text{if } x < 0 \end{cases}$$

find the probability that a randomly selected rat will require more than 3 minutes to traverse the maze.

Customer service
13. The time x (in minutes) that a customer must spend waiting in line at a certain bank is exponentially distributed with density function

$$f(x) = \begin{cases} \frac{1}{4}e^{-x/4} & \text{if } x \geq 0 \\ 0 & \text{if } x < 0 \end{cases}$$

Find the probability that a randomly selected customer at the bank will have to stand in line at least 8 minutes.

Warranty protection
14. The life span x (in months) of a certain electrical appliance is exponentially distributed with density function

$$f(x) = \begin{cases} 0.08e^{-0.08x} & \text{if } x \geq 0 \\ 0 & \text{if } x < 0 \end{cases}$$

The appliance carries a 1-year warranty from the manufacturer. Suppose you purchase one of these appliances, selected at random from the manufacturer's stock. Find the probability that the warranty will expire before your appliance becomes unusable.

15. Find the expected value and variance for the random variable in Problem 1.

16. Find the expected value and variance for the random variable in Problem 2.

17. Find the expected value and variance for the random variable in Problem 3.

18. Find the expected value and variance for the random variable in Problem 5.

19. Find the expected value and variance for the random variable in Problem 6.

Traffic flow   20. Find the average waiting time for cars arriving on red at the traffic light in Problem 9.

Commuting   21. Find the average wait for rush-hour commuters arriving at the station in Problem 10 when no train is at the platform.

Experimental psychology   22. Find the average time required for laboratory rats to traverse the maze in Problem 12.

Customer service   23. Find the average waiting time for customers at the bank in Problem 13.

Expected value   24. Show that the expected value of a uniformly distributed random variable with density function

$$f(x) = \begin{cases} \dfrac{1}{B - A} & \text{if } A \le x \le B \\ 0 & \text{otherwise} \end{cases}$$

is $\dfrac{A + B}{2}$.

Variance   25. Show that the variance of the uniformly distributed random variable in Problem 24 is $\dfrac{(B - A)^2}{12}$.

Expected value   26. Show that the expected value of an exponentially distributed random variable with density function

$$f(x) = \begin{cases} ke^{-kx} & \text{if } x \ge 0 \\ 0 & \text{if } x < 0 \end{cases}$$

is $\dfrac{1}{k}$.

Variance   27. Show that the variance of the exponentially distributed random variable in Problem 26 is $\dfrac{1}{k^2}$.

5   Numerical Integration

In this section you will see some techniques you can use to approximate definite integrals. Numerical methods such as these are needed when the function to be integrated does not have an elementary antiderivative.

Approximation by Rectangles

If $f(x)$ is positive on the interval $a \leq x \leq b$, the definite integral $\int_a^b f(x)\, dx$ is equal to the area under the graph of f between $x = a$ and $x = b$. As you saw in Section 1 of this chapter, one way to approximate this area is to use n rectangles, as shown in Figure 5.1. In particular, you divide the interval $a \leq x \leq b$ into n equal subintervals of width $\Delta x = \dfrac{b - a}{n}$ and let x_j denote the beginning of the jth subinterval. The base of the jth rectangle is the jth

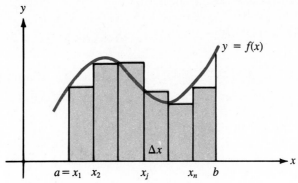

Figure 5.1 Approximation by rectangles.

subinterval, and its height is $f(x_j)$. Hence the area of the jth rectangle is $f(x_j)\, \Delta x$. The sum of the areas of all n rectangles is an approximation to the area under the curve and hence an approximation to the corresponding definite integral. Thus,

$$\int_a^b f(x)\, dx \simeq f(x_1)\, \Delta x + f(x_2)\, \Delta x + \cdots + f(x_n)\, \Delta x$$

This approximation improves as the number of rectangles increases, and you can estimate the integral to any desired degree of accuracy by taking n large enough. However, since fairly large values of n are usually required to achieve reasonable accuracy, approximation by rectangles is rarely used in practice.

Approximation by Trapezoids

The accuracy of the approximation improves significantly if trapezoids are used instead of rectangles. Figure 5.2 shows the area from Figure 5.1 approximated by n trapezoids. Notice how much better the approximation is in this case.

The jth trapezoid is shown in greater detail in Figure 5.3. Notice that it consists of a rectangle with a right triangle on top of it. Since,

$$\text{Area of rectangle} = f(x_{j+1})\, \Delta x$$

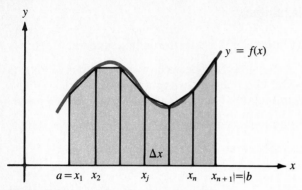

Figure 5.2 Approximation by trapezoids.

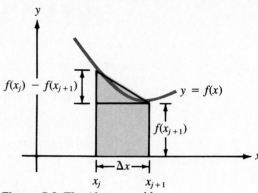

Figure 5.3 The jth trapezoid.

and

$$\text{Area of triangle} = \frac{1}{2}[f(x_j) - f(x_{j+1})] \, \Delta x$$

it follows that

$$\text{Area of trapezoid} = f(x_{j+1}) \, \Delta x + \frac{1}{2}[f(x_j) - f(x_{j+1})] \, \Delta x$$

$$= \frac{1}{2}[f(x_j) + f(x_{j+1})] \, \Delta x$$

The sum of the areas of all n trapezoids is an approximation to the area under the curve and hence an approximation to the corresponding definite integral. Thus,

$$\int_a^b f(x) \, dx \simeq \frac{1}{2}[f(x_1) + f(x_2)] \, \Delta x + \frac{1}{2}[f(x_2) + f(x_3)] \, \Delta x + \cdots$$

$$+ \frac{1}{2}[f(x_n) + f(x_{n+1})] \, \Delta x$$

$$= \frac{\Delta x}{2}[f(x_1) + 2f(x_2) + \cdots + 2f(x_n) + f(x_{n+1})]$$

This approximation formula is known as the **trapezoidal rule** and applies even if the function f is not positive.

The Trapezoidal Rule

$$\int_a^b f(x) \, dx \simeq \frac{\Delta x}{2}[f(x_1) + 2f(x_2) + \cdots + 2f(x_n) + f(x_{n+1})]$$

The use of the trapezoidal rule is illustrated in the following example.

EXAMPLE 5.1 Use the trapezoidal rule with $n = 10$ to approximate $\int_1^2 \frac{1}{x}\, dx$.

SOLUTION
Since

$$\Delta x = \frac{2 - 1}{10} = 0.1$$

the interval $1 \le x \le 2$ is divided into 10 subintervals, as shown in Figure 5.4, by

$$x_1 = 1,\ x_2 = 1.1,\ x_3 = 1.2,\ \ldots,\ x_{10} = 1.9,\ x_{11} = 2$$

Then, by the trapezoidal rule,

$$\int_1^2 \frac{1}{x}\, dx \approx \frac{0.1}{2}\left(\frac{1}{1} + \frac{2}{1.1} + \frac{2}{1.2} + \frac{2}{1.3} + \frac{2}{1.4} + \frac{2}{1.5} + \frac{2}{1.6} + \frac{2}{1.7}\right.$$

$$\left. + \frac{2}{1.8} + \frac{2}{1.9} + \frac{1}{2}\right)$$

$$\approx 0.693771$$

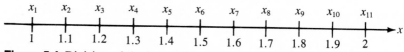

Figure 5.4 Division of the interval $1 \le x \le 2$ into 10 subintervals.

The definite integral in Example 5.1 can be evaluated directly. In particular,

$$\int_1^2 \frac{1}{x}\, dx = \ln |x|\,\Big|_1^2 = \ln 2 \approx 0.693147$$

Thus the approximation of this particular integral by the trapezoidal rule with $n = 10$ is accurate (after round off) to two decimal places.

The Accuracy of the Trapezoidal Rule

The difference between the true value of the integral $\int_a^b f(x)\, dx$ and the approximation generated by the trapezoidal rule when n subintervals are used is denoted by E_n. The following estimate for the absolute value of E_n is proved in more advanced courses.

Error Estimate for the Trapezoidal Rule

If M is the maximum value of $|f''(x)|$ on the interval $a \leq x \leq b$, then

$$|E_n| \leq \frac{M(b - a)^3}{12n^2}$$

The use of this formula is illustrated in the next example.

EXAMPLE 5.2   Estimate the accuracy of the approximation of $\int_1^2 \frac{1}{x}\, dx$ by the trapezoidal rule with $n = 10$.

SOLUTION

Starting with $f(x) = \frac{1}{x}$, compute the derivatives

$$f'(x) = -\frac{1}{x^2} \quad \text{and} \quad f''(x) = \frac{2}{x^3}$$

and observe that the largest value of $|f''(x)|$ for $1 \leq x \leq 2$ is $|f''(1)| = 2$. Apply the error formula with

$$M = 2 \quad a = 1 \quad b = 2 \quad \text{and} \quad n = 10$$

to get

$$|E_{10}| \leq \frac{2(2 - 1)^3}{12(10)^2} \simeq 0.00167$$

That is, the error in the approximation in Example 5.1 is guaranteed to be no greater than 0.00167. (In fact, to five decimal places, the error is 0.00062, as you can see by comparing the approximation obtained in Example 5.1 with the decimal representation of ln 2.)

How to Choose the Number of Subintervals

With the aid of the error estimate you can decide in advance how many subintervals to use to achieve a desired degree of accuracy. Here is an example.

EXAMPLE 5.3   How many subintervals are required to guarantee that the error will be less than 0.00005 in the approximation of $\int_1^2 \frac{1}{x}\, dx$ using the trapezoidal rule?

SOLUTION

From Example 5.2 you know that $M = 2$, $a = 1$, and $b = 2$, so that

$$|E_n| \leq \frac{2(2-1)^3}{12n^2} = \frac{1}{6n^2}$$

The goal is to find the smallest positive integer n for which

$$\frac{1}{6n^2} < 0.00005$$

or, equivalently,

$$n^2 > \frac{1}{6(0.00005)}$$

or

$$n > \sqrt{\frac{1}{6(0.00005)}} \simeq 57.74$$

The smallest such integer is $n = 58$, and so 58 subintervals are required to ensure the desired accuracy.

The relatively large number of subintervals required in Example 5.3 to ensure accuracy to within 0.00005 suggests that approximation by trapezoids may not be efficient enough for some applications. There is another approximation technique, which is no harder to use than the trapezoidal rule, but which requires substantially fewer calculations to achieve a given degree of accuracy. Like the trapezoidal rule, it is based on the approximation of the area under a curve by columns, but unlike the trapezoidal rule, it uses curves rather than lines as the tops of the columns.

Approximation Using Parabolas

The approximation of a definite integral using parabolas is based on the following construction (which is illustrated in Figure 5.5 for $n = 6$). Divide the interval $a \leq x \leq b$ into an even number of subintervals so that adjacent subintervals can be paired with none left over. Approximate the portion of the graph that lies above the first pair of subintervals by the (unique) parabola that passes through the three points $(x_1, f(x_1))$, $(x_2, f(x_2))$, and $(x_3, f(x_3))$, and use the area under this parabola between x_1 and x_3 to approximate the corresponding area under the curve. Do the same for the remaining pairs of subintervals and use the sum of the resulting areas to approximate the total area under the graph. It can be shown that this construction leads to the following approximation scheme known as **Simpson's rule.**

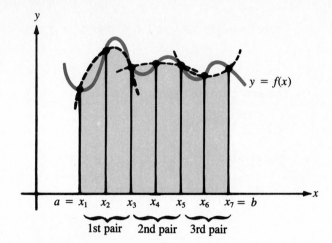

Figure 5.5 Approximation using parabolas.

1st pair   2nd pair   3rd pair

Simpson's Rule

$$\int_a^b f(x)\, dx \simeq \frac{\Delta x}{3}\left[f(x_1) + 4f(x_2) + 2f(x_3) + 4f(x_4) + 2f(x_5)\right.$$
$$\left. + \cdots + 2f(x_{n-1}) + 4f(x_n) + f(x_{n+1})\right]$$

Notice that the first and last function values in the approximating sum in Simpson's rule are multiplied by 1, while the others are multiplied alternately by 4 and 2.

The proof of Simpson's rule is based on the fact that the equation of a parabola is a polynomial of the form $y = Ax^2 + Bx + C$. For each pair of subintervals, the three given points are used to find the coefficients A, B, and C, and the resulting polynomial is then integrated to get the corresponding area. The details of the proof are straightforward but tedious and will be omitted.

EXAMPLE 5.4   Use Simpson's rule with $n = 10$ to approximate $\int_1^2 \frac{1}{x}\, dx$.

SOLUTION
As in Example 5.1, $\Delta x = 0.1$, and hence the interval $1 \le x \le 2$ is divided into the 10 subintervals by

$$x_1 = 1,\ x_2 = 1.1,\ x_3 = 1.2,\ \ldots,\ x_{10} = 1.9,\ x_{11} = 2$$

Then, by Simpson's rule,

$$\int_1^2 \frac{1}{x}\, dx \simeq \frac{0.1}{3}\left(\frac{1}{1} + \frac{4}{1.1} + \frac{2}{1.2} + \frac{4}{1.3} + \frac{2}{1.4} + \frac{4}{1.5} + \frac{2}{1.6} + \frac{4}{1.7}\right.$$

$$+ \frac{2}{1.8} + \frac{4}{1.9} + \frac{1}{2}\Big)$$

$$\simeq 0.693150$$

Notice that this is an excellent approximation to the true value ln 2 = 0.693147.

The Accuracy of Simpson's Rule

The error estimate for Simpson's rule turns out to involve the fourth derivative $f^{(4)}(x)$.

Error Estimate for Simpson's Rule

If M is the maximum value of $|f^{(4)}(x)|$ on the interval $a \leq x \leq b$, then

$$|E_n| \leq \frac{M(b - a)^5}{180n^4}$$

Here is an application of the formula.

EXAMPLE 5.5   Estimate the accuracy of the approximation of $\int_1^2 \frac{1}{x} dx$ by Simpson's rule with $n = 10$.

SOLUTION

Starting with $f(x) = \frac{1}{x}$, compute the derivatives

$$f'(x) = -\frac{1}{x^2} \quad f''(x) = \frac{2}{x^3} \quad f^{(3)}(x) = -\frac{6}{x^4} \quad f^{(4)}(x) = \frac{24}{x^5}$$

and observe that the largest value of $|f^{(4)}(x)|$ on the interval $1 \leq x \leq 2$ is $|f^{(4)}(1)| = 24$.

Now apply the error formula with $M = 24$, $a = 1$, $b = 2$, and $n = 10$ to get

$$|E_{10}| \leq \frac{24(2 - 1)^5}{180(10)^4} \simeq 0.000013$$

That is, the error in the approximation in Example 5.4 is guaranteed to be no greater than 0.000013.

How to Choose the Number of Subintervals

In the next example, the error estimate is used to determine the number of subintervals that are required to ensure a specified degree of accuracy.

EXAMPLE 5.6 How many subintervals are required to ensure accuracy to within 0.00005 in the approximation of $\int_1^2 \frac{1}{x}\,dx$ by Simpson's rule?

SOLUTION
From Example 5.5 you know that $M = 24$, $a = 1$, and $b = 2$. Hence,

$$|E_n| \le \frac{24(2 - 1)^5}{180n^4} = \frac{2}{15n^4}$$

The goal is to find the smallest positive (even) integer for which

$$\frac{2}{15n^4} < 0.00005$$

or, equivalently,

$$n^4 > \frac{2}{15(0.00005)}$$

or

$$n > \left[\frac{2}{15(0.00005)}\right]^{1/4} \simeq 7.19$$

The smallest such (even) integer is $n = 8$, and so eight subintervals are required to ensure the desired accuracy.

Normal Probability Density Functions

In the preceding section you saw how to integrate probability density functions to compute probabilities. The most widely used probability density functions are the **normal density functions,** whose graphs are "bell-shaped" curves. They describe or approximate the distributions of many random variables arising in the social and natural sciences, including heights, weights, test scores, and measurement errors.

The simplest and most important of the normal density functions is the **standard normal density function**

$$f(x) = \frac{1}{\sqrt{2\pi}}e^{-x^2/2}$$

which has mean equal to zero and variance equal to 1 (Figure 5.6). All other normal density functions can be transformed to the standard normal density function by a routine change of variables. Thus, once you know how to compute probabilities using this particular density function, you can compute probabilities associated with any normal density function. (The details can be found in any statistics text.)

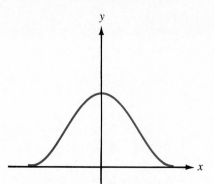

Figure 5.6 The standard normal density function.

Approximation of Areas Under the Standard Normal Curve

If a random variable x has the standard normal distribution, the probability that x lies between a and b is the area under the standard normal curve between $x = a$ and $x = b$ and is given by the integral

$$P(a \leq x \leq b) = \frac{1}{\sqrt{2\pi}} \int_a^b e^{-x^2/2} \, dx$$

Unfortunately, the integrand $e^{-x^2/2}$ does not have an elementary antiderivative and numerical methods must be used to approximate the integral. Using numerical integration and electronic computers, statisticians have compiled highly accurate tables of areas under the standard normal curve. The procedure is illustrated in the next example, in which Simpson's rule is used to approximate an area under this curve.

EXAMPLE 5.7   Use Simpson's rule with $n = 10$ to approximate the probability $P(0 \leq x \leq 1)$, where x is a standard normal random variable.

SOLUTION
The probability $P(0 \leq x \leq 1)$ is the area under the graph of the standard normal density function from $x = 0$ to $x = 1$ (Figure 5.7) and is given by the definite integral

$$\frac{1}{\sqrt{2\pi}} \int_0^1 e^{-x^2/2} \, dx$$

To approximate this integral using 10 subintervals, observe that

$$\Delta x = \frac{1 - 0}{10} = 0.1$$

and that

$$x_1 = 0, \, x_2 = 0.1, \, x_3 = 0.2, \, \ldots, \, x_{10} = 0.9, \, x_{11} = 1$$

Then, by Simpson's rule,

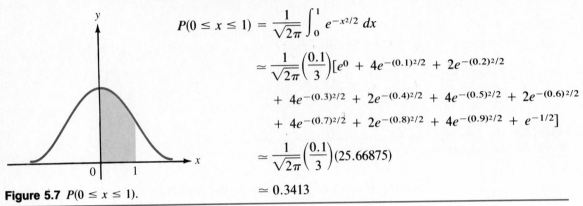

$$P(0 \le x \le 1) = \frac{1}{\sqrt{2\pi}} \int_0^1 e^{-x^2/2} \, dx$$

$$\approx \frac{1}{\sqrt{2\pi}} \left(\frac{0.1}{3}\right)[e^0 + 4e^{-(0.1)^2/2} + 2e^{-(0.2)^2/2}$$

$$+ \ 4e^{-(0.3)^2/2} + 2e^{-(0.4)^2/2} + 4e^{-(0.5)^2/2} + 2e^{-(0.6)^2/2}$$

$$+ \ 4e^{-(0.7)^2/2} + 2e^{-(0.8)^2/2} + 4e^{-(0.9)^2/2} + e^{-1/2}]$$

$$\approx \frac{1}{\sqrt{2\pi}} \left(\frac{0.1}{3}\right)(25.66875)$$

$$\approx 0.3413$$

Figure 5.7 $P(0 \le x \le 1)$.

Problems

In Problems 1 through 8, approximate the given integral using (a) the trapezoidal rule, and (b) Simpson's rule with the specified number of subintervals.

1. $\int_1^2 x^2 \, dx$; $n = 4$

2. $\int_4^6 \frac{1}{\sqrt{x}} \, dx$; $n = 10$

3. $\int_0^1 \frac{1}{1 + x^2} \, dx$; $n = 4$

4. $\int_2^3 \frac{1}{x^2 - 1} \, dx$; $n = 4$

5. $\int_{-1}^0 \sqrt{1 + x^2} \, dx$; $n = 4$

6. $\int_0^3 \sqrt{9 - x^2} \, dx$; $n = 6$

7. $\int_0^1 e^{-x^2} \, dx$; $n = 4$

8. $\int_0^2 e^{x^2} \, dx$; $n = 10$

In Problems 9 through 14, approximate the given integral and estimate the error $|E_n|$ using (a) the trapezoidal rule, and (b) Simpson's rule with the specified number of subintervals.

9. $\int_1^2 \frac{1}{x^2} \, dx$; $n = 4$

10. $\int_0^2 x^3 \, dx$; $n = 8$

11. $\int_1^3 \sqrt{x} \, dx$; $n = 10$

12. $\int_1^2 \ln x \, dx$; $n = 4$

13. $\int_0^1 e^{x^2} \, dx$; $n = 4$

14. $\int_0^{0.6} e^{x^3} \, dx$; $n = 6$

Normal distribution 15. Use Simpson's rule with $n = 8$ to approximate the probability $P(0 \le x \le 0.8)$, where x is a standard normal random variable.

Normal distribution

16. Use Simpson's rule with $n = 8$ to approximate the probability $P(0 \leq x \leq 1.6)$, where x is a standard normal random variable.

In Problems 17 through 22, determine how many subintervals are required to guarantee accuracy to within 0.00005 in the approximation of the given integral by (a) the trapezoidal rule, and (b) Simpson's rule.

17. $\displaystyle\int_1^3 \frac{1}{x}\, dx$

18. $\displaystyle\int_0^4 (x^4 + 2x^2 + 1)\, dx$

19. $\displaystyle\int_1^2 \frac{1}{\sqrt{x}}\, dx$

20. $\displaystyle\int_1^2 \ln(1 + x)\, dx$

21. $\displaystyle\int_{1.2}^{2.4} e^x\, dx$

22. $\displaystyle\int_0^2 e^{x^2}\, dx$

Chapter Summary and Review Problems

Important Terms, Symbols, and Formulas

Fundamental theorem of calculus:

$$\lim_{n\to\infty} \sum_{j=1}^n f(x_j)\, \Delta x = \int_a^b f(x)\, dx = F(b) - F(a)$$

Volume of a solid of revolution:

$$\text{Volume} = \pi \int_a^b [f(x)]^2\, dx$$

$$\text{Average value} = \frac{1}{b - a} \int_a^b f(x)\, dx$$

Improper integrals:

$$\int_a^\infty f(x)\, dx = \lim_{N\to\infty} \int_a^N f(x)\, dx$$

$$\int_{-\infty}^\infty f(x)\, dx = \lim_{N\to\infty} \int_{-N}^0 f(x)\, dx + \lim_{N\to\infty} \int_0^N f(x)\, dx$$

Discrete and continuous random variables

Probability density function:

$$P(a \leq x \leq b) = \int_a^b f(x)\, dx$$

Uniform density function:

$$f(x) = \begin{cases} \dfrac{1}{B - A} & \text{if } A \leq x \leq B \\ 0 & \text{otherwise} \end{cases}$$

Exponential density function:

$$f(x) = \begin{cases} ke^{-kx} & \text{if } x \geq 0 \\ 0 & \text{if } x < 0 \end{cases}$$

Expected value (mean):

$$E(x) = \int_{-\infty}^{\infty} xf(x)\ dx$$

Variance:

$$\text{Var}(x) = \int_{-\infty}^{\infty} [x - E(x)]^2 f(x)\ dx = \int_{-\infty}^{\infty} x^2 f(x)\ dx - [E(x)]^2$$

Trapezoidal rule:

$$\int_a^b f(x)\ dx \simeq \frac{\Delta x}{2} [f(x_1) + 2f(x_2) + \cdots + 2f(x_n) + f(x_{n+1})]$$

Error estimate:

$$|E_n| \leq \frac{M(b-a)^3}{12n^2}$$

where M is the maximum value of $|f''(x)|$ for $a \leq x \leq b$.

Simpson's rule:

$$\int_a^b f(x)\ dx \simeq \frac{\Delta x}{3} [f(x_1) + 4f(x_2) + 2f(x_3) + 4f(x_4) + 2f(x_5) + \cdots$$
$$+ 2f(x_{n-1}) + 4f(x_n) + f(x_{n+1})]$$

Error estimate:

$$|E_n| \leq \frac{M(b-a)^5}{180n^4}$$

where M is the maximum value of $|f^{(4)}(x)|$ for $a \leq x \leq b$.

Review Problems

1. A retailer expects that x months from now consumers will be buying 50 cameras a month at the price of $P(x) = 40 + 3\sqrt{x}$ dollars per camera. Use the characterization of the definite integral as the limit of a sum to find the total revenue the retailer can expect from the sale of the cameras over the next 9 months.

2. Economists predict that x months from now the demand for beef will be $D(x)$ pounds per month and the price will be $P(x)$ dollars per pound. Use the characterization of the definite integral as the limit of a sum to find an expression for the total amount that consumers will spend on beef this year.

In Problems 3 and 4, find the volume of the solid of revolution formed by revolving the region R about the x axis.

3. R is the region under the curve $y = x^2 + 1$ from $x = -1$ to $x = 2$.

4. R is the region under the curve $y = e^{-x/20}$ from $x = 0$ to $x = 10$.

5. Records indicate that t months after the beginning of the year, the price of chicken in local supermarkets was $P(t) = 0.06t^2 - 0.2t + 1.2$ dollars per pound. What was the average price of chicken during the first 6 months of the year?

6. Money is transferred continuously into an account at the constant rate of \$1,200 per year. The account earns interest at the annual rate of 8 percent compounded continuously. How much will be in the account at the end of 5 years?

7. What is the present value of an investment scheme that will generate income continuously at a constant rate of \$1,000 per year for 10 years if the prevailing annual interest rate remains fixed at 7 percent compounded continuously?

8. In a certain community the fraction of the homes placed on the market that remain unsold for at least t weeks is approximately $f(t) = e^{-0.2t}$. If 200 homes are currently on the market and if additional homes are placed on the market at the rate of 8 per week, approximately how many homes will be on the market 10 weeks from now?

9. The population density r miles from the center of a certain city is $D(r) = 6,000e^{-0.1r}$ people per square mile. How many people live between 2 and 3 miles from the center of the city?

In Problems 10 through 18, evaluate the given improper integral.

10. $\int_0^\infty \dfrac{1}{\sqrt[3]{1 + 2x}}\, dx$

11. $\int_0^\infty (1 + 2x)^{-3/2}\, dx$

12. $\int_0^\infty \dfrac{3x}{x^2 + 1}\, dx$

13. $\int_0^\infty 3e^{-5x}\, dx$

14. $\int_0^\infty xe^{-2x}\, dx$

15. $\int_0^\infty 2x^2 e^{-x^3}\, dx$

16. $\int_0^\infty x^2 e^{-2x}\, dx$

17. $\int_2^\infty \dfrac{1}{x(\ln x)^2}\, dx$

18. $\int_0^\infty x^5 e^{-x^3}\, dx$

19. The publishers of a national magazine have found that the fraction of subscribers who remain subscribers for at least t years is $f(t) = e^{-t/10}$. Currently the magazine has 20,000 subscribers and estimates

that new subscriptions will be sold at the rate of 1,000 per year. Approximately how many subscribers will the magazine have in the long run?

20. It is estimated that t years from now a certain investment will be generating income at the rate of $f(t) = 8,000 + 400t$ dollars per year. If the income is generated in perpetuity and the prevailing annual interest rate remains fixed at 10 percent compounded continuously, find the present value of the investment.

21. Demographic studies conducted in a certain city indicate that the fraction of the residents that will remain in the city for at least t years is $f(t) = e^{-t/20}$. The current population of the city is 100,000, and it is estimated that t years from now, new people will be arriving at the rate of $100t$ people per year. If this estimate is correct, what will happen to the population of the city in the long run?

In Problems 22 through 24, integrate the given probability density function to find the indicated probabilities.

22.
$$f(x) = \begin{cases} \dfrac{1}{3} & \text{if } 1 \leq x \leq 4 \\ 0 & \text{otherwise} \end{cases}$$

(a) $P(1 \leq x \leq 4)$ (b) $P(2 \leq x \leq 3)$ (c) $P(x \leq 2)$

23.
$$f(x) = \begin{cases} \dfrac{2}{9}(3 - x) & \text{if } 0 \leq x \leq 3 \\ 0 & \text{otherwise} \end{cases}$$

(a) $P(0 \leq x \leq 3)$ (b) $P(1 \leq x \leq 2)$

24.
$$f(x) = \begin{cases} 0.2e^{-0.2x} & \text{if } x \geq 0 \\ 0 & \text{if } x < 0 \end{cases}$$

(a) $P(x \geq 0)$ (b) $P(1 \leq x \leq 4)$ (c) $P(x \geq 5)$

25. Find the expected value and variance for the random variable in Problem 22.

26. Find the expected value and variance for the random variable in Problem 23.

27. Find the expected value and variance for the random variable in Problem 24.

28. A bakery turns out a fresh batch of chocolate chip cookies every 45 minutes. You arrive (at random) at the bakery, hoping to buy a fresh

cookie. Use an appropriate uniform density function to find the probability that you arrive within 5 minutes (before or after) of the time that the cookies come out of the oven.

29. The time x (in minutes) between the arrivals of successive cars at a toll booth is exponentially distributed with density function

$$f(x) = \begin{cases} 0.5e^{-0.5x} & \text{if } x \geq 0 \\ 0 & \text{if } x < 0 \end{cases}$$

   (a) Find the probability that a randomly selected pair of successive cars will arrive at the toll booth at least 6 minutes apart.
   (b) Find the average time interval between the arrivals of successive cars at the toll booth.

In Problems 30 and 31, approximate the given integral and estimate the error $|E_n|$ using (a) the trapezoidal rule, and (b) Simpson's rule with the specified number of subintervals.

30. $\displaystyle\int_1^3 \frac{1}{x}\, dx;\ n = 10$ 31. $\displaystyle\int_0^2 e^{x^2}\, dx;\ n = 8$

In Problems 32 and 33, determine how many subintervals are required to guarantee accuracy to within 0.00005 in the approximation of the given integral by (a) the trapezoidal rule, and (b) Simpson's rule.

32. $\displaystyle\int_1^3 \sqrt{x}\, dx$ 33. $\displaystyle\int_{0.5}^1 e^{x^2}\, dx$

Chapter

Differential Equations

1   **Elementary Differential Equations**
2   **Separable Differential Equations**
      Chapter Summary and Review Problems

1   Elementary Differential Equations

Any equation that contains a derivative is called a **differential equation.**
For example, the equations

$$\frac{dy}{dx} = 3x^2 + 5 \qquad \frac{dP}{dt} = kP \qquad \text{and} \qquad \left(\frac{dy}{dx}\right)^2 + 3\frac{dy}{dx} + 2y = e^x$$

are all differential equations.

Applications of Differential Equations

Many practical situations, especially those involving rates of change, can
be described mathematically by differential equations. For example, the
assumption that population grows at a rate proportional to its size can be
expressed by the differential equation $\frac{dP}{dt} = kP$, where P denotes the

population size at time t and k is the constant of proportionality. In economics, statements about marginal cost and marginal revenue can be formulated as differential equations.

Here are two examples.

EXAMPLE 1.1 Write a differential equation describing the fact that the rate at which people hear about a new increase in postal rates is proportional to the number of people in the country who have not heard about it.

SOLUTION

Let $Q(t)$ denote the number of people who have heard about the postal rate increase at time t and B the total population of the country. Then,

$$\text{Rate at which people hear about the increase} = \frac{dQ}{dt}$$

and

$$\text{Number of people who have not heard about the increase} = B - Q$$

Hence the desired differential equation is

$$\frac{dQ}{dt} = k(B - Q)$$

where k is the constant of proportionality. Notice that the constant k must be positive because $\frac{dQ}{dt} > 0$ (since Q is an increasing function of t) and $B - Q > 0$ (since $B > Q$).

EXAMPLE 1.2 Write a differential equation describing the fact that when environmental factors impose an upper bound on its size, population grows at a rate that is jointly proportional to its current size and the difference between its upper bound and current size.

SOLUTION

Let $P(t)$ denote the size of the population at time t and B the upper bound imposed on the population by the environment. Then,

$$\text{Rate of population growth} = \frac{dP}{dt}$$

and

$$\text{Difference between upper bound and population} = B - P$$

Since "jointly proportional" means "proportional to the product" (as you saw in Chapter 1, Section 5), it follows that the desired differential equation is

$$\frac{dP}{dt} = kP(B - P)$$

where k is the constant of proportionality. (Convince yourself that the constant k must be positive in the context of this problem.)

General and Particular Solutions

Any function that satisfies a differential equation is said to be a **solution** of that equation. Here is an example to illustrate this concept.

EXAMPLE 1.3 Verify that the function $y = e^x - x$ is a solution of the differential equation $\frac{dy}{dx} - y = x - 1$.

SOLUTION
Substitute

$$y = e^x - x \quad \text{and} \quad \frac{dy}{dx} = e^x - 1$$

in the left-hand side of the differential equation to get

$$\frac{dy}{dx} - y = (e^x - 1) - (e^x - x) = x - 1$$

as required.

You can easily check that the function $y = 3e^x - x$ is also a solution of the differential equation in Example 1.3. In fact, every function of the form $y = Ce^x - x$, where C is a constant, is a solution of this equation. Moreover, it can be shown that *every* solution of this differential equation is of this form. For this reason, the function $y = Ce^x - x$ is said to be the **general solution** of this differential equation. A solution obtained by replacing C by a specific number is sometimes called a **particular solution** of the differential equation.

EXAMPLE 1.4 Find the particular solution of the differential equation $\frac{dy}{dx} - y = x - 1$ that satisfies the condition that $y = 4$ when $x = 0$.

SOLUTION
Use the given condition to determine the numerical value of the constant C in the general solution $y = Ce^x - x$. In particular, substitute $y = 4$ and

$x = 0$ into the general solution to get $C = 4$ and conclude that the desired particular solution is $y = 4e^x - x$.

Solving Differential Equations

There is no single general technique that can be used to solve all differential equations. Instead, there are many specialized techniques (some quite complicated) that apply only to particular types of differential equations. Fortunately, most of the differential equations you will encounter in the social and managerial sciences can be solved by one of two elementary methods. The first of these methods (with which you are already familiar) will be discussed in this section, while the second will be introduced in Section 2.

Differential Equations of the Form $\dfrac{dy}{dx} = g(x)$

The simplest type of differential equation is one of the form

$$\frac{dy}{dx} = g(x)$$

in which the derivative of the quantity in question is given explicitly as a function of the independent variable. The differential equations

$$\frac{dy}{dx} = 3x^2 + 5 \quad \text{and} \quad \frac{dQ}{dt} = 5te^{-2t}$$

are of this form. On the other hand, the differential equation

$$\frac{dP}{dt} = kP$$

describing population growth is not of this form, because the derivative of P is expressed in terms of P itself rather than the independent variable t.

A differential equation of the form $\dfrac{dy}{dx} = g(x)$ is particularly easy to solve. Its general solution is simply $y = G(x) + C$, where G is an antiderivative of g.

Differential Equations of the Form $\dfrac{dy}{dx} = g(x)$

The general solution of the differential equation

$$\frac{dy}{dx} = g(x)$$

is

$$y = \int g(x)\, dx + C$$

You have already solved differential equations of the form $\dfrac{dy}{dx} = g(x)$ in Chapter 5. Here, for purposes of review, are two practical problems involving differential equations of this form.

EXAMPLE 1.5   The resale value of a certain industrial machine decreases over a 10-year period at a rate that depends on the age of the machine. When the machine is x years old, the rate at which its value is changing is $220(x - 10)$ dollars per year. Express the value of the machine as a function of its age and initial value. If the machine was originally worth $12,000, how much will it be worth when it is 10 years old?

SOLUTION

Let $V(x)$ denote the value of the machine when it is x years old. The derivative $\dfrac{dV}{dx}$ is equal to the rate $220(x - 10)$ at which the value of the machine is changing. Hence, you begin with the differential equation

$$\frac{dV}{dx} = 220(x - 10) = 220x - 2{,}200$$

To find V, solve this differential equation by integration.

$$V(x) = \int (220x - 2{,}200)\, dx = 110x^2 - 2{,}200x + C$$

Notice that C is equal to $V(0)$, the initial value of the machine. A more descriptive symbol for this constant is V_0. Using this notation, you can write the general solution as

$$V(x) = 110x^2 - 2{,}200x + V_0$$

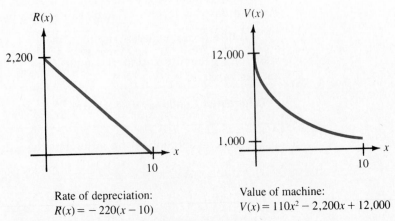

Rate of depreciation:
$R(x) = -220(x - 10)$

Value of machine:
$V(x) = 110x^2 - 2{,}200x + 12{,}000$

Figure 1.1 Depreciation of industrial machinery.

If $V_0 = 12,000$, the corresponding particular solution is

$$V(x) = 110x^2 - 2,200x + 12,000$$

and the value of the machine after 10 years is

$$V(10) = 11,000 - 22,000 + 12,000 = \$1,000$$

Graphs showing the rate of depreciation and the resale value of the machine are sketched in Figure 1.1. Note that the rate of *depreciation* is the negative of the rate of change $\dfrac{dV}{dx} = 220(x - 10)$. (Do you see why?)

EXAMPLE 1.6   An oil well that yields 300 barrels of crude oil a month will run dry in 3 years. It is estimated that t months from now the price of crude oil will be $P(t) = 18 + 0.3\sqrt{t}$ dollars per barrel. If the oil is sold as soon as it is extracted from the ground, what will the total future revenue from the well be?

SOLUTION
Let $R(t)$ denote the revenue generated during the next t months. To construct the relevant differential equation, use the relationship

$$\begin{matrix}\text{Rate of change of revenue} \\ \text{with respect to time}\end{matrix} = \frac{\text{dollars}}{\text{per month}} = \left(\frac{\text{dollars}}{\text{per barrel}}\right)\left(\frac{\text{barrels}}{\text{per month}}\right)$$

Since

$$\begin{matrix}\text{Rate of change of revenue} \\ \text{with respect to time}\end{matrix} = \frac{dR}{dt}$$

Dollars per barrel $= P(t) = 18 + 0.3\sqrt{t}$

and

Barrels per month $= 300$

it follows that

$$\frac{dR}{dt} = P(t)\,(300) \qquad \text{or} \qquad \frac{dR}{dt} = 300(18 + 0.3\sqrt{t}) = 5,400 + 90\sqrt{t}$$

The general solution of this differential equation is

$$R(t) = \int (5,400 + 90\sqrt{t})\, dt = 5,400t + 60t^{3/2} + C$$

Since $R(0) = 0$, it follows that $C = 0$, and so the appropriate particular solution is

$$R(t) = 5,400t + 60t^{3/2}$$

Since the well will run dry in 36 months, the total future revenue will be

$$R(36) = 5,400(36) + 60(216) = \$207,360$$

Problems

In Problems 1 through 12, write a differential equation describing the given situation. Define all variables you introduce. (Do not try to solve the differential equation at this time.)

Growth of bacteria

1. The number of bacteria in a culture grows at a rate that is proportional to the number present.

Radioactive decay

2. A sample of radium decays at a rate that is proportional to its size.

Investment growth

3. An investment grows at a rate equal to 7 percent of its size.

Concentration of drugs

4. The rate at which the concentration of a drug in the bloodstream decreases is proportional to the concentration.

Population growth

5. The population of a certain town increases at the constant rate of 500 people per year.

Marginal cost

6. A manufacturer's marginal cost is $60 per unit.

Temperature change

7. The rate at which the temperature of an object changes is proportional to the difference between its own temperature and the temperature of the surrounding medium.

Dissolution of sugar

8. After being placed in a container of water, sugar dissolves at a rate proportional to the amount of undissolved sugar remaining in the container.

Recall from memory

9. When a person is asked to recall a set of facts, the rate at which the facts are recalled is proportional to the number of relevant facts in the person's memory that have not yet been recalled.

The spread of an epidemic

10. The rate at which an epidemic spreads through a community is jointly proportional to the number of people who have caught the disease and the number who have not.

Corruption in government

11. The rate at which people are implicated in a government scandal is jointly proportional to the number of people already implicated and the number of people involved who have not yet been implicated.

The spread of a rumor

12. The rate at which a rumor spreads through a community is jointly proportional to the number of people in the community who have heard the rumor and the number who have not.

13. Verify that the function $y = Ce^{kx}$ is a solution of the differential equation $\dfrac{dy}{dx} = ky$.

14. Verify that the function $Q = B - Ce^{-kt}$ is a solution of the differential equation $\dfrac{dQ}{dt} = k(B - Q)$.

15. Verify that the function $y = C_1e^x + C_2xe^x$ is a solution of the differential equation $\dfrac{d^2y}{dx^2} - 2\dfrac{dy}{dx} + y = 0$.

16. Verify that the function $y = \dfrac{1}{20}x^4 - \dfrac{C_1}{x} + C_2$ is a solution of the differential equation $x\dfrac{d^2y}{dx^2} + 2\dfrac{dy}{dx} = x^3$.

In Problems 17 through 22, find the general solution of the given differential equation.

17. $\dfrac{dy}{dx} = 3x^2 + 5x - 6$

18. $\dfrac{dP}{dt} = \sqrt{t} + e^{-t}$

19. $\dfrac{dV}{dx} = \dfrac{2}{x+1}$

20. $\dfrac{dA}{dt} = 2te^{t^2+5}$

21. $\dfrac{d^2P}{dt^2} = 50$ (*Hint*: Integrate twice.)

22. $\dfrac{d^2y}{dx^2} = 3x^2 + 5x - 6$

In Problems 23 through 28, find the particular solution of the given differential equation that satisfies the given condition.

23. $\dfrac{dy}{dx} = e^{5x}$; $y = 1$ when $x = 0$

24. $\dfrac{dy}{dx} = 5x^4 - 3x^2 - 2$; $y = 4$ when $x = 1$

25. $\dfrac{dV}{dt} = 16t(t^2 + 1)^3$; $V = 1$ when $t = 0$

26. $\dfrac{d^2y}{dt^2} = 3t^2 + 2t - 1$; $y = 3$ and $\dfrac{dy}{dt} = 0$ when $t = 1$ (*Hint*: Integrate twice.)

27. $\dfrac{d^2A}{dt^2} = e^{-t/2}$; $A = 2$ and $\dfrac{dA}{dt} = 1$ when $t = 0$

28. $\dfrac{d^2H}{dt^2} = -32$; $H = H_0$ and $\dfrac{dH}{dt} = S_0$ when $t = 0$

Depreciation 29. The resale value of a certain industrial machine decreases at a rate that depends on its age. When the machine is t years old, the rate at which its value is changing is $-960e^{-t/5}$ dollars per year.
   (a) Express the value of the machine in terms of its age and initial value.
   (b) If the machine was originally worth $5,200, how much will it be worth when it is 10 years old?

Marginal cost
30. At a certain factory, the marginal cost is $3(q - 4)^2$ dollars per unit when the level of output is q units.
   (a) Express the total production cost in terms of the overhead (the cost of producing no units) and the number of units produced.
   (b) What is the cost of producing 14 units if the overhead is $436?

Population growth
31. Population statistics indicate that x years after 1970 a certain county was growing at a rate of approximately $1,500x^{-1/2}$ people per year. In 1979 the population of the county was 39,000.
   (a) What was the population in 1970?
   (b) If this pattern of population growth continues in the future, how many people will be living in the county in 1995?

Retail prices
32. In a certain section of the country, the price of chicken is currently $3 per kilogram. It is estimated that x weeks from now the price will be increasing at the rate of $3\sqrt{x + 1}$ cents per week. How much will chicken cost 8 weeks from now?

Air pollution
33. In a certain Los Angeles suburb, a reading of air pollution levels taken at 7:00 A.M. shows the ozone level to be 0.25 part per million. A 12-hour forecast of air conditions predicts that t hours later the ozone level will be changing at the rate of $\dfrac{0.24 - 0.03t}{\sqrt{36 + 16t - t^2}}$ parts per million per hour.
   (a) Express the ozone level as a function of t.
   (b) At what time will the peak ozone level occur? What will the ozone level be at this time?

Production of oil
34. A certain oil well that yields 400 barrels of crude oil a month will run dry in 2 years. The price of crude oil is currently $18 per barrel and is expected to rise at the constant rate of 3 cents per barrel per month. If the oil is sold as soon as it is extracted from the ground, what will the total future revenue from the well be?

Farming
35. It is estimated that t days from now a farmer's crop will be increasing at the rate of $0.3t^2 + 0.6t + 1$ bushels per day. By how much will the value of the crop increase during the next 5 days if the market price remains fixed at $3 per bushel?

Water pollution
36. It is estimated that t years from now the population of a certain lakeside community will be changing at the rate of $0.6t^2 + 0.2t + 0.5$ thousand per year. Environmentalists have found that the level of pollution in the lake increases at the rate of approximately 5 units per 1,000 people. If the level of pollution in the lake is currently 60 units, what will the pollution level be 2 years from now?

Stopping distance
37. After its brakes are applied, a certain sports car decelerates at the constant rate of 28 feet per second per second.
   (a) Express the distance the car travels in terms of its speed at the

moment of braking and the amount of time that has elapsed since that moment. [*Hint*: Acceleration is the second derivative of distance. Let $D(t)$ denote the distance the car has traveled after t seconds and solve the differential equation $\dfrac{d^2D}{dt^2} = -28$ by integrating twice.]

(b) Compute the stopping distance if the car was going 60 miles per hour when the brakes were applied. (*Hint*: 60 miles per hour = 88 feet per second.)

Spy story   38. The hero of a popular spy story (who defused the bomb in 5 minutes and survived Problem 18 in Chapter 3, Section 4) is driving the sports car in Problem 37 at a speed of 60 miles per hour on Highway 1 in the remote republic of San Dimas. Suddenly he sees a camel in the road 199 feet in front of him. After a reaction time of 0.7 second, he steps on the brakes. Will he stop before hitting the camel? (See Problem 19 in Chapter 7, Section 2 for the sequel.)

2   Separable Differential Equations

Many useful differential equations can be formally rewritten so that all the terms containing the independent variable appear on one side of the equation and all the terms containing the dependent variable appear on the other. Differential equations with this special property are said to be **separable** and can be solved by the following procedure involving two integrations.

Separable Differential Equations

A differential equation that can be written in the form

$$g(y)\, dy = h(x)\, dx$$

is said to be separable. Its general solution is obtained by integrating both sides of this equation. That is,

$$\int g(y)\, dy = \int h(x)\, dx + C$$

A proof that this procedure works will be given later in this section. First, here are some examples to illustrate how the procedure is used.

EXAMPLE 2.1   Find the general solution of the differential equation $\dfrac{dy}{dx} = \dfrac{2x}{y^2}$.

SOLUTION

To separate the variables, pretend that the derivative $\frac{dy}{dx}$ is actually a quotient and write

$$y^2 \, dy = 2x \, dx$$

Now integrate both sides of this equation to get

$$\int y^2 \, dy = \int 2x \, dx$$

or

$$\frac{1}{3}y^3 = x^2 + C_1 \qquad \text{(where } C_1 \text{ is an arbitrary constant)}$$

and solve for y to conclude that

$$y = (3x^2 + 3C_1)^{1/3}$$

or

$$y = (3x^2 + C)^{1/3} \qquad \text{(where } C = 3C_1 \text{ is an arbitrary constant)}$$

Exponential Models

In the next three examples, you will see how separable differential equations lead to some of the exponential models that were introduced in Chapter 4.

Exponential Growth and Decay

In Chapter 4, Section 4, you saw that if a quantity grows exponentially, its rate of change is proportional to its size. Using separable differential equations we can now establish the converse of this result.

EXAMPLE 2.2   Show that a quantity that grows at a rate proportional to its size grows exponentially.

SOLUTION

Let $Q(t)$ denote the size of the quantity at time t and begin with the differential equation

$$\frac{dQ}{dt} = kQ$$

where k is the constant of proportionality. Separate the variables by writing

$$\frac{1}{Q} \, dQ = k \, dt$$

and integrate both sides of this equation to get

$$\int \frac{1}{Q}\, dQ = \int k\, dt$$

or

$$\ln |Q| = kt + C$$

Solve this equation for $|Q|$ by applying the exponential function to each side to get

$$|Q| = e^{kt+C} = e^C e^{kt}$$

Since $|Q| = \pm Q$, depending on whether Q is positive or negative, you can drop the absolute value sign and write

$$Q = \pm e^C e^{kt}$$

Finally, since the constant $\pm e^C$ is the value of Q when $t = 0$, you can introduce Q_0 to stand for this constant and, using functional notation, write

$$Q(t) = Q_0 e^{kt} \quad \text{(where } Q_0 = \pm e^C)$$

which is precisely the equation describing exponential growth.

You can use a similar argument to show that a quantity that decreases at a rate proportional to its size decreases exponentially. For practice, work out the details.

Exponential Growth and Decay

> A quantity grows (or decays) exponentially if and only if its rate of change is proportional to its size.

A Technical Detail

The solution of Example 2.2 is not quite correct as stated. In particular, the argument that the equation

$$|Q| = e^C e^{kt}$$

can be rewritten as

$$Q = \pm e^C e^{kt}$$

where $\pm e^C$ is a *constant* is valid only if Q never changes sign. If Q were to change sign, the expression $\pm e^C$ would be positive part of the time and negative part of the time and hence would not be constant. Fortunately, it turns out that the continuity of Q guarantees that this cannot happen and that the expression $\pm e^C$ is really a constant. Feel free to ignore this technical point when solving similar separable differential equations and simply assume that the function does not change sign.

Learning Curves

As you saw in Chapter 4, Section 2, the graphs of functions of the form $Q(t) = B - Ae^{-kt}$ are called learning curves because functions of this form often describe the relationship between the efficiency with which an individual performs a task and the amount of training or experience the individual has had. In general, any quantity that grows at a rate that is proportional to the difference between its size and a fixed upper bound can be represented by a function of this form. This is illustrated in the next example.

EXAMPLE 2.3 The rate at which people hear about a new increase in postal rates is proportional to the number of people in the country who have not heard about it. Express the number of people who have heard about the increase as a function of time.

SOLUTION

Let $Q(t)$ denote the number of people who have heard about the increase by time t and B the total population of the country. Then (as you saw in Example 1.1 of this chapter),

$$\frac{dQ}{dt} = k(B - Q)$$

where k is the constant of proportionality. Separate the variables by writing

$$\frac{1}{B - Q} \, dQ = k \, dt$$

and integrate to get

$$\int \frac{1}{B - Q} \, dQ = \int k \, dt$$

or

$$-\ln |B - Q| = kt + C$$

(Be sure you see where the minus sign came from.) This time you can drop the absolute value sign immediately since $B - Q$ cannot be negative in this context. Hence,

$$-\ln (B - Q) = kt + C$$

$$\ln (B - Q) = -kt - C$$

$$B - Q = e^{-kt-C} = e^{-kt}e^{-C}$$

or

$$Q = B - e^{-C}e^{-kt}$$

Denoting the constant e^{-C} by A and using functional notation, you can conclude that

$$Q(t) = B - Ae^{-kt}$$

which is precisely the general equation of a learning curve. For reference, the graph of Q is sketched in Figure 2.1.

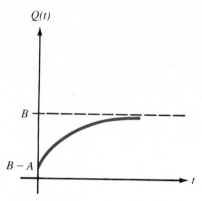

Figure 2.1

A learning curve:

$Q(t) = B - Ae^{-kt}$.

Logistic Curves

In Chapter 4, Section 2, you learned that the graphs of functions of the form $Q(t) = \dfrac{B}{1 + Ae^{-Bkt}}$ are called logistic curves. They are used as models of population growth when environmental factors impose an upper bound on the possible size of the population, and they also describe such phenomena as the spread of rumors and epidemics. The calculation in the next example can be used to show that any quantity that satisfies a differential equation of the form $\dfrac{dQ}{dt} = kQ(B - Q)$ can be represented by a logistic curve.

EXAMPLE 2.4   The rate at which an epidemic spreads through a community is jointly proportional to the number of residents who have been infected and the number of susceptible residents who have not. Express the number of residents who have been infected as a function of time.

SOLUTION

Let $Q(t)$ denote the number of residents who have been infected by time t and B the total number of susceptible residents. Then the number of susceptible residents who have not been infected is $B - Q$, and the differential equation describing the spread of the epidemic is

$$\frac{dQ}{dt} = kQ(B - Q)$$

where k is the constant of proportionality. This is a separable differential equation whose solution is

$$\int \frac{1}{Q(B - Q)} \, dQ = \int k \, dt$$

The trick to finding the integral on the left-hand side is to observe that

$$\frac{1}{Q(B - Q)} = \frac{1}{B}\left[\frac{B}{Q(B - Q)}\right] = \frac{1}{B}\left(\frac{1}{Q} + \frac{1}{B - Q}\right)$$

(which you can verify by adding the fractions in the expression on the right). Hence,

$$\frac{1}{B}\int \frac{1}{Q} \, dQ + \frac{1}{B}\int \frac{1}{B - Q} \, dQ = \int k \, dt$$

$$\frac{1}{B}\ln |Q| - \frac{1}{B}\ln |B - Q| = kt + C$$

$$\frac{1}{B}\ln \left|\frac{Q}{B - Q}\right| = kt + C \qquad \text{since } \ln\frac{u}{v} = \ln u - \ln v$$

or

$$\frac{1}{B}\ln \frac{Q}{B - Q} = kt + C \qquad \text{since } Q > 0 \text{ and } B > Q$$

Solve for Q to get

$$\ln \frac{Q}{B - Q} = Bkt + BC$$

$$\frac{Q}{B - Q} = e^{Bkt + BC} = e^{Bkt}e^{BC} = A_1 e^{Bkt} \qquad \text{where } A_1 = e^{BC}$$

$$Q = (B - Q)A_1 e^{Bkt} = BA_1 e^{Bkt} - QA_1 e^{Bkt}$$

$$Q + QA_1 e^{Bkt} = A_1 B e^{Bkt}$$

$$Q(1 + A_1 e^{Bkt}) = A_1 B e^{Bkt}$$

$$Q = \frac{A_1 B e^{Bkt}}{1 + A_1 e^{Bkt}}$$

To make this formula more attractive, divide the numerator and denominator by $A_1 e^{Bkt}$ to get

$$Q = \frac{B}{(1/A_1)\, e^{-Bkt} + 1}$$

Finally, let A denote the constant $\dfrac{1}{A_1}$ and use functional notation to get

$$Q(t) = \frac{B}{1 + Ae^{-Bkt}}$$

which is, as promised, the general equation of a logistic curve.

For reference, the graph of Q is sketched in Figure 2.2. It is not hard to show that the inflection point occurs when $Q(t) = \dfrac{B}{2}$. (See Problem 30 at the end of this section.) This corresponds to the fact that the epidemic is spreading most rapidly when half of the susceptible residents have been infected.

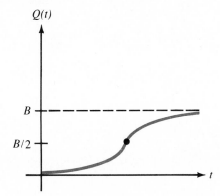

Figure 2.2
A logistic curve:
$Q(t) = \dfrac{B}{1 + Ae^{-Bkt}}.$

Dilution Problem

The next example is typical of an important class of dilution problems that lead to separable differential equations.

EXAMPLE 2.5 The residents of a certain community have voted to discontinue the fluoridation of their water supply. The local reservoir currently holds 200 million gallons of fluoridated water that contains 1,600 pounds of fluoride. The fluoridated water is flowing out of the reservoir at the rate of 4 million gallons per day and is being replaced at the same rate by unfluoridated water. At all times, the remaining fluoride is evenly distributed in the reservoir. Express the amount of fluoride in the reservoir as a function of time.

SOLUTION
Begin with the following relationship:

$$\begin{matrix} \text{Rate of change of fluoride} \\ \text{with respect to time} \end{matrix} = \begin{pmatrix} \text{concentration of} \\ \text{fluoride in water} \end{pmatrix} \begin{pmatrix} \text{rate of flow of} \\ \text{fluoridated water} \end{pmatrix}$$

or

$$\begin{matrix} \text{Pounds} \\ \text{per day} \end{matrix} = \begin{pmatrix} \text{pounds per} \\ \text{million gallons} \end{pmatrix} \begin{pmatrix} \text{million gallons} \\ \text{per day} \end{pmatrix}$$

Let $Q(t)$ denote the number of pounds of fluoride in the reservoir after t days. Then,

$$\begin{matrix} \text{Rate of change of fluoride} \\ \text{with respect to time} \end{matrix} = \frac{dQ}{dt} \quad \text{pounds per day}$$

$$\begin{matrix} \text{Concentration of} \\ \text{fluoride in water} \end{matrix} = \frac{\text{number of pounds of fluoride in reservoir}}{\text{number of million gallons of water in reservoir}}$$

$$= \frac{Q}{200} \quad \text{pounds per million gallons}$$

and

$$\begin{matrix} \text{Rate of flow of} \\ \text{fluoridated water} \end{matrix} = -4 \quad \text{million gallons per day}$$

where the minus sign indicates that the water is leaving the reservoir. Hence,

$$\frac{dQ}{dt} = \frac{Q}{200}(-4) = -\frac{Q}{50}$$

Solving this differential equation by separation of variables, you get

$$\int \frac{1}{Q}\,dQ = -\int \frac{1}{50}\,dt$$

$$\ln Q = -\frac{t}{50} + C$$

or

$$Q = Q_0 e^{-t/50} \quad \text{where } Q_0 = e^C$$

Initially, 1,600 pounds of fluoride were in the reservoir. Replacing Q_0 by 1,600 and using functional notation, you can conclude that

$$Q(t) = 1,600 e^{-t/50}$$

That is, the amount of fluoride in the reservoir decreases exponentially. The situation is illustrated in Figure 2.3.

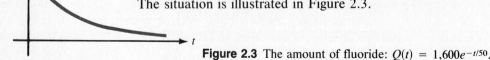

Figure 2.3 The amount of fluoride: $Q(t) = 1,600 e^{-t/50}$.

Why the Method of Separation of Variables Works

It is not hard to see why the method of separation of variables works. Before the variables were separated, the differential equation

$$g(y) \, dy = h(x) \, dx$$

was

$$\frac{dy}{dx} = \frac{h(x)}{g(y)}$$

or, equivalently,

$$g(y) \frac{dy}{dx} - h(x) = 0$$

The left-hand side of this equation can be rewritten in terms of the antiderivatives of g and h. In particular, if G is an antiderivative of g and H an antiderivative of h, it follows from the chain rule that

$$\frac{d}{dx}[G(y) - H(x)] = G'(y)\frac{dy}{dx} - H'(x) = g(y)\frac{dy}{dx} - h(x)$$

Hence the differential equation $g(y) \dfrac{dy}{dx} - h(x) = 0$ says that

$$\frac{d}{dx}[G(y) - H(x)] = 0$$

But constants are the only functions whose derivatives are identically zero, and so

$$G(y) - H(x) = C$$

for some constant C. That is,

$$G(y) = H(x) + C$$

or, equivalently,

$$\int g(y) \, dy = \int h(x) \, dx + C$$

and the proof is complete.

Problems

In Problems 1 through 9, find the general solution of the given separable differential equation.

1. $\dfrac{dy}{dx} = 3y$

2. $\dfrac{dy}{dx} = y^2$

3. $\dfrac{dy}{dx} = e^y$

4. $\dfrac{dy}{dx} = e^{x+y}$

5. $\dfrac{dy}{dx} = \dfrac{x}{y}$ 　　　　　　　　6. $\dfrac{dy}{dx} = \dfrac{y}{x}$

7. $\dfrac{dy}{dx} = y + 10$ 　　　　　　　8. $\dfrac{dy}{dx} = 80 - y$

9. $\dfrac{dy}{dx} = y(1 - 2y)$ 　$\left[Hint: \dfrac{1}{y(1 - 2y)} = \dfrac{1}{y} + \dfrac{2}{1 - 2y} \right]$

In Problems 10 through 14, find the particular solution of the given differential equation that satisfies the given condition.

10. $\dfrac{dy}{dx} = \dfrac{x}{y^2}$; $y = 3$ when $x = 2$

11. $\dfrac{dy}{dx} = 4x^3 y^2$; $y = 2$ when $x = 1$

12. $\dfrac{dy}{dx} = 0.05y$; $y = 500$ when $x = 0$

13. $\dfrac{dy}{dx} = 5(8 - y)$; $y = 6$ when $x = 0$

14. $\dfrac{dy}{dx} = y(1 - y)$; $y = \dfrac{1}{3}$ when $x = 0$

Do Problems 15 through 28 by solving appropriate separable differential equations.

Investment　15. A $1,000 investment grows at a rate equal to 7 percent of its size. Express the value of the investment as a function of time.

Drug concentration　16. The rate at which the concentration of a drug in the bloodstream decreases is proportional to the concentration. Express the concentration of the drug in the bloodstream as a function of time.

Exponential decay　17. Show that a quantity that decays at a rate proportional to its size decays exponentially.

Exponential functions　18. Show that if a differentiable function is equal to its own derivative, then the function must be of the form $y = Ce^x$.

Recall from memory　19. Psychologists believe that when a person is asked to recall a set of facts, the rate at which the facts are recalled is proportional to the number of relevant facts in the subject's memory that have not yet been recalled. Express the number of facts that have been recalled as a function of time and draw the graph.

Dissolution of sugar　20. After being placed in a container of water, sugar dissolves at a rate proportional to the amount of undissolved sugar remaining in the container. Express the amount of sugar that has been dissolved as a function of time and draw the graph.

Newton's law of heating

21. The rate at which the temperature of an object changes is proportional to the difference between its own temperature and that of the surrounding medium. A cold drink is removed from a refrigerator on a hot summer day and placed in an 80° room. Express the temperature of the drink as a function of time (in minutes) if the temperature of the drink was 40° when it left the refrigerator and 50° 20 minutes later.

Newton's law of cooling

22. The rate at which the temperature of an object changes is proportional to the difference between its own temperature and that of the surrounding medium. Express the temperature of the object as a function of time and draw the graph if the temperature of the object is greater than that of the surrounding medium.

Fick's law

23. When a cell is placed in a liquid containing a solute, the solute passes through the cell wall by diffusion. As a result, the concentration of the solute inside the cell changes, increasing if the concentration of the solute outside the cell is greater than the concentration inside and decreasing if the opposite is true. A biological law known as **Fick's law** asserts that the concentration of the solute inside the cell changes at a rate that is jointly proportional to the area of the cell wall and the difference between the concentration of the solute inside and outside the cell. Assuming that the concentration of the solute outside the cell is constant and greater than the concentration inside, derive a formula for the concentration of the solute inside the cell.

Dilution

24. A tank contains 200 gallons of clear water. Brine (salt water) containing 2 pounds of salt per gallon flows into the tank at the rate of 5 gallons per minute, and the mixture, which is stirred so that the salt is evenly distributed at all times, runs out of the tank at the same rate. Express the amount of salt in the tank as a function of time and draw the graph. *Hint*: Let $Q(t)$ denote the number of pounds of salt in the tank after t minutes. Then

$$\frac{dQ}{dt} = \left(\begin{array}{c} \text{rate at which} \\ \text{salt enters tank} \end{array} \right) - \left(\begin{array}{c} \text{rate at which} \\ \text{salt leaves tank} \end{array} \right)$$

Dilution

25. A tank currently holds 200 gallons of brine that contains 3 pounds of salt per gallon. Brine containing 2 pounds of salt per gallon flows into the tank at the rate of 5 gallons per minute, while the mixture, which is kept uniform, runs out of the tank at the same rate. Express the amount of salt in the tank as a function of time and draw the graph.

Air purification

26. A 2,400-cubic-foot room contains an activated charcoal air filter through which air passes at the rate of 400 cubic feet per minute. The ozone in the air is absorbed by the charcoal as the air flows through the filter, and the purified air is recirculated in the room. Assuming that the remaining ozone is evenly distributed throughout the room at all times, determine how long it takes the filter to remove 50 percent of the ozone from the room.

The spread of an epidemic

27. The rate at which an epidemic spreads through a community is jointly proportional to the number of residents who have been infected and the number of susceptible residents who have not. Express the number of residents who have been infected as a function of time (in weeks) if the community has 2,000 susceptible residents, if 500 residents had the disease initially, and if 855 residents had been infected by the end of the 1st week.

Corruption in government

28. The number of people implicated in a certain major government scandal increases at a rate jointly proportional to the number of people already implicated and the number involved who have not yet been implicated. Suppose that 7 people were implicated when a Washington newspaper first made the scandal public, that 9 more were implicated over the next 3 months, and that another 12 were implicated during the following 3 months. Approximately how many people are involved in the scandal? (*Warning*: This problem will test your algebraic ingenuity!)

The spread of an epidemic

29. The rate at which an epidemic spreads through a community is jointly proportional to the number of residents who have been infected and the number of susceptible residents, who have not. Show that the epidemic is spreading most rapidly when one-half of the susceptible residents have been infected. (*Hint*: You do not have to solve a differential equation to do this. Just start with a formula for the *rate* at which the epidemic is spreading and use calculus to maximize this rate.)

Logistic curves

30. Show that if a quantity Q satisfies the differential equation $\dfrac{dQ}{dt} = kQ(B - Q)$, where k and B are positive constants, then the rate of change $\dfrac{dQ}{dt}$ is greatest when $Q(t) = \dfrac{B}{2}$. What does this result tell you about the inflection point of a logistic curve? Explain. (*Hint*: See the hint for Problem 29.)

Chapter Summary and Review Problems

Important Terms, Symbols, and Formulas

Differential equation

General solution; particular solution

Separable differential equation:

$$\text{If } g(y)\, dy = h(x)\, dx, \text{ then } \int g(y)\, dy = \int h(x)\, dx + C$$

Exponential models:

Model	Differential Equation	General Solution	Graph
Exponential growth	$\dfrac{dQ}{dt} = kQ$	$Q(t) = Q_0 e^{kt}$	
Exponential decay	$\dfrac{dQ}{dt} = -kQ$	$Q(t) = Q_0 e^{-kt}$	
Learning curve	$\dfrac{dQ}{dt} = k(B - Q)$	$Q(t) = B - Ae^{-kt}$	
Logistic curve	$\dfrac{dQ}{dt} = kQ(B - Q)$	$Q(t) = \dfrac{B}{1 + Ae^{-Bkt}}$	

Review Problems

In Problems 1 through 4, find the general solution of the given differential equation.

1. $\dfrac{dy}{dx} = x^3 - 3x^2 + 5$ 2. $\dfrac{dy}{dx} = 0.02y$

3. $\dfrac{dy}{dx} = k(80 - y)$ 4. $\dfrac{dy}{dx} = y(1 - y)$

In Problems 5 through 8, find the particular solution of the given differential equation that satisfies the given condition.

5. $\dfrac{dy}{dx} = 5x^4 - 3x^2 - 2$; $y = 4$ when $x = 1$

6. $\dfrac{dy}{dx} = 0.06y$; $y = 100$ when $x = 0$

7. $\dfrac{dy}{dx} = 3 - y$; $y = 2$ when $x = 0$

8. $\dfrac{d^2y}{dx^2} = 2$; $y = 5$ and $\dfrac{dy}{dx} = 3$ when $x = 0$

9. The resale value of a certain industrial machine decreases at a rate proportional to the difference between its current value and its scrap value of $5,000. The machine was bought new for $40,000 and was worth $30,000 after 4 years. How much will it be worth when it is 8 years old?

10. A certain oil well that yields 600 barrels of crude oil per month will run dry in 3 years. The price of crude oil is currently $24 per barrel and is expected to rise at the constant rate of 8 cents per barrel per month. If the oil is sold as soon as it is extracted from the ground, what will the total future revenue from the well be?

11. A tank currently holds 200 gallons of brine that contains 3 pounds of salt per gallon. Clear water flows into the tank at the rate of 4 gallons per minute, while the mixture, which is kept uniform, runs out of the tank at the rate of 5 gallons per minute. How much salt is in the tank at the end of 100 minutes?

12. The rate at which the population of a certain country is growing is jointly proportional to the upper bound of 10 million imposed by environmental factors and the difference between the upper bound and the size of the population. Express the population (in millions) of the country as a function of time (in years measured from 1980) if the population in 1980 was 4 million and the population in 1985 was 4.74 million.

Chapter

9 Functions of Several Variables

1   Functions of Several Variables

In many practical situations, the value of one quantity may depend on the values of two or more others. For example, the amount of water in a reservoir may depend on the amount of rainfall and on the amount of water consumed by local residents. The demand for butter may depend on the price of butter and on the price of margarine. The output at a factory may depend on the amount of capital invested in the plant and on the size of the labor force. Relationships of this sort often can be represented mathematically by functions having more than one independent variable.

In this chapter, we shall restrict our attention to functions of two variables. As you shall see, such functions can be represented geometrically as surfaces in three-dimensional space. We shall use the geometry to gain insight into the new concepts. The theory of functions of three or more variables is similar to that of functions of two variables and is discussed in advanced calculus texts.

Function of Two Variables

A function f of the two variables x and y is a rule that assigns to each ordered pair (x, y) of real numbers in some set, one and only one real number denoted by $f(x, y)$.

The Domain of a Function of Two Variables

The domain of the function $f(x, y)$ is the set of all ordered pairs (x, y) of real numbers for which $f(x, y)$ can be evaluated.

EXAMPLE 1.1   Suppose $f(x, y) = \dfrac{3x^2 + 5y}{x - y}$.

(a) Find the domain of f.
(b) Compute $f(1, -2)$.

SOLUTION
(a) Since division by any real number except zero is possible, the only ordered pairs (x, y) for which f cannot be evaluated are those for which $x = y$. Hence, the domain of f consists of all ordered pairs (x, y) of real numbers for which $x \neq y$.

(b) $f(1, -2) = \dfrac{3(1)^2 + 5(-2)}{1 - (-2)} = \dfrac{3 - 10}{1 + 2} = -\dfrac{7}{3}$

EXAMPLE 1.2   Suppose $f(x, y) = xe^y + \ln x$.

(a) Find the domain of f.
(b) Compute $f(e^2, \ln 2)$

SOLUTION
(a) Since xe^y is defined for all real numbers x and y and since $\ln x$ is defined only for $x > 0$, the domain of f consists of all ordered pairs (x, y) of real numbers for which $x > 0$.

(b) $f(e^2, \ln 2) = e^2 e^{\ln 2} + \ln(e^2) = 2e^2 + 2 = 2(e^2 + 1) \simeq 16.78$

Applications

Here are two elementary applications of functions of two variables to business and economics.

EXAMPLE 1.3 A liquor store in Minneapolis carries two brands of inexpensive white table wine, one from California and the other from New York. The consumer demand for each brand depends not only on its own price but also on the price of the competing brand. Sales figures indicate that if the California wine sells for x dollars per bottle and the New York wine for y dollars per bottle, the demand for the California wine will be

$$D_1 = 300 - 20x + 30y \qquad \text{bottles per month}$$

and the demand for the New York wine will be

$$D_2 = 200 + 40x - 10y \qquad \text{bottles per month}$$

Express the liquor store's total monthly revenue from the sale of these wines as a function of the prices x and y.

SOLUTION

Let R denote the total monthly revenue. Then,

$$R = \text{(number of bottles of California wine sold)(price per bottle)}$$
$$+ \text{(number of bottles of New York wine sold)(price per bottle)}$$

Hence,

$$R(x, y) = (300 - 20x + 30y)(x) + (200 + 40x - 10y)(y)$$
$$= 300x + 200y + 70xy - 20x^2 - 10y^2$$

Cobb-Douglas Production Functions

Output Q at a factory is often regarded as a function of the amount K of capital investment and the size L of the labor force. Output functions of the form

$$Q(K, L) = AK^\alpha L^{1-\alpha}$$

where A and α are positive constants and $0 < \alpha < 1$ have proved to be especially useful in economic analysis. Such functions are known as **Cobb-Douglas production functions.**

EXAMPLE 1.4 Suppose that at a certain factory, output is given by the Cobb-Douglas production function $Q(K, L) = 60K^{1/3}L^{2/3}$ units, where K is the capital investment measured in units of $\$1,000$ and L the size of the labor force measured in worker-hours.

(a) Compute the output if the capital investment is $\$512,000$ and $1,000$ worker-hours of labor are used.

(b) Show that the output in part (a) will double if both the capital investment and the size of the labor force are doubled.

SOLUTION

(a) Evaluate $Q(K, L)$ with $K = 512$ (thousand) and $L = 1,000$ to get

$$Q(512, 1,000) = 60(512)^{1/3}(1,000)^{2/3}$$

$$= 60(8)(100) = 48,000 \text{ units}$$

(b) Evaluate $Q(K, L)$ with $K = 2(512)$ and $L = 2(1,000)$ as follows to get

$$Q[2(512), 2(1,000)] = 60[2(512)]^{1/3}[2(1,000)]^{2/3}$$

$$= 60(2)^{1/3}(512)^{1/3}(2)^{2/3}(1,000)^{2/3}$$

$$= (2)^{1/3 + 2/3}[60(512)^{1/3}(1,000)^{2/3}]$$

$$= 2Q(512, 1,000) = 2(48,000) = 96,000 \text{ units}$$

That is, the output Q when $K = 2(512)$ and $L = 2(1,000)$ is twice the output when $K = 512$ and $L = 1,000$.

Using a calculation similar to the one in part (b) of the preceding example, you can show that if output is related to capital and labor by a Cobb-Douglas production function and if both capital and labor are multiplied by some positive number m, then output will also be multiplied by m. (The details are left as an exercise for you to do.) In economics, production functions with this property are said to have **constant returns to scale**.

Geometric Representation of Functions

Functions of one variable can be represented graphically as curves drawn on a two-dimensional coordinate system. As you will now see, functions of two variables can be represented graphically as **surfaces** drawn on a three-dimensional coordinate system. Unfortunately, there is no analogous way to visualize functions of more than two variables.

A Three-Dimensional Coordinate System

To construct a three-dimensional coordinate system, we adjoin a third axis (the z axis) to the familiar xy coordinate plane as shown in Figure 1.1. Notice that the xy plane is taken as horizontal. The z axis is perpendicular to the xy plane and the upward direction is chosen to be the positive z direction. (For simplicity, only the positive coordinate axes are drawn in Figure 1.1.)

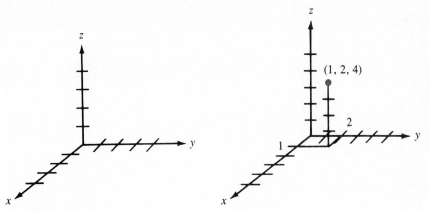

Figure 1.1 A three-dimensional coordinate system.

Figure 1.2 The point (1, 2, 4).

You can describe the location of a point in three-dimensional space by specifying three coordinates. For example, the point in Figure 1.2 that is 4 units above the xy plane and that lies directly over the point $(x, y) = (1, 2)$ is represented by the coordinates $x = 1$, $y = 2$, and $z = 4$ and is denoted by the ordered triple $(x, y, z) = (1, 2, 4)$. Similarly, the ordered triple $(2, -1, -2)$ represents the point that is 2 units below the xy plane and that lies directly under the point $(x, y) = (2, -1)$.

The Graph of $f(x, y)$

To graph a function $f(x, y)$ of the two independent variables x and y, it is customary to introduce the letter z to stand for the dependent variable and to write $z = f(x, y)$. The ordered pairs (x, y) in the domain of f are thought of as points in the xy plane, and the function f assigns a height z to each such point. Thus, if $f(1, 2) = 4$, we would express this fact geometrically by plotting the point $(1, 2, 4)$ on a three-dimensional coordinate system. The graph of f consists of all points (x, y, z) for which $z = f(x, y)$. The function may assign different heights to different points in its domain and, in general, its graph will be a surface in three-dimensional space. The situation is illustrated in Figure 1.3.

In practical work in the social and managerial sciences, you will rarely, if ever, have to graph a function of two variables. Therefore, we shall not take time to develop techniques for sketching surfaces. However, since you may be interested in seeing some surfaces that arise as graphs of actual functions (and because these surfaces will be used to illustrate the theory of maxima and minima in Section 5), portions of two such surfaces are sketched in Figure 1.4.

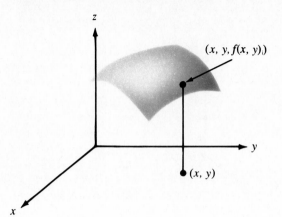

Figure 1.3 The graph of $z = f(x, y)$.

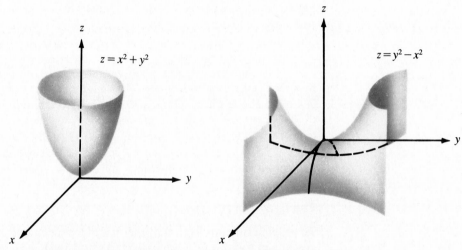

Figure 1.4 The graphs of $z = x^2 + y^2$ and $z = y^2 - x^2$.

Problems

In Problems 1 through 6, specify the domain of f and compute the indicated function values.

1. $f(x, y) = (x - 1)^2 + 2xy^3; f(2, -1), f(1, 2)$

2. $f(x, y) = \dfrac{3x + 2y}{2x + 3y}; f(1, 2), f(-4, 6)$

3. $f(x, y) = \sqrt{y^2 - x^2}; f(4, 5), f(-1, 2)$

4. $f(x, y) = 10x^{1/2}y^{2/3}; f(16, 27), f(4, -1{,}331)$

5. $f(x, y) = \dfrac{x}{\ln y}; f(-1, e^3), f(\ln 9, e^2)$

6. $f(x, y) = \dfrac{e^{xy}}{2 - e^{xy}}; f(1, 0), f(\ln 2, 2)$

Production

7. Using x skilled workers and y unskilled workers, a manufacturer can produce $Q(x, y) = 10x^2y$ units per day. Currently there are 20 skilled workers and 40 unskilled workers on the job.
   (a) How many units are currently being produced each day?
   (b) By how much will the daily production level change if 1 more skilled worker is added to the current work force?
   (c) By how much will the daily production level change if 1 more unskilled worker is added to the current work force?
   (d) By how much will the daily production level change if 1 more skilled worker *and* 1 more unskilled worker are added to the current work force.

Production cost

8. A manufacturer can produce electric typewriters at a cost of $80 apiece and manual typewriters at a cost of $20 apiece.
   (a) Express the manufacturer's total monthly production cost as a function of the number of electric typewriters and the number of manual typewriters produced.
   (b) Compute the total monthly cost if 500 electric and 800 manual typewriters are produced.
   (c) The manufacturer wants to increase the output of electric typewriters by 50 a month from the level in part (b). What corresponding change should be made in the monthly output of manual typewriters so that the total monthly cost will not change?

Retail sales

9. A paint store carries two brands of latex paint. Sales figures indicate that if the first brand is sold for x_1 dollars per gallon and the second for x_2 dollars per gallon, the demand for the first brand will be $D_1(x_1, x_2) = 200 - 10x_1 + 20x_2$ gallons per month, and the demand for the second brand will be $D_2(x_1, x_2) = 100 + 5x_1 - 10x_2$ gallons per month.
   (a) Express the paint store's total monthly revenue from the sale of the paint as a function of the prices x_1 and x_2.
   (b) Compute the revenue in part (a) if the first brand is sold for $6 per gallon and the second for $5 per gallon.

Production

10. The output at a certain factory is $Q(K, L) = 120K^{2/3}L^{1/3}$ units, where K is the capital investment measured in units of $1,000 and L the size of the labor force measured in worker-hours.
    (a) Compute the output if the capital investment is $125,000 and the size of the labor force is 1,331 worker-hours.
    (b) What will happen to the output in part (a) if both the level of capital investment and the size of the labor force are cut in half?

Constant returns to scale

11. Suppose output Q is given by the Cobb-Douglas production function $Q(K, L) = AK^{\alpha}L^{1-\alpha}$, where A and α are positive constants and

$0 < \alpha < 1$. Prove that if K and L are both multiplied by the same positive number m, then the output Q will also be multiplied by m. That is, show that $Q(mK, mL) = mQ(K, L)$.

Constant production curves

12. Using x skilled workers and y unskilled workers, a manufacturer can produce $Q(x, y) = 3x + 2y$ units per day. Currently the work force consists of 10 skilled workers and 20 unskilled workers.

(a) Compute the current daily output.

(b) Find an equation relating the levels of skilled and unskilled labor if the daily output is to remain at its current level.

(c) On a two-dimensional coordinate system, draw the relevant portion of the graph of the equation in part (b). (This graph shows all the combinations of inputs x and y that will result in the current level of output and is known in economics as a **constant production curve**. Constant production curves will be studied in more detail in Section 4.)

(d) What change should be made in the level of unskilled labor y to offset an increase in skilled labor x of 2 workers so that the output will remain at its current level?

2 Partial Derivatives

In many problems involving functions of two variables, the goal is to find the rate of change of the function with respect to one of its variables when the other is held constant. That is, the goal is to differentiate the function with respect to the particular variable in question while keeping the other variable fixed. This process is known as **partial differentiation,** and the resulting derivative is said to be a **partial derivative** of the function.

Partial Derivatives

Suppose $z = f(x, y)$. The partial derivative of f with respect to x is denoted by

$$\frac{\partial z}{\partial x} \quad \text{or} \quad f_x(x, y)$$

and is the function obtained by differentiating f with respect to x treating y as a constant. The partial derivative of f with respect to y is denoted by

$$\frac{\partial z}{\partial y} \quad \text{or} \quad f_y(x, y)$$

and is the function obtained by differentiating f with respect to y treating x as a constant.

Computation of Partial Derivatives

No new rules are needed for the computation of partial derivatives. To compute f_x, simply differentiate f with respect to the single variable x, pretending that y is a constant. To compute f_y, simply differentiate f with respect to y, pretending that x is a constant. Here are some examples.

EXAMPLE 2.1 Find the partial derivatives f_x and f_y if $f(x, y) = x^2 + 2xy^2 + \dfrac{2y}{3x}$.

SOLUTION
To simplify the computation, begin by rewriting the function as

$$f(x, y) = x^2 + 2xy^2 + \frac{2}{3}yx^{-1}$$

To compute f_x, think of f as a function of x and differentiate the sum term by term, treating y as a constant to get

$$f_x(x, y) = 2x + 2y^2 - \frac{2}{3}yx^{-2} = 2x + 2y^2 - \frac{2y}{3x^2}$$

To compute f_y, think of f as a function of y and differentiate term by term, treating x as a constant to get

$$f_y(x, y) = 0 + 4xy + \frac{2}{3}x^{-1} = 4xy + \frac{2}{3x}$$

EXAMPLE 2.2 Find the partial derivatives $\dfrac{\partial z}{\partial x}$ and $\dfrac{\partial z}{\partial y}$ if $z = (x^2 + xy + y)^5$.

SOLUTION
Holding y fixed and using the chain rule to differentiate z with respect to x, you get

$$\frac{\partial z}{\partial x} = 5(x^2 + xy + y)^4(2x + y)$$

Holding x fixed and using the chain rule to differentiate z with respect to y, you get

$$\frac{\partial z}{\partial y} = 5(x^2 + xy + y)^4(x + 1)$$

EXAMPLE 2.3 Find the partial derivatives f_x and f_y if $f(x, y) = xe^{-2xy}$.

SOLUTION
From the product rule,

$$f_x(x, y) = x(-2ye^{-2xy}) + e^{-2xy} = (-2xy + 1)e^{-2xy}$$

and from the constant multiple rule,

$$f_y(x, y) = x(-2xe^{-2xy}) = -2x^2e^{-2xy}$$

Marginal Analysis

In economics, the term **marginal analysis** refers to the practice of using a derivative to estimate the change in the value of a function resulting from a 1-unit increase in one of its variables. In Chapter 2, Section 3, you saw some examples of marginal analysis involving ordinary derivatives of functions of one variable. Here are two examples illustrating marginal analysis for functions of two variables.

Marginal Products of Capital and Labor

EXAMPLE 2.4 Suppose the daily output Q at a factory depends on the amount K of capital (measured in units of $1,000) invested in the plant and equipment, and also on the size L of the labor force (measured in worker-hours). In economics, the partial derivatives $\frac{\partial Q}{\partial K}$ and $\frac{\partial Q}{\partial L}$ are known as the **marginal products** (or marginal productivities) of capital and labor, respectively. Give economic interpretations of these two marginal products.

SOLUTION
The marginal product of labor $\frac{\partial Q}{\partial L}$ is the rate at which output Q changes with respect to labor L for a fixed level K of capital investment. Hence, $\frac{\partial Q}{\partial L}$ is approximately the change in output that will result if capital investment is held fixed and labor is increased by 1 worker-hour.

Similarly, the marginal product of capital $\frac{\partial Q}{\partial K}$ is approximately the change in output that will result if the size of the labor force is held fixed and capital investment is increased by $1,000.

EXAMPLE 2.5 It is estimated that the weekly output at a certain plant is given by the function $Q(x, y) = 1,200x + 500y + x^2y - x^3 - y^2$ units, where x is the number of skilled workers and y the number of unskilled workers employed at the plant. Currently the work force consists of 30 skilled workers and 60 unskilled workers. Use marginal analysis to estimate the change in the weekly output that will result from the addition of 1 more skilled worker if the number of unskilled workers is not changed.

SOLUTION
The partial derivative

$$Q_x(x, y) = 1,200 + 2xy - 3x^2$$

is the rate of change of output with respect to the number of skilled workers. For any values of x and y, this is an approximation of the number of additional units that will be produced each week if the number of skilled workers is increased from x to $x + 1$ while the number of unskilled workers is kept fixed at y. In particular, if the work force is increased from 30 skilled and 60 unskilled workers to 31 skilled and 60 unskilled workers, the resulting change in output is approximately

$$Q_x(30, 60) = 1,200 + 2(30)(60) - 3(30)^2 = 2,100 \text{ units}$$

For practice, compute the change in output exactly by subtracting appropriate values of Q. Is the approximation a good one?

Second-Order Partial Derivatives

Partial derivatives can themselves be differentiated. The resulting functions are called **second-order partial derivatives.** Here is a summary of the definition and notation for the four possible second-order partial derivatives of a function of two variables.

Second-Order Partial Derivatives

If $z = f(x, y)$, the partial derivative of f_x with respect to x is

$$f_{xx} = (f_x)_x \quad \text{or} \quad \frac{\partial^2 z}{\partial x^2} = \frac{\partial}{\partial x}\left(\frac{\partial z}{\partial x}\right)$$

The partial derivative of f_x with respect to y is

$$f_{xy} = (f_x)_y \quad \text{or} \quad \frac{\partial^2 z}{\partial y \partial x} = \frac{\partial}{\partial y}\left(\frac{\partial z}{\partial x}\right)$$

The partial derivative of f_y with respect to x is

$$f_{yx} = (f_y)_x \quad \text{or} \quad \frac{\partial^2 z}{\partial x \partial y} = \frac{\partial}{\partial x}\left(\frac{\partial z}{\partial y}\right)$$

The partial derivative of f_y with respect to y is

$$f_{yy} = (f_y)_y \quad \text{or} \quad \frac{\partial^2 z}{\partial y^2} = \frac{\partial}{\partial y}\left(\frac{\partial z}{\partial y}\right)$$

The computation of second-order partial derivatives is illustrated in the next example.

EXAMPLE 2.6   Compute the four second-order partial derivatives of the function $f(x, y) = xy^3 + 5xy^2 + 2x + 1$.

SOLUTION
Since

$$f_x = y^3 + 5y^2 + 2$$

it follows that

$$f_{xx} = 0 \quad \text{and} \quad f_{xy} = 3y^2 + 10y$$

Since

$$f_y = 3xy^2 + 10xy$$

it follows that

$$f_{yx} = 3y^2 + 10y \quad \text{and} \quad f_{yy} = 6xy + 10x$$

The Mixed Second-Order Partial Derivatives

The two partial derivatives f_{xy} and f_{yx} are sometimes called the **mixed second-order partial derivatives** of f. Notice that the mixed partial derivatives in Example 2.6 were equal. This is not an accident. It turns out that for virtually all the functions you will encounter in practical work, the mixed partial derivatives will be equal. That is, you will get the same answer if you first differentiate f with respect to x and then differentiate the resulting function with respect to y as you would if you perform the differentiation in the opposite order.

An Application to Economics

In the next example, you will see how a second-order partial derivative can convey useful information in a practical situation.

EXAMPLE 2.7   Suppose the output Q at a factory depends on the amount K of capital invested in the plant and equipment, and also on the size L of the labor force, measured in worker-hours. Give an economic interpretation of the sign of the second-order partial derivative $\dfrac{\partial^2 Q}{\partial L^2}$.

SOLUTION

If $\dfrac{\partial^2 Q}{\partial L^2}$ is negative, the marginal product of labor $\dfrac{\partial Q}{\partial L}$ decreases as L increases. This implies that for a fixed level of capital investment, the effect on output of the addition of 1 worker-hour of labor is greater when the work force is small than when the work force is large.

   Similarly, if $\dfrac{\partial^2 Q}{\partial L^2}$ is positive, it follows that for a fixed level of capital investment, the effect on output of the addition of 1 worker-hour of labor is greater when the work force is large than when it is small.

   For most factories operating with adequate work forces, the derivative $\dfrac{\partial^2 Q}{\partial L^2}$ is generally negative. Can you give an economic explanation for this fact?

Geometric Interpretation of Partial Derivatives

As you saw in Section 1, functions of two variables can be represented graphically as surfaces drawn on three-dimensional coordinate systems. In particular, if $z = f(x, y)$, an ordered pair (x, y) in the domain of f can be identified with a point in the xy plane and the corresponding function value $z = f(x, y)$ can be thought of as assigning a height to this point. The graph of f is the surface consisting of all points (x, y, z) in three-dimensional space whose height z is equal to $f(x, y)$.

   The partial derivatives of a function of two variables can be interpreted geometrically as follows. For each fixed number y_0, the points (x, y_0, z) form a vertical plane whose equation is $y = y_0$. If $z = f(x, y)$ and if y is kept fixed at $y = y_0$, then the corresponding points $(x, y_0, f(x, y_0))$ form a curve in three-dimensional space that is the intersection of the surface $z = f(x, y)$ with the plane $y = y_0$. At each point on this curve, the partial derivative $\dfrac{\partial z}{\partial x}$ is simply the slope of the line in the plane $y = y_0$ that is tangent to the curve at the point in question. That is, $\dfrac{\partial z}{\partial x}$ is the slope of the tangent "in the x direction." The situation is illustrated in Figure 2.1a.

   Similarly, if x is kept fixed at $x = x_0$, the corresponding points $(x_0, y, f(x_0, y))$ form a curve that is the intersection of the surface $z = f(x, y)$ with the vertical plane $x = x_0$. At each point on this curve, the partial derivative $\dfrac{\partial z}{\partial y}$ is the slope of the tangent in the plane $x = x_0$. That is, $\dfrac{\partial z}{\partial y}$ is the slope of the tangent "in the y direction." The situation is illustrated in Figure 2.1b.

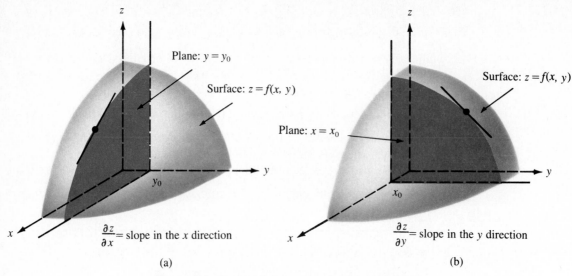

Figure 2.1 Geometric interpretation of partial derivatives.

Problems

In Problems 1 through 12, compute all the first-order partial derivatives of the given function.

1. $f(x, y) = 2xy^5 + 3x^2y + x^2$

2. $z = 5x^2y + 2xy^3 + 3y^2$

3. $z = (3x + 2y)^5$

4. $f(x, y) = (x + xy + y)^3$

5. $f(x, y) = \dfrac{3y}{2x}$

6. $z = \dfrac{x^2}{y^3}$

7. $z = xe^{xy}$

8. $f(x, y) = xye^x$

9. $f(x, y) = \dfrac{2x + 3y}{y - x}$

10. $z = \dfrac{xy^2}{x^2y^3 + 1}$

11. $z = x \ln y$

12. $f(x, y) = x \ln xy$

Marginal analysis 13. At a certain factory, the daily output is $Q(K, L) = 60K^{1/2}L^{1/3}$ units, where K denotes the capital investment measured in units of $1,000 and L the size of the labor force measured in worker-hours. Suppose the current capital investment is $900,000 and that 1,000 worker-hours of labor are used each day. Use marginal analysis to estimate the effect of an additional capital investment of $1,000 on the daily output if the size of the labor force is not changed.

Consumer demand 14. The monthly demand for a certain brand of toasters is given by a function $f(x, y)$, where x is the amount of money (measured in units of $1,000) spent on advertising and y is the selling price (in dollars) of the toasters. Give economic interpretations of the partial derivatives f_x and f_y. Under normal economic conditions, what will the sign of each of these partial derivatives be?

Marginal analysis 15. A grocer's daily profit from the sale of two brands of orange juice is

$$P(x, y) = (x - 30)(70 - 5x + 4y) + (y - 40)(80 + 6x - 7y)$$

cents, where x is the price per can of the first brand and y is the price per can of the second. Currently the first brand sells for 50 cents per can and the second for 52 cents per can. Use marginal analysis to estimate the change in the daily profit that will result if the grocer raises the price of the second brand by 1 cent per can but keeps the price of the first brand unchanged.

Consumer demand 16. Two competing brands of power lawnmowers are sold in the same town. The price of the first brand is x dollars per mower and the price of the second brand is y dollars per mower. The local demand for the first brand of mower is given by a function $D(x, y)$.
(a) How would you expect the demand for the first brand of mower to be affected by an increase in x? By an increase in y?
(b) Translate your answers in part (a) into conditions on the signs of the partial derivatives of D.
(c) If $D(x, y) = a + bx + cy$, what can you say about the signs of the coefficients b and c if your conclusions in parts (a) and (b) are to hold?

Substitute commodities 17. In economics, two commodities are said to be **substitute commodities** if the demand Q_1 for the first increases as the price p_2 of the second increases, and if the demand Q_2 for the second increases as the price p_1 of the first increases.
(a) Give an example of a pair of substitute commodities.
(b) If two commodities are substitutes, what must be true of the partial derivatives $\dfrac{\partial Q_1}{\partial p_2}$ and $\dfrac{\partial Q_2}{\partial p_1}$?
(c) Suppose the demand functions for two commodities are $Q_1 = 3,000 + \dfrac{400}{p_1 + 3} + 50p_2$ and $Q_2 = 2,000 - 100p_1 + \dfrac{500}{p_2 + 4}$. Are the commodities substitutes?

Complementary commodities 18. Two commodities are said to be **complementary commodities** if the demand Q_1 for the first decreases as the price p_2 of the second increases, and if the demand Q_2 for the second decreases as the price p_1 of the first increases.
(a) Give an example of a pair of complementary commodities.
(b) If two commodities are complementary, what must be true of the partial derivatives $\dfrac{\partial Q_1}{\partial p_2}$ and $\dfrac{\partial Q_2}{\partial p_1}$?

(c) Suppose the demand functions for two commodities are $Q_1 = 2,000 + \dfrac{400}{p_1 + 3} - 50p_2$ and $Q_2 = 2,000 - 100p_2 + \dfrac{500}{p_1 + 4}$. Are the commodities complementary?

In Problems 19 through 24, compute all the second-order partial derivatives of the given function.

19. $f(x, y) = 5x^4y^3 + 2xy$

20. $f(x, y) = \dfrac{x + 1}{y - 1}$

21. $f(x, y) = e^{x^2y}$

22. $f(x, y) = \ln (x^2 + y^2)$

23. $f(x, y) = \sqrt{x^2 + y^2}$

24. $f(x, y) = x^2ye^x$

Marginal productivity

25. Suppose the output Q of a factory depends on the amount K of capital investment measured in units of \$1,000 and on the size L of the labor force measured in worker-hours. Give an economic interpretation of the second-order partial derivative $\dfrac{\partial^2 Q}{\partial K^2}$.

Marginal productivity

26. At a certain factory, the output is $Q = 120K^{1/2}L^{1/3}$ units, where K denotes the capital investment measured in units of \$1,000 and L the size of the labor force measured in worker-hours.

(a) Determine the sign of the second-order partial derivative $\dfrac{\partial^2 Q}{\partial L^2}$ and give an economic interpretation.

(b) Determine the sign of the second-order partial derivative $\dfrac{\partial^2 Q}{\partial K^2}$ and give an economic interpretation.

Law of diminishing returns

27. Suppose the daily output Q of a factory depends on the amount K of capital investment and on the size L of the labor force. A **law of diminishing returns** states that in certain circumstances, there is a value L_0 such that the marginal product of labor will be increasing for $L < L_0$ and decreasing for $L > L_0$.

(a) Translate this law of diminishing returns into statements about the sign of a certain second-order partial derivative.

(b) Discuss the economic factors that might account for this phenomenon.

3 The Chain Rule and the Total Differential

In many practical situations, a particular quantity is given as a function of two or more variables, each of which can be thought of as a function of yet another variable, and the goal is to find the rate of change of the quantity with respect to this other variable. For example, the demand for a certain commodity may depend on the price of the commodity itself and on the price of a competing commodity, both of which are increasing with time, and the goal may be to find the rate of change of the demand with respect to time. You can solve problems of this type by using a generalization of the chain rule that was introduced in Chapter 2, Section 5.

The Chain Rule

Recall from Chapter 2 that if z if a function of x, and x is a function of t, then z can be regarded as a function of t, and the rate of change of z with respect to t is given by the chain rule

$$\frac{dz}{dt} = \frac{dz}{dx}\frac{dx}{dt}$$

Here is the corresponding rule for functions of two variables.

Chain Rule for Partial Derivatives

> Suppose z is a function of x and y, each of which is a function of t. Then z can be regarded as a function of t and
>
> $$\frac{dz}{dt} = \frac{\partial z}{\partial x}\frac{dx}{dt} + \frac{\partial z}{\partial y}\frac{dy}{dt}$$

Observe that the expression for $\frac{dz}{dt}$ is the sum of two terms, each of which can be interpreted using the chain rule for a function of one variable. In particular,

$$\frac{\partial z}{\partial x}\frac{dx}{dt} = \text{rate of change of } z \text{ with respect to } t \text{ for fixed } y$$

and

$$\frac{\partial z}{\partial y}\frac{dy}{dt} = \text{rate of change of } z \text{ with respect to } t \text{ for fixed } x$$

The chain rule for partial derivatives says that the total rate of change of z with respect to t is the sum of these two "partial" rates of change.

Here are two examples illustrating the use of the chain rule for partial derivatives.

EXAMPLE 3.1   Find $\frac{dz}{dt}$ if $z = x^2 + 3xy + 1$, $x = 2t + 1$, and $y = t^2$.

SOLUTION
By the chain rule,

$$\frac{dz}{dt} = \frac{\partial z}{\partial x}\frac{dx}{dt} + \frac{\partial z}{\partial y}\frac{dy}{dt} = (2x + 3y)(2) + 3x(2t)$$

which you can rewrite in terms of t by substituting $x = 2t + 1$ and $y = t^2$ to get

$$\frac{dz}{dt} = 4(2t + 1) + 6t^2 + 3(2t + 1)(2t) = 18t^2 + 14t + 4$$

For practice, check this answer by first substituting $x = 2t + 1$ and $y = t^2$ into the formula for z and then differentiating directly with respect to t.

EXAMPLE 3.2   A liquor store carries two brands of inexpensive white wine, one from California and the other from New York. Sales figures indicate that if the California wine is sold for x dollars per bottle and the New York wine for y dollars per bottle, the demand for the California wine will be

$$Q(x, y) = 300 - 20x^2 + 30y \qquad \text{bottles per month}$$

It is estimated that t months from now the price of the California wine will be

$$x = 2 + 0.05t \qquad \text{dollars per bottle}$$

and the price of the New York wine will be

$$y = 2 + 0.1\sqrt{t} \qquad \text{dollars per bottle}$$

At what rate will the demand for the California wine be changing with respect to time 4 months from now?

SOLUTION

Your goal is to find $\dfrac{dQ}{dt}$ when $t = 4$. Using the chain rule you get

$$\frac{dQ}{dt} = \frac{\partial Q}{\partial x}\frac{dx}{dt} + \frac{\partial Q}{\partial y}\frac{dy}{dt} = -40x(0.05) + 30(0.05t^{-1/2})$$

When $t = 4$,

$$x = 2 + 0.05(4) = 2.2$$

and hence

$$\frac{dQ}{dt} = -40(2.2)(0.05) + 30(0.05)(0.5) = -3.65$$

That is, 4 months from now the monthly demand for the California wine will be decreasing at the rate of 3.65 bottles per month.

The Total Differential

In Chapter 2, Section 4, you learned how to use the differential of a function to approximate the change in the function resulting from a small change in its independent variable. In particular, you saw that if y is a function of x,

$$\Delta y \simeq \frac{dy}{dx} \Delta x$$

where Δx is a small change in the variable x and Δy is the corresponding change in the function y. The expression $dy = \frac{dy}{dx} \Delta x$ that was used to approximate Δy was called the differential of y. Here is the analogous approximation formula for functions of two variables.

Approximation Formula

> Suppose z is a function of x and y. If Δx denotes a small change in x and Δy a small change in y, the corresponding change in z is
>
> $$\Delta z \simeq \frac{\partial z}{\partial x} \Delta x + \frac{\partial z}{\partial y} \Delta y$$

Observe that the expression used to approximate Δz is the sum of two terms, each of which is essentially a one-variable differential. In particular,

$$\frac{\partial z}{\partial x} \Delta x \simeq \text{change in } z \text{ due to the change in } x \text{ for fixed } y$$

and

$$\frac{\partial z}{\partial y} \Delta y \simeq \text{change in } z \text{ due to the change in } y \text{ for fixed } x$$

The approximation formula says that the total change in z is approximately equal to the sum of these two partial changes.

The sum of the two one-variable differentials that appears in the approximation formula is called the **total differential** of z and is denoted by the symbol dz. Notice the similarity between the formula for the total differential dz and the chain rule.

The Total Differential

> If z is a function of x and y, the total differential of z is
>
> $$dz = \frac{\partial z}{\partial x} \Delta x + \frac{\partial z}{\partial y} \Delta y$$

The use of the total differential to approximate the change in a function is illustrated in the next example.

EXAMPLE 3.3 At a certain factory, the daily output is $Q = 60K^{1/2}L^{1/3}$ units, where K denotes the capital investment measured in units of \$1,000 and L the size of the labor force measured in worker-hours. The current capital investment is \$900,000 and 1,000 worker-hours of labor are used each day. Estimate the change in output that will result if capital investment is increased by \$1,000 and labor is increased by 2 worker-hours.

SOLUTION

Apply the approximation formula with $K = 900$, $L = 1,000$, $\Delta K = 1$, and $\Delta L = 2$ to get

$$\Delta Q \simeq \frac{\partial Q}{\partial K} \Delta K + \frac{\partial Q}{\partial L} \Delta L$$

$$= 30K^{-1/2}L^{1/3} \Delta K + 20K^{1/2}L^{-2/3} \Delta L$$

$$= 30\left(\frac{1}{30}\right)(10)\,(1)\, + \,20(30)\left(\frac{1}{100}\right)(2)$$

$$= 22 \text{ units}$$

That is, output will increase by approximately 22 units.

Approximation of Percentage Change

The **percentage change** of a quantity expresses the change in the quantity as a percentage of its size prior to the change. In particular,

$$\text{Percentage change} = 100\frac{\text{change in quantity}}{\text{size of quantity}}$$

This formula can be combined with the approximation formula as follows.

Approximation of Percentage Change

Suppose z is a function of x and y. If Δx denotes a small change in x and Δy a small change in y, the corresponding percentage change in z is

$$\text{Percentage change in } z = 100\frac{\Delta z}{z} \simeq 100\frac{\frac{\partial z}{\partial x} \Delta x + \frac{\partial z}{\partial y} \Delta y}{z}$$

EXAMPLE 3.4 Use calculus to approximate the percentage by which the volume of a cylinder increases if the radius increases by 1 percent and the height increases by 2 percent.

SOLUTION

The volume of a cylinder is given by the function $V(r, h) = \pi r^2 h$, where r is the radius and h the height. The fact that r increases by 1 percent means that $\Delta r = 0.01r$, and the fact that h increases by 2 percent means that $\Delta h = 0.02h$. By the approximation formula for percentage change,

$$\text{Percentage change in } V \simeq 100 \frac{\dfrac{\partial V}{\partial r} \Delta r + \dfrac{\partial V}{\partial h} \Delta h}{V}$$

$$= 100 \frac{2\pi rh(0.01r) + \pi r^2(0.02h)}{\pi r^2 h}$$

$$= 100 \frac{0.02 \, \pi r^2 h + 0.02\pi r^2 h}{\pi r^2 h}$$

$$= 100 \frac{0.04\pi r^2 h}{\pi r^2 h} = 4 \text{ percent}$$

Problems

In Problems 1 through 8, use the chain rule to find $\dfrac{dz}{dt}$. Check your answer by writing z explicitly as a function of t and differentiating directly with respect to t.

1. $z = x + 2y$; $x = 3t$, $y = 2t + 1$

2. $z = 3x^2 + xy$; $x = t + 1$, $y = 1 - 2t$

3. $z = \dfrac{x}{y}$; $x = t^2$, $y = 3t$

4. $z = \dfrac{y}{x}$; $x = 2t$, $y = t^3$

5. $z = \dfrac{x + y}{x - y}$; $x = t^3 + 1$, $y = 1 - t^3$

6. $z = (2x + 3y)^2$; $x = 2t$, $y = 3t$

7. $z = (x - y^2)^3$; $x = t^2$, $y = 2t$

8. $z = xy$; $x = e^t$, $y = e^{-t}$

In Problems 9 through 13, use the chain rule to find $\dfrac{dz}{dt}$ for the specified value of t.

9. $z = 2x + 3y$; $x = t^2$, $y = 5t$; $t = 2$

10. $z = x^2 y$; $x = 3t + 1$, $y = t^2 - 1$; $t = 1$

11. $z = \dfrac{3x}{y}; x = t, y = t^2; t = 3$

12. $z = x^{1/2}y^{1/3}; x = 2t, y = 2t^2; t = 2$

13. $z = xy; x = e^{2t}, y = e^{3t}; t = 0$

Consumer demand   14. A paint store carries two brands of latex paint. Sales figures indicate that if the first brand is sold for x dollars per gallon and the second for y dollars per gallon, the demand for the first brand will be $Q(x, y) = 200 - 10x^2 + 20y$ gallons per month. It is estimated that t months from now the price of the first brand will be $x = 5 + 0.02t$ dollars per gallon and the price of the second brand will be $y = 6 + 0.4\sqrt{t}$ dollars per gallon. At what rate will the demand for the first brand of paint be changing with respect to time 9 months from now?

Consumer demand   15. A bicycle dealer has found that if 10-speed bicycles are sold for x dollars apiece and the price of gasoline is y cents per gallon, approximately $f(x, y) = 200 - 24\sqrt{x} + 4(0.1y + 5)^{3/2}$ bicycles will be sold each month. It is estimated that t months from now the bicycles will be selling for $x = 129 + 5t$ dollars apiece and the price of gasoline will be $y = 80 + 10\sqrt{3t}$ cents per gallon. At what rate will the monthly demand for the bicycles be changing with respect to time 3 months from now?

Consumer demand   16. The demand for a certain product is $Q(x, y) = 200 - 10x^2 + 20xy$ units per month, where x is the price of the product and y the price of a competing product. It is estimated that t months from now the price of the product will be $x = 10 + 0.5t$ dollars per unit while the price of the competing product will be $y = 12.8 + 0.2t^2$ dollars per unit.
   (a) At what rate will the demand for the product be changing with respect to time 4 months from now?
   (b) At what percentage rate will the demand for the product be changing with respect to time 4 months from now?

Manufacturing   17. The output at a certain plant is $Q(x, y) = 0.08x^2 + 0.12xy + 0.03y^2$ units per day, where x is the number of hours of skilled labor used and y is the number of hours of unskilled labor used. Currently 80 hours of skilled labor and 200 hours of unskilled labor are used each day. Use the total differential of Q to estimate the change in output that will result if an additional $\dfrac{1}{2}$ hour of skilled labor is used along with an additional 2 hours of unskilled labor.

Manufacturing   18. At a certain factory, the output is $Q = 120K^{1/2}L^{1/3}$ units, where K denotes the capital investment measured in units of $1,000 and L the size of the labor force measured in worker-hours. The current capital investment is $400,000 and 1,000 worker-hours of labor are currently used. Use the total differential of Q to estimate the change in output that will result if capital investment is increased by \$500 and labor is increased by 4 worker-hours.

Retail sales   19. A grocer's daily profit from the sale of two brands of orange juice is
$P(x, y) = (x - 30)(70 - 5x + 4y) + (y - 40)(80 + 6x - 7y)$ cents,
where x is the price per can of the first brand and y is the price per can of the second. Currently the first brand sells for 50 cents per can and the second for 52 cents per can. Use the total differential of P to estimate the change in the daily profit that will result if the grocer raises the price of the first brand by 1 cent per can and raises the price of the second brand by 2 cents per can.

Publishing   20. An editor estimates that if x thousand dollars is spent on development and y thousand on promotion, approximately $Q(x, y) = 20x^{3/2}y$ copies of a new book will be sold. Current plans call for the expenditure of \$36,000 on development and \$25,000 on promotion. Use the total differential of Q to estimate the change in sales that will result if the amount spent on development is increased by \$500 and the amount spent on promotion is decreased by \$500.

Landscaping   21. A rectangular garden that is 30 yards long and 40 yards wide is surrounded by a concrete path that is 0.8 yard wide. Use a total differential to estimate the area of the concrete path.

Packaging   22. A soft-drink can is 12 centimeters tall and has a radius of 3 centimeters. The manufacturer is planning to reduce the height of the can by 0.2 centimeter and the radius by 0.3 centimeter. Use a total differential to estimate how much less drink consumers will find in each can after the new cans are introduced. [*Hint*: Recall that the volume of a cylinder of radius r and height h is $V(r, h) = \pi r^2 h$.]

Volume   23. Use calculus to estimate the percentage by which the volume of a cylinder will change if the radius is increased by 1 percent and the height is decreased by 1.5 percent.

Manufacturing   24. At a certain factory, the daily output is $Q(K, L) = 60K^{1/2}L^{1/3}$ units, where K denotes the capital investment and L the size of the labor force. Use calculus to estimate the percentage by which the daily output will change if capital investment is increased by 1 percent and labor by 2 percent.

Manufacturing   25. At a certain factory, output is given by the Cobb-Douglas production function $Q(K, L) = AK^\alpha L^{1-\alpha}$, where A and α are positive constants with $0 < \alpha < 1$, and where K denotes the capital investment and L the size of the labor force. Use calculus to estimate the percentage by which output will change if both capital and labor are increased by 1 percent.

Manufacturing   26. Using x skilled workers and y unskilled workers, a manufacturer can produce $Q(x, y) = x^2y$ units per day. Currently the work force consists of 16 skilled workers and 32 unskilled workers, and the manufacturer is planning to hire 1 additional skilled worker. Use a total differential to estimate the corresponding change that the manufacturer should make in the level of unskilled labor so that the total output will remain the same. (*Hint*: The goal is to estimate Δy if $\Delta x = 1$ and $\Delta Q = 0$.)

27. Suppose $z = f(x, y)$, $x = at$, and $y = bt$, where a and b are constants. Think of z as a function of t and find an expression for the second derivative $\dfrac{d^2z}{dt^2}$ in terms of the constants a and b and the second-order partial derivatives f_{xx}, f_{yy}, and f_{xy}.

4 Level Curves

There are many situations in which one is interested in the possible combinations of variables x and y for which a function $f(x, y)$ will be equal to a certain constant. For example, a manufacturer whose output depends on the numbers of skilled and unskilled workers in the labor force may wish to determine the possible combinations of skilled and unskilled workers that will result in a certain desired level of output.

The combinations of x and y for which $f(x, y)$ is equal to a fixed number often can be represented geometrically as the points on a curve in the xy plane. Such a curve is said to be a **level curve** of f. If the function f represents the output of a factory and the variables x and y represent inputs (such as skilled and unskilled labor), the level curves of f are sometimes called **constant-production curves** or **isoquants.**

Level Curve

> For any constant C, the points (x, y) for which $f(x, y) = C$ form a curve in the xy plane that is said to be a level curve of f.

Here are two examples.

EXAMPLE 4.1   If $f(x, y) = x^2 - y$, sketch the level curves $f(x, y) = 4$ and $f(x, y) = 9$.

SOLUTION
The level curve $f(x, y) = 4$ consists of all points (x, y) in the xy plane for which

$$x^2 - y = 4 \qquad \text{or} \qquad y = x^2 - 4 = (x - 2)(x + 2)$$

The level curve $f(x, y) = 9$ consists of all points (x, y) in the xy plane for which

$$x^2 - y = 9 \quad \text{or} \quad y = x^2 - 9 = (x - 3)(x + 3)$$

The graphs of these two level curves are sketched in Figure 4.1.

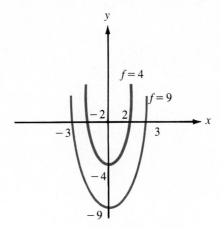

Figure 4.1 The level curves $f = 4$ and $f = 9$ for $f(x, y) = x^2 - y$.

EXAMPLE 4.2 If $f(x, y) = xy$, sketch the level curves $f(x, y) = 1$ and $f(x, y) = 2$.

SOLUTION

The level curve $f(x, y) = 1$ consists of all the points (x, y) in the xy plane for which

$$xy = 1 \quad \text{or} \quad y = \frac{1}{x}$$

The level curve $f(x, y) = 2$ consists of all the points (x, y) in the xy plane for which

$$xy = 2 \quad \text{or} \quad y = \frac{2}{x}$$

The graphs of these two level curves are sketched in Figure 4.2.

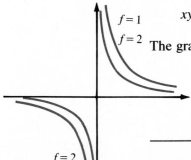

Figure 4.2 The level curves $f = 1$ and $f = 2$ for $f(x, y) = xy$.

Geometric Interpretation of Level Curves

Here is another way you can visualize level curves, which is sometimes useful. Think of $z = f(x, y)$ as the equation of a surface in three-dimensional space. The level curve $f(x, y) = C$ is the projection onto the xy plane of the curve formed by the intersection of the surface $z = f(x, y)$ with the horizontal plane $z = C$. The situation is illustrated in Figure 4.3.

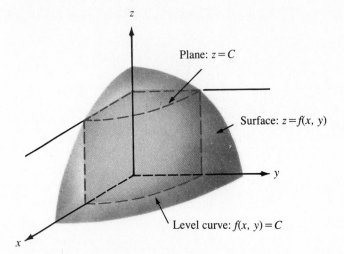

Figure 4.3

Geometric interpretation of a level curve.

The Slope of a Level Curve

The slope of the line that is tangent to the level curve $f(x, y) = C$ at a particular point is given by the derivative $\dfrac{dy}{dx}$. This derivative is the rate of change of y with respect to x on the level curve and hence is approximately the amount by which the y coordinate of a point on the level curve changes when the x coordinate is increased by 1. (See Figure 4.4.) For example, if f represents output and x and y represent the levels of skilled

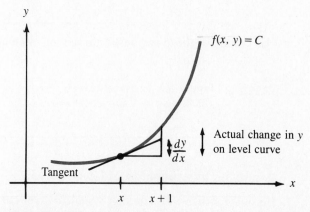

Figure 4.4 The slope of the tangent is approximately the change in y.

and unskilled labor, respectively, the slope $\dfrac{dy}{dx}$ of the tangent to the level curve $f(x,\ y)\ =\ C$ is an approximation to the amount by which the manufacturer should change the level of unskilled labor y to compensate for a 1-unit increase in the level of skilled labor x so that output will remain unchanged.

How to Compute $\dfrac{dy}{dx}$

One way to compute the slope $\dfrac{dy}{dx}$ of (the tangent to) a level curve is to solve the equation $f(x,\ y)\ =\ C$ for y in terms of x, and then differentiate the resulting expression with respect to x. Unfortunately, it is often difficult or even impossible to solve the equation $f(x,\ y)\ =\ C$ explicitly for y. In such cases, you can either differentiate the equation $f(x,\ y)\ =\ C$ implicitly (as in Chapter 2, Section 6) or use the following formula involving the partial derivatives of f.

Formula for the Slope of a Level Curve

> If the level curve $f(x,\ y)\ =\ C$ is the graph of a differentiable function of x, the slope of its tangent is given by the formula
>
> $$\frac{dy}{dx} = -\frac{f_x}{f_y}$$

All three methods are illustrated in the next example.

EXAMPLE 4.3   Find the slope $\dfrac{dy}{dx}$ of the level curve $f(x,\ y)\ =\ C$ if $f(x,\ y)\ =\ xy$ by

(a) Solving for y in terms of x and differentiating the resulting expression.
(b) Differentiating the equation $f(x,\ y)\ =\ C$ implicitly.
(c) Applying the formula.

SOLUTION
(a) Solve the equation $xy\ =\ C$ for y to get

$$y = \frac{C}{x}$$

and differentiate to get

$$\frac{dy}{dx} = -\frac{C}{x^2}$$

(b) Use the product rule to differentiate both sides of the equation

$$xy = C$$

with respect to x, keeping in mind that y is really a function of x. You get

$$x\frac{dy}{dx} + y = 0 \quad \text{or} \quad \frac{dy}{dx} = -\frac{y}{x}$$

which is equivalent to the answer in part (a) since $y = \dfrac{C}{x}$.

(c) Since $f_x = y$ and $f_y = x$, the formula gives

$$\frac{dy}{dx} = -\frac{f_x}{f_y} = -\frac{y}{x}$$

which is the same answer obtained in part (b).

Here is another example illustrating the use of the formula.

EXAMPLE 4.4   Find the slope of the level curve $f(x, y) = 6$ when $x = 2$ if $f(x, y) = x^2y^3 - 5y^3 - x$.

SOLUTION
By the formula,

$$\frac{dy}{dx} = -\frac{f_x}{f_y} = -\frac{2xy^3 - 1}{3x^2y^2 - 15y^2}$$

The desired slope is the value of this derivative when $x = 2$. Before you can compute this slope, you have to find the value of y that corresponds to $x = 2$ on the level curve $f(x, y) = 6$. To do this, substitute $x = 2$ and $f = 6$ into the equation for f to get

$$6 = 4y^3 - 5y^3 - 2 \quad y^3 = -8 \quad \text{or} \quad y = -2$$

Now substitute $x = 2$ and $y = -2$ into the formula for $\dfrac{dy}{dx}$ to get

$$\text{Slope of level curve} = -\frac{2(2)(-2)^3 - 1}{3(2)^2(-2)^2 - 15(-2)^2} = -\frac{-33}{-12} = -\frac{11}{4}$$

Why the Formula Works

The proof that $\dfrac{dy}{dx} = -\dfrac{f_x}{f_y}$ is short, but subtle. Here is the argument.

Suppose the equation $f(x, y) = C$ implicitly defines y as a differentiable function of x. Think of f as a function of the single variable x and differentiate both sides of the equation $f = C$ with respect to x to get

$$\frac{df}{dx} = \frac{dC}{dx} \qquad \text{or} \qquad \frac{df}{dx} = 0 \qquad \text{since } C \text{ is a constant}$$

But, according to the chain rule for partial derivatives,

$$\frac{df}{dx} = \frac{\partial f}{\partial x}\frac{dx}{dx} + \frac{\partial f}{\partial y}\frac{dy}{dx} = \frac{\partial f}{\partial x} + \frac{\partial f}{\partial y}\frac{dy}{dx}$$

Substitute this expression for $\frac{df}{dx}$ into the equation $\frac{df}{dx} = 0$ and conclude, as required, that

$$\frac{\partial f}{\partial x} + \frac{\partial f}{\partial y}\frac{dy}{dx} = 0 \qquad \text{or} \qquad \frac{dy}{dx} = -\frac{\partial f/\partial x}{\partial f/\partial y} = -\frac{f_x}{f_y}$$

Applications

In each of the following two examples from economics, the slope of a level curve is used in the approximation of the change in one variable that is needed to offset a small change in the other variable so that the value of the function will remain unchanged. The procedure is based on the one-variable approximation formula from Chapter 2, Section 4, which is restated here for level curves using the partial derivative formula for $\frac{dy}{dx}$.

Approximation on Level Curves

The change in y needed to offset a small change Δx in x so that the value of the function $f(x, y)$ will remain unchanged is

$$\Delta y \simeq \frac{dy}{dx}\,\Delta x = -\frac{f_x}{f_y}\,\Delta x$$

Constant-Production Curves

The next example illustrates how the slope of a constant-production curve can be used to make decisions concerning the allocation of labor.

EXAMPLE 4.5 Using x worker-hours of skilled labor and y worker-hours of unskilled labor, a manufacturer can produce $f(x, y) = x^2y$ units. Currently 16 worker-hours of skilled labor and 32 worker-hours of unskilled labor are used, and the manufacturer is planning to increase skilled labor by $\frac{1}{4}$ worker-hour. Use calculus to estimate the corresponding change that the manufacturer should make in the level of unskilled labor so that the total output will remain the same.

SOLUTION

The present output is $f(16, 32) = 8,192$ units. The combinations of x and y for which output will remain at this level are the coordinates of the points that lie on the constant-production curve $f(x, y) = 8,192$ (Figure 4.5).

By the approximation formula, the change Δy in unskilled labor that should be made to offset a change of $\Delta x = \frac{1}{4}$ in skilled labor so that output will remain unchanged is

$$\Delta y \simeq \frac{dy}{dx} \Delta x = \left(-\frac{f_x}{f_y}\right)\left(\frac{1}{4}\right) = \left(-\frac{2xy}{x^2}\right)\left(\frac{1}{4}\right) = -\frac{y}{2x}$$

Evaluation at the current levels $x = 16$ and $y = 32$ gives

$$\Delta y \simeq -\frac{32}{2(16)} = -1$$

That is, to compensate for the proposed increase of $\frac{1}{4}$ worker-hour of skilled labor, the manufacturer should reduce the level of unskilled labor by approximately 1 worker-hour.

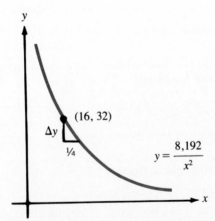

Figure 4.5
The constant-production
curve $f = 8,192$,
where $f(x, y) = x^2y$.

Indifference Curves

A classic problem in economic theory concerns a consumer who is considering the purchase of a number of units of each of two commodities. Associated with the consumer is a **utility function** $U(x, y)$, which measures the total satisfaction (or **utility**) the consumer derives from having x units of the first commodity and y units of the second. The problem is to decide how many units of each commodity the consumer should buy to maximize utility while staying within a fixed budget.

In Section 6, you shall see how to solve this maximization problem. The solution involves the notion of an **indifference curve,** which is simply a level curve $U(x, y) = C$ of the utility function. That is, an indifference curve gives all the combinations of x and y that lead to the same level of consumer satisfaction. To make this concept more concrete, here is a typical example involving the slope of an indifference curve.

EXAMPLE 4.6 Suppose that the utility derived by a consumer from x units of one commodity and y units of a second commodity is given by the utility function $U(x, y) = x^{3/2}y$. The consumer currently owns $x = 16$ units of the first commodity and $y = 20$ units of the second. Use calculus to estimate how many units of the second commodity the consumer could substitute for 1 unit of the first commodity without changing the total utility.

SOLUTION

The current level of utility is $U(16, 20) = 1{,}280$. The corresponding indifference curve is sketched in Figure 4.6.

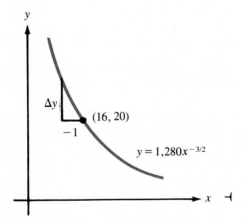

Figure 4.6 The indifference curve $U = 1{,}280$, where $U(x, y) = x^{3/2}y$.

The goal is to estimate the change Δy required to offset a change of $\Delta x = -1$ so that utility will remain at its current level. The approximation formula

$$\Delta y \approx \frac{dy}{dx}\,\Delta x = -\frac{dy}{dx} = \frac{U_x}{U_y} = \frac{\frac{3}{2}x^{1/2}y}{x^{3/2}} = \frac{3y}{2x}$$

with $x = 16$ and $y = 20$ gives

$$\Delta y \simeq \frac{3(20)}{2(16)} = 1.875 \text{ units}$$

That is, the consumer could substitute 1.875 units of the second commodity for 1 unit of the first commodity without affecting the level of utility.

Problems

In Problems 1 through 7, sketch the indicated level curves.

1. $f(x, y) = x + 2y; f = 1, f = 2, f = 3$

2. $f(x, y) = x^2 + y; f = 0, f = 4, f = 9$

3. $f(x, y) = x^2 - 4x - y; f = -4, f = 5$

4. $f(x, y) = \frac{x}{y}; f = -2, f = 2$

5. $f(x, y) = xy; f = 1, f = -1, f = 2, f = -2$

6. $f(x, y) = ye^x; f = 0, f = 1$

7. $f(x, y) = xe^y; f = 1, f = e$

In Problems 8 through 13, use the formula $\dfrac{dy}{dx} = -\dfrac{f_x}{f_y}$ to find the slope of the level curve $f = C$, where C is a constant. Check your answer by differentiating implicitly.

8. $f(x, y) = x^2 - 4x - y$

9. $f(x, y) = \dfrac{x}{y}$

10. $f(x, y) = xe^y$

11. $f(x, y) = x^2y + 2y^3 - 3x - 2y$

12. $f(x, y) = x + 1 - 2xy^2$

13. $f(x, y) = x \ln y$

In Problems 14 through 18, find the slope of the indicated level curve at the specified value of x for the given function.

14. $f = 1; x = 3; f(x, y) = x^2 - y^3$

15. $f = 8; x = 0; f(x, y) = x^2 + xy + y^3$

16. $f = 9; x = 1; f(x, y) = x^2y - 3xy + 5$

17. $f = 8; x = -1; f(x, y) = (x^2 + y)^3$

18. $f = 2; x = \dfrac{1}{2}; f(x, y) = \dfrac{e^y}{x}$

Allocation of labor 19. Using x hours of skilled labor and y hours of unskilled labor, a manufacturer can produce $f(x, y) = 10xy^{1/2}$ units. Currently the manufacturer uses 30 hours of skilled labor and 36 hours of unskilled labor and is planning to use 1 additional hour of skilled labor. Use calculus to estimate the corresponding change that the manufacturer should make in the level of unskilled labor so that the total output will remain the same.

Allocation of labor 20. Suppose the manufacturer in Problem 19 currently uses 30 hours of skilled labor and 36 hours of unskilled labor and is planning to use 1 additional hour of *unskilled* labor. Use calculus to estimate the corresponding change that should be made in the level of skilled labor so that the total output will remain the same. (*Hint*: Use the derivative $\frac{dx}{dy}$.)

Allocation of resources 21. At a certain factory, the daily output is $Q = 200K^{1/2}L^{1/3}$ units, where K denotes the capital investment measured in units of $1,000 and L the size of the labor force measured in worker-hours. The current level of capital investment is $60,000 and the current size of the labor force is 10,000 worker-hours. The manufacturer is planning to increase the capital investment by $500. Use calculus to estimate the corresponding decrease that the manufacturer can make in the size of the labor force without affecting the daily output.

Allocation of resources 22. At a certain factory, output Q is related to inputs x and y by the function $Q = 2x^3 + 3x^2y + y^3$. If the current levels of input are $x = 20$ and $y = 10$, use calculus to estimate the change in input x that should be made to offset an increase of 0.5 unit in input y so that output will be maintained at its current level.

Utility 23. Suppose that the utility derived by a consumer from x units of one commodity and y units of a second commodity is given by the utility function $U(x, y) = 2x^3y^2$. The consumer currently owns $x = 5$ units of the first commodity and $y = 4$ units of the second. Use calculus to estimate how many units of the second commodity the consumer could substitute for 1 unit of the first commodity without affecting total utility.

Utility 24. Suppose that the utility derived by a consumer from x units of one commodity and y units of a second commodity is given by the utility function $U(x, y) = (x + 1)(y + 2)$. The consumer currently owns $x = 25$ units of the first commodity and $y = 8$ units of the second. Use calculus to estimate how many units of the first commodity the consumer could substitute for 1 unit of the second commodity without affecting total utility.

5   Relative Maxima and Minima

In this section you will learn how to use partial derivatives to find the relative maxima and minima of functions of two variables.

In geometric terms, a **relative maximum** of a function $f(x, y)$ is a peak, a point on the surface $z = f(x, y)$ that is higher than any nearby point on the surface. A **relative minimum** is the bottom of a valley, a point that is lower than any nearby point on the surface. For example, when $(x, y) = (a, b)$, the function sketched in Figure 5.1a has a relative minimum while the function sketched in Figure 5.1b has a relative maximum.

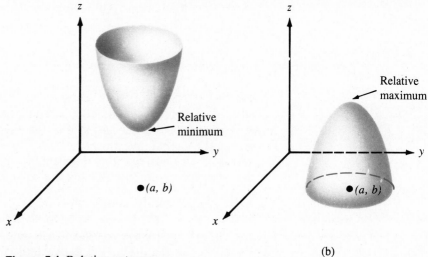

(b)

Figure 5.1 Relative extrema.

Critical Points

The points (a, b) in the domain of $f(x, y)$ for which both $f_x(a, b) = 0$ and $f_y(a, b) = 0$ are said to be **critical points** of f. Like the critical points for functions of one variable, these critical points play an important role in the study of relative maxima and minima.

To see the connection between critical points and relative extrema, suppose $f(x, y)$ has a relative maximum at (a, b). Then the curve formed by intersecting the surface $z = f(x, y)$ with the vertical plane $y = b$ has a relative maximum and hence a horizontal tangent when $x = a$. (See Figure 5.2a.) Since the partial derivative $f_x(a, b)$ is the slope of this tangent, it follows that $f_x(a, b) = 0$. Similarly, the curve formed by intersecting the surface $z = f(x, y)$ with the plane $x = a$ has a relative maximum when $y = b$ (see Figure 5.2b), and so $f_y(a, b) = 0$. This shows that a point at which a function of two variables has a relative maximum must be a critical point. A similar argument shows that a point at which a

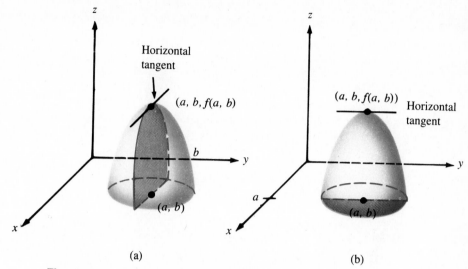

(a)   (b)

Figure 5.2 The partial derivatives are zero at a relative extremum.

function of two variables has a relative minimum must also be a critical point.

Here is a more precise statement of the situation.

Critical Points and Relative Extrema

A point (a, b) in the domain of $f(x, y)$ for which both

$$f_x(a, b) = 0 \quad \text{and} \quad f_y(a, b) = 0$$

is said to be a critical point of f.

If the first-order partial derivatives of f are defined at all points in some region in the xy plane, then the relative extrema of f in the region can occur only at critical points.

Although all the relative extrema of a function must occur at critical points, not every critical point of a function is necessarily a relative extremum. Consider, for example, the function $f(x, y) = y^2 - x^2$, whose graph, which resembles a saddle, is sketched in Figure 5.3. In this case, $f_x(0, 0) = 0$ because the surface has a relative *maximum* (and hence a horizontal tangent) "in the x direction," and $f_y(0, 0) = 0$ because the surface has a relative *minimum* (and hence a horizontal tangent) "in the y direction." Hence $(0, 0)$ is a critical point of f, but it is not a relative extremum. For a critical point to be a relative extremum, the nature of the extremum must be the same in *all directions*. A critical point that is neither a relative maximum nor a relative minimum is called a **saddle point**.

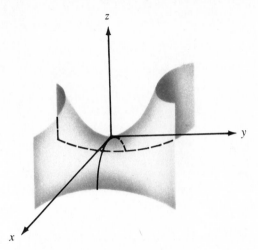

Figure 5.3
The surface $z = y^2 - x^2$.

The Second-Derivative Test

Here is a procedure involving second-order partial derivatives that you can use to decide whether a given critical point is a relative maximum, a relative minimum, or a saddle point. This procedure is the two-variable version of the second-derivative test for functions of a single variable that you saw in Chapter 3, Section 2.

The Second-Derivative Test

Suppose that (a, b) is a critical point of the function $f(x, y)$. Let

$$D = f_{xx}(a, b)f_{yy}(a, b) - [f_{xy}(a, b)]^2$$

If $D < 0$, then f has a saddle point at (a, b).

If $D > 0$ and $f_{xx}(a, b) < 0$, then f has a relative maximum at (a, b).

If $D > 0$ and $f_{xx}(a, b) > 0$, then f has a relative minimum at (a, b).

If $D = 0$, the test is inconclusive and f may have either a relative extremum or a saddle point at (a, b).

The sign of the quantity D that appears in the second-derivative test tells you whether the function has a relative extremum at (a, b): if D is positive, (a, b) is a relative extremum and if D is negative, (a, b) is a saddle point.

Moreover, if D is positive [guaranteeing that (a, b) is a relative extremum], the function has either a relative maximum *in all directions* or a relative minimum *in all directions*. To decide which, you can restrict your attention to any *one* direction (say the x direction) and use the one-variable second-derivative test to conclude that f has a relative maximum at (a, b) if $f_{xx}(a, b)$ is negative and a relative minimum at (a, b) if $f_{xx}(a, b)$ is positive.

The proof of the second-derivative test involves ideas that are beyond the scope of this text and will be omitted. Here are some examples illustrating the use of this test.

EXAMPLE 5.1　Classify the critical point of the function $f(x, y) = x^2 + y^2$.

SOLUTION
Since

$$f_x = 2x \quad \text{and} \quad f_y = 2y$$

the only critical point of f is $(0, 0)$. To test this point, use the second-order partial derivatives

$$f_{xx} = 2 \quad f_{yy} = 2 \quad \text{and} \quad f_{xy} = 0$$

to get

$$D(x, y) = f_{xx}f_{yy} - (f_{xy})^2 = 2(2) - 0 = 4$$

That is, $D(x, y) = 4$ for *all* points (x, y) and, in particular,

$$D(0, 0) = 4 > 0$$

Hence, f has a relative extremum at $(0, 0)$. Moreover, since

$$f_{xx}(0, 0) = 2 > 0$$

it follows that the relative extremum at $(0, 0)$ is a relative minimum.
For reference, the graph of f is sketched in Figure 5.4.

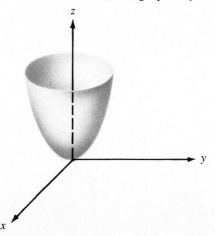

Figure 5.4
The surface $z = x^2 + y^2$ with a relative minimum at $(0, 0)$.

EXAMPLE 5.2   Classify the critical points of the function $f(x, y) = y^2 - x^2$.

SOLUTION
Since

$$f_x = -2x \qquad \text{and} \qquad f_y = 2y$$

the only critical point of f is $(0, 0)$. To test this point, use the second-order partial derivatives

$$f_{xx} = -2 \quad f_{yy} = 2 \qquad \text{and} \qquad f_{xy} = 0$$

to get

$$D(x, y) = f_{xx}f_{yy} - (f_{xy})^2 = -2(2) - 0 = -4$$

That is, $D(x, y) = -4$ for all points (x, y) and, in particular,

$$D(0, 0) = -4 < 0$$

It follows that f must have a saddle point at $(0, 0)$.
The graph of f is shown in Figure 5.5.

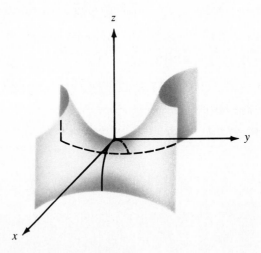

Figure 5.5 The surface $z = y^2 - x^2$ with a saddle point at $(0, 0)$.

Solving the equations $f_x = 0$ and $f_y = 0$ simultaneously to find the critical points of a function of two variables is rarely as simple as in Examples 5.1 and 5.2. The algebra in the next example is more typical. Before proceeding, you may wish to refer to the algebra review in the appendix in which techniques for solving systems of two equations in two unknowns are discussed.

EXAMPLE 5.3   Classify the critical points of the function $f(x, y) = x^3 - y^3 + 6xy$.

SOLUTION
Since

$$f_x = 3x^2 + 6y \quad \text{and} \quad f_y = -3y^2 + 6x$$

you find the critical points of f by solving simultaneously the two equations

$$3x^2 + 6y = 0 \quad \text{and} \quad -3y^2 + 6x = 0$$

From the first equation, you get

$$y = -\frac{x^2}{2}$$

which you can substitute into the second equation to get

$$-\frac{3x^4}{4} + 6x = 0 \quad \text{or} \quad -x(x^3 - 8) = 0$$

The solutions of this equation are $x = 0$ and $x = 2$. These are the x coordinates of the critical points of f. To get the corresponding y coordinates, substitute these values of x into the equation $y = -\frac{x^2}{2}$ (or into either one of the two original equations). You will find that $y = 0$ when $x = 0$ and $y = -2$ when $x = 2$. It follows that the critical points of f are $(0, 0)$ and $(2, -2)$.

The second-order partial derivatives of f are

$$f_{xx} = 6x \quad f_{yy} = -6y \quad \text{and} \quad f_{xy} = 6$$

Hence,

$$D(x, y) = f_{xx}f_{yy} - (f_{xy})^2 = -36xy - 36 = -36(xy + 1)$$

Since

$$D(0, 0) = -36[0(0) + 1] = -36 < 0$$

it follows that f has a saddle point at $(0, 0)$. Since

$$D(2, -2) = -36[2(-2) + 1] = 108 > 0$$

and

$$f_{xx}(2, -2) = 6(2) = 12 > 0$$

it follows that f has a relative minimum at $(2, -2)$.

Practical Optimization Problems

In the next example, you will see how to apply the theory of relative extrema to solve an optimization problem from economics. Actually, you will be trying to find the *absolute* maximum of a certain function. It turns out, however, that the absolute and relative maxima of this function coincide. In fact, in the majority of two-variable optimization problems in the social sciences, the relative extrema and absolute extrema coincide. For this reason, the theory of absolute extrema for functions of two variables will not be developed in this text, and you may assume that the relative extremum you find as the solution of a practical optimization problem is actually the absolute extremum.

Maximization of Profit

EXAMPLE 5.4   The only grocery store in a small rural community carries two brands of frozen orange juice, a local brand that it obtains at the cost of 30 cents per can and a well-known national brand that it obtains at the cost of 40 cents per can. The grocer estimates that if the local brand is sold for x cents per can and the national brand for y cents per can, approximately $70 - 5x + 4y$ cans of the local brand and $80 + 6x - 7y$ cans of the national brand will be sold each day. How should the grocer price each brand to maximize the profit from the sale of the juice? (Assume that the absolute maximum and the relative maximum of the profit function are the same.)

SOLUTION
Since

$$\text{Profit} = \frac{\text{profit from the sale}}{\text{of the local brand}} + \frac{\text{profit from the sale}}{\text{of the national brand}}$$

it follows that the total daily profit from the sale of the juice is given by the function

$$f(x, y) = (x - 30)(70 - 5x + 4y) + (y - 40)(80 + 6x - 7y)$$

$$= -5x^2 + 10xy - 20x - 7y^2 + 240y - 5{,}300$$

Compute the partial derivatives

$$f_x = -10x + 10y - 20 \quad \text{and} \quad f_y = 10x - 14y + 240$$

set them equal to zero to get

$$-10x + 10y - 20 = 0 \quad \text{and} \quad 10x - 14y + 240 = 0$$

or

$$-x + y = 2 \quad \text{and} \quad 5x - 7y = -120$$

and solve these equations simultaneously to get

$$x = 53 \quad \text{and} \quad y = 55$$

It follows that (53, 55) is the only critical point of f.

Now use the second-order partial derivatives

$$f_{xx} = -10 \qquad f_{yy} = -14 \qquad \text{and} \qquad f_{xy} = 10$$

to get

$$D(x, y) = f_{xx}f_{yy} - (f_{xy})^2 = -10(-14) - 100 = 40$$

Since

$$D(53, 55) = 40 > 0 \qquad \text{and} \qquad f_{xx}(53, 55) = -10 < 0$$

it follows that f has a (relative) maximum when $x = 53$ and $y = 55$. That is, the grocer can maximize profit by selling the local brand of juice for 53 cents per can and the national brand for 55 cents per can.

Problems

In Problems 1 through 7, find the critical points of the given function and classify them as relative maxima, relative minima, or saddle points.

1. $f(x, y) = 5 - x^2 - y^2$

2. $f(x, y) = 2x^2 - 3y^2$

3. $f(x, y) = xy$

4. $f(x, y) = xy + \dfrac{8}{x} + \dfrac{8}{y}$

5. $f(x, y) = 2x^3 + y^3 + 3x^2 - 3y - 12x - 4$

6. $f(x, y) = (x - 1)^2 + y^3 - 3y^2 - 9y + 5$

7. $f(x, y) = x^3 + y^2 - 6xy + 9x + 5y + 2$

In Problems 8 through 12, assume that the absolute maximum and the relative maximum of the profit function are the same.

Retail sales 8. A liquor store carries two competing brands of inexpensive wine, one from California and the other from New York. The owner of the store can obtain both wines at a cost of \$2 per bottle and estimates that if the California wine is sold for x dollars per bottle and the New York wine for y dollars per bottle, consumers will buy approximately $40 - 50x + 40y$ bottles of the California wine and $20 + 60x - 70y$ bottles of the New York wine each day. How should the owner price the wines to generate the largest possible profit?

Pricing 9. The telephone company is planning to introduce two new types of executive communications systems that it hopes to sell to its largest commercial customers. It is estimated that if the first type of system is priced at x hundred dollars per system and the second type at y

hundred dollars per system, approximately $40 - 8x + 5y$ consumers will buy the first type and $50 + 9x - 7y$ will buy the second type. If the cost of manufacturing the first type is \$1,000 per system and the cost of manufacturing the second type is \$3,000 per system, how should the telephone company price the systems to generate the largest possible profit?

Allocation of funds 10. A manufacturer is planning to sell a new product at the price of \$150 per unit and estimates that if x thousand dollars is spent on development and y thousand dollars is spent on promotion, consumers will buy approximately $\dfrac{320y}{y + 2} + \dfrac{160x}{x + 4}$ units of the product. If manufacturing costs for this product are \$50 per unit, how much should the manufacturer spend on development and how much on promotion to generate the largest possible profit from the sale of this product? [*Hint*: Profit = (number of units)(price per unit − cost per unit) − total amount spent on development and promotion.]

Profit under monopoly 11. A manufacturer with exclusive rights to a sophisticated new industrial machine is planning to sell a limited number of the machines to both foreign and domestic firms. The price the manufacturer can expect to receive for the machines will depend on the number of machines made available. (For example, if only a few of the machines are placed on the market, competitive bidding among prospective purchasers will tend to drive the price up.) It is estimated that if the manufacturer supplies x machines to the domestic market and y machines to the foreign market, the machines will sell for $60 - \dfrac{x}{5} + \dfrac{y}{20}$ thousand dollars apiece at home and for $50 - \dfrac{y}{10} + \dfrac{x}{20}$ thousand dollars apiece abroad. If the manufacturer can produce the machines at the cost of \$10,000 apiece, how many should be supplied to each market to generate the largest possible profit?

Profit under monopoly 12. A manufacturer with exclusive rights to a new industrial machine is planning to sell a limited number of them and estimates that if x machines are supplied to the domestic market and y to the foreign market, the machines will sell for $150 - \dfrac{x}{6}$ thousand dollars apiece at home and for $100 - \dfrac{y}{10}$ thousand dollars apiece abroad.

(a) How many machines should the manufacturer supply to the domestic market to generate the largest possible profit at home?
(b) How many machines should the manufacturer supply to the foreign market to generate the largest possible profit abroad?
(c) How many machines should the manufacturer supply to each market to generate the largest possible *total* profit?

(d) Is the relationship between the answers in parts (a), (b), and (c) accidental? Explain. Does a similar relationship hold in Problem 11? What accounts for the difference between these two problems in this respect?

Level curves 13. Sometimes one can classify the critical points of a function by inspecting its level curves. In each of the following cases, determine the nature of the critical point of f at $(0, 0)$.

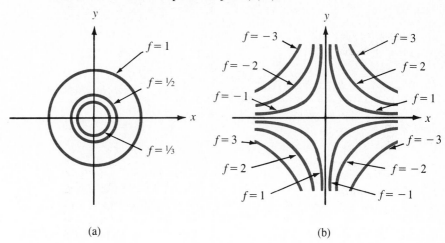

(a)   (b)

14. Let $f(x, y) = x^2 + y^2 - 4xy$. Show that f does *not* have a relative minimum at its critical point $(0, 0)$, even though it does have a relative minimum at $(0, 0)$ in both the x and y directions. (*Hint*: Consider the direction defined by the line $y = x$. That is, substitute x for y in the formula for f and analyze the resulting function of x.)

6 Lagrange Multipliers

Constrained Optimization Problems

In many applied problems, a function of two variables is to be optimized subject to a restriction or **constraint** on the variables. For example, an editor, constrained to stay within a fixed budget of $60,000, may wish to decide how to divide this money between development and promotion in order to maximize the future sales of a new book. If x denotes the amount of money allocated to development, y the amount allocated to promotion, and $f(x, y)$ the corresponding number of books that will be sold, the editor would like to maximize the sales function $f(x, y)$ subject to the budgetary constraint that $x + y = 60,000$.

For a geometric interpretation of the process of optimizing a function of two variables subject to a constraint, think of the function itself as a surface in three-dimensional space and of the constraint (which is an equation involving x and y) as a curve in the xy plane. When you find the maximum or minimum of the function subject to the given constraint, you are restricting your attention to the portion of the surface that lies directly above the constraint curve. The highest point on this portion of the surface is the constrained maximum, and the lowest point is the constrained minimum. The situation is illustrated in Figure 6.1.

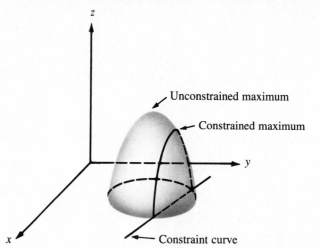

Figure 6.1
Constrained and unconstrained extrema.

You have already seen some constrained optimization problems in Chapter 3 of this text. (See Chapter 3, Section 4, Example 4.2, for instance.) The technique you used in Chapter 3 to solve such a problem was to reduce it to a problem involving only one variable by solving the constraint equation for one of the variables and then substituting the resulting expression into the function that was to be optimized. The success of this technique depended on solving the constraint equation for one of the variables, which, in many problems, is difficult or even impossible to do. In this section you will see a more versatile technique called the **method of Lagrange multipliers,** in which the introduction of a *third* variable allows you to solve constrained optimization problems without first solving the constraint equation for one of the variables. Here is a statement of the method.

The Method of Lagrange Multipliers

Suppose $f(x, y)$ and $g(x, y)$ are functions whose first-order partial derivatives exist. To find the relative maximum and relative minimum of $f(x, y)$ subject to the constraint that $g(x, y) = k$ for some constant k, introduce a new variable λ (the Greek letter lambda) and solve the following three equations simultaneously:

$$f_x(x, y) = \lambda g_x(x, y) \qquad f_y(x, y) = \lambda g_y(x, y) \qquad g(x, y) = k$$

> The desired relative extrema will be found among the resulting points (x, y).

In the next example, the method of Lagrange multipliers will be used to solve the problem from Chapter 3, Section 4, Example 4.2.

EXAMPLE 6.1   The highway department is planning to build a picnic area for motorists along a major highway. It is to be rectangular with an area of 5,000 square yards and is to be fenced off on the three sides not adjacent to the highway. What is the least amount of fencing that will be needed to complete the job?

SOLUTION

Label the sides of the picnic area as indicated in Figure 6.2 and let f denote the amount of fencing required. Then,

$$f(x, y) = x + 2y$$

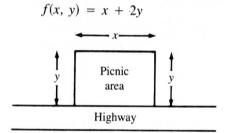

Figure 6.2 Rectangular picnic area.

The goal is to minimize f subject to the constraint that the area must be 5,000 square yards; that is, subject to the constraint

$$xy = 5,000$$

Let $g(x, y) = xy$ and use the partial derivatives

$$f_x = 1 \qquad f_y = 2 \qquad g_x = y \qquad \text{and} \qquad g_y = x$$

to write the three Lagrange equations

$$1 = \lambda y \qquad 2 = \lambda x \qquad \text{and} \qquad xy = 5,000$$

From the first and second equations you get

$$\lambda = \frac{1}{y} \qquad \text{and} \qquad \lambda = \frac{2}{x}$$

(since $y \neq 0$ and $x \neq 0$), which implies that

$$\frac{1}{y} = \frac{2}{x} \qquad \text{or} \qquad x = 2y$$

Now substitute $x = 2y$ into the third Lagrange equation to get

$$2y^2 = 5,000 \quad \text{or} \quad y = \pm 50$$

and finally use $y = 50$ in the equation $x = 2y$ to get $x = 100$. It follows that $x = 100$ and $y = 50$ are the values that minimize the function $f(x, y) = x + 2y$ subject to the constraint that $xy = 5,000$. That is, the optimal picnic area is 100 yards wide (along the highway), extends 50 yards back from the road, and requires $100 + 50 + 50 = 200$ yards of fencing.

Strictly speaking the solution of the preceding example was incomplete since no attempt was made to verify that the optimal values $x = 100$ and $y = 50$ actually minimized the function f subject to the given constraint. There is a version of the second-derivative test that could have been used to show that the point (100, 50) was indeed a constrained relative minimum of the function f. And, with a little more work, it could have been shown that this relative minimum was really the absolute constrained minimum. The techniques needed to carry out this analysis are discussed in more advanced texts. In this text, you may assume that the constrained maximum or minimum you get using the method of Lagrange multipliers is the extremum you are seeking.

Before we see why the method of Lagrange multipliers works, here is one more example illustrating its use.

EXAMPLE 6.2   Find the maximum and minimum values of the function $f(x, y) = xy$ subject to the constraint $x^2 + y^2 = 8$.

SOLUTION
Let $g(x, y) = x^2 + y^2$ and use the partial derivatives

$$f_x = y \quad f_y = x \quad g_x = 2x \quad \text{and} \quad g_y = 2y$$

to get the three Lagrange equations

$$y = 2\lambda x \quad x = 2\lambda y \quad \text{and} \quad x^2 + y^2 = 8$$

Neither x nor y can be zero if all three of these equations are to hold (do you see why?), and so you can rewrite the first two equations as

$$2\lambda = \frac{y}{x} \quad \text{and} \quad 2\lambda = \frac{x}{y}$$

which implies that

$$\frac{y}{x} = \frac{x}{y} \quad \text{or} \quad x^2 = y^2$$

Now substitute $x^2 = y^2$ into the third equation to get

$$2x^2 = 8 \quad \text{or} \quad x = \pm 2$$

If $x = 2$, it follows from the equation $x^2 = y^2$ that $y = 2$ or $y = -2$. Similarly, if $x = -2$, it follows that $y = 2$ or $y = -2$. Hence, the four points at which the constrained extrema can occur are $(2, 2)$, $(2, -2)$, $(-2, 2)$, and $(-2, -2)$. Since

$$f(2, 2) = f(-2, -2) = 4 \quad \text{and} \quad f(2, -2) = f(-2, 2) = -4$$

it follows that when $x^2 + y^2 = 8$, the maximum value of $f(x, y)$ is 4, which occurs at the points $(2, 2)$ and $(-2, -2)$, and the minimum value is -4, which occurs at $(2, -2)$ and $(-2, 2)$.

For practice, check these answers by solving the optimization problem using the methods of Chapter 3.

A Word of Advice

Notice that in each of the preceding examples, the first two Lagrange equations were used to eliminate the new variable λ, and then the resulting expression relating x and y was substituted into the third Lagrange equation. For most constrained optimization problems you will encounter, this particular sequence of steps will lead you quickly to the desired solution.

Why the Method of Lagrange Multipliers Works

Although a rigorous explanation of why the method of Lagrange multipliers works involves advanced ideas beyond the scope of this text, there is a rather simple geometric argument that you should find convincing. Suppose the constraint curve $g(x, y) = k$ and the level curves $f(x, y) = C$ are drawn in the xy plane as shown in Figure 6.3.

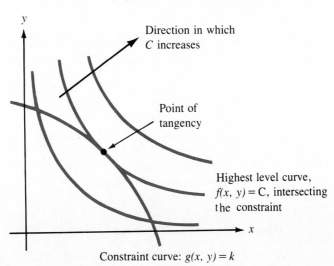

Figure 6.3
Increasing level curves and the constraint curve.

To maximize $f(x, y)$ subject to the constraint $g(x, y) = k$, you must find the highest level curve of f that intersects the constraint curve. As the sketch in Figure 6.3 suggests, this critical intersection will occur at a point at which the constraint curve is tangent to a level curve; that is, at which the slope of the constraint curve $g(x, y) = k$ is equal to the slope of a level curve $f(x, y) = C$. According to the formula you learned in Section 4 of this chapter,

$$\text{Slope of constraint curve} = -\frac{g_x}{g_y}$$

and

$$\text{Slope of level curve} = -\frac{f_x}{f_y}$$

Hence the condition that the slopes be equal can be expressed by the equation

$$-\frac{f_x}{f_y} = -\frac{g_x}{g_y} \quad \text{or, equivalently,} \quad \frac{f_x}{g_x} = \frac{f_y}{g_y}$$

If you let λ denote this common ratio, you have

$$\lambda = \frac{f_x}{g_x} \quad \text{and} \quad \lambda = \frac{f_y}{g_y}$$

from which you get the first two Lagrange equations

$$f_x = \lambda g_x \quad \text{and} \quad f_y = \lambda g_y$$

The third Lagrange equation

$$g(x, y) = k$$

is simply a statement of the fact that the point in question actually lies on the constraint curve.

Suppose all three of the Lagrange equations are satisfied at a certain point (a, b). Then f will reach its constrained *maximum* at (a, b) if the *highest* level curve that intersects the constraint curve does so at this point. On the other hand, if the *lowest* level curve that intersects the constraint curve does so at (a, b), then f will achieve its constrained *minimum* at this point. The situation is illustrated further in Figure 6.4, which shows the constraint curve $x^2 + y^2 = 8$ and the optimal level curves $xy = 4$ and $xy = -4$ from Example 6.2. Notice that in this case there are four points at which the constraint curve is tangent to a level curve. Two of these points, $(2, 2)$ and $(-2, -2)$ maximize f subject to the given constraint while the other two, $(2, -2)$ and $(-2, 2)$, minimize f subject to the constraint.

Applications to Economics

In the next two examples, the method of Lagrange multipliers is used to solve constrained optimization problems from economics.

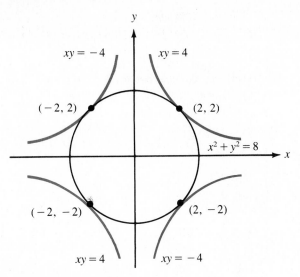

Figure 6.4 Constraint curve and optimal level curves.

Allocation of Funds

EXAMPLE 6.3 An editor has been allotted $60,000 to spend on the development and promotion of a new book. It is estimated that if x thousand dollars is spent on development and y thousand on promotion, approximately $f(x, y) = 20x^{3/2}y$ copies of the book will be sold. How much money should the editor allocate to development and how much to promotion in order to maximize sales?

SOLUTION
The goal is to maximize the function $f(x, y) = 20x^{3/2}y$ subject to the constraint $g(x, y) = 60$, where $g(x, y) = x + y$. The corresponding Lagrange equations are

$$30x^{1/2}y = \lambda \qquad 20x^{3/2} = \lambda \qquad \text{and} \qquad x + y = 60$$

From the first two equations you get

$$30x^{1/2}y = 20x^{3/2}$$

Since the maximum value of f clearly does not occur when $x = 0$, you may assume that $x \neq 0$ and divide both sides of this equation by $30x^{1/2}$ to get

$$y = \frac{2}{3}x$$

Substituting this expression into the third equation, you get

$$x + \frac{2}{3}x = 60 \qquad \text{or} \qquad \frac{5}{3}x = 60$$

from which it follows that

$$x = 36 \quad \text{and} \quad y = 24$$

That is, to maximize sales, the editor should spend $36,000 on development and $24,000 on promotion. If this is done, approximately $f(36, 24) = 103,680$ copies of the book will be sold.

A graph showing the relationship between the budgetary constraint and the level curve for optimal sales is sketched in Figure 6.5.

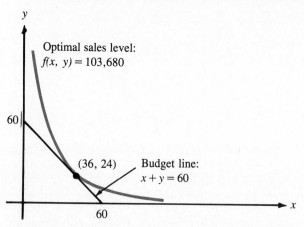

Figure 6.5 Budgetary constraint and optimal sales level.

Maximization of Utility

In Section 4 you were introduced to a classic problem in economics in which a consumer is considering the purchase of a number of units of each of two commodities. Associated with the consumer is a **utility function** $U(x, y)$, which measures the total satisfaction or **utility** the consumer receives from having x units of the first commodity and y units of the second. The problem is to determine how many units of each commodity the consumer should buy to maximize utility while staying within a fixed budget. The application of the method of Lagrange multipliers to the utility problem is illustrated in the next example.

EXAMPLE 6.4 A consumer has $600 to spend on two commodities, the first of which costs $20 per unit and the second $30 per unit. Suppose that the utility derived by the consumer from x units of the first commodity and y units of the second commodity is given by the **Cobb-Douglas utility function** $U(x, y) = 10x^{0.6}y^{0.4}$. How many units of each commodity should the consumer buy to maximize utility?

SOLUTION

The total cost of buying x units of the first commodity at \$20 per unit and y units of the second at \$30 per unit is $20x + 30y$. Since the consumer has only \$600 to spend, the goal is to maximize utility $U(x, y)$ subject to the budgetary constraint that $20x + 30y = 600$.

The three Lagrange equations are

$$6x^{-0.4}y^{0.4} = 20\lambda \qquad 4x^{0.6}y^{-0.6} = 30\lambda \qquad \text{and} \qquad 20x + 30y = 600$$

From the first two equations you get

$$\frac{6x^{-0.4}y^{0.4}}{20} = \frac{4x^{0.6}y^{-0.6}}{30}$$

$$9x^{-0.4}y^{0.4} = 4x^{0.6}y^{-0.6}$$

$$9y = 4x \qquad \text{or} \qquad y = \frac{4}{9}x$$

Substituting this into the third equation you get

$$20x + 30\left(\frac{4}{9}x\right) = 600$$

from which it follows that

$$x = 18 \qquad \text{and} \qquad y = 8$$

That is, to maximize utility, the consumer should buy 18 units of the first commodity and 8 units of the second.

Recall from Section 4 that the level curves of a utility function are known as **indifference curves.** A graph showing the relationship between the optimal indifference curve $U(x, y) = C$, where $C = U(18, 8)$ and the budgetary constraint $20x + 30y = 600$ is sketched in Figure 6.6.

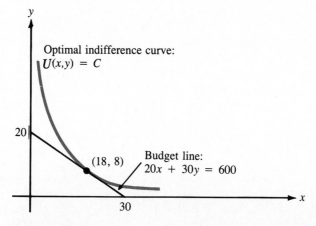

Figure 6.6 Budgetary constraint and optimal indifference curve.

The Significance of the Lagrange Multiplier λ

You can solve most constrained optimization problems by the method of Lagrange multipliers without actually obtaining a numerical value for the **Lagrange multiplier** λ. In some problems, however, you may want to compute λ. This is because λ has the following useful interpretation.

The Lagrange Multiplier

Suppose M is the maximum (or minimum) value of $f(x, y)$ subject to the constraint $g(x, y) = k$. The Lagrange multiplier λ is the rate of change of M with respect to k. That is,

$$\lambda = \frac{dM}{dk}$$

Hence,

$\lambda \simeq$ change in M resulting from a 1-unit increase in k

The proof of this fact will be given after the following examples illustrating its use.

EXAMPLE 6.5   Suppose the editor in Example 6.3 is allotted \$60,200 instead of \$60,000 to spend on the development and promotion of the new book. Estimate how the additional \$200 will affect the maximum sales level.

SOLUTION
In Example 6.3, you solved the three Lagrange equations

$$30x^{1/2}y = \lambda \qquad 20x^{3/2} = \lambda \qquad \text{and} \qquad x + y = 60$$

to conclude that the maximum value M of $f(x, y)$ subject to the constraint $x + y = 60$ occurred when $x = 36$ and $y = 24$. To find λ, substitute these values of x and y into the first or second Lagrange equations. Using the second equation, you get

$$\lambda = 20(36)^{3/2} = 4{,}320$$

The goal is to estimate the change ΔM in the maximal sales that will result from an increase of $\Delta k = 0.2$ (thousand dollars) in the available funds. Since $\lambda = \dfrac{dM}{dk}$, the one-variable approximation formula from Chapter 2, Section 4 gives

$$\Delta M \simeq \frac{dM}{dk} \, \Delta k = \lambda \, \Delta k = 4{,}320(0.2) = 864$$

That is, maximal sales of the book will increase by approximately 864 copies if the budget is increased from $60,000 to $60,200 and the money is allocated optimally.

In the context of the maximization of utility subject to a budgetary constraint, the Lagrange multiplier λ is the approximate change in maximum utility resulting from a 1-unit increase in the budget and is known as the **marginal utility of money.** Here is an example.

EXAMPLE 6.6 Suppose the consumer in Example 6.4 has $601 instead of $600 to spend on the two commodities. Estimate how the additional $1 will affect the maximum utility.

SOLUTION
From the first Lagrange equation

$$6x^{-0.4}y^{0.4} = 20\lambda$$

in Example 6.4 it follows that

$$\lambda = \frac{6x^{-0.4}y^{0.4}}{20} = \frac{3y^{0.4}}{10x^{0.4}} = 0.3\left(\frac{y}{x}\right)^{0.4}$$

Since the maximum value M of utility when $600 was available occurred when $x = 18$ and $y = 8$, substitute these values into the formula for λ to get

$$\lambda = 0.3\left(\frac{8}{18}\right)^{0.4} \simeq 0.22$$

which is approximately the increase ΔM in maximum utility resulting from the $1 increase in available funds.

Proof that $\lambda = \dfrac{dM}{dk}$

The proof that $\lambda = \dfrac{dM}{dk}$ is based on the chain rule for partial derivatives from Section 3. Here is the argument.

Suppose M is the optimal value of f subject to the constraint $g = k$. Then, $M = f(x, y)$ for some ordered pair (x, y) that satisfies the three Lagrange equations:

$$f_x = \lambda g_x \qquad f_y = \lambda g_y \qquad \text{and} \qquad g = k$$

Moreover, the coordinates of the optimal ordered pair (x, y) depend on k (since different constraint levels will generally lead to different optimal combinations of x and y). Thus,

$$M = f(x, y) \qquad \text{(where } x \text{ and } y \text{ are functions of } k\text{)}$$

By the chain rule for partial derivatives,

$$\frac{dM}{dk} = \frac{\partial M}{\partial x}\frac{dx}{dk} + \frac{\partial M}{\partial y}\frac{dy}{dk}$$

$$= f_x\frac{dx}{dk} + f_y\frac{dy}{dk} \qquad \text{since } M = f(x, y)$$

$$= \lambda g_x\frac{dx}{dk} + \lambda g_y\frac{dy}{dk} \qquad \text{by the first two Lagrange equations}$$

$$= \lambda\left(g_x\frac{dx}{dk} + g_y\frac{dy}{dk}\right)$$

$$= \lambda\frac{dg}{dk} \qquad \text{by the chain rule applied to } g(x, y)$$

$$= \lambda(1) = \lambda \qquad \text{since } g = 1 \cdot \lambda \text{ by the third Lagrange equation}$$

and the proof is complete.

Problems

Use the method of Lagrange multipliers to solve the following problems.

1. Find the maximum value of the function $f(x, y) = xy$ subject to the constraint $x + y = 1$.

2. Find the maximum and minimum values of the function $f(x, y) = xy$ subject to the constraint $x^2 + y^2 = 1$.

3. Find the minimum value of the function $f(x, y) = x^2 + y^2$ subject to the constraint $xy = 1$.

4. Find the minimum value of the function $f(x, y) = x^2 + 2y^2 - xy$ subject to the constraint $2x + y = 22$.

5. Find the minimum value of the function $f(x, y) = x^2 - y^2$ subject to the constraint $x^2 + y^2 = 4$.

6. Find the maximum and minimum values of the function $f(x, y) = 8x^2 - 24xy + y^2$ subject to the constraint $x^2 + y^2 = 1$.

Construction   7. A farmer wishes to fence off a rectangular pasture along the bank of a river. The area of the pasture is to be 3,200 square meters, and no fencing is needed along the river bank. Find the dimensions of the pasture that will require the least amount of fencing.

Construction   8. There are 320 meters of fencing available to enclose a rectangular field. How should the fencing be used so that the enclosed area is as large as possible?

Postal service

9. According to postal regulations, the girth plus length of parcels sent by fourth-class mail may not exceed 72 inches. What is the largest possible volume of a rectangular parcel with two square sides that can be sent by fourth-class mail?

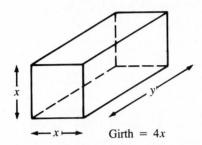

Girth = $4x$

Packaging

10. Use the fact that 12 fluid ounces is (approximately) 6.89π cubic inches to find the dimensions of the 12-ounce beer can that can be constructed using the least amount of metal. (Recall that the volume of a cylinder of radius r and height h is $\pi r^2 h$, that the circumference of a circle of radius r is $2\pi r$, and that the area of a circle of radius r is πr^2.)

Packaging

11. A cylindrical can is to hold 4π cubic inches of frozen orange juice. The cost per square inch of constructing the metal top and bottom is twice the cost per square inch of constructing the cardboard side. What are the dimensions of the least expensive can?

Allocation of funds

12. A manufacturer has $8,000 to spend on the development and promotion of a new product. It is estimated that if x thousand dollars is spent on development and y thousand is spent on promotion, sales will be approximately $f(x, y) = 50x^{1/2}y^{3/2}$ units. How much money should the manufacturer allocate to development and how much to promotion to maximize sales?

Allocation of funds

13. If x thousand dollars is spent on labor and y thousand dollars is spent on equipment, the output at a certain factory will be $Q(x, y) = 60x^{1/3}y^{2/3}$ units. If $120,000 is available, how should this be allocated between labor and equipment to generate the largest possible output?

Allocation of funds

14. A manufacturer is planning to sell a new product at the price of $150 per unit and estimates that if x thousand dollars is spent on development and y thousand dollars is spent on promotion, approximately

$$\frac{320y}{y + 2} + \frac{160x}{x + 4}$$

units of the product will be sold. The cost of manufacturing the product is $50 per unit. If the manufacturer has a total of $8,000 to spend on development and promotion, how should this money be allocated to generate the largest possible profit. [*Hint*:

Profit = (number of units)(price per unit − cost per unit) − total amount spent on development and promotion.]

Marginal analysis
15. Use the Lagrange multiplier λ to estimate the change in the maximum output of the factory in Problem 13 that would result if the money available for labor and equipment was increased by $1,000.

Marginal analysis
16. Suppose the manufacturer in Problem 14 decides to spend $8,100 instead of $8,000 on the development and promotion of the new product. Use the Lagrange multiplier λ to estimate how this change will affect the maximum possible profit.

Allocation of unrestricted funds
17. (a) If unlimited funds are available, how much should the manufacturer in Problem 14 spend on development and how much on promotion in order to generate the largest possible profit? (*Hint*: Use the methods of Section 5.)
   (b) What is the value of the Lagrange multiplier λ that corresponds to the optimal budget in part (a)? Explain your answer in light of the interpretation of λ as $\dfrac{dM}{dk}$.
   (c) Your answer to part (b) should suggest another method for solving the problem in part (a). Solve the problem using this new method.

Utility
18. A consumer has $280 to spend on two commodities, the first of which costs $2 per unit and the second $5 per unit. Suppose that the utility derived by the consumer from x units of the first commodity and y units of the second commodity is $U(x, y) = 100x^{0.25}y^{0.75}$.
   (a) How many units of each commodity should the consumer buy to maximize utility?
   (b) Compute the marginal utility of money and interpret the result in economic terms.

Utility
19. A consumer has k dollars to spend on two commodities, the first of which costs a dollars per unit and the second b dollars per unit. Suppose that the utility derived by the consumer from x units of the first commodity and y units of the second commodity is given by the Cobb-Douglas utility function $U(x, y) = x^{\alpha}y^{\beta}$, where $0 < \alpha < 1$ and $\alpha + \beta = 1$. Show that utility will be maximized when $x = \dfrac{k\alpha}{a}$ and $y = \dfrac{k\beta}{b}$.

7 The Method of Least Squares

Throughout this text, you have seen examples in which functions relating two or more variables were differentiated or integrated to obtain useful information about practical situations. In this section, you will see one of the techniques that researchers use to determine such functions from observed data.

Suppose data consisting of n points $(x_1, y_1), (x_2, y_2), \ldots, (x_n, y_n)$ are known and the goal is to find a function $y = f(x)$ that fits the data reasonably well. The first step is to decide what type of function to try. Sometimes this can be done by a theoretical analysis of the underlying practical situation and sometimes by inspection of the graph of the n points. Two sets of data are plotted in Figure 7.1. In Figure 7.1a, the points lie roughly along a straight line, so a linear function $y = mx + b$ would be an appropriate choice in this case. In Figure 7.1b, the points appear to follow an exponential curve, and a function of the form $y = Ae^{-kx}$ would be reasonable.

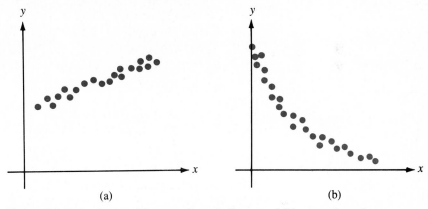

(a)   (b)

Figure 7.1 Data that are (a) approximately linear and (b) approximately exponential.

The Least-Squares Criterion

Once the type of function has been chosen, the next step is to determine the particular function of this type whose graph is "closest" to the given set of points. A convenient way to measure how close a curve is to a set of points is to compute the sum of the squares of the vertical distances from the points to the curve. In Figure 7.2, for example, this is the sum $d_1^2 +$

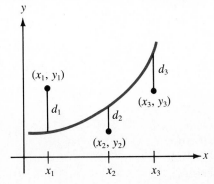

Figure 7.2 Sum of the squares of the vertical distances: $d_1^2 + d_2^2 + d_3^2$.

$d_2^2 + d_3^2$. The closer the curve is to the points, the smaller this sum will be, and the curve for which this sum is smallest is said to be **closest** to the set of points **according to the least-squares criterion.**

The use of the least-squares criterion to fit a linear function to a set of points is illustrated in the following example. The computation involves the technique from Section 5 for minimizing a function of two variables.

EXAMPLE 7.1   Use the least-squares criterion to find the equation of the line that is closest to the three points $(1, 1)$, $(2, 3)$, and $(4, 3)$.

SOLUTION
As indicated in Figure 7.3, the sum of the squares of the vertical distances from the three given points to the line $y = mx + b$ is

$$d_1^2 + d_2^2 + d_3^2 = (m + b - 1)^2 + (2m + b - 3)^2 + (4m + b - 3)^2$$

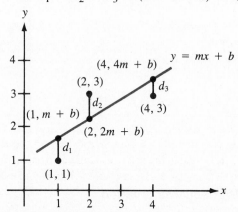

Figure 7.3 $d_1^2 + d_2^2 + d_3^2 = (m + b - 1)^2 + (2m + b - 3)^2 + (4m + b - 3)^2$.

This sum depends on the coefficients m and b that define the line, and so the sum can be thought of as a function $S(m, b)$ of the two variables m and b. The goal, therefore, is to find the values of m and b that minimize the function

$$S(m, b) = (m + b - 1)^2 + (2m + b - 3)^2 + (4m + b - 3)^2$$

You do this by setting the partial derivatives $\dfrac{\partial S}{\partial m}$ and $\dfrac{\partial S}{\partial b}$ equal to zero to get

$$\frac{\partial S}{\partial m} = 2(m + b - 1) + 4(2m + b - 3) + 8(4m + b - 3)$$

$$= 42m + 14b - 38 = 0$$

and

$$\frac{\partial S}{\partial b} = 2(m + b - 1) + 2(2m + b - 3) + 2(4m + b - 3)$$

$$= 14m + 6b - 14 = 0$$

and solving the resulting simplified equations

$$21m + 7b = 19$$

$$7m + 3b = 7$$

simultaneously for m and b to conclude that

$$m = \frac{4}{7} \quad \text{and} \quad b = 1$$

It can be shown that the critical point $(m, b) = \left(\frac{4}{7}, 1\right)$ does indeed minimize the function $S(m, b)$, and so it follows that

$$y = \frac{4}{7}x + 1$$

is the equation of the line that is closest to the three given points.

The Least-Squares Line

The line that is closest to a set of points according to the least-squares criterion is called the **least-squares line** for the points. (The term **regression line** is also used, especially in statistical work.) The procedure used in Example 7.1 can be generalized to give the following formulas for the slope m and the y intercept b of the least-squares line for an arbitrary set of n points (x_1, y_1), (x_2, y_2), . . . , (x_n, y_n). The formulas involve sums of the x and y values. All the sums run from $j = 1$ to $j = n$. To simplify the notation, the indices are omitted and, for example, Σx is used instead of $\sum_{j=1}^{n} x_j$.

The Least-Squares Line

> The equation of the least-squares line for the n points (x_1, y_1), (x_2, y_2), . . . , (x_n, y_n) is $y = mx + b$, where
>
> $$m = \frac{n\Sigma xy - \Sigma x \Sigma y}{n\Sigma x^2 - (\Sigma x)^2} \quad \text{and} \quad b = \frac{\Sigma x^2 \Sigma y - \Sigma x \Sigma xy}{n\Sigma x^2 - (\Sigma x)^2}$$

EXAMPLE 7.2 Use the formulas to find the least-squares line for the points $(1, 1)$, $(2, 3)$, and $(4, 3)$ from Example 7.1.

SOLUTION

Arrange your calculations as follows:

x	y	xy	x^2
1	1	1	1
2	3	6	4
4	3	12	16
$\Sigma x = 7$	$\Sigma y = 7$	$\Sigma xy = 19$	$\Sigma x^2 = 21$

Then use the formulas with $n = 3$ to get

$$m = \frac{3(19) - 7(7)}{3(21) - (7)^2} = \frac{4}{7} \quad \text{and} \quad b = \frac{21(7) - 7(19)}{3(21) - (7)^2} = 1$$

from which it follows that the equation of the least-squares line is

$$y = \frac{4}{7}x + 1$$

Least-Squares Prediction

The least-squares line (or curve) that fits data collected in the past can be used to make rough predictions about the future. This is illustrated in Figure 7.4, which shows the least-squares line for a company's annual sales for its first 4 years of operation. A reasonable estimate of fifth-year sales is the value of y obtained from the equation of the line when $x = 5$.

Least-squares prediction is illustrated further in the next example. To keep the calculations relatively simple, an unrealistically small set of data is used.

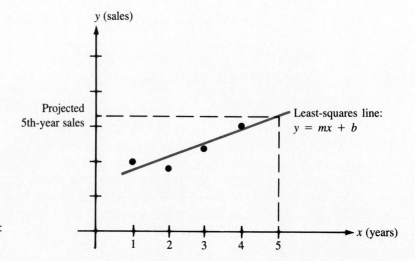

Figure 7.4
Projected fifth-year sales:
$y = 5m + b$.

EXAMPLE 7.3 A college admissions officer has compiled the following data relating students' high-school and college grade-point averages:

High-school GPA	2.0	2.5	3.0	3.0	3.5	3.5	4.0	4.0
College GPA	1.5	2.0	2.5	3.5	2.5	3.0	3.0	3.5

Find the equation of the least-squares line for these data and use it to predict the college GPA of a student whose high-school GPA is 3.75.

SOLUTION

Let x denote the high-school GPA and y the college GPA, and arrange the calculations as follows:

x	y	xy	x^2
2.0	1.5	3.0	4.0
2.5	2.0	5.0	6.25
3.0	2.5	7.5	9.0
3.0	3.5	10.5	9.0
3.5	2.5	8.75	12.25
3.5	3.0	10.5	12.25
4.0	3.0	12.0	16.0
4.0	3.5	14.0	16.0
$\Sigma x = 25.5$	$\Sigma y = 21.5$	$\Sigma xy = 71.25$	$\Sigma x^2 = 84.75$

Use the least-squares formula with $n = 8$ to get

$$m = \frac{8(71.25) - 25.5(21.5)}{8(84.75) - (25.5)^2} \simeq 0.78$$

and

$$b = \frac{84.75(21.5) - 25.5(71.25)}{8(84.75) - (25.5)^2} \simeq 0.19$$

The equation of the least-squares line is therefore

$$y = 0.78x + 0.19$$

To predict the college GPA y of a student whose high-school GPA x is 3.75, substitute $x = 3.75$ into the equation of the least-squares line. This gives

$$y = 0.78(3.75) + 0.19 = 3.12$$

which suggests that the student's college GPA might be about 3.1.

A graph of the original data and of the corresponding least-squares line is shown in Figure 7.5. Actually, in practice, it is a good idea to plot the data *before* proceeding with the calculations. By looking at the graph you will usually be able to tell whether approximation by a straight line is

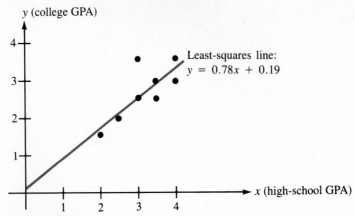

Figure 7.5 The least-squares line for high-school and college GPAs.

appropriate or whether a curve of some other shape should be used instead.

___ rve Fitting

receding examples, the least-squares criterion was used to
on to a set of data. With appropriate modifications, the
be used to fit nonlinear functions to data. For exam-
dratic function $y = Ax^2 + Bx + C$ whose graph is
nts, you proceed as in Example 7.1 and form the sum
ertical distances from the given points to the graph.
a function $S(A, B, C)$ of the coefficients A, B, and
ction of three variables, you set all three of its
ves equal to zero and solve the resulting system
e unknowns. A computer is often used to
lations.

Pro

In Prob.   plot the given points, and use the method of
Example   corresponding least-squares line.

1. $(0, 1), (2, \quad , 2)$
2. $(1, 1), (2, 2), (6, 0)$

3. $(1, 2), (2, 4), \quad , 4), (5, 2)$
4. $(1, 5), (2, 4), (3, 2), (6, 0)$

In Problems 5 through 8, plot the given points, and use the formula to find the corresponding least-squares line.

5. $(1, 2), (2, 2), (2, 3), (5, 5)$

6. $(-4, -1), (-3, 0), (-1, 0), (0, 1), (1, 2)$

7. $(-2, 5), (0, 4), (2, 3), (4, 2), (6, 1)$

8. $(-6, 2), (-3, 1), (0, 0), (0, -3), (1, -1), (3, -2)$

College admissions   9. Over the past 4 years, a college admissions officer has compiled the following data (measured in units of 1,000) relating the number of college catalogs requested by high-school students by December 1 to the number of completed applications received by March 1:

Catalogs requested	4.5	3.5	4.0	5.0
Applications received	1.0	0.8	1.0	1.5

(a) Plot these data on a graph.
(b) Find the equation of the least-squares line.
(c) Use the least-squares line to predict the number of completed applications that will be received by March 1 if 4,800 catalogs are requested by December 1.

Sales   10. A company's annual sales (in units of 1 billion dollars) for its first 5 years of operation are shown in the following table:

Year	1	2	3	4	5
Sales	0.9	1.5	1.9	2.4	3.0

(a) Plot these data on a graph.
(b) Find the equation of the least-squares line.
(c) Use the least-squares line to predict the company's sixth-year sales.

Voter turnout   11. On election day, the polls in a certain state open at 8:00 A.M. Every 2 hours after that, an election official determines what percentage of the registered voters have already cast their ballots. The data through 6:00 P.M. are shown below:

Time	10:00	12:00	2:00	4:00	6:00
Percentage turnout	12	19	24	30	37

(a) Plot these data on a graph.
(b) Find the equation of the least-squares line. (Let x denote the number of hours after 8:00 A.M.)
(c) Use the least-squares line to predict what percentage of the registered voters will have cast their ballots by the time the polls close at 8:00 P.M.

Public health   12. In a study of five industrial areas, a researcher obtained the following data relating the average number of units of a certain pollutant in the air and the incidence (per 100,000 people) of a certain disease:

Units of pollutant	3.4	4.6	5.2	8.0	10.7
Incidence of disease	48	52	58	76	96

(a) Plot these data on a graph.

(b) Find the equation of the least-squares line.

(c) Use the least-squares line to estimate the incidence of the disease in an area with an average pollution level of 7.3 units.

Chapter Summary and Review Problems

Important Terms, Symbols, and Formulas

Function of two variables: $z = f(x, y)$

Partial derivatives:

$$f_x = \frac{\partial z}{\partial x} \qquad f_y = \frac{\partial z}{\partial y}$$

Second-order partial derivatives:

$$f_{xx} = \frac{\partial^2 z}{\partial x^2} \qquad f_{xy} = \frac{\partial^2 z}{\partial y \partial x} \qquad f_{yx} = \frac{\partial^2 z}{\partial x \partial y} \qquad f_{yy} = \frac{\partial^2 z}{\partial y^2}$$

Mixed second-order partial derivatives: $f_{xy} = f_{yx}$

Chain rule: $\dfrac{dz}{dt} = \dfrac{\partial z}{\partial x}\dfrac{dx}{dt} + \dfrac{\partial z}{\partial y}\dfrac{dy}{dt}$

Approximation formula; total differential:

$$\Delta z \simeq \frac{\partial z}{\partial x}\Delta x + \frac{\partial z}{\partial y}\Delta y = dz$$

Percentage change in $z = 100\dfrac{\Delta z}{z} \simeq 100\dfrac{\dfrac{\partial z}{\partial x}\Delta x + \dfrac{\partial z}{\partial y}\Delta y}{z}$

Level curve: $f(x, y) = C$

Slope of a level curve: $\dfrac{dy}{dx} = -\dfrac{f_x}{f_y}$

Approximation on level curves: $\Delta y \simeq \dfrac{dy}{dx}\Delta x = -\dfrac{f_x}{f_y}\Delta x$

Constant-production curve; indifference curve

Relative maximum; relative minimum; saddle point

Critical point: $f_x = f_y = 0$

Second-derivative test at a critical point: Let $D = f_{xx}f_{yy} - (f_{xy})^2$

If $D < 0$, f has a saddle point.

If $D > 0$ and $f_{xx} < 0$, f has a relative maximum.

If $D > 0$ and $f_{xx} > 0$, f has a relative minimum.

Method of Lagrange multipliers: To find the relative extremum of $f(x, y)$ subject to $g(x, y) = k$, solve the equations

$$f_x = \lambda g_x \qquad f_y = \lambda g_y \qquad \text{and} \qquad g = k$$

The Lagrange multiplier: $\lambda = \dfrac{dM}{dk} \approx$ change in M resulting from a 1-unit increase in k, where M is the optimal value of $f(x, y)$ subject to $g(x, y) = k$.

Least-squares criterion

Least-squares line: $y = mx + b$, where

$$m = \frac{n\Sigma xy - \Sigma x \Sigma y}{n\Sigma x^2 - (\Sigma x)^2} \qquad \text{and} \qquad b = \frac{\Sigma x^2 \Sigma y - \Sigma x \Sigma xy}{n\Sigma x^2 - (\Sigma x)^2}$$

Review Problems

1. For each of the following functions, compute the first-order partial derivatives f_x and f_y:

   (a) $f(x, y) = 2x^3y + 3xy^2 + \dfrac{y}{x}$

   (b) $f(x, y) = (xy^2 + 1)^5$

   (c) $f(x, y) = xye^{xy}$

2. For each of the following functions, compute the second-order partial derivatives f_{xx}, f_{yy}, f_{xy}, and f_{yx}.

   (a) $f(x, y) = x^2 + y^3 - 2xy^2$

   (b) $f(x, y) = e^{x^2 + y^2}$

   (c) $f(x, y) = x \ln y$

3. At a certain factory, the daily output is approximately $40K^{1/3}L^{1/2}$ units, where K denotes the capital investment measured in units of $\$1,000$ and L denotes the size of the labor force measured in worker-hours. Suppose the current capital investment is $\$125,000$ and that 900 worker-hours of labor are used each day. Use marginal analysis to estimate the effect that an additional capital investment of $\$1,000$ will have on the daily output if the size of the labor force is not changed.

4. In economics, the marginal product of labor is the rate at which output Q changes with respect to labor L for a fixed level of capital investment K. An economic law states that, under certain circumstances, the marginal product of labor increases as the level of capital investment increases. Translate this law into a mathematical statement involving a second-order partial derivative.

5. Use the chain rule to find $\dfrac{dz}{dt}$.

   (a) $z = x^3 - 3xy^2; x = 2t, y = t^2$

   (b) $z = x \ln y; x = 2t, y = e^t$

6. A grocery store carries two brands of diet cola. Sales figures indicate that if the first brand is sold for x cents per can and the second brand for y cents per can, consumers will buy $Q(x, y) = 240 + 0.1y^2 - 0.2x^2$ cans of the first brand per week. Currently the first brand sells for 45 cents per can, and the second brand sells for 48 cents per can. Use the total differential to estimate how the demand for the first brand of diet cola will change if the price of the first brand is increased by 2 cents per can while the price of the second brand is decreased by 1 cent per can.

7. At a certain factory the daily output is $Q(K, L) = 120K^{1/3}L^{2/3}$ units, where K denotes the capital investment and L the size of the labor force. Use calculus to estimate the percentage by which the daily output will change if capital investment is increased by 2 percent and labor by 1 percent.

8. Suppose that when apples sell for x cents per pound and bakers earn y dollars per hour, the price of apple pies at the local supermarket is $p(x, y) = \dfrac{1}{2}x^{1/3}y^{1/2}$ dollars per pie. Suppose also that t months from now, the price of apples will be $x = 23 + \sqrt{8t}$ cents per pound and bakers' wages will be $y = 3.96 + 0.02t$ dollars per hour. If the supermarket can sell $Q(p) = \dfrac{3,600}{p}$ pies per week when the price is p dollars per pie, at what rate will the weekly demand Q be changing with respect to time 2 months from now?

9. For each of the following functions, sketch the indicated level curves.
   (a) $f(x, y) = x^2 - y; f = 2, f = -2$
   (b) $f(x, y) = 6x + 2y; f = 0, f = 1, f = 2$

10. For each of the following functions, find the slope of the indicated level curve at the specified value of x.
    (a) $f(x, y) = x^2 - y^3; f = 2; x = 1$
    (b) $f(x, y) = xe^y; f = 2; x = 2$

11. Using x skilled workers and y unskilled workers, a manufacturer can produce $Q(x, y) = 60x^{1/3}y^{2/3}$ units per day. Currently the manufacturer employs 10 skilled workers and 40 unskilled workers and is planning to hire 1 additional skilled worker. Use calculus to estimate the corresponding change that the manufacturer should make in the level of unskilled labor so that the total output will remain the same.

12. Find the critical points of each of the following functions and classify them as relative maxima, relative minima, or saddle points.
    (a) $f(x, y) = x^3 + y^3 + 3x^2 - 18y^2 + 81y + 5$
    (b) $f(x, y) = x^2 + y^3 + 6xy - 7x - 6y$

13. Use the method of Lagrange multipliers to find the maximum and minimum values of the function $f(x, y) = x^2 + 2y^2 + 2x + 3$ subject to the constraint $x^2 + y^2 = 4$.

14. Use the method of Lagrange multipliers to prove that of all rectangles with a given perimeter the square has the largest area.

15. A manufacturer is planning to sell a new product at the price of $350 per unit and estimates that if x thousand dollars is spent on development and y thousand dollars is spent on promotion, consumers will buy approximately $\dfrac{250y}{y + 2} + \dfrac{100x}{x + 5}$ units of the product. If manufacturing costs for this product are $150 per unit, how much should the manufacturer spend on development and how much on promotion to generate the largest possible profit if unlimited funds are available?

16. Suppose the manufacturer in Problem 15 has only $11,000 to spend on the development and promotion of the new product. How should this money be allocated to generate the largest possible profit?

17. Suppose the manufacturer in Problem 16 decides to spend $12,000 instead of $11,000 on the development and promotion of the new product. Use the Lagrange multiplier λ to estimate how this change will affect the maximum possible profit.

18. Plot the points (1, 1), (1, 2), (3, 2), and (4, 3), and use partial derivatives to find the corresponding least-squares line.

19. The marketing manager for a certain company has compiled the following data relating monthly advertising expenditure and monthly sales (both measured in units of $1,000):

Advertising	3	4	7	9	10
Sales	78	86	138	145	156

    (a) Plot these data on a graph.
    (b) Find the least-squares line.
    (c) Use the least-squares line to predict monthly sales if the monthly advertising expenditure is $5,000.

10 Double Integrals

1   Double Integrals

This chapter is about definite integrals of functions of two variables. They are known as **double integrals** and are evaluated by a process involving repeated partial antidifferentiation. In this first section you will see how to evaluate double integrals once their limits of integration are known. In Section 2 you will learn how to determine the limits of integration. And in Section 3 you will see a sampling of applications, all of which are generalizations of applications of definite integrals of functions of one variable with which you are already familiar.

The discussion in this chapter will be kept at an introductory level. Additional detail and theory can be found in more advanced calculus texts.

Evaluation of Double Integrals

The symbol

$$\int_a^b \int_c^d f(x, y) \, dy \, dx$$

is called a double integral (or **iterated integral**) and is an abbreviation for

$$\int_a^b \left[\int_c^d f(x, y) \, dy \right] dx$$

To evaluate a double integral you first compute the inner definite integral

$$\int_c^d f(x, y) \, dy$$

taking the antiderivative of f with respect to y while keeping x fixed. The result will be a function of the single variable x, which you then integrate with respect to x between $x = a$ and $x = b$. Here is an example.

EXAMPLE 1.1   Evaluate $\int_0^1 \int_{-1}^2 xy^2 \, dy \, dx$.

SOLUTION

First perform the inner integration with respect to y, treating x as a constant to get

$$\int_{-1}^2 xy^2 \, dy = \frac{1}{3}xy^3 \Big|_{y=-1}^{y=2} = \frac{8}{3}x + \frac{1}{3}x = 3x$$

(Note that the limits of integration -1 and 2 refer to the variable y.) Now integrate the result of this calculation with respect to x from $x = 0$ to $x = 1$ to conclude that

$$\int_0^1 \int_{-1}^2 xy^2 \, dy \, dx = \int_0^1 3x \, dx = \frac{3}{2}x^2 \Big|_0^1 = \frac{3}{2}$$

To give your solution a more professional appearance, arrange your work compactly as follows:

$$\int_0^1 \int_{-1}^2 xy^2 \, dy \, dx = \int_0^1 \left(\frac{1}{3}xy^3 \Big|_{y=-1}^{y=2} \right) dx = \int_0^1 3x \, dx = \frac{3}{2}x^2 \Big|_0^1 = \frac{3}{2}$$

Double Integrals with Variable Limits of Integration

In the preceding example, all four of the limits of integration were constants. In the next example, the limits of integration on the inner integral are not constants but functions of x. The technique for evaluating the integral remains the same.

EXAMPLE 1.2　Evaluate $\displaystyle\int_0^1 \int_{x^2}^{\sqrt{x}} 160xy^3 \, dy \, dx$.

SOLUTION

$$\int_0^1 \int_{x^2}^{\sqrt{x}} 160xy^3 \, dy \, dx = \int_0^1 \left(40xy^4 \Big|_{y=x^2}^{y=\sqrt{x}} \right) dx \qquad \text{since } x \text{ is treated as a constant}$$

$$= \int_0^1 [40x(\sqrt{x})^4 - 40x(x^2)^4] \, dx$$

$$= \int_0^1 (40x^3 - 40x^9) \, dx$$

$$= (10x^4 - 4x^{10}) \Big|_0^1$$

$$= 6$$

In Example 1.2, the variable limits of integration on the inner integral were functions of x. These functions, x^2 and $\sqrt{x}$, were substituted for y during the first integration, resulting in an expression containing only the variable x, which was then integrated between the constant limits $x = 0$ and $x = 1$. In general, the outer limits of integration must be constants (so that the final answer will be a constant), while the inner limits may be functions of the variable with respect to which the second integration is to be performed.

In the next example, x is the first variable of integration and y is the second. Notice that in this case the variable limit of integration on the inner integral is a function of y.

EXAMPLE 1.3　Evaluate $\displaystyle\int_0^1 \int_0^y y^2 e^{xy} \, dx \, dy$.

SOLUTION

$$\int_0^1 \int_0^y y^2 e^{xy} \, dx \, dy = \int_0^1 \left(ye^{xy} \Big|_{x=0}^{x=y} \right) dy \qquad \text{since } \int e^{xy} \, dx = \frac{1}{y}e^{xy}$$

$$= \int_0^1 (ye^{y^2} - y) \, dy$$

$$= \left(\frac{1}{2}e^{y^2} - \frac{1}{2}y^2 \right) \Big|_0^1$$

$$= \left(\frac{1}{2}e - \frac{1}{2} \right) - \left(\frac{1}{2} - 0 \right)$$

$$= \frac{1}{2}e - 1$$

Techniques of Integration

In performing the repeated antidifferentiation required to evaluate a double integral, you may have to use one or more of the special techniques of integration you learned in Chapter 5. The next example involves integration by parts.

EXAMPLE 1.4 Evaluate $\int_0^1 \int_{1-y}^1 e^y \, dx \, dy$.

SOLUTION

$$\int_0^1 \int_{1-y}^1 e^y \, dx \, dy = \int_0^1 \left(xe^y \Big|_{x=1-y}^{x=1} \right) dy$$

$$= \int_0^1 [e^y - (1 - y)e^y] \, dy$$

$$= \int_0^1 ye^y \, dy$$

$$= ye^y \Big|_0^1 - \int_0^1 e^y \, dy \qquad \text{integration by parts}$$

$$= (ye^y - e^y) \Big|_0^1 = 1$$

Problems

Evaluate the following double integrals.

1. $\int_0^1 \int_1^2 x^2 y \, dx \, dy$

2. $\int_1^2 \int_0^1 x^2 y \, dy \, dx$

3. $\int_0^{\ln 2} \int_{-1}^0 2xe^y \, dx \, dy$

4. $\int_2^3 \int_{-1}^1 (x + 2y) \, dy \, dx$

5. $\int_1^3 \int_0^1 \frac{2xy}{x^2 + 1} \, dx \, dy$

6. $\int_0^1 \int_0^1 x^2 e^{xy} \, dy \, dx$

7. $\int_0^4 \int_0^{\sqrt{x}} x^2 y \, dy \, dx$

8. $\int_0^1 \int_1^5 y\sqrt{1 - y^2} \, dx \, dy$

9. $\int_0^1 \int_{y-1}^{1-y} (2x + y) \, dx \, dy$

10. $\int_0^1 \int_{x^2}^x 2xy \, dy \, dx$

11. $\int_0^2 \int_{x^2}^4 xe^y \, dy \, dx$

12. $\int_0^1 \int_{ey}^e \frac{1}{x} \, dx \, dy$

13. $\int_0^1 \int_0^x \frac{2y}{x^3 + 1} \, dy \, dx$

14. $\int_0^2 \int_0^{\sqrt{4-y^2}} y \, dx \, dy$

15. $\int_0^1 \int_0^x xe^y \, dy \, dx$

16. $\int_1^2 \int_x^{x^2} e^{y/x} \, dy \, dx$

17. $\int_0^1 \int_y^1 ye^{x+y} \, dx \, dy$

18. $\int_1^e \int_0^{\ln x} 2 \, dy \, dx$

2 Finding Limits of Integration

The limits of integration for the definite integral of a function of one variable come from the inequality defining a corresponding interval on the x axis. In particular, the integral of $f(x)$ over the interval $a \leq x \leq b$ is $\int_a^b f(x) \, dx$. A double integral of a function $f(x, y)$ of two variables is an integral associated with a region R in the xy plane. The notation

$$\iint_R f(x, y) \, dA$$

is sometimes used to denote such an integral. It can be shown that the corresponding limits of integration come from inequalities involving x and y that define the region R.

In this section, the procedure for finding limits of integration for double integrals will be discussed for regions of two fundamental types. More complicated regions of integration can usually be divided into two or more of these elementary regions.

Regions Described in Terms of Vertical Cross Sections

The region R in Figure 2.1 is bounded below by the graph of the function $y = g_1(x)$ and above by the graph of the function $y = g_2(x)$, and it extends from $x = a$ on the left to $x = b$ on the right. The inequalities

$$a \leq x \leq b \qquad \text{and} \qquad g_1(x) \leq y \leq g_2(x)$$

can be used to describe such a region. The first inequality specifies the interval in which x must lie, and the second indicates the lower and upper bounds of the vertical cross section of R for each x in this interval. Roughly speaking, the inequalities state that "y goes from $g_1(x)$ to $g_2(x)$ for each x between a and b."

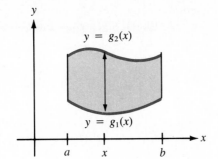

Figure 2.1 Region described by $a \le x \le b$ and $g_1(x) \le y \le g_2(x)$.

EXAMPLE 2.1 Let R be the region bounded by the curve $y = x^2$ and the line $y = 2x$. Use inequalities to describe R in terms of its vertical cross sections.

SOLUTION

Begin with a sketch of the curve and line as shown in Figure 2.2, identify the region R, and, for reference, draw a vertical cross section. Solve the equations $y = x^2$ and $y = 2x$ simultaneously to find the points of intersection, $(0, 0)$ and $(2, 4)$. Observe that in the region R, the variable x takes on all values from $x = 0$ to $x = 2$ and that for each such value of x, the vertical cross section is bounded below by $y = x^2$ and above by $y = 2x$. Hence R can be described by the inequalities

$$0 \le x \le 2 \qquad \text{and} \qquad x^2 \le y \le 2x$$

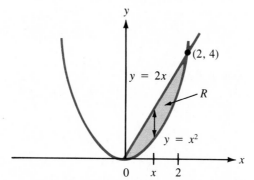

Figure 2.2 Region bounded by $y = x^2$ and $y = 2x$.

Regions Described in Terms of Horizontal Cross Sections

The region R in Figure 2.3 is bounded on the left by the graph of $x = h_1(y)$ and on the right by the graph of $x = h_2(y)$, and it extends from $y = c$ on the bottom to $y = d$ on top. It can be described by the inequalities

$$c \le y \le d \qquad \text{and} \qquad h_1(y) \le x \le h_2(y)$$

where the first inequality specifies the interval in which y must lie, and the second indicates the left-hand and right-hand bounds of a horizontal cross section. Roughly speaking, the inequalities state that "x goes from $h_1(y)$ to $h_2(y)$ for each y between c and d."

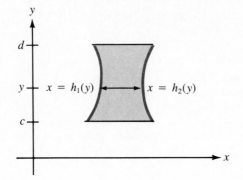

Figure 2.3 Region bounded by $c \le y \le d$ and $h_1(y) \le x \le h_2(y)$.

EXAMPLE 2.2   Let R be the region from Example 2.1 bounded by the curve $y = x^2$ and the line $y = 2x$. Use inequalities to describe R in terms of its horizontal cross sections.

SOLUTION
As in Example 2.1, sketch the region and find the points of intersection of the line and curve, but this time draw a horizontal cross section (Figure 2.4).

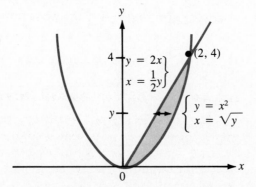

Figure 2.4 Region bounded by $y = x^2$ and $y = 2x$.

In the region R, the variable y takes on all values from $y = 0$ to $y = 4$, and for each such value of y, the horizontal cross section extends from the line $y = 2x$ on the left to the curve $y = x^2$ on the right. Since the equation of the line can be rewritten as $x = \frac{1}{2}y$ and the equation of the curve as $x = \sqrt{y}$, the inequalities describing R in terms of its horizontal cross sections are

$$0 \leq y \leq 4 \quad \text{and} \quad \frac{1}{2}y \leq x \leq \sqrt{y}$$

Limits of Integration

To evaluate a double integral over a region of one of these types, you use an iterated integral whose limits of integration come from the inequalities describing the region. Here is a more precise statement of the procedure.

Limits of Integration for Double Integrals

If R can be described by the inequalities

$$a \leq x \leq b \quad \text{and} \quad g_1(x) \leq y \leq g_2(x)$$

then

$$\iint\limits_{R} f(x, y) \, dA = \int_{a}^{b} \int_{g_1(x)}^{g_2(x)} f(x, y) \, dy \, dx$$

If R can be described by the inequalities

$$c \leq y \leq d \quad \text{and} \quad h_1(y) \leq x \leq h_2(y)$$

then

$$\iint\limits_{R} f(x, y) \, dA = \int_{c}^{d} \int_{h_1(y)}^{h_2(y)} f(x, y) \, dx \, dy$$

Notice that in each case, the limits of integration on the inner integral are functions of the *second* variable of integration. (In some cases, one or both of these functions may be constant.) The limits on the outer integral are always constants.

Here are some examples.

EXAMPLE 2.3 Evaluate $\iint\limits_{R} 40x^2y \, dA$, where R is the region bounded by the curve $y = \sqrt{x}$ and the line $y = x$.

SOLUTION
From the sketch in Figure 2.5, observe that R can be described in terms of vertical cross sections by the inequalities

$$0 \leq x \leq 1 \quad \text{and} \quad x \leq y \leq \sqrt{x}$$

Hence

$$\iint_R 40x^2y \, dA = \int_0^1 \int_x^{\sqrt{x}} 40x^2y \, dy \, dx$$

$$= \int_0^1 \left(20x^2y^2 \Big|_{y=x}^{y=\sqrt{x}}\right) dx$$

$$= \int_0^1 (20x^3 - 20x^4) \, dx$$

$$= (5x^4 - 4x^5)\Big|_0^1$$

$$= 1$$

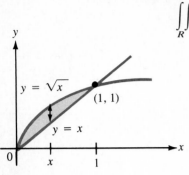

Figure 2.5 Region bounded by $y = \sqrt{x}$ and $y = x$.

Observe that R can also be described in terms of horizontal cross sections by the inequalities

$$0 \le y \le 1 \qquad \text{and} \qquad y^2 \le x \le y$$

The corresponding double integral is

$$\int_0^1 \int_{y^2}^y 40x^2y \, dx \, dy$$

For practice (and to check your answer), evaluate this integral. The answer, of course, should also be 1.

EXAMPLE 2.4   Evaluate $\displaystyle\iint_R (x + y) \, dA$, where R is the triangle with vertices $(0, 0)$, $(0, 1)$, and $(1, 1)$.

SOLUTION
The triangle is sketched in Figure 2.6. It is bounded above by the horizontal line $y = 1$ and below by the line $y = x$ and can be described in terms of vertical cross sections by the inequalities

$$0 \le x \le 1 \qquad \text{and} \qquad x \le y \le 1$$

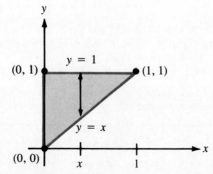

Figure 2.6 Triangle with vertices $(0, 0)$, $(0, 1)$, and $(1, 1)$.

Hence,

$$\iint_R (x + y)\, dA = \int_0^1 \int_x^1 (x + y)\, dy\, dx$$

$$= \int_0^1 \left[\left(xy + \frac{1}{2} y^2 \right) \Big|_{y = x}^{y = 1} \right] dx$$

$$= \int_0^1 \left[\left(x + \frac{1}{2} \right) - \left(x^2 + \frac{1}{2} x^2 \right) \right] dx$$

$$= \int_0^1 \left(\frac{1}{2} + x - \frac{3}{2} x^2 \right) dx$$

$$= \left(\frac{1}{2} x + \frac{1}{2} x^2 - \frac{1}{2} x^3 \right) \Big|_0^1$$

$$= \frac{1}{2}$$

For practice, check this answer by evaluating the corresponding double integral with the order of integration reversed.

Integration over More Complex Regions

Many regions that are not of one of the fundamental types can be divided into two or more subregions that are. The next example illustrates how to integrate over such a region.

EXAMPLE 2.5   Evaluate $\iint_R 1\, dA$, where R is the region in the first quadrant that lies under the curve $y = \dfrac{1}{x}$ and is bounded by this curve and the lines $y = x$, $y = 0$, and $x = 2$.

SOLUTION
The region is sketched in Figure 2.7. Observe that to the left of $x = 1$, vertical cross sections of R are bounded above by the line $y = x$, while to the right of $x = 1$, they are bounded above by the curve $y = \dfrac{1}{x}$. This prevents a simple description of R in terms of vertical cross sections and suggests that R be broken into two subregions, R_1 and R_2, as shown in Figure 2.7.

R_1 can be described by the inequalities

$$0 \le x \le 1 \quad \text{and} \quad 0 \le y \le x$$

while R_2 can be described by the inequalities

$$1 \le x \le 2 \qquad \text{and} \qquad 0 \le y \le \frac{1}{x}$$

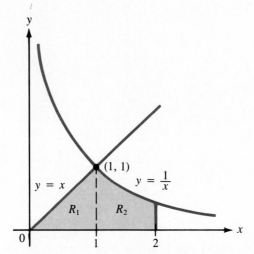

Figure 2.7 Region in the first quadrant bounded by $y = \dfrac{1}{x}$, $y = x$, $y = 0$, and $x = 2$.

Hence,

$$\iint\limits_{R} 1 \, dA = \iint\limits_{R_1} 1 \, dA + \iint\limits_{R_2} 1 \, dA$$

$$= \int_0^1 \int_0^x 1 \, dy \, dx + \int_1^2 \int_0^{1/x} 1 \, dy \, dx$$

$$= \int_0^1 x \, dx + \int_1^2 \frac{1}{x} \, dx$$

$$= \frac{1}{2}x^2 \Big|_0^1 + \ln x \Big|_1^2$$

$$= \frac{1}{2} + \ln 2$$

For practice, do the problem again, this time using horizontal cross sections. Notice that this solution also involves the use of two iterated integrals.

Selecting the Order of Integration

Before evaluating a double integral, it is a good idea to take a moment to decide which of the two possible orders of integration will lead to the simpler calculation. A wise choice of order can result in substantial savings of time and effort.

EXAMPLE 2.6 Set up the iterated integral you would prefer to use to evaluate $\iint\limits_R y\, dA$, where R is the region bounded by $y = \ln x$, $y = 1$, $x = 0$, and $y = 0$.

SOLUTION
The region is sketched in Figure 2.8. To the left of $x = 1$, vertical cross sections are bounded below by the x axis ($y = 0$), while to the right of $x = 1$, they are bounded below by the curve $y = \ln x$. As a result, the use of vertical cross sections will involve the two iterated integrals

$$\iint\limits_R y\, dA = \int_0^1 \int_0^1 y\, dy\, dx + \int_1^e \int_{\ln x}^1 y\, dy\, dx$$

the second of which will eventually require the integration of $(\ln x)^2$, which is not easy.

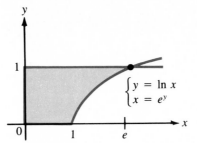

Figure 2.8 Region bounded by $y = \ln x$, $y = 1$, $x = 0$, and $y = 0$.

$$\begin{cases} y = \ln x \\ x = e^y \end{cases}$$

On the other hand, the horizontal cross sections are always bounded on the left by the y axis ($x = 0$) and on the right by the curve $y = \ln x$, whose equation can be rewritten as $x = e^y$. The inequalities that describe R in terms of its horizontal cross sections are

$$0 \le y \le 1 \quad \text{and} \quad 0 \le x \le e^y$$

and the corresponding integral is

$$\iint\limits_R y\, dA = \int_0^1 \int_0^{e^y} y\, dx\, dy$$

Clearly, this second integral, in which x is the first variable of integration, is preferable to the pair of integrals that is needed if y is to be the first variable of integration.

For practice, evaluate $\iint\limits_R y\, dA$ using both methods. Each calculation will involve integration by parts. The answer is 1.

In the next example, the choice of the order of integration is critical. The wrong choice leads to an integral that is impossible to evaluate by elementary methods.

EXAMPLE 2.7 Evaluate $\iint\limits_{R} e^{y^2}\, dA$, where R is the triangle bounded by the lines $y = \frac{1}{2}x$, $y = 1$, and $x = 0$.

SOLUTION
The region is sketched in Figure 2.9. The two possible iterated integrals are

$$\int_0^2 \int_{x/2}^1 e^{y^2}\, dy\, dx \qquad \text{and} \qquad \int_0^1 \int_0^{2y} e^{y^2}\, dx\, dy$$

The first integral, in which y is the first variable of integration, requires the evaluation of $\int e^{y^2}\, dy$, which cannot be done by elementary methods. The second, on the other hand, is easy,

$$\int_0^1 \int_0^{2y} e^{y^2}\, dx\, dy = \int_0^1 \left(xe^{y^2}\Big|_{x=0}^{x=2y} \right) dy$$

$$= \int_0^1 2ye^{y^2}\, dy = e^{y^2}\Big|_0^1 = e - 1$$

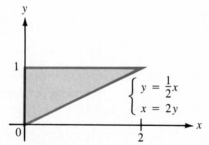

Figure 2.9 Region bounded by $y = \frac{1}{2}x$, $y = 1$, and $x = 0$.

Problems

In Problems 1 through 6, use inequalities to describe R in terms of its vertical and horizontal cross sections.

1. R is the region bounded by $y = x^2$ and $y = 3x$.

2. R is the region bounded by $y = \sqrt{x}$ and $y = x^2$.

3. R is the rectangle with vertices $(-1, 1)$, $(2, 1)$, $(2, 2)$, and $(-1, 2)$.

4. R is the triangle with vertices $(1, 0)$, $(1, 1)$, and $(2, 0)$.

5. R is the region bounded by $y = \ln x$, $y = 0$, and $x = e$.

6. R is the region bounded by $y = e^x$, $y = 2$, and $x = 0$.

In Problems 7 through 22, evaluate the given double integral for the specified region R.

7. $\iint\limits_R 3xy^2 \, dA$, where R is the rectangle bounded by the lines $x = -1$, $x = 2$, $y = -1$, and $y = 0$.

8. $\iint\limits_R (x + 2y) \, dA$, where R is the triangle with vertices $(0, 0)$, $(1, 0)$, and $(0, 2)$.

9. $\iint\limits_R xe^y \, dA$, where R is the triangle with vertices $(0, 0)$, $(1, 0)$, and $(1, 1)$.

10. $\iint\limits_R 48xy \, dA$, where R is the region bounded by $y = x^3$ and $y = \sqrt{x}$.

11. $\iint\limits_R (2y - x) \, dA$, where R is the region bounded by $y = x^2$ and $y = 2x$.

12. $\iint\limits_R 12x \, dA$, where R is the region bounded by $y = x^2$ and $y = 6 - x$.

13. $\iint\limits_R 2 \, dA$, where R is the region bounded by $y = \dfrac{16}{x}$, $y = x$, and $x = 2$.

14. $\iint\limits_R y \, dA$, where R is the region bounded by $y = \sqrt{x}$, $y = 2 - x$, and $y = 0$.

15. $\iint\limits_R 4x \, dA$, where R is the region in the first quadrant bounded by $y = 4 - x^2$, $y = 3x$, and $x = 0$.

16. $\iint\limits_R 4x \, dA$, where R is the region in the first quadrant bounded by $y = 4 - x^2$, $y = 3x$, and $y = 0$.

17. $\iint\limits_R (2x + 1) \, dA$, where R is the triangle with vertices $(-1, 0)$, $(1, 0)$, and $(0, 1)$.

18. $\iint\limits_R 2x \, dA$, where R is the region bounded by $y = \dfrac{1}{x^2}$, $y = x$, $x = 2$, and $y = 0$.

19. $\iint\limits_R \dfrac{1}{y^2 + 1} \, dA$, where R is the triangle bounded by the lines $y = \dfrac{1}{2}x$, $y = -x$, and $y = 2$.

20. $\iint\limits_R e^{y^3}\, dA$, where R is the region bounded by $y = \sqrt{x}$, $y = 1$, and $x = 0$.

21. $\iint\limits_R 12x^2 e^{y^2}\, dA$, where R is the region in the first quadrant bounded by $y = x^3$ and $y = x$.

22. $\iint\limits_R y\, dA$, where R is the region bounded by $y = \ln x$, $y = 0$, and $x = e$.

In Problems 23 through 34, sketch the region of integration for the given integral, and set up an equivalent integral with the order of integration reversed. (In some cases, two or more integrals may be needed.)

23. $\displaystyle\int_0^2 \int_0^{4-x^2} f(x, y)\, dy\, dx$

24. $\displaystyle\int_0^1 \int_0^{2y} f(x, y)\, dx\, dy$

25. $\displaystyle\int_0^1 \int_{x^3}^{\sqrt{x}} f(x, y)\, dy\, dx$

26. $\displaystyle\int_0^4 \int_{y/2}^{\sqrt{y}} f(x, y)\, dx\, dy$

27. $\displaystyle\int_1^{e^2} \int_{\ln x}^2 f(x, y)\, dy\, dx$

28. $\displaystyle\int_0^{\ln 3} \int_{e^x}^3 f(x, y)\, dy\, dx$

29. $\displaystyle\int_{-1}^1 \int_{x^2+1}^2 f(x, y)\, dy\, dx$

30. $\displaystyle\int_{-1}^1 \int_{-\sqrt{y+1}}^{\sqrt{y+1}} f(x, y)\, dx\, dy$

31. $\displaystyle\int_0^1 \int_x^{2-x} f(x, y)\, dy\, dx$

32. $\displaystyle\int_0^3 \int_{y/3}^{\sqrt{4-y}} f(x, y)\, dx\, dy$

33. $\displaystyle\int_{-3}^2 \int_{x^2}^{6-x} f(x, y)\, dy\, dx$

34. $\displaystyle\int_2^3 \int_x^{16/x} f(x, y)\, dy\, dx$

3 Applications of Double Integrals

In this section, you will see a sampling of applications of double integrals, most of which are generalizations of applications of definite integrals of functions of one variable with which you are already familiar.

The Area of a Plane Region

The area of a region R in the xy plane can be computed as the double integral over R of the constant function $f(x, y) = 1$.

Area Formula

The area of a region R in the xy plane is given by the formula

$$\text{Area of } R = \iint_R 1 \, dA$$

To get a feel for why the area formula holds, consider the elementary region R shown in Figure 3.1, which is bounded above by the curve $y = g_2(x)$ and below by the curve $y = g_1(x)$, and which extends from $x = a$ to $x = b$. According to the double-integral formula for area,

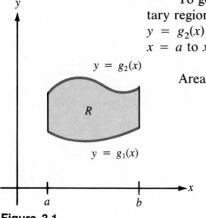

Figure 3.1

Area of $R = \displaystyle\iint_R 1 \, dA.$

$$\text{Area of } R = \iint_R 1 \, dA$$

$$= \int_a^b \int_{g_1(x)}^{g_2(x)} 1 \, dy \, dx$$

$$= \int_a^b \left[y \Big|_{y=g_1(x)}^{y=g_2(x)} \right] dx$$

$$= \int_a^b \left[g_2(x) - g_1(x) \right] dx$$

which is precisely the formula for the area between two curves that you saw in Chapter 6, Section 2.

Here are two examples illustrating the use of the area formula.

EXAMPLE 3.1 Find the area of the region R bounded by the curves $y = x^3$ and $y = x^2$.

SOLUTION

The region is shown in Figure 3.2. Using the area formula you get

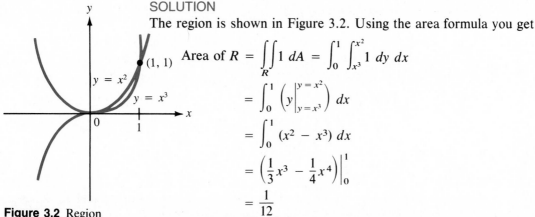

Figure 3.2 Region bounded by $y = x^3$ and $y = x^2$.

$$\text{Area of } R = \iint_R 1 \, dA = \int_0^1 \int_{x^3}^{x^2} 1 \, dy \, dx$$

$$= \int_0^1 \left(y \Big|_{y=x^3}^{y=x^2} \right) dx$$

$$= \int_0^1 (x^2 - x^3) \, dx$$

$$= \left(\frac{1}{3} x^3 - \frac{1}{4} x^4 \right) \Big|_0^1$$

$$= \frac{1}{12}$$

The area calculated in Example 3.1 using a double integral is the same as that calculated in Example 2.6 of Chapter 6 using a single integral. Compare the two solutions. Observe, in particular, that after the integration with respect to y, the double integral in Example 3.1 reduces to the single integral that was used in Example 2.6.

EXAMPLE 3.2   Find the area of the region R in the first quadrant bounded by $y = x^3$, $y = x - 2$, $y = 0$, and $y = 1$.

SOLUTION

From the sketch in Figure 3.3 you see that the use of vertical cross sections to describe R will lead to three iterated integrals. To avoid this, use horizontal cross sections.

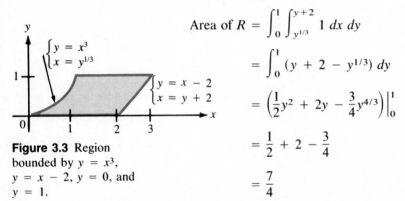

$$\text{Area of } R = \int_0^1 \int_{y^{1/3}}^{y+2} 1 \; dx \; dy$$

$$= \int_0^1 (y + 2 - y^{1/3}) \; dy$$

$$= \left(\frac{1}{2}y^2 + 2y - \frac{3}{4}y^{4/3} \right)\Big|_0^1$$

$$= \frac{1}{2} + 2 - \frac{3}{4}$$

$$= \frac{7}{4}$$

Figure 3.3 Region bounded by $y = x^3$, $y = x - 2$, $y = 0$, and $y = 1$.

The Volume under a Surface

Recall from Chapter 6, Section 2, that the definite integral $\int_a^b f(x) \; dx$ of a nonnegative function $f(x)$ of one variable gives the area of the region under the graph of f and above the interval $a \le x \le b$. It can be shown that the double integral $\iint\limits_R f(x, y) \; dA$ of a nonnegative function $f(x, y)$ gives the **volume of the solid** under the graph of f and above the region R.

Volume Formula

If $f(x, y) \ge 0$ for all points (x, y) in R, and if S is the solid bounded above by the surface $z = f(x, y)$ and below by R, then

$$\text{Volume of } S = \iint\limits_R f(x, y) \; dA$$

Here is an example illustrating the use of this formula.

EXAMPLE 3.3 Find the volume under the surface $z = e^{-x}e^{-y}$ and above the triangle with vertices $(0, 0)$, $(1, 0)$, and $(0, 1)$.

SOLUTION

The triangle is shown in Figure 3.4a. The surface under which you are to find the volume is the graph of the function $f(x, y) = e^{-x}e^{-y}$, which is nonnegative for all values of x and y. Hence,

$$\text{Volume} = \int_0^1 \int_0^{1-x} e^{-x}e^{-y}\, dy\, dx$$

$$= \int_0^1 \left(-e^{-x}e^{-y}\Big|_{y=0}^{y=1-x} \right) dx$$

$$= \int_0^1 \left(-e^{-x}e^{x-1} + e^{-x} \right) dx$$

$$= \int_0^1 \left(-e^{-1} + e^{-x} \right) dx$$

$$= \left(-xe^{-1} - e^{-x} \right)\Big|_0^1$$

$$= 1 - 2e^{-1}$$

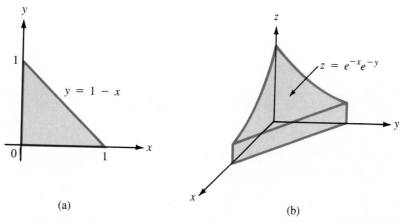

Figure 3.4 Triangle and solid for Example 3.3.

(a)

(b)

Notice that you do not have to draw the three-dimensional solid to compute its volume (although you do have to check that the function is nonnegative for all points in the region of integration). The picture of the solid in Figure 3.4b is shown here simply to help you visualize the situation.

The Average Value of a Function

In Chapter 7, Section 2, you saw that the average value of a continuous function $f(x)$ over an interval $a \leq x \leq b$ is given by the integral formula

$$\text{Average value} = \frac{1}{b-a} \int_a^b f(x) \, dx$$

That is, to find the average value of a function of one variable over an interval, you integrate the function over the interval and divide by the length of the interval. The two-variable procedure is similar. In particular, to find the average value of a function of two variables over a region, you integrate the function over the region and divide by the area of the region.

Average-Value Formula

> The average value of a continuous function $f(x, y)$ over a region R is given by the formula
>
> $$\text{Average value} = \frac{1}{\text{area of } R} \iint_R f(x, y) \, dA$$

EXAMPLE 3.4 Suppose R represents the surface of a lake and $f(x, y)$ the depth of the lake at the point (x, y). Find an expression for the average depth of the lake.

SOLUTION
According to the average-value formula,

$$\text{Average depth} = \frac{1}{\text{area of } R} \iint_R f(x, y) \, dA$$

Moreover, the area of R is itself a double integral:

$$\text{Area of } R = \iint_R 1 \, dA$$

Hence,

$$\text{Average depth} = \frac{\displaystyle\iint_R f(x, y) \, dA}{\displaystyle\iint_R 1 \, dA}$$

EXAMPLE 3.5 Find the average value of the function $f(x, y) = xe^y$ on the triangle with vertices $(0, 0)$, $(1, 0)$, and $(1, 1)$.

SOLUTION
The triangle is shown in Figure 3.5. Its area can be calculated without integration using the familiar formula from plane geometry:

$$\text{Area} = \frac{1}{2}(\text{base})(\text{height}) = \frac{1}{2}(1)(1) = \frac{1}{2}$$

Hence,

$$\text{Average value} = \frac{1}{1/2} \int_0^1 \int_0^x xe^y \, dy \, dx$$

$$= 2 \int_0^1 \left(xe^y \Big|_{y=0}^{y=x} \right) dx$$

$$= 2 \int_0^1 (xe^x - x) \, dx$$

$$= 2\left(xe^x - e^x - \frac{1}{2}x^2 \right)\Big|_0^1$$

$$= 2\left[\left(e - e - \frac{1}{2} \right) - (-1) \right]$$

$$= 1$$

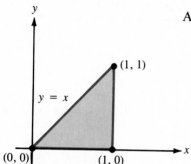

Figure 3.5 Triangle with vertices $(0, 0)$, $(1, 0)$, and $(1, 1)$.

Joint Probability Density Functions

One of the most important applications of integration in the social, managerial, and life sciences is the computation of probabilities. The technique of integrating probability density functions to find probabilities was introduced in Chapter 6, Section 2, and discussed in more detail in Chapter 7, Section 4. Recall that a probability density function for a single variable x is a nonnegative function $f(x)$ such that the probability that x is between a and b is given by the formula

$$P(a \le x \le b) = \int_a^b f(x) \, dx$$

In geometric terms, the probability $P(a \le x \le b)$ is the area under the graph of f from $x = a$ to $x = b$ (Figure 3.6a).

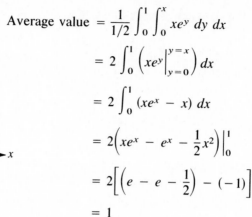

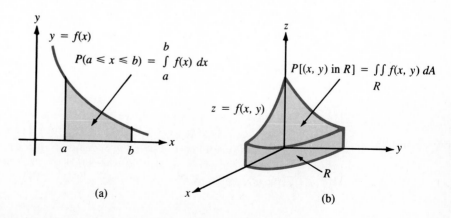

Figure 3.6 Probability as area and volume.

(a)

(b)

In situations involving two variables, you compute probabilities by evaluating double integrals of a two-variable density function. In particular, you integrate a **joint probability density function,** which is a non-negative function $f(x, y)$ such that the probability that x is between a and b and y is between c and d is given by the formula

$$P(a \leq x \leq b \text{ and } c \leq y \leq d) = \int_{c}^{d} \int_{a}^{b} f(x, y) \, dx \, dy$$

$$= \int_{a}^{b} \int_{c}^{d} f(x, y) \, dy \, dx$$

More generally, the probability that the ordered pair (x, y) lies in a region R is given by the formula

$$P[(x, y) \text{ in } R] = \iint_{R} f(x, y) \, dA$$

In geometric terms, the probability that (x, y) is in R is the volume under the graph of f above the region R (Figure 3.6b).

The techniques for constructing joint probability density functions from experimental data are beyond the scope of this book and are discussed in most probability and statistics texts. The use of double integrals to compute probabilities once the appropriate density functions are known is illustrated in the next two examples.

EXAMPLE 3.6   Smoke detectors manufactured by a certain firm contain two independent circuits, one manufactured at the firm's California plant and the other at the firm's plant in Ohio. Reliability studies suggest that if x denotes the life span (in years) of a randomly selected circuit from the California plant and y the life span (in years) of a randomly selected circuit from the Ohio plant, the joint probability density function for x and y is $f(x, y) = e^{-x}e^{-y}$. If the smoke detector will operate as long as either of its circuits is operating, find the probability that a randomly selected smoke detector will fail within 1 year.

SOLUTION

Since the smoke detector will operate as long as either of its circuits is operating, it will fail within 1 year if and only if *both* of its circuits fail within 1 year. The desired probability is therefore the probability that both $0 \leq x \leq 1$ and $0 \leq y \leq 1$. The points (x, y) for which both these inequalities hold form the square R shown in Figure 3.7. The corresponding probability is the double integral of the density function f over this region R. That is,

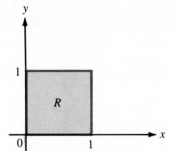

Figure 3.7 Square consisting of all points (x, y) for which $0 \le x \le 1$ and $0 \le y \le 1$.

$$P(0 \le x \le 1 \text{ and } 0 \le y \le 1) = \int_0^1 \int_0^1 e^{-x} e^{-y} \, dx \, dy$$

$$= \int_0^1 \left(-e^{-x} e^{-y} \Big|_{x=0}^{x=1} \right) dy$$

$$= \int_0^1 -(e^{-1} - 1) e^{-y} \, dy$$

$$= (e^{-1} - 1) e^{-y} \Big|_0^1$$

$$= (e^{-1} - 1)^2$$

$$\approx 0.3996$$

EXAMPLE 3.7   Suppose x denotes the time (in minutes) that a person sits in the waiting room of a certain dentist and y the time (in minutes) that a person stands in line at a certain bank. You have an appointment with the dentist, after which you are planning to cash a check at the bank. If $f(x, y)$ is the joint probability density function for x and y, write down a double integral that gives the probability that your *total* waiting time will be no more than 20 minutes.

SOLUTION
The goal is to find the probability that $x + y \le 20$. The points (x, y) for which $x + y \le 20$ lie on or below the line $x + y = 20$. Moreover, since x and y stand for nonnegative quantities, only those points in the first quadrant are meaningful in this particular context. The problem, then, is to find the probability that a randomly selected point (x, y) lies in R, where R is the region in the first quadrant bounded by the line $x + y = 20$ and the coordinate axes (Figure 3.8). This probability is given by the double integral of the density function f over the region R. That is,

$$P[(x, y) \text{ in } R] = \iint_R f(x, y) \, dA = \int_0^{20} \int_0^{20-x} f(x, y) \, dy \, dx$$

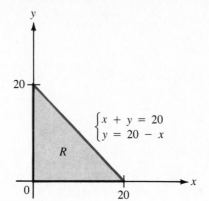

Figure 3.8 Triangle consisting of all points (x, y) for which $x + y \leq 20$.

Problems

In Problems 1 through 10, use a double integral to find the area of R.

1. R is the triangle with vertices $(-4, 0)$, $(2, 0)$, and $(2, 6)$.

2. R is the triangle with vertices $(0, -1)$, $(-2, 1)$, and $(2, 1)$.

3. R is the region bounded by $y = \frac{1}{2}x^2$ and $y = 2x$.

4. R is the region bounded by $y = \sqrt{x}$ and $y = x^2$.

5. R is the region bounded by $y = x^2 - 4x + 3$ and the x axis.

6. R is the region bounded by $y = x^2 + 6x + 5$ and the x axis.

7. R is the region bounded by $y = \ln x$, $y = 0$, and $x = e$.

8. R is the region bounded by $y = x$, $y = \ln x$, $y = 0$, and $y = 1$.

9. R is the region in the first quadrant bounded by $y = 4 - x^2$, $y = 3x$, and $y = 0$.

10. R is the region in the first quadrant that lies under the curve $y = \dfrac{16}{x}$ and is bounded by this curve and the lines $y = x$, $y = 0$, and $x = 8$.

In Problems 11 through 16, find the volume of the solid bounded above by the surface $z = f(x, y)$ and below by the region R.

11. $f(x, y) = 6 - 2x - y$; R is the rectangle with vertices $(0, 0)$, $(1, 0)$, $(0, 2)$, and $(1, 2)$.

12. $f(x, y) = 2x + y$; R is the triangle bounded by $y = x$, $y = 2 - x$, and the x axis.

13. $f(x, y) = e^{y^2}$; R is the triangle bounded by $y = \frac{1}{2}x$, $x = 0$, and $y = 1$.

14. $f(x, y) = x + 1$; R is the region in the first quadrant bounded by $y = 8 - x^2$, $y = x^2$, and the y axis.

15. $f(x, y) = e^{2x+y}$; R is the triangle with vertices $(0, 0)$, $(1, 0)$, and $(0, 1)$.

16. $f(x, y) = 4xe^y$; R is the triangle bounded by $y = 2x$, $y = 2$, and $x = 0$.

Air pollution 17. Suppose R is the region within the boundary of Los Angeles County and $f(x, y)$ is the number of units of carbon monoxide in the air at noon at the point (x, y) in R. Find an expression for the average level of carbon monoxide in the county at noon.

Elevation 18. Suppose the rectangle R in the first quadrant bounded by the coordinate axes and the lines $x = 3$ and $y = 1$ represents the region inside the boundary of a certain national park. Suppose the elevation above sea level at any point (x, y) in the park is $f(x, y) = 900(2x + y^2)$ feet. Find the average elevation in the park.

Property value 19. Suppose the triangle R with vertices $(0, 0)$, $(1, 0)$, and $(1, 1)$ represents the region inside the boundary of a certain rural congressional district. Suppose the property value at any point (x, y) in the district is $f(x, y) = 400xe^{-y}$ dollars per acre. Find the average property value in the district.

In Problems 20 through 25, find the average value of the given function over the specified region R.

20. $f(x, y) = 6xy$; R is the triangle with vertices $(0, 0)$, $(0, 1)$, and $(3, 1)$.

21. $f(x, y) = 3y$; R is the triangle with vertices $(0, 0)$, $(4, 0)$, and $(2, 2)$.

22. $f(x, y) = e^{x^2}$; R is the triangle with vertices $(0, 0)$, $(1, 0)$, and $(1, 1)$.

23. $f(x, y) = x$; R is the region bounded by $y = 4 - x^2$ and the x axis.

24. $f(x, y) = e^{x^3}$; R is the region in the first quadrant bounded by $y = x^2$, $y = 0$, and $x = 1$.

25. $f(x, y) = e^x y^{-1/2}$; R is the region in the first quadrant bounded by $y = x^2$, $x = 0$, and $y = 1$.

Probability 26. Suppose the joint probability density function for the nonnegative variables x and y is $f(x, y) = 2e^{-2x}e^{-y}$. Find the probability that $0 \le x \le 1$ and $1 \le y \le 2$.

Probability 27. Suppose the joint probability density function for the nonnegative variables x and y is $f(x, y) = xe^{-x}e^{-y}$. Find the probability that $0 \le x \le 1$ and $0 \le y \le 2$.

Probability 28. Suppose the joint probability density function for the nonnegative variables x and y is $f(x, y) = 2e^{-2x}e^{-y}$. Find the probability that $x + y \le 1$.

Probability 29. Suppose the joint probability density function for the nonnegative variables x and y is $f(x, y) = xe^{-x}e^{-y}$. Find the probability that $x + y \leq 1$.

Health care   30. Suppose x is the length of time (in days) that a person stays in the hospital after abdominal surgery and y the length of time (in days) that a person stays in the hospital after orthopedic surgery. On Monday, the patient in bed 107A undergoes an emergency appendectomy (abdominal surgery), while the patient's roommate in bed 107B undergoes (orthopedic) surgery for the repair of torn knee cartilage. If the joint probability density function for x and y is $f(x, y) = \frac{1}{12}e^{-x/4}e^{-y/3}$, find the probability that both patients will be discharged from the hospital within 3 days.

Warranty protection   31. A certain appliance consisting of two independent electronic components will be usable as long as either one of its components is still operating. The appliance carries a warranty from the manufacturer guaranteeing replacement if the appliance becomes unusable within 1 year of the date of purchase. Let x denote the life span (in years) of the first component and y the life span (in years) of the second, and suppose that the joint probability density function for x and y is $f(x, y) = \frac{1}{4}e^{-x/2}e^{-y/2}$. You purchase one of these appliances, selected at random from the manufacturer's stock. Find the probability that the warranty will expire before your appliance becomes unusable. (*Hint*: You want the probability that the appliance does *not* fail during the first year, which you can compute as 1 minus the probability that it *does* fail during this period.)

Customer service   32. Suppose x denotes the time (in minutes) that a person stands in line at a certain bank and y the duration (in minutes) of a routine transaction at the teller's window. You arrive at the bank to deposit a check. If the joint probability density function for x and y is $f(x, y) = \frac{1}{8}e^{-x/4}e^{-y/2}$, find the probability that you will complete your business at the bank within 8 minutes.

Insurance sales   33. Suppose x denotes the time (in minutes) that a person spends with an agent choosing a life insurance policy and y the time (in minutes) that the agent spends doing the paperwork once the client has decided. You arrange to meet with an insurance agent to buy a life insurance policy. If the joint probability density function for x and y is $f(x, y) = \frac{1}{300}e^{-x/30}e^{-y/10}$, find the probability that the entire transaction will take more than half an hour. (*Hint*: The probability you want is 1 minus the probability that the transaction will be completed within 30 minutes.)

Chapter Summary and Review Problems

Important Terms, Symbols, and Formulas

Double integrals: If R is described by vertical cross sections,

$$\iint_R f(x, y)\, dA = \int_a^b \int_{g_1(x)}^{g_2(x)} f(x, y)\, dy\, dx$$

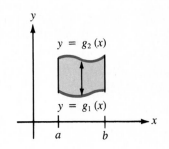

If R is described by horizontal cross sections,

$$\iint_R f(x, y)\, dA = \int_c^d \int_{h_1(y)}^{h_2(y)} f(x, y)\, dx\, dy$$

Area of $R = \iint_R 1\, dA$

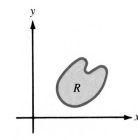

Volume of $S = \iint_R f(x, y)\, dA$

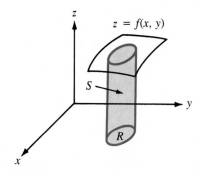

Average value of f over $R = \dfrac{1}{\text{area of } R} \iint_R f(x, y)\, dA$

Joint probability density function:

$$P[(x, y) \text{ in } R] = \iint_R f(x, y)\, dA$$

Review Problems

In Problems 1 through 5, evaluate the given double integral.

1. $\int_0^1 \int_{-2}^0 (2x + 3y)\, dy\, dx$

2. $\int_0^1 \int_0^y x\sqrt{1-y^3}\, dx\, dy$

3. $\int_0^1 \int_{-x}^x \frac{6xy^2}{x^5 + 1}\, dy\, dx$

4. $\int_0^1 \int_0^2 e^{-x-y}\, dy\, dx$

5. $\int_0^1 \int_0^x xe^{2y}\, dy\, dx$

In Problems 6 through 10, evaluate the given double integral for the specified region R.

6. $\iint_R 6x^2y\, dA$, where R is the rectangle with vertices $(-1, 0)$, $(2, 0)$, $(2, 3)$, and $(-1, 3)$.

7. $\iint_R (x + 2y)\, dA$, where R is the triangle with vertices $(0, 0)$, $(0, 3)$, and $(1, 1)$.

8. $\iint_R 40x^2y\, dA$, where R is the region bounded by $y = \sqrt{x}$ and $y = \frac{1}{2}x$.

9. $\iint_R 6y^2e^{x^2}\, dA$, where R is the region in the first quadrant bounded by $y = \sqrt[3]{x}$ and $y = x$.

10. $\iint_R 32xy^3\, dA$, where R is the region bounded by $y = \frac{8}{x}$, $y = \sqrt{x}$, and $y = 1$.

In Problems 11 through 14, sketch the region of integration for the given iterated integral, and set up an equivalent iterated integral (or integrals) with the order of integration reversed.

11. $\int_0^4 \int_{x^2/8}^{\sqrt{x}} f(x, y)\, dy\, dx$

12. $\int_0^2 \int_1^{e^y} f(x, y)\, dx\, dy$

13. $\int_0^2 \int_{x^2}^{8-2x} f(x, y)\, dy\, dx$

14. $\int_{-3}^1 \int_{x^2}^{6-x} f(x, y)\, dy\, dx$

15. Find the area of the region bounded by $y = x^2 - 4$ and $y = x - 2$.

16. Find the area of the region bounded by $y = \ln x$, $y = 0$, $y = 1$, and $x = 0$.

17. Find the area of the region in the first quadrant bounded by $y = x^2$, $y = 2 - x$, and $y = 0$.

18. Find the volume under the surface $z = 2xy$ and above the triangle with vertices $(0, 0)$, $(2, 0)$, and $(0, 1)$.

19. Find the volume under the surface $z = xe^y$ and above the region bounded by $y = x$ and $y = x^2$.

20. Find the average value of $f(x, y) = 2xy$ over the region bounded by $y = x^2$ and $y = x$.

21. Suppose R is the region inside the boundary of a certain county in the midwest. After a winter storm, the depth of snow at the point (x, y) in R was $f(x, y)$ feet.
    (a) Assuming x and y are measured in feet, find an expression for the number of cubic feet of snow that fell on the county.
    (b) Find an expression for the average depth of snow in the county.

22. Suppose the joint probability density function for the nonnegative variables x and y is $f(x, y) = 6e^{-2x}e^{-3y}$.
    (a) Find the probability that $0 \le x \le 1$ and $0 \le y \le 2$.
    (b) Find the probability that $x + y \le 2$.

23. Suppose x denotes the time (in minutes) that a person spends in a certain doctor's waiting room and y the duration (in minutes) of a complete physical examination. You arrive at the doctor's office for a physical 50 minutes before you are due to leave for a meeting. If the joint probability density function for x and y is $f(x, y) = \dfrac{1}{500} e^{-x/10}e^{-y/50}$, find the probability that you will be late leaving for your meeting.

Chapter

Infinite Series and Taylor Approximation

1   Infinite Series

The sum of infinitely many numbers may be finite. This statement, which may seem paradoxical at first, plays a central role in mathematics and has a variety of important applications. The purpose of this chapter is to explore its meaning and some of its consequences.

You are already familiar with the phenomenon of a finite-valued infinite sum. You know, for example, that the repeating decimal $0.333\ldots$ stands for the infinite sum $\dfrac{3}{10} + \dfrac{3}{100} + \dfrac{3}{1,000} + \cdots$ and that its value is the finite number $\dfrac{1}{3}$. What may not be familiar, however, is exactly what it means to say that this infinite sum "adds up to" $\dfrac{1}{3}$. The situation will be clarified in this introductory section.

Infinite Series

An expression of the form

$$a_1 + a_2 + \cdots + a_n + \cdots$$

is called an **infinite series.** It is customary to use summation notation to write series compactly as follows:

$$a_1 + a_2 + \cdots + a_n + \cdots = \sum_{n=1}^{\infty} a_n$$

The use of this notation is illustrated in the following example.

EXAMPLE 1.1   (a) Write out some representative terms of the series $\sum_{n=1}^{\infty} \dfrac{1}{2^n}$.

(b) Use summation notation to write the series $1 - \dfrac{1}{4} + \dfrac{1}{9} - \dfrac{1}{16} + \cdots$ in compact form.

SOLUTION

(a) $\displaystyle\sum_{n=1}^{\infty} \frac{1}{2^n} = \frac{1}{2} + \frac{1}{4} + \frac{1}{8} + \cdots + \frac{1}{2^n} + \cdots$

(b) The nth term of this series is

$$\frac{(-1)^{n+1}}{n^2}$$

where the factor $(-1)^{n+1}$ generates the alternating signs starting with a plus sign when $n = 1$. Thus,

$$1 - \frac{1}{4} + \frac{1}{9} - \frac{1}{16} + \cdots = \sum_{n=1}^{\infty} \frac{(-1)^{n+1}}{n^2}$$

Applications of Infinite Series

Infinite series arise in a variety of practical situations. In Section 2 you will see applications of series to such fields as economics, medicine, and probability. In Section 3 you will see how infinite series are used in the approximation of functions that may be difficult or impossible to evaluate directly. As a preview of these applications, and to give you a more concrete feel for how series arise in practice, here is an investment problem that leads to an infinite series.

EXAMPLE 1.2 Find an expression for the amount of money you should invest today at an annual interest rate of 10 percent compounded continuously so that, starting next year, you can make annual withdrawals of $400 in perpetuity.

SOLUTION

The amount you should invest today to generate the desired sequence of withdrawals is the sum of the present values of the individual withdrawals. You compute the present value of each withdrawal using the formula $P = Be^{-rt}$ from Chapter 4, Section 5, with $B = 400$, $r = 0.1$, and t being the time (in years) at which the withdrawal is made. Thus,

$$P_1 = \text{present value of 1st withdrawal} = 400e^{-0.1(1)} = 400e^{-0.1}$$

$$P_2 = \text{present value of 2nd withdrawal} = 400e^{-0.1(2)} = 400e^{-0.2}$$

$$\vdots$$

$$P_n = \text{present value of } n\text{th withdrawal} = 400e^{-0.1(n)}$$

and so on. The situation is illustrated in Figure 1.1. The amount that you should invest today is the sum of these (infinitely many) present values.

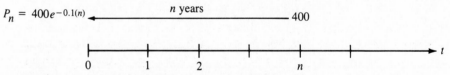

Figure 1.1 The present value of the nth withdrawal.

That is,

$$\text{Amount to be invested} = \sum_{n=1}^{\infty} P_n = \sum_{n=1}^{\infty} 400e^{-0.1(n)}$$

In Section 2 you will learn how to show that this infinite series "adds up to" $3,808.33. As an experiment, use your calculator to add the first several terms of this series and see what happens as you include more and more terms.

Convergence and Divergence of Infinite Series

Roughly speaking, an infinite series is said to **converge** if it "adds up to" a finite number and to **diverge** if it does not. A more precise statement of the criterion for convergence involves the sum

$$S_n = a_1 + a_2 + \cdots + a_n$$

of the first n terms of the series. This (finite) sum is called the **nth partial sum** of the series, and its behavior as n approaches infinity determines the convergence or divergence of the series. Here is a summary of the criterion.

Convergence and Divergence

An infinite series $\displaystyle\sum_{n=1}^{\infty} a_n$ with nth partial sum

$$S_n = a_1 + a_2 + \cdots + a_n$$

is said to converge if there is a (finite) number S such that

$$\lim_{n \to \infty} S_n = S$$

In this case, S is said to be the sum of the series, and one writes

$$\sum_{n=1}^{\infty} a_n = S$$

If S_n does not have a finite limit as n approaches infinity, the series is said to diverge.

According to this criterion, to determine the convergence or divergence of an infinite series, you start by finding an expression for the sum of the first n terms of the series and then take the limit of this finite sum as n approaches infinity. This procedure should be reminiscent of the one that was used in Chapter 7, Section 3, to determine the convergence or divergence of an improper integral.

The notions of convergence and divergence of infinite series are illustrated in the following examples.

EXAMPLE 1.3   Investigate the possible convergence of the series $\displaystyle\sum_{n=1}^{\infty} \frac{1}{n(n+1)}$.

SOLUTION
To get a feel for the series, you might begin by computing a few of its partial sums.

$$S_1 = \frac{1}{1 \cdot 2} = \frac{1}{2}$$

$$S_2 = \frac{1}{1 \cdot 2} + \frac{1}{2 \cdot 3} = \frac{1}{2} + \frac{1}{6} = \frac{2}{3}$$

$$S_3 = \left(\frac{1}{1 \cdot 2} + \frac{1}{2 \cdot 3}\right) + \frac{1}{3 \cdot 4} = \frac{2}{3} + \frac{1}{12} = \frac{3}{4}$$

$$S_4 = \left(\frac{1}{1 \cdot 2} + \frac{1}{2 \cdot 3} + \frac{1}{3 \cdot 4}\right) + \frac{1}{4 \cdot 5} = \frac{3}{4} + \frac{1}{20} = \frac{4}{5}$$

$$S_5 = \left(\frac{1}{1 \cdot 2} + \frac{1}{2 \cdot 3} + \frac{1}{3 \cdot 4} + \frac{1}{4 \cdot 5}\right) + \frac{1}{5 \cdot 6} = \frac{4}{5} + \frac{1}{30} = \frac{5}{6}$$

These calculations *suggest* that $S_n = \dfrac{n}{n + 1}$ and that the sum of the series is 1. However, since the reason for the observed pattern among the partial sums is probably not clear, it is advisable to seek an algebraic solution to the problem.

The trick is to use the fact that

$$\frac{1}{n(n + 1)} = \frac{1}{n} - \frac{1}{n + 1}$$

to write the series as

$$\sum_{n=1}^{\infty} \frac{1}{n(n + 1)} = \sum_{n=1}^{\infty} \left(\frac{1}{n} - \frac{1}{n + 1}\right)$$

and the nth partial sum as

$$S_n = \left(\frac{1}{1} - \frac{1}{2}\right) + \left(\frac{1}{2} - \frac{1}{3}\right) + \left(\frac{1}{3} - \frac{1}{4}\right) + \cdots + \left(\frac{1}{n} - \frac{1}{n + 1}\right)$$

Since all but the first and last terms in this sum cancel, S_n can be rewritten as

$$S_n = 1 - \frac{1}{n + 1} = \frac{n}{n + 1}$$

Finally, since

$$\lim_{n \to \infty} S_n = \lim_{n \to \infty} \frac{n}{n + 1} = 1$$

it follows that the series converges to 1. That is,

$$\sum_{n=1}^{\infty} \frac{1}{n(n + 1)} = 1$$

The series in the preceding example is known as a **telescoping series** because of the cancellation that takes place within the nth partial sum.

The series in the next example is the one that corresponds to the decimal representation of the fraction $\frac{1}{3}$. It is an example of an important type of series that will be examined in more detail in Section 2.

EXAMPLE 1.4 Investigate the possible convergence of the series $\sum_{n=1}^{\infty} \frac{3}{10^n}$.

SOLUTION
Notice that each term of the series

$$\sum_{n=1}^{\infty} \frac{3}{10^n} = \frac{3}{10} + \frac{3}{10^2} + \frac{3}{10^3} + \cdots$$

is $\frac{1}{10}$ times the preceding term. This leads to the following trick for finding a compact formula for S_n.

Begin with the nth partial sum in the form

$$S_n = \frac{3}{10} + \frac{3}{10^2} + \frac{3}{10^3} + \cdots + \frac{3}{10^n}$$

and multiply both sides of this equation by $\frac{1}{10}$ to get

$$\frac{1}{10} S_n = \frac{3}{10^2} + \frac{3}{10^3} + \cdots + \frac{3}{10^n} + \frac{3}{10^{n+1}}$$

Now subtract the expression for $\frac{1}{10} S_n$ from the expression for S_n. Most of the terms cancel, and you are left with

$$\frac{9}{10} S_n = \frac{3}{10} - \frac{3}{10^{n+1}} = \frac{3}{10}\left(1 - \frac{1}{10^n}\right)$$

or

$$S_n = \frac{1}{3}\left(1 - \frac{1}{10^n}\right)$$

Then,

$$\lim_{n \to \infty} S_n = \lim_{n \to \infty} \frac{1}{3}\left(1 - \frac{1}{10^n}\right) = \frac{1}{3}$$

and so

$$\sum_{n=1}^{\infty} \frac{3}{10^n} = \frac{1}{3}$$

as expected.

A Necessary Condition for Convergence

The individual terms a_n of a convergent series $\sum_{n=1}^{\infty} a_n$ must approach zero as n increases without bound. To see this, write a_n as the difference of two partial sums as follows:

$$a_n = (a_1 + a_2 + \cdots + a_n) - (a_1 + a_2 + \cdots + a_{n-1})$$

$$= S_n - S_{n-1}$$

Then, if $\sum\limits_{n=1}^{\infty} a_n$ converges to some number S, both partial sums S_n and S_{n-1} approach S as n approaches infinity, and so

$$\lim_{n \to \infty} a_n = \lim_{n \to \infty} S_n - \lim_{n \to \infty} S_{n-1} = S - S = 0$$

This observation can be rephrased as a test for *divergence* as follows.

Test for Divergence

> If $\lim\limits_{n \to \infty} a_n \neq 0$, then $\sum\limits_{n=1}^{\infty} a_n$ diverges.

The use of this test is illustrated in the next example.

EXAMPLE 1.5   Investigate the possible convergence of the series $\sum\limits_{n=1}^{\infty} \dfrac{n+3}{2n+1}$.

SOLUTION
The limit of the nth term is

$$\lim_{n \to \infty} a_n = \lim_{n \to \infty} \frac{n+3}{2n+1} = \frac{1}{2}$$

which is not equal to zero. Hence the series cannot converge and must diverge.

Look again at the series in Example 1.5. Its individual terms a_n decrease as n increases. In particular,

$$a_1 = \frac{4}{3} \simeq 1.3333$$

$$a_2 = \frac{5}{5} = 1$$

$$a_3 = \frac{6}{7} \simeq 0.8571$$

$$a_4 = \frac{7}{9} \simeq 0.7778$$

$$a_5 = \frac{8}{11} \simeq 0.7273$$

.
.
.

$$a_{50} = \frac{53}{101} \simeq 0.5248$$

.
.
.

$$a_{100} = \frac{103}{201} \simeq 0.5124$$

and so on. However, it is because the terms do not decrease *to zero* (but approach $\frac{1}{2}$ instead) that the series $\sum\limits_{n=1}^{\infty} a_n$ cannot converge.

Warning Against Misuse of the Divergence Test

The test for divergence states that a series whose individual terms do not approach zero must diverge. The converse of this test is *not* true. That is, the fact that the individual terms of a series approach zero does not guarantee that the series converges. Here is an example.

EXAMPLE 1.6 Investigate the possible convergence of the series $\sum\limits_{n=1}^{\infty} \dfrac{1}{n}$.

SOLUTION
Since

$$\lim_{n \to \infty} a_n = \lim_{n \to \infty} \frac{1}{n} = 0$$

it is possible, but not guaranteed, that the series converges.

It is natural to proceed by computing a few of the partial sums to get a better feel for the series.

$$S_1 = 1$$

$$S_2 = 1 + \frac{1}{2} = \frac{3}{2} = 1.5$$

$$S_3 = \left(1 + \frac{1}{2}\right) + \frac{1}{3} = \frac{3}{2} + \frac{1}{3} = \frac{11}{6} \simeq 1.8333$$

$$S_4 = \left(1 + \frac{1}{2} + \frac{1}{3}\right) + \frac{1}{4} = \frac{11}{6} + \frac{1}{4} = \frac{25}{12} \simeq 2.0833$$

$$S_5 = \left(1 + \frac{1}{2} + \frac{1}{3} + \frac{1}{4}\right) + \frac{1}{5} = \frac{25}{12} + \frac{1}{5} = \frac{137}{60} \simeq 2.2833$$

$$S_6 = \left(1 + \frac{1}{2} + \frac{1}{3} + \frac{1}{4} + \frac{1}{5}\right) + \frac{1}{6} = \frac{137}{60} + \frac{1}{6} = \frac{147}{60} \doteq 2.45$$

Unfortunately, there is no obvious pattern among these partial sums, nor is it even clear whether or not they are approaching a finite limit.

Actually, this series diverges. To see this, group the terms as follows:

$$1 + \underbrace{\frac{1}{2}}_{} + \underbrace{\frac{1}{3} + \frac{1}{4}}_{2 \text{ terms}} + \underbrace{\frac{1}{5} + \frac{1}{6} + \frac{1}{7} + \frac{1}{8}}_{2^2 \text{ terms}} + \underbrace{\frac{1}{9} + \cdots + \frac{1}{16}}_{2^3 \text{ terms}} + \underbrace{\frac{1}{17} + \cdots}_{}$$

The sum of the terms in each group is greater than or equal to $\frac{1}{2}$. For example,

$$\frac{1}{3} + \frac{1}{4} > \frac{1}{4} + \frac{1}{4} = \frac{1}{2}$$

and

$$\frac{1}{5} + \frac{1}{6} + \frac{1}{7} + \frac{1}{8} > \frac{1}{8} + \frac{1}{8} + \frac{1}{8} + \frac{1}{8} = \frac{1}{2}$$

By including sufficiently many of these groups, you can make the corresponding partial sum as large as you like, from which it follows that the infinite series must diverge.

The series $\sum\limits_{n=1}^{\infty} \frac{1}{n}$ in Example 1.6 is known as the **harmonic series** and is one of the most well-known examples of a divergent series whose individual terms approach zero.

It is instructive to compare the convergent series $\sum\limits_{n=1}^{\infty} \frac{1}{n(n+1)}$ from Example 1.3 with the divergent harmonic series $\sum\limits_{n=1}^{\infty} \frac{1}{n}$ from Example 1.6. Each is a series of positive terms that approach zero as n increases. Yet one converges and has a finite sum, while the other diverges and has an "infinite sum." The reason for the difference is that, as n increases, the terms $\frac{1}{n(n+1)}$ approach zero more quickly than the terms $\frac{1}{n}$ do. The situation is illustrated in the following table and should be reminiscent of the difference between convergent and divergent improper integrals.

n	1	2	3	4	5	10	50
$\dfrac{1}{n}$	1	0.5	0.3333	0.25	0.2	0.1	0.02
$\dfrac{1}{n(n+1)}$	0.5	0.1667	0.0833	0.05	0.0333	0.0091	0.0004

A Useful Property of Convergent Series

Like finite sums, convergent infinite series obey a distributive law with respect to multiplication. According to this law, if you multiply each term of a convergent series by a constant c, you get a new series whose sum is c times the sum of the original series.

The Distributive Law for Series

For any constant c,

$$\sum_{n=1}^{\infty} ca_n = c \sum_{n=1}^{\infty} a_n$$

This law follows from the fact that the sum of an infinite series is determined by its nth partial sum, which, being a finite sum, obeys the familiar distributive law from elementary algebra. In particular,

$$\sum_{n=1}^{\infty} ca_n = \lim_{n \to \infty} (ca_1 + ca_2 + \cdots + ca_n)$$

$$= \lim_{n \to \infty} c(a_1 + a_2 + \cdots + a_n)$$

$$= c \lim_{n \to \infty} (a_1 + a_2 + \cdots + a_n)$$

$$= c \sum_{n=1}^{\infty} a_n$$

The distributive law also holds for divergent series in the sense that if $\sum_{n=1}^{\infty} a_n$ diverges, so does $\sum_{n=1}^{\infty} ca_n$ (provided, of course that $c \neq 0$).

The distributive law allows you to factor constants out of infinite series. Such factorization will be used several times in the next section to simplify calculations.

Problems

In Problems 1 through 6, use summation notation to write the given series in compact form.

1. $\dfrac{1}{3} + \dfrac{1}{9} + \dfrac{1}{27} + \dfrac{1}{81} + \cdots$

2. $1 + \dfrac{1}{8} + \dfrac{1}{27} + \dfrac{1}{64} + \cdots$

3. $\dfrac{1}{2} + \dfrac{2}{3} + \dfrac{3}{4} + \dfrac{4}{5} + \cdots$

4. $\dfrac{1}{3} + \dfrac{2}{5} + \dfrac{3}{7} + \dfrac{4}{9} + \cdots$

5. $\dfrac{1}{2} - \dfrac{4}{3} + \dfrac{9}{4} - \dfrac{16}{5} + \cdots$

6. $-3 + \dfrac{9}{4} - \dfrac{27}{9} + \dfrac{81}{16} - \cdots$

In Problems 7 through 10, find the fourth partial sum S_4 of the given series.

7. $\displaystyle\sum_{n=1}^{\infty} \dfrac{1}{2^n}$

8. $\displaystyle\sum_{n=1}^{\infty} \dfrac{n}{n+1}$

9. $\displaystyle\sum_{n=1}^{\infty} \dfrac{(-1)^n}{n}$

10. $\displaystyle\sum_{n=1}^{\infty} \dfrac{(-1)^{n+1}}{n(n+1)}$

In Problems 11 through 18, find the sum of the given convergent series by taking the limit of a compact expression for the nth partial sum.

11. $\displaystyle\sum_{n=1}^{\infty} \left(\dfrac{1}{n+3} - \dfrac{1}{n+4} \right)$

12. $\displaystyle\sum_{n=1}^{\infty} \left(\dfrac{1}{n} - \dfrac{1}{n+2} \right)$

13. $\displaystyle\sum_{n=1}^{\infty} \dfrac{1}{(n+1)(n+2)}$

14. $\displaystyle\sum_{n=1}^{\infty} \dfrac{1}{(n+2)(n+3)}$

15. $\displaystyle\sum_{n=1}^{\infty} \dfrac{6}{10^n}$

16. $\displaystyle\sum_{n=1}^{\infty} \dfrac{1}{2^n}$

17. $\displaystyle\sum_{n=1}^{\infty} \dfrac{4}{3^n}$

18. $\displaystyle\sum_{n=1}^{\infty} \left(-\dfrac{1}{5} \right)^n$

In Problems 19 through 25, determine whether or not the given series converges, and if it does, find its sum.

19. $\displaystyle\sum_{n=1}^{\infty} \dfrac{n}{2n+1}$

20. $\displaystyle\sum_{n=1}^{\infty} \dfrac{1}{n+1}$

21. $\displaystyle\sum_{n=1}^{\infty} \left(\dfrac{1}{n+1} - \dfrac{1}{n} \right)$

22. $\displaystyle\sum_{n=1}^{\infty} \dfrac{n^2}{50n^2+1}$

23. $\displaystyle\sum_{n=1}^{\infty} \left(-\dfrac{2}{3} \right)^n$

24. $\displaystyle\sum_{n=1}^{\infty} \left(-\dfrac{3}{2} \right)^n$

25. $\displaystyle\sum_{n=1}^{\infty} \dfrac{1}{\sqrt{n}}$ (*Hint:* Show that $S_n \geq \dfrac{n}{\sqrt{n}} = \sqrt{n}$.)

2 The Geometric Series

A **geometric series** is an infinite series in which the ratio of successive terms is constant. For example, the series

$$3 + \frac{3}{2} + \frac{3}{4} + \frac{3}{8} + \cdots$$

is geometric because each term is one-half of the preceding term. In general, a geometric series is a series of the form

$$\sum_{n=m}^{\infty} ar^n = ar^m + ar^{m+1} + ar^{m+2} + \cdots$$

The constant r, which is the ratio of the successive terms of a geometric series, is known as the **ratio** of the series. The ratio of a geometric series may be positive or negative. For example,

$$\sum_{n=0}^{\infty} \frac{2}{(-3)^n} = 2 - \frac{2}{3} + \frac{2}{9} - \frac{2}{27} + \cdots$$

is a geometric series with ratio $r = -\frac{1}{3}$.

Elementary Geometric Series

The most elementary geometric series are those for which $m = 0$, and $a = 1$, that is, series of the form

$$\sum_{n=0}^{\infty} r^n = 1 + r + r^2 + \cdots$$

Because of the distributive law, every geometric series can be written as a multiple of one of these elementary geometric series. In particular,

$$\sum_{n=m}^{\infty} ar^n = ar^m + ar^{m+1} + ar^{m+2} + \cdots$$

$$= ar^m(1 + r + r^2 + \cdots) \qquad \text{by the distributive law}$$

$$= ar^m \sum_{n=0}^{\infty} r^n$$

For this reason, it suffices to study the elementary geometric series.

Convergence of Geometric Series

A geometric series $\sum_{n=0}^{\infty} r^n$ converges if the absolute value of its ratio r is less than 1 and diverges otherwise. The divergence when $|r| \geq 1$ follows from the fact that, in this case, the individual terms r^n do not approach zero as n increases. When $|r| < 1$, you can establish the convergence of the series and derive a formula for its sum by means of a calculation similar to the one you saw in Example 1.4 of the preceding section. In particular, begin with the nth partial sum in the form

$$S_n = 1 + r + r^2 + \cdots + r^{n-1}$$

(Notice that the power of r in the nth term is only $n - 1$ because the first term in the sum is $1 = r^0$.) Now multiply both sides of this equation by r to get

$$rS_n = r + r^2 + \cdots + r^{n-1} + r^n$$

and subtract the expression for rS_n from the expression for S_n to get

$$(1 - r)S_n = 1 - r^n$$

or

$$S_n = \frac{1 - r^n}{1 - r}$$

Since $|r| < 1$,

$$\lim_{n \to \infty} r^n = 0$$

and so

$$\lim_{n \to \infty} S_n = \lim_{n \to \infty} \frac{1 - r^n}{1 - r} = \frac{1}{1 - r}$$

Hence the series converges, and its sum is $\frac{1}{1 - r}$.

The Sum of a Geometric Series

If $|r| < 1$,

$$\sum_{n=0}^{\infty} r^n = \frac{1}{1 - r}$$

Notice that this formula for the sum of a geometric series is valid only for elementary series of the form

$$\sum_{n=0}^{\infty} r^n = 1 + r + r^2 + \cdots$$

that start with the term $r^0 = 1$. To apply the formula to any other geometric series, you will first have to rewrite the series as a constant multiple of an elementary one. The use of the formula is illustrated in the following examples.

EXAMPLE 2.1 Find the sum of the series $\sum\limits_{n=0}^{\infty} \left(-\dfrac{2}{3}\right)^n$.

SOLUTION
This is an elementary geometric series with $r = -\dfrac{2}{3}$. Hence,

$$\sum_{n=0}^{\infty} \left(-\frac{2}{3}\right)^n = \frac{1}{1 - (-2/3)} = \frac{1}{5/3} = \frac{3}{5}$$

EXAMPLE 2.2 Find the sum of the series $\sum\limits_{n=0}^{\infty} \dfrac{3}{2^n}$.

SOLUTION
$$\sum_{n=0}^{\infty} \frac{3}{2^n} = 3 \sum_{n=0}^{\infty} \left(\frac{1}{2}\right)^n = 3\left(\frac{1}{1 - 1/2}\right) = 3\left(\frac{1}{1/2}\right) = 6$$

EXAMPLE 2.3 Find the sum of the series $\sum\limits_{n=2}^{\infty} \dfrac{2}{5^n}$.

SOLUTION

$$\sum_{n=2}^{\infty} \frac{2}{5^n} = \frac{2}{5^2} + \frac{2}{5^3} + \frac{2}{5^4} + \cdots$$

$$= \frac{2}{5^2}\left(1 + \frac{1}{5} + \frac{1}{5^2} + \cdots\right) \qquad \text{by the distributive law}$$

$$= \frac{2}{25} \sum_{n=0}^{\infty} \left(\frac{1}{5}\right)^n$$

$$= \frac{2}{25}\left(\frac{1}{1 - 1/5}\right) = \frac{2}{25}\left(\frac{5}{4}\right) = \frac{1}{10}$$

EXAMPLE 2.4 Find the sum of the series $\sum\limits_{n=1}^{\infty} \dfrac{(-2)^n}{3^{n+1}}$.

SOLUTION

$$\sum_{n=1}^{\infty} \frac{(-2)^n}{3^{n+1}} = \frac{-2}{3^2} + \frac{(-2)^2}{3^3} + \frac{(-2)^3}{3^4} + \cdots$$

$$= \frac{-2}{3^2}\left[1 + \left(-\frac{2}{3}\right) + \left(-\frac{2}{3}\right)^2 + \cdots\right]$$

$$= -\frac{2}{9} \sum_{n=0}^{\infty} \left(-\frac{2}{3}\right)^n$$

$$= -\frac{2}{9}\left(\frac{1}{1 + 2/3}\right) = -\frac{2}{9}\left(\frac{3}{5}\right) = -\frac{2}{15}$$

Applications of Geometric Series

Geometric series arise in many branches of mathematics and in a variety of applications. Here are four illustrations taken from number theory, economics, medicine, and probability.

Repeating Decimals

A repeating decimal is a geometric series, and its value is the sum of the series. Here is an example.

EXAMPLE 2.5   Express the repeating decimal 0.232323 . . . as a fraction.

SOLUTION
Write the decimal as a geometric series as follows:

$$0.232323 \ldots = \frac{23}{100} + \frac{23}{10,000} + \frac{23}{1,000,000} + \cdots$$

$$= \frac{23}{100}\left[1 + \frac{1}{100} + \left(\frac{1}{100}\right)^2 + \cdots\right]$$

$$= \frac{23}{100} \sum_{n=0}^{\infty} \left(\frac{1}{100}\right)^n$$

$$= \frac{23}{100}\left(\frac{1}{1 - 1/100}\right) = \frac{23}{100}\left(\frac{100}{99}\right) = \frac{23}{99}$$

The Multiplier Effect in Economics

A tax rebate that returns a certain amount of money to taxpayers can result in spending that is many times this amount. This phenomenon is known in economics as the **multiplier effect.** It occurs because the portion of the rebate that is spent by one individual becomes income for one or more others who, in turn, spend some of it again, creating income for yet other individuals to spend. If the fraction of income that is saved remains constant as this process continues indefinitely, the total amount spent as a result of the rebate is the sum of a geometric series. Here is an example.

EXAMPLE 2.6   Suppose that nationwide, approximately 90 percent of all income is spent and 10 percent saved. How much additional spending will be generated by a 40 billion dollar tax rebate if savings habits do not change?

SOLUTION

The amount (in billions) spent by original recipients of the rebate is

$$0.9(40)$$

This becomes new income, of which 90 percent or

$$0.9[0.9(40)] = (0.9)^2(40)$$

is spent. This, in turn, generates additional spending of

$$0.9[(0.9)^2(40)] = (0.9)^3(40)$$

and so on. The total amount spent if this process continues indefinitely is

$$0.9(40) + (0.9)^2(40) + (0.9)^3(40) + \cdots$$

$$= 0.9(40)[1 + 0.9 + (0.9)^2 + \cdots]$$

$$= 36 \sum_{n=0}^{\infty} (0.9)^n = 36\left(\frac{1}{1 - 0.9}\right) = 360 \text{ billion}$$

Accumulation of Medication in the Body

In Example 3.5 of Chapter 7, a nuclear power plant generated radioactive waste, which decayed exponentially. The problem was to determine the long-run accumulation of waste from the plant. Because the waste was being generated *continuously*, the solution involved an (improper) integral. The situation in the next example is similar. A patient receives medication which is eliminated from the body exponentially, and the goal is to determine the long-run accumulation of medication in the patient's body. In this case, however, the medication is being administered in *discrete* doses, and the solution to the problem involves an infinite series rather than an integral.

EXAMPLE 2.7   A patient is given an injection of 10 units of a certain drug every 24 hours. The drug is eliminated exponentially so that the fraction that remains in the patient's body after t days is $f(t) = e^{-t/5}$. If the treatment is continued indefinitely, approximately how many units of the drug will eventually be in the patient's body just prior to an injection?

SOLUTION

Of the original dose of 10 units, only $10e^{-1/5}$ units are left in the patient's body after 1 day (just prior to the second injection). That is,

$$\text{Amount in body after 1 day} = S_1 = 10e^{-1/5}$$

The medication in the patient's body after 2 days consists of what remains from the first *two* doses. Of the original dose, only $10e^{-2/5}$ units are left (since 2 days have elapsed), and of the second dose, $10e^{-1/5}$ units remain. (See Figure 2.1.) Hence,

$$\text{Amount in body after 2 days} = S_2 = 10e^{-1/5} + 10e^{-2/5}$$

Similarly,

$$\begin{matrix}\text{Amount in body} \\ \text{after } n \text{ days}\end{matrix} = S_n = 10e^{-1/5} + 10e^{-2/5} + \cdots + 10e^{-n/5}$$

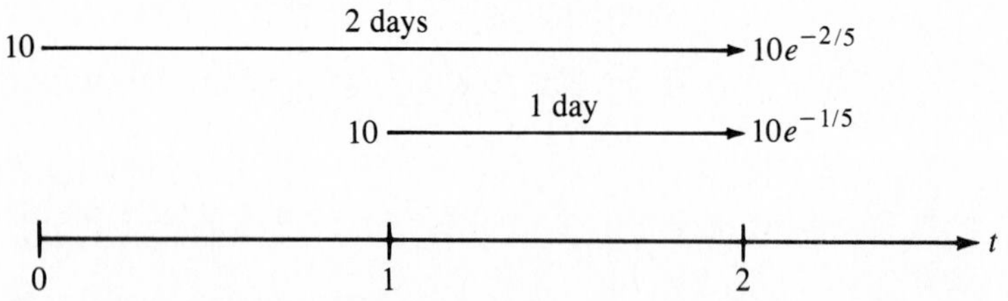

Figure 2.1 Amount of medication in the body after 2 days.

The amount S of medication in the patient's body in the long run is the limit of S_n as n approaches infinity. That is,

$$S = \lim_{n \to \infty} S_n = \sum_{n=1}^{\infty} 10e^{-n/5}$$

$$= \sum_{n=1}^{\infty} 10(e^{-1/5})^n = 10e^{-1/5} \sum_{n=0}^{\infty} (e^{-1/5})^n$$

$$= 10e^{-1/5}\left(\frac{1}{1 - e^{-1/5}}\right) \approx 45.17 \text{ units}$$

Geometric Random Variables

Suppose a random experiment with two possible outcomes (traditionally called "success" and "failure") is performed repeatedly until the first success occurs. The number x of repetitions required to produce the first success is known as a **geometric random variable.**

A geometric random variable is a discrete random variable that can take on any positive integer value. For any positive integer n, the probability that $x = n$ is the probability that the first $n - 1$ trials of the experiment result in failure and the nth in success:

$$\underbrace{F \quad F \quad F \cdots F}_{n-1 \text{ failures}} \quad \underbrace{S}_{1\text{st success}}$$

If, on each trial, the probability of success is p, then the probability of failure must be $1 - p$, and the probability of $n - 1$ failures followed by 1 success is the product

$$(1 - p)(1 - p)(1 - p) \cdots (1 - p)p$$

That is,

$$P(x = n) = (1 - p)^{n-1}p$$

If $p \neq 0$, success will eventually occur if the experiment is repeated indefinitely. Hence, $P(x \geq 1)$ ought to be 1. To compute this probability, you find the sum of a certain geometric series as follows:

$$
\begin{aligned}
P(x \geq 1) &= \sum_{n=1}^{\infty} P(x = n) \\
&= \sum_{n=1}^{\infty} (1 - p)^{n-1}p \\
&= p + (1 - p)p + (1 - p)^2 p + \cdots \\
&= p[1 + (1 - p) + (1 - p)^2 + \cdots] \\
&= p \sum_{n=0}^{\infty} (1 - p)^n \\
&= \frac{p}{1 - (1 - p)} = \frac{p}{p} = 1
\end{aligned}
$$

A simple application of geometric random variables is given in the next example.

EXAMPLE 2.8 In a certain two-person board game played with 1 die, each player needs a six to start play. You and your opponent take turns rolling the die. Find the probability that you will get the first six if you roll first.

SOLUTION
In this context, "success" is getting a six and "failure" is getting any of the other 5 possible numbers. Hence, the probability of success is $p = \dfrac{1}{6}$ and the probability of failure is $1 - p = \dfrac{5}{6}$.

Let x denote the number of the roll on which the first six comes up. If x is odd, the first six will occur on one of your rolls, while if x is even, the first six will occur on one of your opponent's rolls. Hence, the probability that you get the first six is

$$P(x \text{ is odd}) = P(x = 1) + P(x = 3) + P(x = 5) + \cdots$$

$$= \frac{1}{6} + \left(\frac{5}{6}\right)^2 \left(\frac{1}{6}\right) + \left(\frac{5}{6}\right)^4\left(\frac{1}{6}\right) + \cdots$$

$$= \frac{1}{6} + \left(\frac{25}{36}\right)\left(\frac{1}{6}\right) + \left(\frac{25}{36}\right)^2\left(\frac{1}{6}\right) + \cdots$$

$$= \frac{1}{6} \sum_{n=0}^{\infty} \left(\frac{25}{36}\right)^n$$

$$= \frac{1}{6}\left(\frac{1}{1 - 25/36}\right) = \frac{1}{6}\left(\frac{36}{11}\right) = \frac{6}{11} \approx 0.5455$$

Problems

In Problems 1 through 14, determine whether the given geometric series converges, and if so, find its sum.

1. $\displaystyle\sum_{n=0}^{\infty} \left(\frac{4}{5}\right)^n$

2. $\displaystyle\sum_{n=0}^{\infty} \left(-\frac{4}{5}\right)^n$

3. $\displaystyle\sum_{n=0}^{\infty} \frac{2}{3^n}$

4. $\displaystyle\sum_{n=0}^{\infty} \frac{2}{(-3)^n}$

5. $\displaystyle\sum_{n=1}^{\infty} \left(\frac{3}{2}\right)^n$

6. $\displaystyle\sum_{n=1}^{\infty} \frac{3}{2^n}$

7. $\displaystyle\sum_{n=2}^{\infty} \frac{3}{(-4)^n}$

8. $\displaystyle\sum_{n=2}^{\infty} \left(-\frac{4}{3}\right)^n$

9. $\displaystyle\sum_{n=1}^{\infty} 5(0.9)^n$

10. $\displaystyle\sum_{n=1}^{\infty} e^{-0.2n}$

11. $\displaystyle\sum_{n=1}^{\infty} \frac{3^n}{4^{n+2}}$

12. $\displaystyle\sum_{n=2}^{\infty} \frac{(-2)^{n-1}}{3^{n+1}}$

13. $\displaystyle\sum_{n=0}^{\infty} \frac{4^{n+1}}{5^{n-1}}$

14. $\displaystyle\sum_{n=2}^{\infty} (-1)^n \frac{2^{n+1}}{3^{n-3}}$

In Problems 15 through 18, express the given decimal as a fraction.

15. $0.3333\ldots$

16. $0.5555\ldots$

17. $0.252525\ldots$

18. $1.405405405\ldots$

Multiplier effect 19. Suppose that nationwide, approximately 92 percent of all income is spent and 8 percent saved. How much additional spending will be generated by a 50 billion dollar tax cut if savings habits do not change?

Present value
20. An investment guarantees annual payments of $1,000 in perpetuity, with the payments beginning immediately. Find the present value of this investment if the prevailing annual interest rate remains fixed at 12 percent compounded continuously. (*Hint:* The present value of the investment is the sum of the present values of the individual payments.)

Present value
21. How much should you invest today at an annual interest rate of 15 percent compounded continuously so that, starting next year, you can make annual withdrawals of $2,000 in perpetuity?

Accumulation of medication
22. A patient is given an injection of 20 units of a certain drug every 24 hours. The drug is eliminated exponentially so that the fraction that remains in the patient's body after t days is $f(t) = e^{-t/2}$. If the treatment is continued indefinitely, approximately how many units of the drug will eventually be in the patient's body just prior to an injection?

Group membership
23. Each January first, the administration of a certain private college adds 6 new members to its board of trustees. If the fraction of trustees who remain active for at least t years is $f(t) = e^{-0.2t}$, approximately how many active trustees will the college have on December thirty-first in the long run?

Games of chance
24. You and a friend take turns rolling a die until one of you wins by getting a three or a four. If your friend rolls first, find the probability that you will win.

Numismatics
25. In 1959, the design on the reverse side of U.S. pennies was changed from one depicting a pair of wheat stalks to one depicting the Lincoln Memorial. Today, very few of the old pennies are still in circulation. If only 0.2 percent of the pennies now in circulation were minted before 1959, find the probability that you will have to examine at least 100 pennies to find one with the old design. (Incidentally, there is an easy way to do this problem that does not involve infinite series. For practice, do it both ways.)

Quality control
26. Three inspectors take turns checking electronic components as they come off an assembly line. If 10 percent of all the components produced on the assembly line are defective, find the probability that the inspector who checks the first component will be the one who finds the first defective component.

3   Taylor Approximation

The purpose of this section is to develop a general procedure for finding infinite series that converge to given functions and to illustrate how the partial sums of these series can be used to approximate the functions.

Even if the original functions are difficult or impossible to evaluate directly, the partial sums of the corresponding infinite series turn out to be polynomials and can be evaluated with ease. The approximation of functions by the partial sums of their infinite series is how electronic computers and calculators are programmed to generate values of functions such as e^x, ln x, and the trigonometric functions.

Power Series

The infinite series that are used in the approximation of functions are series of the form

$$\sum_{n=0}^{\infty} a_n x^n = a_0 + a_1 x + a_2 x^2 + \cdots$$

in which each term is a constant times a power of x. Series of this type are called **power series.** Notice that the nth partial sum of such a power series is a polynomial

$$S_n = a_0 + a_1 x + a_2 x^2 + \cdots + a_{n-1} x^{n-1}$$

Convergence of Power Series

If the variable x in a power series is given a specific numerical value, the series becomes a series of constant terms, which either converges or diverges. A power series may converge for some values of x and diverge for other values of x. It can be shown that to each power series there corresponds a symmetric interval of the form $-R \leq x \leq R$ in the interior of which the series converges and outside of which it diverges (Figure 3.1). (At the endpoints $x = -R$ and $x = R$, convergence or divergence is possible.) The number R is called the **radius of convergence** of the series, and the set of all points for which the series converges is called its **interval of convergence.** Thus the interval of convergence of a power series consists of the interval $-R < x < R$, where R is the radius of convergence, and possibly one or both of the endpoints $x = -R$ and $x = R$. (If the series happens to converge for all values of x, the radius of convergence is said to be ∞.)

Figure 3.1 The interval of convergence for a power series.

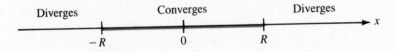

Factorial Notation

The construction of power series that converge to given functions involves products of the form

$$n(n - 1)(n - 2) \cdots 3 \cdot 2 \cdot 1$$

It is customary to abbreviate such a product using the symbol $n!$, which is read "*n* factorial." For example, $3! = 3 \cdot 2 \cdot 1 = 6$. Moreover, $0!$ is *defined* to be 1 (because, as it turns out, this definition leads to compact representations of certain formulas involving factorials).

Factorial Notation

If n is a positive integer,

$$n! = n(n - 1)(n - 2) \cdots 3 \cdot 2 \cdot 1$$

Moreover,

$$0! = 1$$

Power Series That Converge to Given Functions

Imagine that a function $f(x)$ is given and that you would like to find the corresponding coefficients a_n such that the power series $\sum\limits_{n=0}^{\infty} a_n x^n$ converges to $f(x)$ on some interval. To discover what these coefficients might be, suppose that

$$f(x) = \sum_{n=0}^{\infty} a_n x^n = a_0 + a_1 x + a_2 x^2 + a_3 x^3 + \cdots + a_n x^n + \cdots$$

If $x = 0$, only the first term in the sum is nonzero and so

$$a_0 = f(0)$$

Now differentiate the series term by term. It can be shown that if the original power series converges to $f(x)$ on the interval $-R < x < R$, then the differentiated series converges to $f'(x)$ on this interval. Hence,

$$f'(x) = a_1 + 2a_2 x + 3a_3 x^2 + \cdots + na_n x^{n-1} + \cdots$$

and, if $x = 0$, it follows that

$$a_1 = f'(0)$$

Differentiate again to get

$$f''(x) = 2a_2 + 3 \cdot 2a_3 x + \cdots + n(n - 1)a_n x^{n-2} + \cdots$$

and let $x = 0$ to conclude that

$$f''(0) = 2a_2 \quad \text{or} \quad a_2 = \frac{f''(0)}{2}$$

Similarly, the third derivative of f is

$$f^{(3)}(x) = 3 \cdot 2a_3 + \cdots + n(n - 1)(n - 2)x^{n-3} + \cdots$$

and

$$f^{(3)}(0) = 3 \cdot 2a_3 \quad \text{or} \quad a_3 = \frac{f^{(3)}(0)}{3!}$$

and so on. In general,

$$f^{(n)}(0) = n!a_n \quad \text{or} \quad a_n = \frac{f^{(n)}(0)}{n!}$$

where $f^{(n)}(0)$ denotes the nth derivative of f evaluated at $x = 0$, and $f^{(0)}(0) = f(0)$.

The preceding argument shows that *if* there is any power series that converges to $f(x)$, it must be the one whose coefficients are obtained from the derivatives of f by the formula

$$a_n = \frac{f^{(n)}(0)}{n!}$$

This series is known as the **Taylor series of f** (about $x = 0$), and the corresponding coefficients a_n are called the **Taylor coefficients of f.**

Taylor Series

The Taylor series of $f(x)$ about $x = 0$ is the power series $\sum\limits_{n=0}^{\infty} a_n x^n$, where

$$a_n = \frac{f^{(n)}(0)}{n!}$$

Convergence of Taylor Series

Although the Taylor series of a function is the only power series that can possibly converge to the function, it need not actually do so. There are, in fact, some functions whose Taylor series converge, but not to the values of the function. Fortunately, such functions rarely arise in practice. Most of the functions you are likely to encounter are equal to their Taylor series wherever the series converge.

Calculation of Taylor Series

The use of the formula for the Taylor coefficients is illustrated in the following example.

The Taylor Series for e^x

EXAMPLE 3.1 Find the Taylor series of the function $f(x) = e^x$ about $x = 0$.

SOLUTION
Compute the Taylor coefficients as follows:

$$f(x) = e^x \qquad f(0) = 1 \qquad a_0 = \frac{f(0)}{0!} = \frac{1}{0!}$$

$$f'(x) = e^x \qquad f'(0) = 1 \qquad a_1 = \frac{f'(0)}{1!} = \frac{1}{1!}$$

$$f''(x) = e^x \qquad f''(0) = 1 \qquad a_2 = \frac{f''(0)}{2!} = \frac{1}{2!}$$

$$f^{(3)}(x) = e^x \qquad f^{(3)}(0) = 1 \qquad a_3 = \frac{f^{(3)}(0)}{3!} = \frac{1}{3!}$$

$$\begin{matrix} \cdot \\ \cdot \\ \cdot \end{matrix} \qquad\qquad \begin{matrix} \cdot \\ \cdot \\ \cdot \end{matrix} \qquad\qquad \begin{matrix} \cdot \\ \cdot \\ \cdot \end{matrix}$$

$$f^{(n)}(x) = e^x \qquad f^{(n)}(0) = 1 \qquad a_n = \frac{f^{(n)}(0)}{n!} = \frac{1}{n!}$$

The corresponding Taylor series is

$$\sum_{n=0}^{\infty} a_n x^n = \sum_{n=0}^{\infty} \frac{x^n}{n!}$$

which can be shown to converge to e^x for all x. That is,

$$e^x = \sum_{n=0}^{\infty} \frac{x^n}{n!} = 1 + x + \frac{x^2}{2!} + \frac{x^3}{3!} + \cdots \qquad \text{for all } x$$

The Series for $f(x) = \dfrac{1}{1 - x}$

Recall from the discussion of the geometric series in Section 2 that if $|r| < 1$, $\sum_{n=0}^{\infty} r^n = \dfrac{1}{1 - r}$. It should come as no surprise, therefore, that the Taylor series for the function $f(x) = \dfrac{1}{1 - x}$ is $\sum_{n=0}^{\infty} x^n$. Here is the calculation.

EXAMPLE 3.2 Find the Taylor series of the function $f(x) = \dfrac{1}{1 - x}$ about $x = 0$.

SOLUTION
Compute the Taylor coefficients as follows:

$$f(x) = \frac{1}{1 - x} \qquad f(0) = 1 \qquad a_0 = \frac{f(0)}{0!} = \frac{1}{0!} = 1$$

$$f'(x) = \frac{1}{(1 - x)^2} \qquad f'(0) = 1 \qquad a_1 = \frac{f'(0)}{1!} = \frac{1}{1!} = 1$$

$$f''(x) = \frac{2 \cdot 1}{(1 - x)^3} \qquad f''(0) = 2! \qquad a_2 = \frac{f''(0)}{2!} = \frac{2!}{2!} = 1$$

$$f^{(3)}(x) = \frac{3 \cdot 2 \cdot 1}{(1 - x)^4} \qquad f^{(3)}(0) = 3! \qquad a_3 = \frac{f^{(3)}(0)}{3!} = \frac{3!}{3!} = 1$$

. . .

. . .

. . .

$$f^{(n)}(x) = \frac{n!}{(1 - x)^{n+1}} \qquad f^{(n)}(0) = n! \qquad a_n = \frac{f^{(n)}(0)}{n!} = \frac{n!}{n!} = 1$$

The corresponding Taylor series is $\sum_{n=0}^{\infty} x^n$, which, as you already know, converges to $f(x) = \frac{1}{1 - x}$ for $|x| < 1$. Thus,

$$\frac{1}{1 - x} = \sum_{n=0}^{\infty} x^n = 1 + x + x^2 + \cdots \qquad \text{for } |x| < 1$$

Modification of Taylor Series

As you have seen in the preceding two examples, the use of the formula for the Taylor coefficients to find the Taylor series of a function can be somewhat tedious. Sometimes you will be able to avoid this procedure and find the Taylor series of a function by modifying another Taylor series that you already know. One technique that is particularly useful is to replace x in the series that is known by a more complicated expression, say $g(x)$. In particular, if $f(x) = \sum_{n=0}^{\infty} a_n x^n$ for $|x| < R$ and $g(x)$ is any function of x, you get the Taylor series of the composite function $f[g(x)]$ by replacing x by $g(x)$ in the series for $f(x)$. The new series will converge for any value of x for which $|g(x)| < R$.

Substitution in Taylor Series

If $f(x) = \sum_{n=0}^{\infty} a_n x^n$ for $|x| < R$, then

$$f[g(x)] = \sum_{n=0}^{\infty} a_n [g(x)]^n \qquad \text{for } |g(x)| < R$$

Here is an example.

EXAMPLE 3.3 Starting with an appropriate geometric series, find the Taylor series for the function $\dfrac{x}{1 + 3x^2}$ and specify the interval of convergence.

SOLUTION
Start with the fact that

$$\frac{1}{1 - x} = \sum_{n=0}^{\infty} x^n \qquad \text{for } |x| < 1$$

and replace x by $-3x^2$ to get

$$\frac{1}{1 + 3x^2} = \sum_{n=0}^{\infty} (-3x^2)^n = \sum_{n=0}^{\infty} (-3)^n x^{2n}$$

The new interval of convergence is $|-3x^2| < 1$, which can be rewritten as $|x^2| < \dfrac{1}{3}$ or $|x| < \dfrac{1}{\sqrt{3}}$.

Now multiply by x (which does not affect the interval of convergence) to get

$$\frac{x}{1 + 3x^2} = \sum_{n=0}^{\infty} (-3)^n x^{2n+1} \qquad \text{for } |x| < \frac{1}{\sqrt{3}}$$

Taylor Series about $x = a$

In certain situations it will be inappropriate (or impossible) to represent a given function by a power series of the form $\sum_{n=0}^{\infty} a_n x^n$, and you will have to use a series of the form $\sum_{n=0}^{\infty} a_n (x - a)^n$ instead, where a is a constant. This is the case, for example, for the function $f(x) = \ln x$, which is undefined for $x \leq 0$ and so cannot possibly be the sum of a series of the form $\sum_{n=0}^{\infty} a_n x^n$ on a symmetric interval about $x = 0$.

Using an argument similar to the one at the beginning of this section, you can show that if $f(x) = \sum_{n=0}^{\infty} a_n (x - a)^n$, the coefficients of the series are related to the derivatives of f by the formula

$$a_n = \frac{f^{(n)}(a)}{n!}$$

The power series in $(x - a)$ with these coefficients is called the **Taylor series of $f(x)$ about $x = a$**. It can be shown that series of this type converge on symmetric intervals of the form $a - R < x < a + R$, including possibly one or both of the endpoints $x = a - R$ and $x = a + R$. Notice that the Taylor series about $x = 0$ is simply a special case of the more general Taylor series about $x = a$.

Taylor Series about $x = a$

The Taylor series of $f(x)$ about $x = a$ is the power series

$$\sum_{n=0}^{\infty} a_n(x - a)^n$$

where

$$a_n = \frac{f^{(n)}(a)}{n!}$$

Here is an example.

The Taylor Series for ln x

EXAMPLE 3.4   Find the Taylor series of the function $f(x) = \ln x$ about $x = 1$.

SOLUTION
Compute the Taylor coefficients as follows:

$$f(x) = \ln x \qquad f(1) = 0 \qquad a_0 = \frac{f(1)}{0!} = 0$$

$$f'(x) = \frac{1}{x} \qquad f'(1) = 1 \qquad a_1 = \frac{f'(1)}{1!} = 1$$

$$f''(x) = \frac{-1}{x^2} \qquad f''(1) = -1 \qquad a_2 = \frac{f''(1)}{2!} = -\frac{1}{2}$$

$$f^{(3)}(x) = \frac{2!}{x^3} \qquad f^{(3)}(1) = 2! \qquad a_3 = \frac{f^{(3)}(1)}{3!} = \frac{1}{3}$$

$$f^{(4)}(x) = \frac{-3!}{x^4} \qquad f^{(4)}(1) = -3! \qquad a_4 = \frac{f^{(4)}(1)}{4!} = -\frac{1}{4}$$

$$\begin{matrix} \cdot \\ \cdot \\ \cdot \end{matrix} \qquad \begin{matrix} \cdot \\ \cdot \\ \cdot \end{matrix} \qquad \begin{matrix} \cdot \\ \cdot \\ \cdot \end{matrix}$$

$$f^{(n)}(x) = (-1)^{n+1}\frac{(n - 1)!}{x^n} \qquad f^{(n)}(1) = (-1)^{n+1}(n - 1)! \qquad a_n = \frac{(-1)^{n+1}}{n}$$

The corresponding Taylor series is

$$\sum_{n=1}^{\infty} \frac{(-1)^{n+1}}{n}(x-1)^n$$

which can be shown to converge to $\ln x$ for $0 < x \le 2$. That is,

$$\ln x = \sum_{n=1}^{\infty} \frac{(-1)^{n+1}}{n}(x-1)^n$$

$$= (x-1) - \frac{1}{2}(x-1)^2 + \frac{1}{3}(x-1)^3 - \cdots \qquad \text{for } 0 < x \le 2$$

Taylor Polynomials

The partial sums of a Taylor series of a function are polynomials that can be used to approximate the function. In general, the more terms there are in the partial sum, the better the approximation will be. It is customary to let $P_n(x)$ denote the $(n+1)$st partial sum, which is a polynomial of degree (at most) n. In particular,

$$P_n(x) = f(a) + f'(a)(x-a) + \frac{f''(a)}{2!}(x-a)^2 + \cdots + \frac{f^{(n)}(a)}{n!}(x-a)^n$$

The polynomial $P_n(x)$ is known as the **nth Taylor polynomial** of $f(x)$ about $x = a$.

Approximation by Taylor Polynomials

The approximation of $f(x)$ by its Taylor polynomials $P_n(x)$ is most accurate near $x = a$ and for large values of n. Here is a geometric argument that should give you some additional insight into the situation.

Observe first that at $x = a$, f and P_n are equal, as are, respectively, their first n derivatives. For example,

$$P_2(x) = f(a) + f'(a)(x-a) + \frac{f''(a)}{2!}(x-a)^2 \quad \text{and} \quad P_2(a) = f(a)$$

$$P_2'(x) = f'(a) + f''(a)(x-a) \quad \text{and} \quad P_2'(a) = f'(a)$$

$$P_2''(x) = f''(a) \quad \text{and} \quad P_2''(a) = f''(a)$$

That fact that $P_n(a) = f(a)$ implies that the graphs of P_n and f intersect at $x = a$. The fact that $P_n'(a) = f'(a)$ implies further that the graphs have the same slope at $x = a$. The fact that $P_n''(a) = f''(a)$ imposes the further restriction that the graphs have the same concavity at $x = a$. In general, as n increases, the number of matching derivatives increases, and the graph of P_n approximates more closely that of f near $x = a$. The situation is illustrated in Figure 3.2, which shows the graphs of e^x and its first three Taylor polynomials.

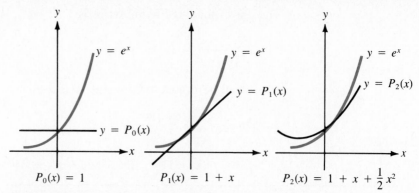

$$P_0(x) = 1 \qquad\qquad P_1(x) = 1 + x \qquad\qquad P_2(x) = 1 + x + \frac{1}{2}x^2$$

Figure 3.2 The graphs of e^x and its first three Taylor polynomials.

The use of Taylor polynomials to approximate functions is illustrated in the following examples.

EXAMPLE 3.5   Use a Taylor polynomial of degree 6 to approximate e.

SOLUTION
From Example 3.1

$$e^x = \sum_{n=0}^{\infty} \frac{x^n}{n!} \qquad \text{for all } x$$

Hence,

$$P_6(x) = 1 + x + \frac{x^2}{2!} + \frac{x^3}{3!} + \frac{x^4}{4!} + \frac{x^5}{5!} + \frac{x^6}{6!}$$

and

$$e = e^1 \simeq P_6(1) = 1 + 1 + \frac{1}{2!} + \frac{1}{3!} + \frac{1}{4!} + \frac{1}{5!} + \frac{1}{6!}$$

$$= 1 + 1 + \frac{1}{2} + \frac{1}{6} + \frac{1}{24} + \frac{1}{120} + \frac{1}{720} \simeq 2.71806$$

By analysis of the error that results from this sort of approximation (based on techniques beyond the scope of this text), it can be shown that the actual value of e, rounded off to five decimal places, is 2.71828.

EXAMPLE 3.6   Use an appropriate Taylor polynomial of degree 3 to approximate $\sqrt{4.1}$.

SOLUTION
The goal is to estimate $f(x) = \sqrt{x}$ when $x = 4.1$. Since 4.1 is close to 4 and since the values of f and its derivatives at $x = 4$ are easy to compute, it is natural to use a Taylor polynomial about $x = 4$.

Compute the necessary Taylor coefficients about $x = 4$ as follows:

$$f(x) = \sqrt{x} \qquad f(4) = 2 \qquad a_0 = \frac{2}{0!} = 2$$

$$f'(x) = \frac{1}{2}x^{-1/2} \qquad f'(4) = \frac{1}{4} \qquad a_1 = \frac{1/4}{1!} = \frac{1}{4}$$

$$f''(x) = -\frac{1}{4}x^{-3/2} \qquad f''(4) = -\frac{1}{32} \qquad a_2 = \frac{-1/32}{2!} = -\frac{1}{64}$$

$$f^{(3)}(x) = \frac{3}{8}x^{-5/2} \qquad f^{(3)}(4) = \frac{3}{256} \qquad a_3 = \frac{3/256}{3!} = \frac{1}{512}$$

The corresponding Taylor polynomial of degree 3 is

$$P_3(x) = 2 + \frac{1}{4}(x - 4) - \frac{1}{64}(x - 4)^2 + \frac{1}{512}(x - 4)^3$$

and so

$$\sqrt{4.1} \simeq P_3(4.1) = 2 + \frac{1}{4}(0.1) - \frac{1}{64}(0.1)^2 + \frac{1}{512}(0.1)^3 \simeq 2.02485$$

Incidentally, all five decimal places of this estimate are correct. Rounded off to seven decimal places, the true value of $\sqrt{4.1}$ is 2.0248457.

Approximation of Definite Integrals

In the next example, a Taylor polynomial is used to estimate a definite integral whose exact value cannot be calculated by elementary methods.

EXAMPLE 3.7   Use a Taylor polynomial of degree 8 to approximate $\int_0^1 e^{-x^2}\, dx$.

SOLUTION
The easiest way to get the Taylor series of e^{-x^2} is to start with the series

$$e^x = \sum_{n=0}^{\infty} \frac{x^n}{n!}$$

for e^x and replace x by $-x^2$ to get

$$e^{-x^2} = \sum_{n=0}^{\infty} \frac{(-1)^n x^{2n}}{n!}$$

Thus,

$$e^{-x^2} \simeq 1 - x^2 + \frac{1}{2}x^4 - \frac{1}{6}x^6 + \frac{1}{24}x^8$$

and so

$$\int_0^1 e^{-x^2} \, dx \approx \int_0^1 \left(1 - x^2 + \frac{1}{2}x^4 - \frac{1}{6}x^6 + \frac{1}{24}x^8\right) dx$$

$$= \left(x - \frac{1}{3}x^3 + \frac{1}{10}x^5 - \frac{1}{42}x^7 + \frac{1}{216}x^9\right)\Big|_0^1$$

$$= 1 - \frac{1}{3} + \frac{1}{10} - \frac{1}{42} + \frac{1}{216} \approx 0.7475$$

Problems

In Problems 1 through 8, use the formula for the Taylor coefficients to find the Taylor series of f about $x = 0$.

1. $f(x) = e^{3x}$

2. $f(x) = e^{-2x}$

3. $f(x) = \ln(1 + x)$

4. $f(x) = \dfrac{1}{2 - x}$

5. $f(x) = \dfrac{e^x + e^{-x}}{2}$

6. $f(x) = \dfrac{e^x - e^{-x}}{2}$

7. $f(x) = (1 + x)e^x$

8. $f(x) = (3x + 2)e^x$

In Problems 9 through 14, use the formula for the Taylor coefficients to find the Taylor series of f about $x = a$.

9. $f(x) = e^{2x}$; $a = 1$

10. $f(x) = e^{-3x}$; $a = -1$

11. $f(x) = \dfrac{1}{x}$; $a = 1$

12. $f(x) = \ln 2x$; $a = \dfrac{1}{2}$

13. $f(x) = \dfrac{1}{2 - x}$; $a = 1$

14. $f(x) = \dfrac{1}{1 + x}$; $a = 2$

In Problems 15 through 18, modify one of the Taylor series you already know to find the Taylor series of the given function and specify an interval on which the series converges.

15. $f(x) = \dfrac{x}{1 - 5x}$

16. $f(x) = \dfrac{1}{1 + 8x^3}$

17. $f(x) = e^{-x/2}$

18. $f(x) = 2xe^{x^2/2}$

In Problems 19 through 24, use a Taylor polynomial of degree n to approximate the given number.

19. $\sqrt{3.8}$; $n = 3$

20. $\sqrt{1.2}$; $n = 3$

21. $\ln 1.1$; $n = 5$

22. $\ln 0.7$; $n = 5$

23. $e^{0.3}$; $n = 4$

24. $\dfrac{1}{\sqrt{e}}$; $n = 4$

In Problems 25 through 28, use a Taylor polynomial of degree n to approximate the given integral.

25. $\displaystyle\int_0^{1/2} e^{-x^2}\, dx$; $n = 6$

26. $\displaystyle\int_{-0.2}^{0.1} e^{-x^2}\, dx$; $n = 4$

27. $\displaystyle\int_0^{0.1} \dfrac{1}{1 + x^2}\, dx$; $n = 4$

28. $\displaystyle\int_{-1/2}^{0} \dfrac{1}{1 - x^3}\, dx$; $n = 9$

4 Newton's Method

The Roots of an Equation

A value of a variable that satisfies an equation is said to be a **solution** or **root** of the equation. In this section, you will see a technique known as **Newton's method** that can be used to approximate roots of equations of the form $f(x) = 0$. (As you will see, the method is related to the approximation of functions by their Taylor polynomials, which you studied in the preceding section.) In geometric terms, a root of an equation $f(x) = 0$ is an x intercept of the graph of f.

Like most techniques for approximating a root of an equation, Newton's method requires that you start with a rough estimate of the location of the root. The following property of continuous functions is often used to obtain this initial estimate.

The Intermediate Value Theorem

If $f(x)$ is continuous on the interval $a \leq x \leq b$, and if $f(a)$ and $f(b)$ have opposite signs, then the equation $f(x) = 0$ has at least one root between $x = a$ and $x = b$.

In geometric terms, the intermediate value theorem says that if f is continuous and if the points $(a, f(a))$ and $(b, f(b))$ lie on opposite sides of the x axis, then the graph of f (which is an unbroken curve) must cross the x axis somewhere between $x = a$ and $x = b$. The situation is illustrated in Figure 4.1.

The use of the intermediate value theorem to estimate the location of roots is illustrated in the following example.

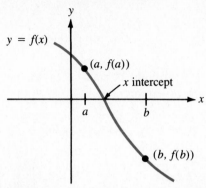

Figure 4.1 The intermediate value theorem.

EXAMPLE 4.1   Make a rough sketch of the function $f(x) = x^3 - x^2 - 1$, and use the intermediate value theorem to estimate the location of the roots of the equation $f(x) = 0$.

SOLUTION
The graph (obtained with the aid of the first derivative) is sketched in Figure 4.2. It indicates that the equation $f(x) = 0$ has only one root. Moreover, since

$$f(1) = -1 < 0 \qquad \text{and} \qquad f(2) = 3 > 0$$

it follows from the intermediate value theorem that the root lies between $x = 1$ and $x = 2$.

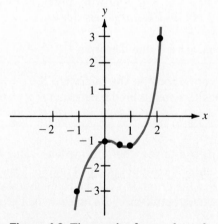

Figure 4.2 The graph of $y = x^3 - x^2 - 1$.

Newton's Method

The basic idea behind Newton's method is illustrated in Figure 4.3a, in which r is a root of the equation $f(x) = 0$, x_0 is an initial approximation to r, and x_1 is a better approximation obtained by taking the x intercept of the line that is tangent to the graph of f at $(x_0, f(x_0))$.

[In the terminology of the preceding section, the line that is tangent to the graph of f at $(x_0, f(x_0))$ is the graph of the first Taylor polynomial of $f(x)$ about $x = x_0$. Hence, Newton's method can be viewed as an application of the approximation of a function by its Taylor polynomials.]

To find a formula for the improved approximation x_1, recall that the slope of the tangent line through $(x_0, f(x_0))$ is the derivative $f'(x_0)$. Hence,

$$f'(x_0) = \text{slope} = \frac{\Delta y}{\Delta x} = \frac{f(x_0) - 0}{x_0 - x_1}$$

or, equivalently,

$$x_1 = x_0 - \frac{f(x_0)}{f'(x_0)}$$

If the procedure is repeated using x_1 as the initial approximation, an even better approximation is obtained (Figure 4.3b). This approximation, x_2, is related to x_1 as x_1 was to x_0. That is,

$$x_2 = x_1 - \frac{f(x_1)}{f'(x_1)}$$

The process can be continued until the desired degree of accuracy is obtained. In general, the nth approximation x_n is related to the $(n - 1)$st by the formula

$$x_n = x_{n-1} - \frac{f(x_{n-1})}{f'(x_{n-1})}$$

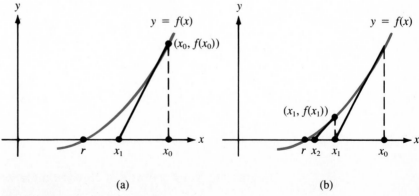

(a)   (b)

Figure 4.3 Newton's method.

Newton's Method

To approximate a root of the equation $f(x) = 0$, start with an initial estimate x_0 and generate a sequence of increasingly accurate approximations $x_1, x_2, x_3, \ldots$ using the formula

$$x_n = x_{n-1} - \frac{f(x_{n-1})}{f'(x_{n-1})}$$

The use of Newton's method is illustrated in the following example.

EXAMPLE 4.2   Use three repetitions of Newton's method to approximate $\sqrt{5}$.

SOLUTION
The goal is to find the root of the equation $f(x) = 0$, where

$$f(x) = x^2 - 5$$

The derivative of f is $f'(x) = 2x$, and so

$$x - \frac{f(x)}{f'(x)} = x - \frac{x^2 - 5}{2x} = \frac{x^2 + 5}{2x}$$

Thus, for $n = 1, 2, 3, \ldots$

$$x_n = \frac{x_{n-1}^2 + 5}{2x_{n-1}}$$

A convenient choice for the initial estimate is $x_0 = 2$ (since $2^2 = 4$, which is close to 5). Then,

$$x_1 = \frac{x_0^2 + 5}{2x_0} = 2.250 \qquad \text{using } x_0 = 2$$

$$x_2 = \frac{x_1^2 + 5}{2x_1} \simeq 2.2361111 \qquad \text{using } x_1 = 2.250$$

$$x_3 = \frac{x_2^2 + 5}{2x_2} \simeq 2.2360680 \qquad \text{using } x_2 = 2.2361111$$

Thus, $\sqrt{5}$ is approximately 2.236. (Actually, rounded off to seven decimal places, $\sqrt{5} = 2.2360680$, which is exactly the value obtained by Newton's method after only three repetitions!)

Estimating the Accuracy of Newton's Method

A systematic analysis of the error involved in approximation by Newton's method is beyond the scope of this text. However, there is a simple rule of thumb, which states that, to use Newton's method to approximate a root to a given number of decimal places, round off all calculations to one more decimal place and stop when there is no change from one approximation to the next. This is illustrated in the following examples.

EXAMPLE 4.3 Use Newton's method to approximate the roots of the equation $x^3 - x^2 - 1 = 0$ to three decimal places.

SOLUTION

Let $f(x) = x^3 - x^2 - 1$. Then,

$$x - \frac{f(x)}{f'(x)} = x - \frac{x^3 - x^2 - 1}{3x^2 - 2x} = \frac{2x^3 - x^2 + 1}{3x^2 - 2x}$$

and so, for $n = 1, 2, 3, \ldots$

$$x_n = \frac{2x_{n-1}^3 - x_{n-1}^2 + 1}{3x_{n-1}^2 - 2x_{n-1}}$$

The goal is to find the roots of the equation $f(x) = 0$. As you saw in Example 4.1, there is only one such root, and it must lie between 1 and 2. Either of these two values is a reasonable choice for the initial estimate x_0. If you take $x_0 = 1$, you get

$$x_1 = \frac{2x_0^3 - x_0^2 + 1}{3x_0^2 - 2x_0} = 2 \qquad \text{using } x_0 = 1$$

$$x_2 = \frac{2x_1^3 - x_1^2 + 1}{3x_1^2 - 2x_1} = 1.6250 \qquad \text{using } x_1 = 2$$

$$x_3 = \frac{2x_2^3 - x_2^2 + 1}{3x_2^2 - 2x_2} \approx 1.4858 \qquad \text{using } x_2 = 1.6250$$

$$x_4 = \frac{2x_3^3 - x_3^2 + 1}{3x_3^2 - 2x_3} \approx 1.4660 \qquad \text{using } x_3 = 1.4858$$

$$x_5 = \frac{2x_4^3 - x_4^2 + 1}{3x_4^2 - 2x_4} \approx 1.4656 \qquad \text{using } x_4 = 1.4660$$

$$x_6 = \frac{2x_5^3 - x_5^2 + 1}{3x_5^2 - 2x_5} \approx 1.4656 \qquad \text{using } x_5 = 1.4656$$

Since x_5 and x_6 are identical (to four decimal places), no further computation is necessary. You now simply round off the number 1.4656 to conclude that, to three-decimal-place accuracy, the desired root is 1.466.

EXAMPLE 4.4 Use Newton's method to approximate the solution of the equation $e^x = -x$ to three decimal places.

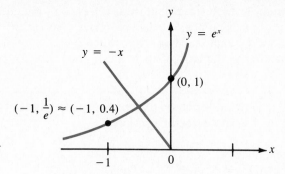

Figure 4.4 The graphs of $y = e^x$ and $y = -x$.

SOLUTION

From the graphs of $y = e^x$ and $y = -x$ in Figure 4.4, you see that the equation has a solution somewhere between $x = -1$ and $x = 0$.

Let $f(x) = e^x + x$ so that the desired solution is the root of the equation $f(x) = 0$. Then,

$$x - \frac{f(x)}{f'(x)} = x - \frac{e^x + x}{e^x + 1} = \frac{(x - 1)e^x}{e^x + 1}$$

and so, for $n = 1, 2, 3, \ldots,$

$$x_n = \frac{(x_{n-1} - 1)e^{x_{n-1}}}{e^{x_{n-1}} + 1}$$

Taking $x_0 = -0.5$ (as suggested by the graph), you get

$$x_1 = \frac{(x_0 - 1)e^{x_0}}{e^{x_0} + 1} \simeq -0.5663$$

$$x_2 = \frac{(x_1 - 1)e^{x_1}}{e^{x_1} + 1} \simeq -0.5671$$

$$x_3 = \frac{(x_2 - 1)e^{x_2}}{e^{x_2} + 1} \simeq -0.5671$$

Hence, to three-decimal-place accuracy, the solution of the equation $e^x = -x$ is -0.567.

Convergence of Newton's Method

It can be shown that if the approximations $x_1, x_2, x_3, \ldots$ generated by Newton's method converge to some (finite) number, this number must be a root of the equation $f(x) = 0$. However, there are cases in which the approximations do not converge, even though the equation has a root. Roughly speaking, this occurs if the initial estimate x_0 is not "close enough" to the root. This phenomenon is illustrated in Figure 4.5, which shows successive "approximations" that fail to converge to the root but

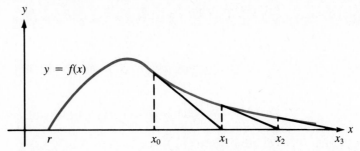

Figure 4.5 Newton's method fails because x_0 is not close enough to r.

instead increase without bound. In such cases, it is usually possible to achieve convergence by reworking the problem using a better initial estimate.

Problems

In Problems 1 through 8, use Newton's method to find all the roots of the given equation to three-decimal-place accuracy.

1. $x^2 - 12 = 0$

2. $x^3 + 49 = 0$

3. $x^3 + x^2 - 1 = 0$

4. $x^3 + 2x^2 - x + 1 = 0$

5. $e^x = -2x$

6. $e^{-x} = x - 1$

7. $x^2 - 5x + 1 = 0$ (*Hint*: There are two roots.)

8. $x^4 - 4x^3 + 10 = 0$ (*Hint*: There are two roots.)

In Problems 9 through 12, use Newton's method to compute the given number to five decimal places.

9. $\sqrt{2}$ 10. $\sqrt{7}$

11. $\sqrt[3]{9}$ 12. $\sqrt[3]{-20}$

Chapter Summary and Review Problems

Important Terms, Symbols, and Formulas

Infinite series: $\displaystyle\sum_{n=1}^{\infty} a_n = a_1 + a_2 + a_3 + \cdots$

nth partial sum: $S_n = a_1 + a_2 + \cdots + a_n$

Convergence and divergence:

$$\sum_{n=1}^{\infty} a_n \text{ converges to } S \text{ if and only if } \lim_{n \to \infty} S_n = S$$

Test for divergence: If $\lim_{n \to \infty} a_n \neq 0$, then $\displaystyle\sum_{n=1}^{\infty} a_n$ diverges.

Distributive law: $\displaystyle\sum_{n=1}^{\infty} ca_n = c \sum_{n=1}^{\infty} a_n$

Harmonic series: $\displaystyle\sum_{n=1}^{\infty} \frac{1}{n}$ diverges (even though $\lim_{n \to \infty} a_n = 0$).

Geometric series: If $|r| < 1$, $\displaystyle\sum_{n=0}^{\infty} r^n = \frac{1}{1 - r}$

Power series: $\displaystyle\sum_{n=0}^{\infty} a_n x^n; \sum_{n=0}^{\infty} a_n (x - a)^n$

Radius of convergence; interval of convergence

Factorial notation: $n! = n(n - 1)(n - 2) \cdots 3 \cdot 2 \cdot 1; 0! = 1$

Taylor series of $f(x)$ about $x = a$:

$$\sum_{n=0}^{\infty} a_n (x - a)^n \quad \text{where } a_n = \frac{f^{(n)}(a)}{n!}$$

Important Taylor series:

$$e^x = \sum_{n=0}^{\infty} \frac{x^n}{n!} \quad \text{for all } x$$

$$\frac{1}{1 - x} = \sum_{n=0}^{\infty} x^n \quad \text{for } |x| < 1$$

$$\ln x = \sum_{n=1}^{\infty} \frac{(-1)^{n+1}}{n}(x - 1)^n \quad \text{for } 0 < x \leq 2$$

Substitution in power series

nth Taylor polynomial: $P_n(x)$

Root of an equation

Intermediate value theorem

Newton's method: $x_n = x_{n-1} - \dfrac{f(x_{n-1})}{f'(x_{n-1})}$

Review Problems

In Problems 1 through 3, find the sum of the given convergent series by taking the limit of a compact expression for the nth partial sum.

1. $\displaystyle\sum_{n=1}^{\infty} \left(\frac{1}{n + 1} - \frac{1}{n + 3}\right)$

2. $\displaystyle\sum_{n=2}^{\infty} \frac{1}{n(n - 1)}$

3. $\displaystyle\sum_{n=1}^{\infty} \frac{2}{(-3)^n}$

In Problems 4 through 7, determine whether the given geometric series converges, and, if so, find its sum.

4. $\displaystyle\sum_{n=1}^{\infty} \frac{3}{(-5)^n}$

5. $\displaystyle\sum_{n=0}^{\infty} \left(-\frac{3}{2}\right)^n$

6. $\displaystyle\sum_{n=1}^{\infty} e^{-0.5n}$

7. $\displaystyle\sum_{n=2}^{\infty} \frac{2^{n+1}}{3^{n-3}}$

8. Express the repeating decimal $1.545454\ldots$ as a fraction.

9. A ball is dropped from a height of 6 feet and allowed to bounce indefinitely. How far will the ball travel if it always rebounds to 80 percent of its previous height?

10. How much should you invest today at an annual interest rate of 12 percent compounded continuously so that, starting next year, you can make annual withdrawals of $500 in perpetuity?

11. A patient is given an injection of 10 units of a certain drug every 24 hours. The drug is eliminated exponentially so that the fraction that remains in the patient's body after t days is $f(t) = e^{-0.8t}$. If the treatment is continued indefinitely, approximately how many units of the drug will eventually be in the patient's body immediately following an injection?

12. You and two friends take turns rolling a die until one of you wins by getting a six. Find the probability that you will win if you roll second.

In Problems 13 through 16, use the formula for the Taylor coefficients to find the Taylor series of f about $x = a$.

13. $f(x) = \dfrac{1}{x + 3}; \ a = 0$

14. $f(x) = e^{-3x}; \ a = 0$

15. $f(x) = (1 + 2x)e^x; \ a = 0$

16. $f(x) = \dfrac{1}{(1 + x)^2}; \ a = -2$

17. Modify a geometric series to find the Taylor series about $x = 0$ for the function $f(x) = \dfrac{4x}{1 + 4x^2}$.

18. Use a Taylor polynomial of degree 3 to approximate $\sqrt{0.9}$.

19. Use a Taylor polynomial of degree 10 to approximate $\displaystyle\int_0^{1/2} \dfrac{x}{1 + x^3}\, dx$.

20. Use Newton's method to find $\sqrt{55}$ to five decimal places.

In Problems 21 and 22, use Newton's method to find all the roots of the given equation to three decimal places.

21. $x^3 + 3x^2 + 1 = 0$ 22. $e^{-x} = 3x$

12 Trigonometric Functions

1   The Trigonometric Functions

In this chapter you will be introduced to some functions that are widely used in the natural sciences to study periodic or rhythmic phenomena such as oscillations, the periodic motion of planets, and the respiratory cycle and heartbeat of animals. These functions are also related to the measurement of angles and hence play an important role in such fields as architecture, navigation, and surveying.

Angles

An **angle** is formed when one line segment in the plane is rotated into another about their common endpoint. The resulting angle is said to be a **positive angle** if the rotation is in a counterclockwise direction and a **negative angle** if the rotation is in a clockwise direction. The situation is illustrated in Figure 1.1.

Positive angle   Negative angle

Figure 1.1 Positive and negative angles.

Measurement of Angles

You are probably already familiar with the use of degrees to measure angles. A **degree** is the amount by which a line segment must be rotated so that its free endpoint traces out $\frac{1}{360}$ of a circle. Thus, for example, a complete counterclockwise rotation generating an entire circle contains 360°, one-half of a complete counterclockwise rotation contains 180°, and one-sixth of a complete counterclockwise rotation contains 60°. Some important angles are shown in Figure 1.2.

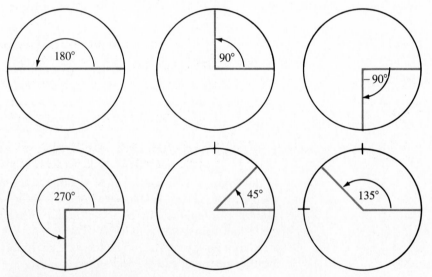

Figure 1.2 Angles measured in degrees.

Radian Measure

Although measurement of angles in degrees is convenient for many geometric applications, there is another unit of angle measurement called a **radian** that leads to simpler rules for the differentiation and integration of trigonometric functions. One radian is defined to be the amount by which

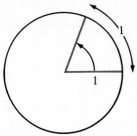

Figure 1.3 An angle of 1 radian.

a line segment of length 1 must be rotated so that its free endpoint traces out a circular arc of length 1. The situation is illustrated in Figure 1.3.

The total circumference of a circle of radius 1 is 2π. Hence 2π radians is equal to 360°, π radians is equal to 180°, and, in general, the relationship between radians and degrees is given by the following proportion.

Conversion Formula

$$\frac{\text{Degrees}}{180} = \frac{\text{radians}}{\pi}$$

Here is an example.

EXAMPLE 1.1   (a) Convert 45° to radians.

(b) Convert $\dfrac{\pi}{6}$ radians to degrees.

SOLUTION
(a) From the proportion

$$\frac{45}{180} = \frac{\text{radians}}{\pi}$$

it follows that

$$\text{Radians} = \frac{\pi}{4}$$

That is, 45° equals $\dfrac{\pi}{4}$ radians.

(b) From the proportion

$$\frac{\text{Degrees}}{180} = \frac{\pi/6}{\pi}$$

it follows that

$$\text{Degrees} = 30$$

That is, $\dfrac{\pi}{6}$ radians equals 30°.

For reference, six of the most important angles are shown in Figure 1.4 along with their measurements in degrees and in radians.

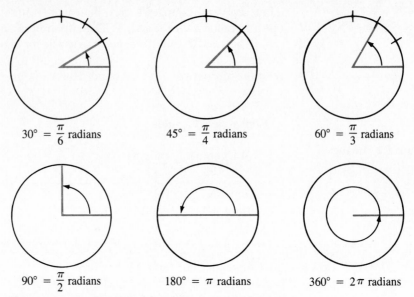

$$30° = \frac{\pi}{6} \text{ radians} \qquad 45° = \frac{\pi}{4} \text{ radians} \qquad 60° = \frac{\pi}{3} \text{ radians}$$

$$90° = \frac{\pi}{2} \text{ radians} \qquad 180° = \pi \text{ radians} \qquad 360° = 2\pi \text{ radians}$$

Figure 1.4 Six important angles in degrees and radians.

The Sine and the Cosine

Suppose the line segment joining the points $(0, 0)$ and $(1, 0)$ on the x axis is rotated through an angle of θ radians so that the free endpoint of the segment moves from $(1, 0)$ to a point (x, y) as in Figure 1.5. The x and y coordinates of the point (x, y) are known, respectively, as the **cosine** and **sine** of the angle θ. The symbol $\cos \theta$ is used to denote the cosine of θ, and $\sin \theta$ is used to denote its sine.

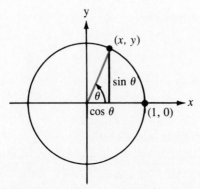

Figure 1.5 The sine and cosine of an angle.

The Sine and Cosine

For any angle θ,

$$\cos \theta = x \qquad \text{and} \qquad \sin \theta = y$$

where (x, y) is the point to which $(1, 0)$ is carried by a rotation of θ radians about the origin.

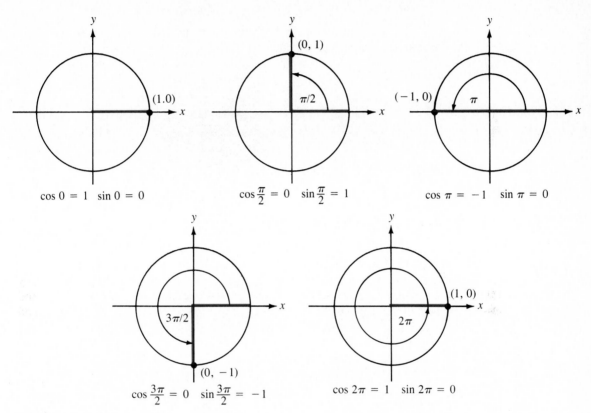

Figure 1.6 The sine and cosine of multiples of $\frac{\pi}{2}$.

The values of the sine and cosine for multiples of $\frac{\pi}{2}$ can be read from Figure 1.6 and are summarized in the following table.

θ	0	$\frac{\pi}{2}$	π	$\frac{3\pi}{2}$	2π
cos θ	1	0	-1	0	1
sin θ	0	1	0	-1	0

The sine and cosine of a few other important angles are also easy to obtain geometrically, as you will see shortly. For other angles, you can use Table III at the back of the book or your calculator to find the sine and cosine.

Elementary Properties of the Sine and Cosine

Since there are 2π radians in a complete rotation, it follows that

$$\sin(\theta + 2\pi) = \sin\theta \quad \text{and} \quad \cos(\theta + 2\pi) = \cos\theta$$

That is, the sine and cosine functions are **periodic** with period 2π. The situation is illustrated in Figure 1.7.

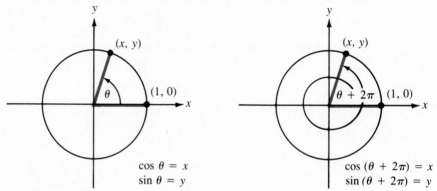

Figure 1.7 The periodicity of $\sin\theta$ and $\cos\theta$.

Since negative angles correspond to clockwise rotations, it follows that

$$\sin(-\theta) = -\sin\theta \quad \text{and} \quad \cos(-\theta) = \cos\theta$$

This is illustrated in Figure 1.8.

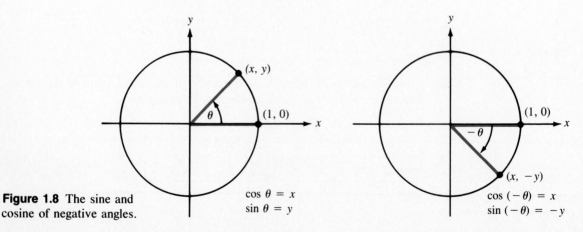

Figure 1.8 The sine and cosine of negative angles.

Properties of the Sine and Cosine

$$\sin (\theta + 2\pi) = \sin \theta \quad \text{and} \quad \cos (\theta + 2\pi) = \cos \theta$$

$$\sin (-\theta) = -\sin \theta \quad \text{and} \quad \cos (-\theta) = \cos \theta$$

The use of these properties is illustrated in the following example.

EXAMPLE 1.2 Evaluate the following expressions.

(a) $\cos (-\pi)$ (b) $\sin \left(-\dfrac{\pi}{2}\right)$ (c) $\cos 3\pi$ (d) $\sin \dfrac{5\pi}{2}$

SOLUTION

(a) Since $\cos \pi = -1$, it follows that

$$\cos (-\pi) = \cos \pi = -1$$

(b) Since $\sin \dfrac{\pi}{2} = 1$, it follows that

$$\sin \left(-\frac{\pi}{2}\right) = -\sin \frac{\pi}{2} = -1$$

(c) Since $3\pi = \pi + 2\pi$ and $\cos \pi = -1$, it follows that

$$\cos 3\pi = \cos (\pi + 2\pi) = \cos \pi = -1$$

(d) Since $\dfrac{5\pi}{2} = \dfrac{\pi}{2} + 2\pi$ and $\sin \dfrac{\pi}{2} = 1$, it follows that

$$\sin \frac{5\pi}{2} = \sin \left(\frac{\pi}{2} + 2\pi\right) = \sin \frac{\pi}{2} = 1$$

The Graphs of sin θ and cos θ

It is obvious from the definitions of the sine and cosine that as θ goes from 0 to 2π, the function $\sin \theta$ oscillates between 1 and -1, starting with $\sin 0 = 0$, and the function $\cos \theta$ oscillates between 1 and -1, starting with $\cos 0 = 1$. This observation, together with the elementary properties previously derived, suggests that the graphs of the functions $\sin \theta$ and $\cos \theta$ resemble the curves in Figures 1.9 and 1.10, respectively.

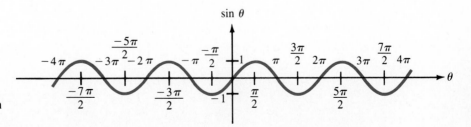

Figure 1.9 The graph of sin θ.

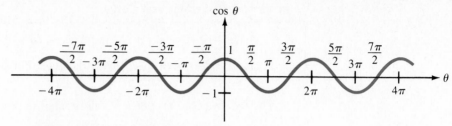

Figure 1.10 The graph of cos θ.

Other Trigonometric Functions

Four other useful trigonometric functions can be defined in terms of the sine and cosine as follows.

The Tangent, Cotangent, Secant, and Cosecant

For any angle θ,

$$\tan \theta = \frac{\sin \theta}{\cos \theta} \qquad \cot \theta = \frac{1}{\tan \theta} = \frac{\cos \theta}{\sin \theta}$$

$$\sec \theta = \frac{1}{\cos \theta} \qquad \csc \theta = \frac{1}{\sin \theta}$$

provided the denominators are not zero.

EXAMPLE 1.3 Evaluate the following expressions.

(a) $\tan \pi$ (b) $\cot \dfrac{\pi}{2}$ (c) $\sec(-\pi)$ (d) $\csc\left(-\dfrac{5\pi}{2}\right)$

SOLUTION

(a) Since $\sin \pi = 0$ and $\cos \pi = -1$, it follows that

$$\tan \pi = \frac{\sin \pi}{\cos \pi} = \frac{0}{-1} = 0$$

(b) Since $\sin \dfrac{\pi}{2} = 1$ and $\cos \dfrac{\pi}{2} = 0$, it follows that

$$\cot \frac{\pi}{2} = \frac{\cos(\pi/2)}{\sin(\pi/2)} = \frac{0}{1} = 0$$

(c) Since $\cos \pi = -1$, it follows that

$$\sec(-\pi) = \frac{1}{\cos(-\pi)} = \frac{1}{\cos \pi} = -1$$

(d) Since $\dfrac{5\pi}{2} = \dfrac{\pi}{2} + 2\pi$ and $\sin\dfrac{\pi}{2} = 1$, it follows that

$$\csc\left(-\frac{5\pi}{2}\right) = \frac{1}{\sin\,(-5\pi/2)} = \frac{1}{-\sin\,(5\pi/2)} = \frac{1}{-\sin\,(\pi/2)} = -1$$

Right Triangles

If you had a high-school course in trigonometry, you may remember the following definitions of the sine, cosine, and tangent involving the sides of a right triangle like the one in Figure 1.11.

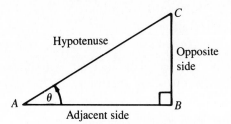

Figure 1.11 Triangle used to define trigonometric functions.

The Trigonometry of Right Triangles

$$\sin\theta = \frac{\text{opposite}}{\text{hypotenuse}} \qquad \cos\theta = \frac{\text{adjacent}}{\text{hypotenuse}} \qquad \tan\theta = \frac{\text{opposite}}{\text{adjacent}}$$

The definitions of trigonometric functions that you have seen in this section involving the coordinates of points on a circle of radius 1 are equivalent to the definitions from high-school trigonometry. To see this, superimpose an xy coordinate system over the triangle ABC as shown in Figure 1.12, and draw the circle of radius 1 that is centered at the origin.

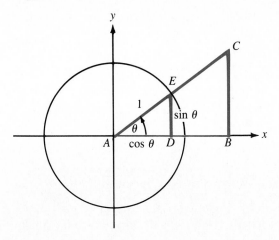

Figure 1.12 Similar triangles.

Since ABC and ADE are similar triangles, it follows that

$$\frac{AD}{AE} = \frac{AB}{AC}$$

But the length AD is the x coordinate of point E on the circle, and so, by the definition of the cosine,

$$AD = \cos\theta$$

Moreover,

$$AE = 1 \qquad AB = \text{adjacent side} \qquad \text{and} \qquad AC = \text{hypotenuse}$$

Hence,

$$\frac{\cos\theta}{1} = \frac{\text{adjacent}}{\text{hypotenuse}} \qquad \text{or} \qquad \cos\theta = \frac{\text{adjacent}}{\text{hypotenuse}}$$

For practice, convince yourself that similar arguments lead to the formulas for the sine and tangent.

Calculations with Right Triangles

Many calculations involving trigonometric functions can be performed easily and quickly with the aid of appropriate right triangles. For example, from the well-known 30-60-90° triangle in Figure 1.13, you see immediately that

$$\sin 30° = \frac{1}{2} \qquad \text{and} \qquad \cos 30° = \frac{\sqrt{3}}{2}$$

or, using radian measure,

$$\sin\frac{\pi}{6} = \frac{1}{2} \qquad \text{and} \qquad \cos\frac{\pi}{6} = \frac{\sqrt{3}}{2}$$

Here is an example that further illustrates the use of right triangles.

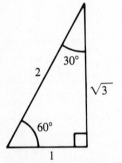

Figure 1.13 A 30-60-90° triangle.

EXAMPLE 1.4   Find $\tan\theta$ if $\sec\theta = \frac{3}{2}$.

SOLUTION
Since

$$\sec\theta = \frac{1}{\cos\theta} = \frac{\text{hypotenuse}}{\text{adjacent}}$$

begin by drawing a right triangle (Figure 1.14) in which the hypotenuse has length 3 and the side adjacent to the angle θ has length 2.

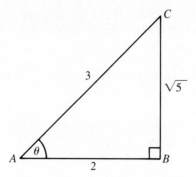

Figure 1.14 Right triangle for Example 1.4.

By the pythagorean theorem,

$$\text{Length of } BC = \sqrt{3^2 - 2^2} = \sqrt{5}$$

and so

$$\tan \theta = \frac{\text{opposite}}{\text{adjacent}} = \frac{\sqrt{5}}{2}$$

Important Angles

Using appropriate right triangles and the definitions and elementary properties of the sine and cosine, you should now be able to find the sine and cosine of the important angles listed in the following table. Before proceeding, take a few minutes to fill in the table so that you will have this information summarized for easy reference in the future.

A Table of Important Values

θ	0	$\frac{\pi}{6}$	$\frac{\pi}{4}$	$\frac{\pi}{3}$	$\frac{\pi}{2}$	$\frac{2\pi}{3}$	$\frac{3\pi}{4}$	$\frac{5\pi}{6}$	π
sin θ									
cos θ									

Trigonometric Identities

An **identity** is an equation that holds for all values of its variable or variables. Identities involving trigonometric functions are called **trigonometric identities** and can be used to simplify trigonometric expressions. You have already seen some elementary trigonometric identities in this section. For example, the formulas

$$\cos(-\theta) = \cos\theta \quad \text{and} \quad \sin(-\theta) = -\sin\theta$$

are identities because they hold for all values of θ. Most trigonometry texts list dozens of trigonometric identities. Fortunately, only a few of them will be needed in this book.

One of the most important and well-known trigonometric identity is a simple consequence of the pythagorean theorem. It states that

$$\sin^2 \theta + \cos^2 \theta = 1$$

where $\sin^2 \theta$ stands for $(\sin \theta)^2$ and $\cos^2 \theta$ stands for $(\cos \theta)^2$. To see why this identity holds, look at Figure 1.15, which shows a point (x, y) on a circle of radius 1. By definition,

$$y = \sin \theta \quad \text{and} \quad x = \cos \theta$$

and, by the pythagorean theorem,

$$y^2 + x^2 = 1$$

from which the identity follows immediately.

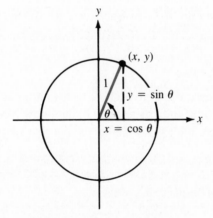

Figure 1.15 $\sin^2 \theta + \cos^2 \theta = 1$.

The Pythagorean Identity

For any angle θ,

$$\sin^2 \theta + \cos^2 \theta = 1$$

The following formulas for the cosine and sine of the sum of two angles are particularly useful. Proofs of these identities can be found in most trigonometry texts and will be omitted here.

The Addition Formulas

For any angles a and b,

$$\cos (a + b) = \cos a \cos b - \sin a \sin b$$

and

$$\sin (a + b) = \sin a \cos b + \cos a \sin b$$

If the angles a and b in the addition formulas are equal, the identities can be rewritten as follows.

The Double-Angle Formulas

For any angle θ,

$$\cos 2\theta = \cos^2 \theta - \sin^2 \theta$$

and

$$\sin 2\theta = 2 \sin \theta \cos \theta$$

Some of these trigonometric identities will be used in Section 2 to derive the formulas for the derivatives of the sine and cosine functions. They can also be used in the solution of certain trigonometric equations.

Trigonometric Equations

The following examples illustrate the use of trigonometric identities to solve trigonometric equations.

EXAMPLE 1.5   Find all the values of θ in the interval $0 \leq \theta \leq \pi$ that satisfy the equation $\sin 2\theta = \sin \theta$.

SOLUTION
Use the double-angle formula

$$\sin 2\theta = 2 \sin \theta \cos \theta$$

to rewrite the given equation as

$$2 \sin \theta \cos \theta = \sin \theta$$

or

$$2 \sin \theta \cos \theta - \sin \theta = 0$$

and factor to get

$$\sin \theta (2 \cos \theta - 1) = 0$$

which is satisfied if

$$\sin \theta = 0 \qquad \text{or} \qquad 2 \cos \theta - 1 = 0$$

From the graphs of the sine and cosine functions and from the table of important values, which you completed earlier in this section, it should be clear that the only solutions of

$$\sin \theta = 0$$

in the interval $0 \le \theta \le \pi$ are

$$\theta = 0 \quad \text{and} \quad \theta = \pi$$

and that the only solution of

$$2 \cos \theta - 1 = 0 \quad \text{or} \quad \cos \theta = \frac{1}{2}$$

in the interval $0 \le \theta \le \pi$ is

$$\theta = \frac{\pi}{3}$$

Hence, the only solutions of the original equation in the specified interval are $\theta = 0$, $\theta = \frac{\pi}{3}$, and $\theta = \pi$.

EXAMPLE 1.6 Find all the values of θ in the interval $0 \le \theta \le 2\pi$ that satisfy the equation $\sin^2 \theta - \cos^2 \theta + \sin \theta = 0$.

SOLUTION
Use the pythagorean identity

$$\sin^2 \theta + \cos^2 \theta = 1 \quad \text{or} \quad \cos^2 \theta = 1 - \sin^2 \theta$$

to rewrite the original equation without any cosine terms as

$$\sin^2 \theta - (1 - \sin^2 \theta) + \sin \theta = 0$$

or

$$2 \sin^2 \theta + \sin \theta - 1 = 0$$

and factor to get

$$(2 \sin \theta - 1)(\sin \theta + 1) = 0$$

The only solutions of

$$2 \sin \theta - 1 = 0 \quad \text{or} \quad \sin \theta = \frac{1}{2}$$

in the interval $0 \le \theta \le 2\pi$ are

$$\theta = \frac{\pi}{6} \quad \text{and} \quad \theta = \frac{5\pi}{6}$$

and the only solution of

$$\sin \theta + 1 = 0 \quad \text{or} \quad \sin \theta = -1$$

in the interval is

$$\theta = \frac{3\pi}{2}$$

Hence, the solutions of the original equation in the specified interval are $\theta = \frac{\pi}{6}$, $\theta = \frac{5\pi}{6}$, and $\theta = \frac{3\pi}{2}$.

Problems

In Problems 1 through 6, specify the number of degrees in the given angle.

1.

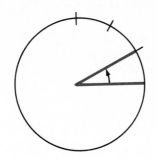

2.

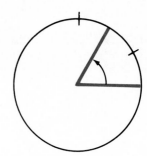

3.

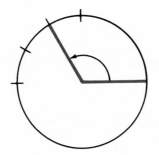

4.

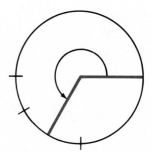

5.

6.

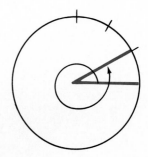

In Problems 7 through 12, show the given angle graphically.

7. 60°   8. −45°

9. 240°   10. 120°

11. −150°   12. 405°

In Problems 13 through 18, convert from degrees to radians.

13. 15°   14. 270°

15. −150°   16. −240°

17. 540°   18. 1°

In Problems 19 through 24, convert from radians to degrees.

19. $\dfrac{5\pi}{6}$ 20. $\dfrac{2\pi}{3}$

21. $\dfrac{3\pi}{2}$ 22. $-\dfrac{\pi}{12}$

23. $-\dfrac{3\pi}{4}$ 24. 1

In Problems 25 through 30, specify the number of radians in the given angle.

25. 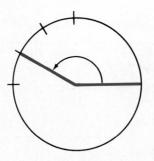   26.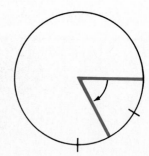

27. 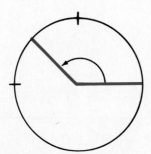   28.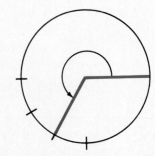

29. 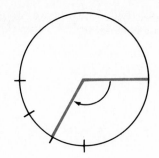   30.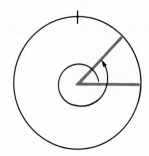

In Problems 31 through 36, show graphically the angles with the given radian measure.

31. $\dfrac{3\pi}{4}$

32. $-\dfrac{\pi}{6}$

33. $\dfrac{2\pi}{3}$

34. $\dfrac{5\pi}{6}$

35. $-\dfrac{5\pi}{6}$

36. $\dfrac{3\pi}{2}$

In Problems 37 through 44, evaluate the given expression without using Table III or your calculator.

37. $\cos \dfrac{7\pi}{2}$

38. $\sin 5\pi$

39. $\cos \left(-\dfrac{5\pi}{2}\right)$

40. $\sin \left(-\dfrac{7\pi}{2}\right)$

41. $\cot \dfrac{5\pi}{2}$

42. $\sec 3\pi$

43. $\tan (-\pi)$

44. $\csc \left(-\dfrac{7\pi}{2}\right)$

45. Use a right triangle to find $\cos \dfrac{\pi}{3}$ and $\sin \dfrac{\pi}{3}$.

46. Use a right triangle to find $\cos \dfrac{\pi}{4}$ and $\sin \dfrac{\pi}{4}$.

47. Complete the following table:

θ	$\dfrac{7\pi}{6}$	$\dfrac{5\pi}{4}$	$\dfrac{4\pi}{3}$	$\dfrac{3\pi}{2}$	$\dfrac{5\pi}{3}$	$\dfrac{7\pi}{4}$	$\dfrac{11\pi}{6}$	2π
sin θ								
cos θ								

In Problems 48 through 58, evaluate the given expression without using Table III or your calculator.

48. $\cos\left(-\dfrac{2\pi}{3}\right)$

49. $\sin\left(-\dfrac{\pi}{6}\right)$

50. $\sin\left(-\dfrac{3\pi}{4}\right)$

51. $\cos\dfrac{7\pi}{3}$

52. $\sin\left(-\dfrac{7\pi}{6}\right)$

53. $\tan\dfrac{\pi}{6}$

54. $\cot\dfrac{\pi}{3}$

55. $\sec\dfrac{\pi}{3}$

56. $\csc\dfrac{2\pi}{3}$

57. $\tan\left(-\dfrac{\pi}{4}\right)$

58. $\cot\left(-\dfrac{5\pi}{3}\right)$

59. Find $\tan\theta$ if $\sec\theta = \dfrac{5}{4}$.

60. Find $\tan\theta$ if $\cos\theta = \dfrac{3}{5}$.

61. Find $\tan\theta$ if $\csc\theta = \dfrac{5}{3}$.

62. Find $\sin\theta$ if $\tan\theta = \dfrac{2}{3}$.

63. Find $\cos\theta$ if $\tan\theta = \dfrac{3}{4}$.

64. Find $\sec\theta$ if $\cot\theta = \dfrac{4}{3}$.

In Problems 65 through 73, find all the values of θ in the specified interval that satisfy the given equation.

65. $\cos\theta - \sin 2\theta = 0;\ 0 \le \theta \le 2\pi$

66. $3\sin^2\theta + \cos 2\theta = 2;\ 0 \le \theta \le 2\pi$

67. $\sin 2\theta = \sqrt{3}\cos\theta;\ 0 \le \theta \le \pi$

68. $\cos 2\theta = \cos\theta;\ 0 \le \theta \le \pi$

69. $2\cos^2\theta = \sin 2\theta;\ 0 \le \theta \le \pi$

70. $3\cos^2\theta - \sin^2\theta = 2;\ 0 \le \theta \le \pi$

71. $2\cos^2\theta + \sin\theta = 1;\ 0 \le \theta \le \pi$

72. $\cos^2\theta - \sin^2\theta + \cos\theta = 0;\ 0 \le \theta \le \pi$

73. $\sin^2\theta - \cos^2\theta + 3\sin\theta = 1;\ 0 \le \theta \le \pi$

74. Starting with the identity $\sin^2\theta + \cos^2\theta = 1$, derive the identity $1 + \tan^2\theta = \sec^2\theta$.

75. Starting with the addition formula

$$\sin(a + b) = \sin a \cos b + \cos a \sin b$$

   derive the subtraction formula

$$\sin(a - b) = \sin a \cos b - \cos a \sin b$$

76. Starting with the addition formula

$$\cos(a + b) = \cos a \cos b - \sin a \sin b$$

   derive the subtraction formula

$$\cos(a - b) = \cos a \cos b + \sin a \sin b$$

77. Starting with the addition formulas for the sine and cosine, derive the addition formula

$$\tan(a + b) = \frac{\tan a + \tan b}{1 - \tan a \tan b}$$

   for the tangent.

78. Use the addition formula for the sine to derive the identity

$$\sin\left(\frac{\pi}{2} - \theta\right) = \cos \theta$$

79. Use the addition formula for the cosine to derive the identity

$$\cos\left(\frac{\pi}{2} - \theta\right) = \sin \theta$$

80. Give geometric arguments involving the coordinates of points on a circle of radius 1 to show why the identities in Problems 78 and 79 are valid.

2   Differentiation of Trigonometric Functions

The trigonometric functions are relatively easy to differentiate. In this section you will learn how. In the following section you will see a variety of rate-of-change and optimization problems that can be solved using derivatives of trigonometric functions.

The Derivatives of the Sine and Cosine

Here are the formulas for the derivatives of the sine and cosine functions. The proofs will be discussed later in this section, after you have had a chance to practice using the formulas.

The Derivatives of sin θ and cos θ

If θ is measured in radians,

$$\frac{d}{d\theta} \sin \theta = \cos \theta \qquad \text{and} \qquad \frac{d}{d\theta} \cos \theta = -\sin \theta$$

These formulas are usually used in conjunction with the chain rule in the following forms:

Chain Rule for Sine and Cosine

If h is a differentiable function of θ and θ is measured in radians,

$$\frac{d}{d\theta} \sin h(\theta) = h'(\theta) \cos h(\theta)$$

and

$$\frac{d}{d\theta} \cos h(\theta) = -h'(\theta) \sin h(\theta)$$

Here are some examples.

EXAMPLE 2.1 Differentiate the function $f(\theta) = \sin (3\theta + 1)$.

SOLUTION
Using the chain rule for the sine function with $h(\theta) = 3\theta + 1$, you get

$$f'(\theta) = 3 \cos (3\theta + 1)$$

EXAMPLE 2.2 Differentiate the function $f(\theta) = \cos^2 \theta$.

SOLUTION
Since

$$\cos^2 \theta = (\cos \theta)^2$$

you use the chain rule for powers and the formula for the derivative of the cosine to get

$$f'(\theta) = 2 \cos \theta \, (-\sin \theta) = -2 \cos \theta \sin \theta$$

The Derivatives of Other Trigonometric Functions

The differentiation formulas for the sine and cosine can be used to obtain differentiation formulas for the other trigonometric functions. The procedure is illustrated in the next example.

EXAMPLE 2.3 Show that $\dfrac{d}{d\theta}\tan\theta = \sec^2\theta$.

SOLUTION
Write

$$\tan\theta = \frac{\sin\theta}{\cos\theta}$$

and use the quotient rule to get

$$\frac{d}{d\theta}\tan\theta = \frac{\cos\theta\,(\cos\theta) - \sin\theta\,(-\sin\theta)}{\cos^2\theta}$$

$$= \frac{\cos^2\theta + \sin^2\theta}{\cos^2\theta}$$

$$= \frac{1}{\cos^2\theta} \qquad \text{since } \sin^2\theta + \cos^2\theta = 1$$

$$= \sec^2\theta \qquad \text{since } \sec\theta = \frac{1}{\cos\theta}$$

The Derivative of tan θ

If θ is measured in radians,

$$\frac{d}{d\theta}\tan\theta = \sec^2\theta$$

Chain Rule for the Tangent

If h is a differentiable function of θ and θ is measured in radians,

$$\frac{d}{d\theta}\tan h(\theta) = h'(\theta)\sec^2 h(\theta)$$

Here is an example illustrating the use of the chain rule for the tangent.

EXAMPLE 2.4 Differentiate the function $f(\theta) = \tan\theta^2$.

SOLUTION
Using the chain rule for tangents, you get

$$f'(\theta) = 2\theta\sec^2\theta^2$$

The derivations of the differentiation formulas for the other trigonometric functions are left as exercises for you to do. (See Problems 21, 22, and 23 at the end of this section.) Only the formulas for the derivatives of the sine, cosine, and tangent will be needed for the applications in Section 3.

The Proof that $\frac{d}{d\theta} \sin \theta = \cos \theta$

Since the function $f(\theta) = \sin \theta$ is unrelated to any of the functions whose derivatives were obtained in previous chapters, it will be necessary to return to the *definition* of the derivative (from Chapter 2, Section 1) to find the derivative in this case. In terms of θ, the definition states that

$$f'(\theta) = \lim_{\Delta\theta \to 0} \frac{f(\theta + \Delta\theta) - f(\theta)}{\Delta\theta}$$

In performing the required calculations when $f(\theta) = \sin \theta$, you will need to use the addition formula

$$\sin (a + b) = \sin a \cos b + \cos a \sin b$$

from the preceding section, as well as the following two facts about the behavior of $\sin \theta$ and $\cos \theta$ as θ approaches zero.

Two Important Limits

If θ is measured in radians,

$$(1) \quad \lim_{\theta \to 0} \frac{\sin \theta}{\theta} = 1$$

and

$$(2) \quad \lim_{\theta \to 0} \frac{\cos \theta - 1}{\theta} = 0$$

Proofs of the Limits

To see why the first of these limits holds, look at Figure 2.1. From the definition of the sine of an angle, it follows that

$$\sin \theta = y = \text{length of segment } PQ$$

and since θ is measured in radians, it follows that

$$\theta = \text{length of arc } PR$$

(Recall that 1 radian is the amount by which a line segment of length 1 must be rotated so that its free endpoint traces out a circular arc of length

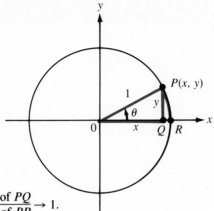

Figure 2.1 $\dfrac{\sin \theta}{\theta} = \dfrac{\text{length of } PQ}{\text{length of } PR} \to 1.$

1. Hence θ radians is the amount by which a line segment of length 1 must be rotated so that its free endpoint traces out a circular arc of length θ.) Thus,

$$\frac{\sin \theta}{\theta} = \frac{\text{length of segment } PQ}{\text{length of arc } PR}$$

The picture suggests that as θ approaches zero, the length of the segment PQ and that of the arc PR will get closer and closer to one another so that their ratio will approach 1. (A formal proof of this fact can be found in more advanced texts.) Hence,

$$\lim_{\theta \to 0} \frac{\sin \theta}{\theta} = 1$$

and the first limit is established.

The second limit,

$$\lim_{\theta \to 0} \frac{\cos \theta - 1}{\theta} = 0$$

can be derived algebraically from the first one. (See Problem 24 at the end of this section.) It also has the following simple geometric interpretation, which should give you additional insight into why it holds.

Figure 2.2 shows a portion of the graph of the function $\cos \theta$ near $\theta = 0$. The graph has a relative maximum when $\theta = 0$, and so, assuming that $\cos \theta$ is differentiable at $\theta = 0$ (which we have not yet proved), it follows that the derivative of $\cos \theta$ at $\theta = 0$ is zero. But the difference quotient you would use to find this derivative is

$$\frac{\cos (0 + \Delta\theta) - \cos 0}{\Delta\theta} = \frac{\cos \Delta\theta - 1}{\Delta\theta}$$

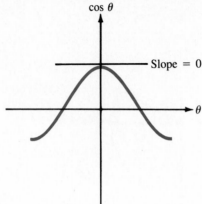

Figure 2.2

$\lim\limits_{\theta \to 0} \dfrac{\cos\theta - 1}{\theta} =$

(slope of tangent at 0) = 0.

Hence,

$$\lim_{\Delta\theta \to 0} \frac{\cos \Delta\theta - 1}{\Delta\theta} = 0$$

which, except for the symbol used to denote the variable ($\Delta\theta$ instead of θ), is precisely the limit under consideration.

Proof of the Derivative Formula

With these preliminaries out of the way, the proof of the formula for the derivative of $\sin\theta$ is not hard. Here are the details.

Let $f(\theta) = \sin\theta$. Then,

$$\frac{f(\theta + \Delta\theta) - f(\theta)}{\Delta\theta} = \frac{\sin(\theta + \Delta\theta) - \sin\theta}{\Delta\theta}$$

$$= \frac{\sin\theta \cos\Delta\theta + \cos\theta \sin\Delta\theta - \sin\theta}{\Delta\theta}$$

by the addition formula for the sine

$$= \frac{\sin\theta \cos\Delta\theta - \sin\theta}{\Delta\theta} + \frac{\cos\theta \sin\Delta\theta}{\Delta\theta}$$

$$= \sin\theta \frac{\cos\Delta\theta - 1}{\Delta\theta} + \cos\theta \frac{\sin\Delta\theta}{\Delta\theta}$$

Now let $\Delta\theta$ approach zero to get the derivative. Since

$$\lim_{\Delta\theta \to 0} \frac{\cos\Delta\theta - 1}{\Delta\theta} = 0 \qquad \text{by limit (2)}$$

and

$$\lim_{\Delta\theta \to 0} \frac{\sin\Delta\theta}{\Delta\theta} = 1 \qquad \text{by limit (1)}$$

it follows that

$$f'(\theta) = \lim_{\Delta\theta \to 0} \sin\theta \frac{\cos\Delta\theta - 1}{\Delta\theta} + \lim_{\Delta\theta \to 0} \cos\theta \frac{\sin\Delta\theta}{\Delta\theta}$$

$$= (\sin\theta)(0) + (\cos\theta)(1) = \cos\theta$$

That is,

$$\frac{d}{d\theta} \sin\theta = \cos\theta$$

The Proof that $\dfrac{d}{d\theta} \cos\theta = -\sin\theta$

The formula for the derivative of $\cos\theta$ can be obtained from the definition of the derivative by a calculation similar to the one just given for $\sin\theta$. (Problem 25 at the end of this section will suggest that you try it.) Here is an attractive alternative derivation that takes advantage of the formula for the derivative of the sine function and uses the identities

$$\cos\theta = \sin\left(\frac{\pi}{2} - \theta\right) \qquad \text{and} \qquad \sin\theta = \cos\left(\frac{\pi}{2} - \theta\right)$$

which relate the sine and cosine functions. (These identities are simple consequences of the addition formulas for the sine and cosine. They can also be obtained geometrically, directly from the definitions of the sine and cosine. See Problems 78, 79, and 80 of the preceding section.)

To obtain the derivative of $\cos\theta$, start with the identity

$$\cos\theta = \sin\left(\frac{\pi}{2} - \theta\right)$$

and differentiate using the chain rule for the sine to get

$$\frac{d}{d\theta} \cos\theta = (-1)\cos\left(\frac{\pi}{2} - \theta\right)$$

Then apply the identity

$$\sin\theta = \cos\left(\frac{\pi}{2} - \theta\right)$$

to the right-hand side to conclude that

$$\frac{d}{d\theta} \cos\theta = -\sin\theta$$

Problems

In Problems 1 through 20, differentiate the given function.

1. $f(\theta) = \sin 3\theta$

2. $f(\theta) = \cos 2\theta$

3. $f(\theta) = \sin (1 - 2\theta)$

4. $f(\theta) = \sin \theta^2$

5. $f(\theta) = \cos (\theta^3 + 1)$

6. $f(\theta) = \sin^2 \theta$

7. $f(\theta) = \cos^2 \left(\dfrac{\pi}{2} - \theta\right)$

8. $f(\theta) = \sin (2\theta + 1)^2$

9. $f(\theta) = \cos (1 + 3\theta)^2$

10. $f(\theta) = e^{-\theta} \sin \theta$

11. $f(\theta) = e^{-\theta/2} \cos 2\pi\theta$

12. $f(\theta) = \dfrac{\cos \theta}{1 - \cos \theta}$

13. $f(\theta) = \dfrac{\sin \theta}{1 + \sin \theta}$

14. $f(\theta) = \tan (5\theta + 2)$

15. $f(\theta) = \tan (1 - \theta^5)$

16. $f(\theta) = \tan^2 \theta$

17. $f(\theta) = \tan^2 \left(\dfrac{\pi}{2} - 2\pi\theta\right)$

18. $f(\theta) = \tan (\pi - 4\theta)^2$

19. $f(\theta) = \ln \sin^2 \theta$

20. $f(\theta) = \ln \tan^2 \theta$

21. Show that $\dfrac{d}{d\theta} \sec \theta = \sec \theta \tan \theta$.

22. Show that $\dfrac{d}{d\theta} \csc \theta = -\csc \theta \cot \theta$.

23. Show that $\dfrac{d}{d\theta} \cot \theta = -\csc^2 \theta$.

24. Use the fact that

$$\lim_{\theta \to 0} \frac{\sin \theta}{\theta} = 1$$

to prove that

$$\lim_{\theta \to 0} \frac{\cos \theta - 1}{\theta} = 0$$

Hint: Write

$$\frac{\cos \theta - 1}{\theta} = \frac{\cos \theta - 1}{\theta} \cdot \frac{\cos \theta + 1}{\cos \theta + 1}$$

and multiply out the terms on the right-hand side. Simplify the resulting expression using an appropriate trigonometric identity, and then let θ approach zero, keeping in mind that

$$\lim_{\theta \to 0} \sin \theta = 0 \quad \text{and} \quad \lim_{\theta \to 0} \cos \theta = 1$$

25. Use the definition of the derivative to prove that $\dfrac{d}{d\theta} \cos \theta = -\sin \theta$.

3   Applications of Trigonometric Functions

In this section, you will see a sampling of applications involving trigonometric functions and their derivatives. The first example is a rate-of-change problem, and the subsequent examples are optimization problems.

Related Rates

In the first example, the goal is to find the rate of change of an angle θ with respect to time t. An explicit formula relating θ and t is not given. However, the relationship between θ and another variable x is known, as is the rate of change of x with respect to t. Using the chain rule and implicit differentiation, you will be able to put the information together to find the rate of change of θ with respect to t. Problems of this type are known as **related rates problems**.

EXAMPLE 3.1   An observer watches a plane approach at a speed of 400 miles per hour and at an altitude of 4 miles. At what rate is the angle of elevation of the observer's line of sight changing with respect to time when the horizontal distance between the plane and the observer is 3 miles?

SOLUTION
Let x denote the horizontal distance between the plane and the observer, let t denote time (in hours), and draw a diagram representing the situation as in Figure 3.1.

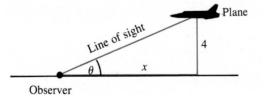

Figure 3.1 Observation of an approaching plane.

You know that $\dfrac{dx}{dt} = -400$ (the minus sign indicating that the distance x is decreasing), and your goal is to find $\dfrac{d\theta}{dt}$ when $x = 3$. From the right triangle in Figure 3.1 you see that

$$\tan \theta = \frac{4}{x}$$

Differentiate both sides of this equation with respect to t (remembering to use the chain rule) to get

$$\sec^2 \theta \, \frac{d\theta}{dt} = -\frac{4}{x^2} \frac{dx}{dt}$$

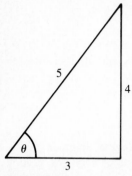

Figure 3.2 Triangle for the computation of cos θ when $x = 3$.

or

$$\frac{d\theta}{dt} = -\frac{4}{x^2} \cos^2 \theta \frac{dx}{dt}$$

The values of x and $\dfrac{dx}{dt}$ are given. You get the value of $\cos^2 \theta$ when $x = 3$ from the right triangle in Figure 3.2. In particular, when $x = 3$,

$$\cos^2 \theta = \left(\frac{3}{5}\right)^2 = \frac{9}{25}$$

Substituting $x = 3$, $\dfrac{dx}{dt} = -400$, and $\cos^2 \theta = \dfrac{9}{25}$ into the formula for $\dfrac{d\theta}{dt}$, you get

$$\frac{d\theta}{dt} = -\frac{4}{9}\left(\frac{9}{25}\right)(-400) = 64 \text{ radians per hour}$$

Optimization Problems

The next two examples are optimization problems involving trig-onometric functions. To solve problems of this type, proceed as in Chapter 3, Section 4. In particular: (1) construct a function representing the quantity to be optimized in terms of a convenient variable; (2) identify an interval on which the function has a practical interpretation; (3) differentiate the function, and set the derivative equal to zero to find the critical points; and (4) compare the values of the function at the critical points in the interval and at the endpoints (or use the second-derivative test) to verify that the desired absolute extremum has been found.

EXAMPLE 3.2 Two sides of a triangle are 4 inches long. What should the angle between these sides be to make the area of the triangle as large as possible?

SOLUTION

The triangle is shown in Figure 3.3. In general, the area of a triangle is given by the formula

$$\text{Area} = \frac{1}{2}(\text{base})(\text{height})$$

In this case,

$$\text{Base} = 4$$

and, since $\sin \theta = \dfrac{h}{4}$,

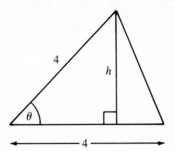

Figure 3.3 Triangle for Example 3.2.

Height $= h = 4 \sin \theta$

Hence the area of the triangle is given by the function

$$A(\theta) = \frac{1}{2}(4)(4 \sin \theta) = 8 \sin \theta$$

Since only values of θ between $\theta = 0$ and $\theta = \pi$ radians are meaningful in the context of this problem, the goal is to find the absolute maximum of the function $A(\theta)$ on the interval $0 \leq \theta \leq \pi$.
The derivative of $A(\theta)$ is

$$A'(\theta) = 8 \cos \theta$$

which is zero on the interval $0 \leq \theta \leq \pi$ only when $\theta = \frac{\pi}{2}$. Comparing

$$A(0) = 8 \sin 0 = 0$$

$$A\left(\frac{\pi}{2}\right) = 8 \sin \frac{\pi}{2} = 8(1) = 8$$

and

$$A(\pi) = 8 \sin \pi = 0$$

you conclude that the area is maximized when the angle θ measures $\frac{\pi}{2}$ radians (or 90°); that is, when the triangle is a right triangle.

Illumination from a Light Source

EXAMPLE 3.3   A lamp with adjustable height hangs directly above the center of a circular kitchen table that is 8 feet in diameter. The illumination at the edge of the table is directly proportional to the cosine of the angle θ and inversely proportional to the square of the distance d, where θ and d are as shown in Figure 3.4. How close to the table should the lamp be pulled to maximize the illumination at the edge of the table?

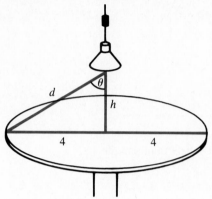

Figure 3.4 Hanging lamp and table for Example 3.3.

SOLUTION

Let L denote the illumination at the edge of the table. Then,

$$L = \frac{k \cos \theta}{d^2}$$

where k is a (positive) constant of proportionality. Moreover, from the right triangle in Figure 3.4,

$$\sin \theta = \frac{4}{d} \quad \text{or} \quad d = \frac{4}{\sin \theta}$$

Hence,

$$L(\theta) = \frac{k \cos \theta}{(4/\sin \theta)^2} = \frac{k}{16} \cos \theta \sin^2 \theta$$

Only values of θ between 0 and $\frac{\pi}{2}$ radians are meaningful in the context of this problem. Hence, the goal is to find the absolute maximum of the function $L(\theta)$ on the interval $0 \le \theta \le \frac{\pi}{2}$.

Using the product rule and chain rule to differentiate $L(\theta)$, you get

$$L'(\theta) = \frac{k}{16}[\cos \theta \, (2 \sin \theta \cos \theta) + \sin^2 \theta \, (-\sin \theta)]$$

$$= \frac{k}{16}(2 \cos^2 \theta \sin \theta - \sin^3 \theta)$$

$$= \frac{k}{16} \sin \theta \, (2 \cos^2 \theta - \sin^2 \theta)$$

which is zero when

$$\sin \theta = 0 \quad \text{or} \quad 2 \cos^2 \theta - \sin^2 \theta = 0$$

The only value of θ in the interval $0 \leq \theta \leq \dfrac{\pi}{2}$ for which $\sin \theta = 0$ is the endpoint $\theta = 0$. To solve the equation

$$2 \cos^2 \theta - \sin^2 \theta = 0$$

rewrite it as

$$2 \cos^2 \theta = \sin^2 \theta$$

divide both sides by $\cos^2 \theta$

$$2 = \frac{\sin^2 \theta}{\cos^2 \theta} = \left(\frac{\sin \theta}{\cos \theta}\right)^2$$

and take the square root of each side to get

$$\pm\sqrt{2} = \frac{\sin \theta}{\cos \theta} = \tan \theta$$

Since both $\sin \theta$ and $\cos \theta$ are nonnegative on the interval $0 \leq \theta \leq \dfrac{\pi}{2}$, you may discard the negative square root and conclude that the critical point occurs when

$$\tan \theta = \sqrt{2}$$

Although it would not be hard to use your calculator (or Table III at the back of the book) to find (or, at least, to estimate) the angle θ for which $\tan \theta = \sqrt{2}$, it is not necessary to do so. Instead, look at the right triangle in Figure 3.5 in which

$$\tan \theta = \frac{\text{opposite}}{\text{adjacent}} = \frac{\sqrt{2}}{1} = \sqrt{2}$$

and observe that

$$\sin \theta = \frac{\text{opposite}}{\text{hypotenuse}} = \frac{\sqrt{2}}{\sqrt{3}} \qquad \text{and} \qquad \cos \theta = \frac{\text{adjacent}}{\text{hypotenuse}} = \frac{1}{\sqrt{3}}$$

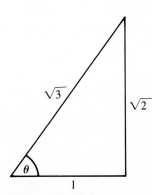

Figure 3.5 Right triangle with $\tan \theta = \sqrt{2}$.

Substitute these values into the equation for $L(\theta)$ to conclude that, at the critical point,

$$L(\theta) = \frac{k}{16} \cos \theta \sin^2 \theta = \frac{k}{16}\left(\frac{1}{\sqrt{3}}\right)\left(\frac{\sqrt{2}}{\sqrt{3}}\right)^2 = \frac{k}{24\sqrt{3}}$$

Compare this value with the values of $L(\theta)$ at the endpoints $\theta = 0$ and $\theta = \dfrac{\pi}{2}$, which are

$$L(0) = \frac{k}{16}(\cos 0)(\sin^2 0) = 0 \qquad \text{since } \sin 0 = 0$$

and

$$L\left(\frac{\pi}{2}\right) = \frac{k}{16}\left(\cos \frac{\pi}{2}\right)\left(\sin^2 \frac{\pi}{2}\right) = 0 \qquad \text{since } \cos \frac{\pi}{2} = 0$$

The largest of these possible maximum values is $\dfrac{k}{24\sqrt{3}}$, which occurs when $\tan \theta = \sqrt{2}$.

Finally, to find the height h that maximizes the illumination L, observe from Figure 3.4 that

$$\tan \theta = \frac{4}{h}$$

Since $\tan \theta = \sqrt{2}$, it follows that

$$h = \frac{4}{\tan \theta} = \frac{4}{\sqrt{2}} \approx 2.83 \text{ feet}$$

Minimization of Travel Time

In the next example, calculus is used to establish a general principle about the minimization of travel time. As you will see subsequently, the principle can be rephrased to give a well-known law about the reflection of light.

EXAMPLE 3.4   Two off-shore oil wells are, respectively, a and b miles out to sea. A motorboat that travels at a constant speed s transports workers from the first well to the shore and then proceeds to the second well. Show that the total travel time is minimized if the angle α between the motorboat's path of arrival and the shoreline is equal to the angle β between the shoreline and the motorboat's path of departure.

SOLUTION

Even though the goal is to prove that two angles are equal, it turns out that the easiest way to solve this problem is to let the variable x represent a convenient distance and to introduce trigonometry only at the end.

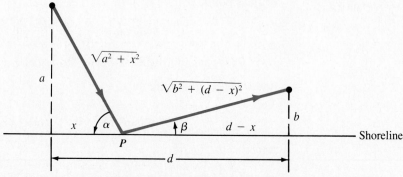

Figure 3.6 Path of motorboat from oil well to shore to oil well.

Begin with a sketch of the situation as in Figure 3.6, and define the variable x and the constant distance d as indicated. Then, by the pythagorean theorem,

$$\text{Distance from first well to } P = \sqrt{a^2 + x^2}$$

and

$$\text{Distance from second well to } P = \sqrt{b^2 + (d - x)^2}$$

Since (for constant speeds),

$$\text{Time} = \frac{\text{distance}}{\text{speed}}$$

the total travel time is given by the function

$$T(x) = \frac{\sqrt{a^2 + x^2}}{s} + \frac{\sqrt{b^2 + (d - x)^2}}{s}$$

where $0 \leq x \leq d$.

Then,

$$T'(x) = \frac{x}{s\sqrt{a^2 + x^2}} - \frac{d - x}{s\sqrt{b^2 + (d - x)^2}}$$

If you set $T'(x)$ equal to zero and do a little algebra, you get

$$\frac{x}{\sqrt{a^2 + x^2}} = \frac{d - x}{\sqrt{b^2 + (d - x)^2}}$$

Now look again at the right triangles in Figure 3.6 and observe that

$$\frac{x}{\sqrt{a^2 + x^2}} = \cos \alpha \qquad \text{and} \qquad \frac{d - x}{\sqrt{b^2 + (d - x)^2}} = \cos \beta$$

Hence, $T'(x) = 0$ implies that

$$\cos \alpha = \cos \beta$$

In the context of this problem, both α and β must measure between 0 and $\frac{\pi}{2}$ radians, and, for angles in this range, equality of the cosines implies equality of the angles themselves. (See, for example, the graph of the cosine function in Figure 1.10.) Hence, $T'(x) = 0$ implies that

$$\alpha = \beta$$

Moreover, it should be clear from the geometry of Figure 3.6 that no matter what the relative sizes of a, b, and d may be, there is a (unique) point P for which $\alpha = \beta$. It follows that the function $T(x)$ has a critical point in the interval $0 \leq x \leq d$ and that at this critical point, $\alpha = \beta$.

To verify that this critical point is really the absolute minimum, you can use the second-derivative test. A routine calculation gives

$$T''(x) = \frac{a^2}{s(a^2 + x^2)^{3/2}} + \frac{b^2}{s[b^2 + (d - x)^2]^{3/2}}$$

(For practice, check this calculation.) Since $T''(x)$ is positive for all values of x, and since there is only one critical point in the interval $0 \le x \le d$, it follows that this critical point does indeed correspond to the absolute minimum of the total time T on this interval.

Reflection of Light

According to **Fermat's principle** in optics, light traveling from one point to another takes the path that requires the least amount of time. Suppose (as illustrated in Figure 3.7) that a ray of light is transmitted from a source at a point A, strikes a reflecting surface (such as a mirror) at point P, and is subsequently received by an observer at point B.

Since, by Fermat's principle, the path from A to P to B minimizes time, it follows from the calculation in Example 3.4 that the angles α and β are equal. This, in turn, implies the **law of reflection,** which states that the **angle of incidence** (θ_1 in Figure 3.7) must be equal to the **angle of reflection** (θ_2 in Figure 3.7).

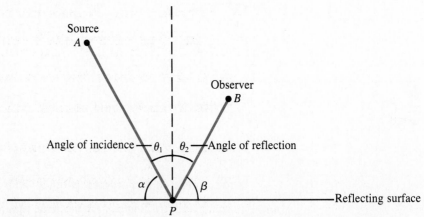

Figure 3.7 The reflection of light: $\theta_1 = \theta_2$.

Problems

Related rates

1. An observer watches a plane approach at a speed of 500 miles per hour and at an altitude of 3 miles. At what rate is the angle of elevation of the observer's line of sight changing with respect to time when the horizontal distance between the plane and the observer is 4 miles?

Related rates

2. A person 6 feet tall is watching a streetlight 18 feet high while walking toward it at a speed of 5 feet per second. At what rate is the angle of elevation of the person's line of sight changing with respect to time when the person is 9 feet from the base of the light?

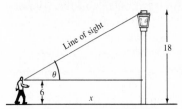

Related rates

3. An attendant is standing at the end of a pier 12 feet above the water and is pulling a rope attached to a rowboat at the rate of 4 feet of rope per minute. At what rate is the angle that the rope makes with the surface of the water changing with respect to time when the boat is 16 feet from the pier?

Related rates

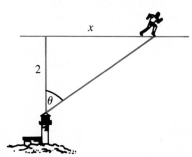

4. A revolving searchlight in a lighthouse 2 miles off shore is following a jogger running along the shore. When the jogger is 1 mile from the point on the shore that is closest to the lighthouse, the searchlight is turning at the rate of 0.25 revolution per hour. How fast is the jogger running at that moment? (*Hint*: Since 0.25 revolution per hour is $\frac{\pi}{2}$ radians per hour, the problem is to find $\frac{dx}{dt}$ for the accompanying picture when $x = 1$ and $\frac{d\theta}{dt} = \frac{\pi}{2}$.)

Maximization of volume

5. You have a piece of metal that is 20 meters long and 6 meters wide, which you are going to bend to form a trough as indicated in the following diagram. At what angle should the sides meet so that the volume of the trough is as large as possible? (*Hint*: The volume is the length of the trough times the area of its triangular cross section.)

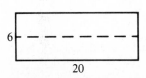

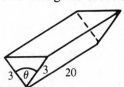

Maximization of area

6. Prove that of all isosceles triangles whose equal sides are of a specified length, the triangle of greatest area is the right triangle.

Maximization of area

7. The two sides and the base of an isosceles trapezoid are each 5 inches long. At what angle should the sides meet the horizontal top to maximize the area of the trapezoid? [*Hints*: (1) As indicated in the accompanying picture, the area of the trapezoid is the sum of the areas of two right triangles and the area of a rectangle. (2) You should be able to factor the derivative of the area function if you first simplify it using an appropriate trigonometric identity.]

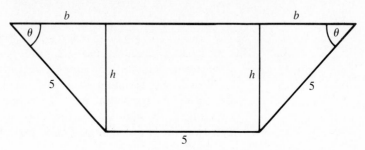

Illumination from a
light source

8. In Example 3.3, an alternative characterization of the illumination is that it is directly proportional to the sine of the angle ϕ and inversely proportional to the square of the distance d, where ϕ is the angle at which the ray of light meets the table. (That is, ϕ is the angle in the left-hand corner of the triangle in Figure 3.4 between the hypotenuse labeled d and the base labeled 4.) For practice, do Example 3.3 again, this time using this alternative characterization of the illumination. (Your final answer, of course, should be the same as before.)

Minimization of
horizontal clearance

9. Find the length of the longest pipe that can be carried horizontally around a corner joining two corridors that are $2\sqrt{2}$ feet wide. [*Hint*: Show that the horizontal clearance $C = x + y$ can be written as

$$C(\theta) = \frac{2\sqrt{2}}{\sin \theta} + \frac{2\sqrt{2}}{\cos \theta}$$

and find the absolute *minimum* of $C(\theta)$ on an appropriate interval.]

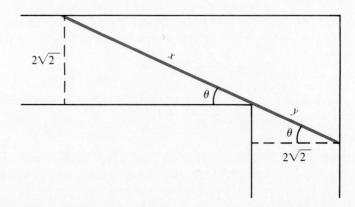

Refraction of light 10. The accompanying figure shows a ray of light emitted from a source at point A under water and subsequently received by an observer at point B above the surface of the water. If v_1 is the speed of light in water and v_2 the speed of light in air, show that

$$\frac{\sin \theta_1}{\sin \theta_2} = \frac{v_1}{v_2}$$

where θ_1 is the **angle of incidence** and θ_2 the **angle of refraction**. (*Hint*: Apply Fermat's principle, which states that light takes the path requiring the least amount of time, and use the solution to Example 3.4 as a guide. You may assume without proof that the total time is minimized when its derivative is equal to zero.)

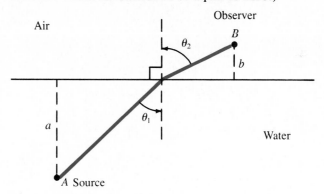

Chapter Summary and Review Problems

Important Terms, Symbols, and Formulas

Angle measurement: $\dfrac{\text{degrees}}{180} = \dfrac{\text{radians}}{\pi}$

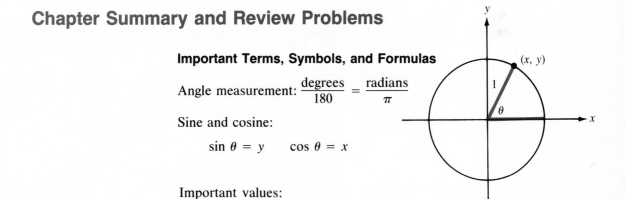

Sine and cosine:

$$\sin \theta = y \qquad \cos \theta = x$$

Important values:

θ	0	$\frac{\pi}{6}$	$\frac{\pi}{4}$	$\frac{\pi}{3}$	$\frac{\pi}{2}$	$\frac{2\pi}{3}$	$\frac{3\pi}{4}$	$\frac{5\pi}{6}$	π
sin θ	0	$\frac{1}{2}$	$\frac{\sqrt{2}}{2}$	$\frac{\sqrt{3}}{2}$	1	$\frac{\sqrt{3}}{2}$	$\frac{\sqrt{2}}{2}$	$\frac{1}{2}$	0
cos θ	1	$\frac{\sqrt{3}}{2}$	$\frac{\sqrt{2}}{2}$	$\frac{1}{2}$	0	$-\frac{1}{2}$	$-\frac{\sqrt{2}}{2}$	$-\frac{\sqrt{3}}{2}$	-1

Other trigonometric functions:

$$\tan \theta = \frac{\sin \theta}{\cos \theta} \qquad \cot \theta = \frac{\cos \theta}{\sin \theta}$$

$$\sec \theta = \frac{1}{\cos \theta} \qquad \csc \theta = \frac{1}{\sin \theta}$$

Trigonometric identities:

Periodicity: $\sin (\theta + 2\pi) = \sin \theta \qquad \cos (\theta + 2\pi) = \cos \theta$

Negative angles: $\sin (-\theta) = -\sin \theta \qquad \cos (-\theta) = \cos \theta$

Pythagorean identity: $\sin^2 \theta + \cos^2 \theta = 1$

Addition formulas:

$$\cos (a + b) = \cos a \cos b - \sin a \sin b$$

$$\sin (a + b) = \sin a \cos b + \cos a \sin b$$

Double-angle formulas:

$$\cos 2\theta = \cos^2 \theta - \sin^2 \theta$$

$$\sin 2\theta = 2 \sin \theta \cos \theta$$

Right triangles:

$$\sin \theta = \frac{\text{opposite}}{\text{hypotenuse}}$$

$$\cos \theta = \frac{\text{adjacent}}{\text{hypotenuse}}$$

$$\tan \theta = \frac{\text{opposite}}{\text{adjacent}}$$

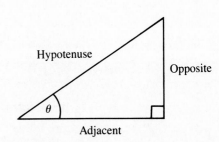

Differentiation formulas:

$$\frac{d}{d\theta} \sin \theta = \cos \theta \qquad \frac{d}{d\theta} \cos \theta = -\sin \theta \qquad \frac{d}{d\theta} \tan \theta = \sec^2 \theta$$

Chain rules:

$$\frac{d}{d\theta} \sin h(\theta) = h'(\theta) \cos h(\theta)$$

$$\frac{d}{d\theta} \cos h(\theta) = -h'(\theta) \sin h(\theta)$$

$$\frac{d}{d\theta} \tan h(\theta) = h'(\theta) \sec^2 h(\theta)$$

Review Problems

1. Specify the radian measurement and degree measurement for each of the following angles.

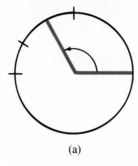

(a)

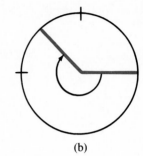

(b)

2. Convert $50°$ to radians.

3. Convert 0.25 radian to degrees.

4. Evaluate the given expressions without using Table III or your calculator.

   (a) $\sin\left(-\dfrac{5\pi}{3}\right)$ (b) $\cos\dfrac{15\pi}{4}$

   (c) $\sec\dfrac{7\pi}{3}$ (d) $\cot\dfrac{2\pi}{3}$

5. Find $\tan\theta$ if $\sin\theta = \dfrac{4}{5}$.

6. Find $\csc\theta$ if $\cot\theta = \dfrac{\sqrt{5}}{2}$.

In Problems 7 through 10, find all the values of θ in the specified interval that satisfy the given equation.

7. $2\cos\theta + \sin 2\theta = 0;\ 0 \le \theta \le 2\pi$

8. $3\sin^2\theta - \cos^2\theta = 2;\ 0 \le \theta \le \pi$

9. $2\sin^2\theta = \cos 2\theta;\ 0 \le \theta \le \pi$

10. $5\sin\theta - 2\cos^2\theta = 1;\ 0 \le \theta \le 2\pi$

11. Starting with the identity $\sin^2\theta + \cos^2\theta = 1$, derive the identity $1 + \cot^2\theta = \csc^2\theta$.

12. Starting with the double-angle formulas for the sine and cosine, derive the double-angle formula

    $$\tan 2\theta = \frac{2\tan\theta}{1 - \tan^2\theta}$$

    for the tangent.

13. (a) Starting with the addition formulas for the sine and cosine, derive the identities

$$\cos\left(\frac{\pi}{2} + \theta\right) = -\sin\theta \quad \text{and} \quad \sin\left(\frac{\pi}{2} + \theta\right) = \cos\theta$$

(b) Give geometric arguments to establish the identities in part (a).

In Problems 14 through 19, differentiate the given function.

14. $f(\theta) = \sin(3\theta + 1)^2$

15. $f(\theta) = \cos^2(3\theta + 1)$

16. $f(\theta) = \tan(3\theta^2 + 1)$

17. $f(\theta) = \tan^2(3\theta^2 + 1)$

18. $f(\theta) = \dfrac{\sin\theta}{1 - \cos\theta}$

19. $f(\theta) = \ln\cos^2\theta$

20. Show that $\dfrac{d}{d\theta}\sin\theta\cos\theta = \cos 2\theta$.

21. On New Year's Eve, a holiday reveler is watching the descent of a lighted ball from atop a tall building that is 600 feet away. The ball is falling at the rate of 20 feet per minute. At what rate is the angle of elevation of the reveler's line of sight changing with respect to time when the ball is 800 feet from the ground?

22. A trough 9 meters long is to have a cross section consisting of an isosceles trapezoid in which the base and two sides are all 4 meters long. At what angle should the sides of the trapezoid meet the horizontal top to maximize the capacity of the trough?

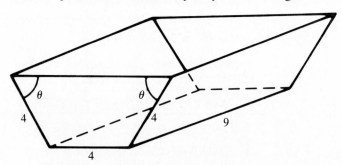

23. Find the length of the longest pipe that can be carried horizontally around a corner joining a corridor that is 8 feet wide and one that is $5\sqrt{5}$ feet wide.

24. A cable is run in a straight line from a power plant on one side of a river 900 meters wide to a point P on the other side and then along the river bank to a factory, 3,000 meters downstream from the power plant. The cost of running the cable under the water is $5 per meter, while the cost over land is $4 per meter. If θ is the (smaller) angle

between the segment of cable under the river and the opposite bank, show that $\cos \theta = \frac{4}{5}$ (the ratio of the per-meter costs) for the route that minimizes the total installation cost. (You may assume without proof that the absolute minimum occurs when the derivative of the cost function is zero.)

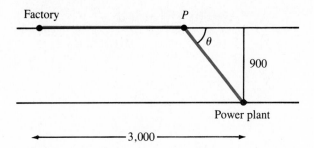

Appendix

A   Algebra Review

This section contains a review of certain algebraic topics that are used throughout the book. A set of practice problems appears at the end of the section, and the answers to the odd-numbered problems are given at the back of the book.

The Real Numbers

An **integer** is a "whole number," either positive or negative. For example, 1, 2, 875, -15, -83, and 0 are integers, while $\frac{2}{3}$, 8.71, and $\sqrt{2}$ are not.

    A **rational number** is a number that can be expressed as the quotient

604

$\frac{a}{b}$ of two integers, where $b \neq 0$. For example,

$$\frac{2}{3} \qquad \frac{8}{3} \qquad -6\frac{1}{2} = \frac{-13}{2} \qquad \text{and} \qquad 0.25 = \frac{25}{100} = \frac{1}{4}$$

are rational numbers. Every integer is a rational number since it can be expressed as itself divided by 1. When expressed in decimal form, rational numbers are either terminating or infinitely repeating decimals. For example,

$$\frac{5}{8} = 0.625 \qquad \frac{1}{3} = 0.333 \ldots \qquad \text{and} \qquad \frac{13}{11} = 1.181818 \ldots$$

A number that cannot be expressed as the quotient of two integers is called an **irrational number**. For example,

$$\sqrt{2} \simeq 1.41421356 \qquad \text{and} \qquad \pi \simeq 3.14159265$$

are irrational numbers.

The rational numbers and irrational numbers form the **real numbers** and can be visualized geometrically as points on a **number line** as illustrated in Figure A.1.

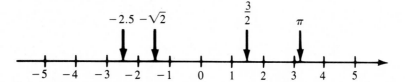

Figure A.1 The number line.

Inequalities

If a and b are real numbers and a is to the right of b on the number line, we say that **a is greater than b** and write $a > b$. If a is to the left of b, we say that **a is less than b** and write $a < b$. (Figure A.2.) For example,

$$5 > 2 \qquad -12 < 0 \qquad \text{and} \qquad -8.2 < -2.4$$

Figure A.2 Inequalities.

Moreover,

$$\frac{6}{7} < \frac{7}{8}$$

as you can see by writing

$$\frac{6}{7} = \frac{48}{56} \qquad \text{and} \qquad \frac{7}{8} = \frac{49}{56}$$

The symbol ≥ stands for **greater than or equal to,** and the symbol ≤ stands for **less than or equal to.** Thus, for example,

$$-3 \geq -4 \qquad -3 \geq -3 \qquad -4 \leq -3 \qquad \text{and} \qquad -4 \leq -4$$

Intervals

A set of real numbers that can be represented on the number line by a line segment is called an **interval.** Inequalities can be used to describe intervals. For example, the interval $a \leq x < b$ consists of all real numbers x that are between a and b, including a but excluding b. This interval is shown in Figure A.3. The numbers a and b are known as the **endpoints** of the interval. The square bracket at a indicates that a is included in the interval, while the rounded bracket at b indicates that b is excluded.

Figure A.3 The interval $a \leq x < b$.

Intervals may be finite or infinite in extent and may or may not contain any endpoints. The possibilities (including customary notation and terminology) are illustrated in Figure A.4.

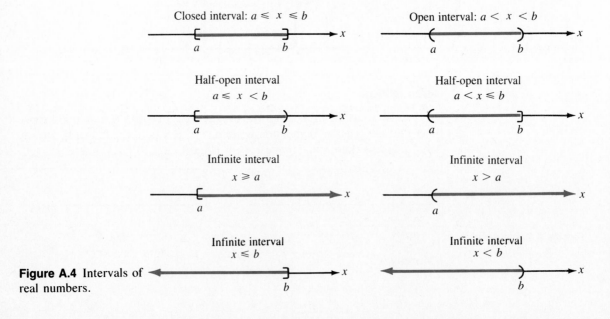

Figure A.4 Intervals of real numbers.

Closed interval: $a \leq x \leq b$

Open interval: $a < x < b$

Half-open interval $a \leq x < b$

Half-open interval $a < x \leq b$

Infinite interval $x \geq a$

Infinite interval $x > a$

Infinite interval $x \leq b$

Infinite interval $x < b$

EXAMPLE A.1 Use inequalities to describe the following intervals.

(a)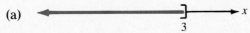

(b)
   -2

(c) ![number line with open circle at -2 and bracket at 3]
   -2 3

SOLUTION

(a) $x \leq 3$ (b) $x > -2$ (c) $-2 < x \leq 3$

EXAMPLE A.2   Represent each of the following intervals as a line segment on a number line.

(a) $x < -1$ (b) $-1 \leq x \leq 2$ (c) $x > 2$

SOLUTION

(a)
   -1

(b) ![number line with bracket at -1 and bracket at 2]
   -1 2

(c) ![number line with open circle at 2]
   2

Absolute Value

The **absolute value** of a real number is its nonnegative value or magnitude. Thus, the absolute value of a nonnegative number is the number itself, while the absolute value of a negative number is the number without the minus sign. The absolute value of a number x is denoted by $|x|$. Here is the formal definition.

Absolute Value

For any real number x, the absolute value of x is

$$|x| = \begin{cases} x & \text{if } x \geq 0 \\ -x & \text{if } x < 0 \end{cases}$$

For example,

$$|-3| = 3 \qquad |3| = 3 \qquad |0| = 0 \qquad \text{and} \qquad |5 - 9| = |-4| = 4$$

Distance

The absolute value of a number can be interpreted geometrically as its distance from zero on the number line. More generally, the distance between any two numbers is the absolute value of their difference (taken in either order). The situation is illustrated in Figure A.5.

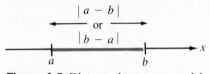

Figure A.5 Distance between a and $b = |a - b| = |b - a|$.

EXAMPLE A.3   Find the distance on the number line between -2 and 3.

SOLUTION
The distance between two numbers is the absolute value of their difference. Hence,

$$\text{Distance} = |-2 - 3| = |-5| = 5$$

The situation is illustrated in Figure A.6.

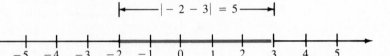

Figure A.6 Distance between -2 and 3.

Absolute Values and Intervals

The geometric interpretation of absolute value as distance can be used to simplify certain algebraic inequalities involving absolute values. Here is an example.

EXAMPLE A.4   Find the interval consisting of all real numbers x such that $|x - 1| \le 3$.

SOLUTION
In geometric terms, the numbers x for which $|x - 1| \le 3$ are those whose distance from 1 is less than or equal to 3. As illustrated in Figure A.7, these are the numbers in the interval $-2 \le x \le 4$.

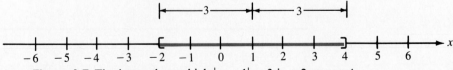

Figure A.7 The interval on which $|x - 1| \le 3$ is $-2 \le x \le 4$.

To find this interval algebraically, without relying on the geometry, rewrite the inequality

$$|x - 1| \le 3$$

as

$$-3 \le x - 1 \le 3$$

and add 1 to each term to get

$$-3 + 1 \le x - 1 + 1 \le 3 + 1$$

or

$$-2 \le x \le 4$$

Exponential Notation

The following rules define the expression a^x for $a > 0$ and all rational values of x.

Definition of a^x for $x > 0$

Integer powers: If n is a positive integer,

$$a^n = a \cdot a \cdots a$$

where the product $a \cdot a \cdots a$ contains n factors.

Fractional powers: If n and m are positive integers,

$$a^{n/m} = (\sqrt[m]{a})^n$$

where $\sqrt[m]{}$ denotes the positive mth root.

Negative powers: $a^{-x} = \dfrac{1}{a^x}$

Zero power: $a^0 = 1$

EXAMPLE A.5 Evaluate the following expressions (without using your calculator).

(a) $9^{1/2}$ (b) $27^{2/3}$ (c) $8^{-1/3}$ (d) $\left(\dfrac{1}{100}\right)^{-3/2}$ (e) 5^0

SOLUTION
(a) $9^{1/2} = \sqrt{9} = 3$

(b) $27^{2/3} = (\sqrt[3]{27})^2 = 3^2 = 9$

(c) $8^{-1/3} = \dfrac{1}{8^{1/3}} = \dfrac{1}{\sqrt[3]{8}} = \dfrac{1}{2}$

(d) $\left(\dfrac{1}{100}\right)^{-3/2} = 100^{3/2} = (\sqrt{100})^3 = 10^3 = 1,000$

(e) $5^0 = 1$

Laws of Exponents

Exponents obey the following useful laws.

Laws of Exponents

> The product law: $a^r a^s = a^{r+s}$
>
> The quotient law: $\dfrac{a^r}{a^s} = a^{r-s}$
>
> The power law: $(a^r)^s = a^{rs}$

The use of these laws is illustrated in the next two examples.

EXAMPLE A.6 Evaluate the following expressions (without using a calculator).

(a) $(2^{-2})^3$ (b) $\dfrac{3^3}{3^{1/3}(3^{2/3})}$ (c) $2^{7/4}(8^{-1/4})$

SOLUTION

(a) $(2^{-2})^3 = 2^{-6} = \dfrac{1}{2^6} = \dfrac{1}{64}$

(b) $\dfrac{3^3}{3^{1/3}(3^{2/3})} = \dfrac{3^3}{3^{1/3+2/3}} = \dfrac{3^3}{3^1} = 3^2 = 9$

(c) $2^{7/4}(8^{-1/4}) = 2^{7/4}(2^3)^{-1/4} = 2^{7/4}(2^{-3/4}) = 2^{7/4-3/4} = 2^1 = 2$

EXAMPLE A.7 Solve each of the following equations for n.

(a) $\dfrac{a^5}{a^2} = a^n$ (b) $(a^n)^5 = a^{20}$

SOLUTION

(a) Since $\dfrac{a^5}{a^2} = a^{5-2} = a^3$, it follows that $n = 3$.

(b) Since $(a^n)^5 = a^{5n}$, it follows that $5n = 20$ or $n = 4$.

Factoring

To factor an expression is to write it as a product of two or more terms, called **factors.** Factoring is used to simplify complicated expressions and to solve equations, and is based on the distributive law for addition and multiplication.

The Distributive Law

For any real numbers a, b, and c,

$$ab + ac = a(b + c)$$

The factoring techniques you will need in this book are illustrated in the following examples.

Factoring Out Common Terms

EXAMPLE A.8   Factor the expression $3x^4 - 6x^3$.

SOLUTION
Since $3x^3$ is a factor of each of the terms in this expression, you can use the distributive law to "factor out" $3x^3$ and write

$$3x^4 - 6x^3 = 3x^3(x - 2)$$

EXAMPLE A.9   Factor the expression $1,000 - 200(x - 5)$.

SOLUTION
The greatest factor common to the terms $1,000$ and $200(x - 5)$ is 200. Factor this out to get

$$\begin{aligned} 1,000 - 200(x - 5) &= 200[5 - (x - 5)] \\ &= 200(5 - x + 5) \\ &= 200(10 - x) \end{aligned}$$

EXAMPLE A.10   Simplify the expression $10(1 - x)^4(x + 1)^4 + 8(x + 1)^5(1 - x)^3$.

SOLUTION
The greatest common factor is $2(1 - x)^3(x + 1)^4$. Factor this out to get

$$10(1 - x)^4(x + 1)^4 + 8(x + 1)^5(1 - x)^3$$
$$= 2(1 - x)^3(x + 1)^4[5(1 - x) + 4(x + 1)]$$

Since no further factorization is possible, do the multiplication in the square brackets and combine the resulting terms to conclude that

$$10(1 - x)^4(x + 1)^4 + 8(x + 1)^5(1 - x)^3$$

$$= 2(1 - x)^3(x + 1)^4(5 - 5x + 4x + 4)$$

$$= 2(1 - x)^3(x + 1)^4(9 - x)$$

Simplification of Quotients by Factoring and Canceling

The next example illustrates how you can combine factoring and canceling to simplify certain types of quotients that arise frequently in calculus.

EXAMPLE A.11   Simplify the quotient

$$\frac{4(x + 3)^4(x - 2)^2 - 6(x + 3)^3(x - 2)^3}{(x + 3)(x - 2)^3}$$

SOLUTION
First simplify the numerator to get

$$\frac{4(x + 3)^4(x - 2)^2 - 6(x + 3)^3(x - 2)^3}{(x + 3)(x - 2)^3}$$

$$= \frac{2(x + 3)^3(x - 2)^2[2(x + 3) - 3(x - 2)]}{(x + 3)(x - 2)^3}$$

$$= \frac{2(x + 3)^3(x - 2)^2(2x + 6 - 3x + 6)}{(x + 3)(x - 2)^3}$$

$$= \frac{2(x + 3)^3(x - 2)^2(12 - x)}{(x + 3)(x - 2)^3}$$

and then "cancel" the common factor of $(x + 3)(x - 2)^2$ from the numerator and denominator to conclude that

$$\frac{4(x + 3)^4(x - 2)^2 - 6(x + 3)^3(x - 2)^3}{(x + 3)(x - 2)^3} = \frac{2(x + 3)^2(12 - x)}{x - 2}$$

Polynomials

A **polynomial** is an expression of the form

$$a_0 + a_1x + a_2x^2 + \cdots + a_nx^n$$

where n is a nonnegative integer and $a_0, a_1, a_2, \ldots, a_n$ are real numbers known as the **coefficients** of the polynomial. If $a_n \neq 0$, n is said to be the

degree of the polynomial. For example, $3x^5 - 7x^2 + 12$ is a polynomial of degree 5.

Factoring Polynomials with Integer Coefficients

Many of the polynomials that arise in practice have integer coefficients (or are closely related to polynomials that do). Techniques for factoring polynomials with integer coefficients are illustrated in the following examples. In each, the goal will be to rewrite the given polynomial as a product of polynomials of lower degree that also have integer coefficients.

EXAMPLE A.12 Factor the polynomial $x^2 - 2x - 3$ using integer coefficients.

SOLUTION
The goal is to write the polynomial as a product of the form

$$x^2 - 2x - 3 = (x + a)(x + b)$$

where a and b are integers. The distributive law implies that

$$(x + a)(x + b) = x^2 + (a + b)x + ab$$

Hence, the goal is to find integers a and b such that

$$x^2 - 2x - 3 = x^2 + (a + b)x + ab$$

or, equivalently, such that

$$a + b = -2 \quad \text{and} \quad ab = -3$$

From the list

$$1, -3 \quad \text{and} \quad -1, 3$$

of pairs of integers whose product is -3, choose $a = -3$ and $b = 1$ as the only pair whose sum is -2. It follows that

$$x^2 - 2x - 3 = (x - 3)(x + 1)$$

which you should check by multiplying out the right-hand side.

EXAMPLE A.13 Factor the polynomial $x^3 - 8$ using integer coefficients.

SOLUTION
The fact that $2^3 = 8$ tells you that $x - 2$ must be a factor of this expression. That is, there are integers a and b for which

$$x^3 - 8 = (x - 2)(x^2 + ax + b)$$

Since

$$(x - 2)(x^2 + ax + b) = x^3 + (a - 2)x^2 + (b - 2a)x - 2b$$

(which you should check for yourself), the goal is to find integers a and b for which

$$a - 2 = 0 \qquad b - 2a = 0 \qquad \text{and} \qquad 2b = 8$$

Clearly, the only such integers are $a = 2$ and $b = 4$. Hence,

$$x^3 - 8 = (x - 2)(x^2 + 2x + 4)$$

Convince yourself by examining pairs of integers whose product is 4 that the polynomial $x^2 + 2x + 4$ cannot be factored further with integer coefficients.

The Difference of Two Squares

Here is the formula (which you can verify by multiplication) for the factorization of the difference between two perfect squares. It is particularly useful, and you should memorize it.

> **The Difference of Two Squares**
>
> For any real numbers a and b,
> $$a^2 - b^2 = (a + b)(a - b)$$

The use of this formula is illustrated in the next example.

EXAMPLE A.14   Factor the polynomial $x^5 - 4x^3$ using integer coefficients.

SOLUTION
First factor out x^3 to get

$$x^5 - 4x^3 = x^3(x^2 - 4)$$

and then factor $x^2 - 4$ (which is the difference of two squares) to conclude that

$$x^5 - 4x^3 = x^3(x + 2)(x - 2)$$

The Solution of Equations by Factoring

The **solutions** of an equation are the values of the variable that make the equation true. For example, $x = 2$ is a solution of the equation

$$x^3 - 6x^2 + 12x - 8 = 0$$

because substitution of 2 for x gives

$$2^3 - 6(2^2) + 12(2) - 8 = 8 - 24 + 24 - 8 = 0$$

In the following two examples, you will see how factoring can be used to solve certain equations. The technique is based on the fact that if the product of two (or more) terms is equal to zero, then at least one of the terms must be equal to zero. For example, if $ab = 0$, then either $a = 0$ or $b = 0$ (or both).

EXAMPLE A.15   Solve the equation $x^2 - 3x = 10$.

SOLUTION
First subtract 10 from both sides to get

$$x^2 - 3x - 10 = 0$$

and then factor the resulting polynomial on the left-hand side to get

$$(x - 5)(x + 2) = 0$$

Since the product $(x - 5)(x + 2)$ can be zero only if one (or both) of its factors is zero, it follows that the solutions are $x = 5$ (which makes the first factor zero) and $x = -2$ (which makes the second factor zero).

EXAMPLE A.16   Solve the equation $1 - \dfrac{1}{x} - \dfrac{2}{x^2} = 0$.

SOLUTION
Put the fractions on the left-hand side over a common denominator and add to get

$$\frac{x^2}{x^2} - \frac{x}{x^2} - \frac{2}{x^2} = 0$$

or

$$\frac{x^2 - x - 2}{x^2} = 0$$

Now factor the polynomial in the numerator to get

$$\frac{(x + 1)(x - 2)}{x^2} = 0$$

Since a quotient is zero only if its numerator is zero, it follows that the solutions are $x = -1$ and $x = 2$.

The Solution of Quadratic Equations

An equation of the form

$$ax^2 + bx + c = 0 \qquad \text{(for } a \neq 0\text{)}$$

is said to be a **quadratic equation.** A quadratic equation can have at most two solutions. As you have seen, one way to find the solutions is to factor the equation. When the factors are not obvious or when the equation cannot be factored at all, you can use the following special formula to solve quadratic equations.

The Quadratic Formula

The solutions of the quadratic equation

$$ax^2 + bx + c = 0 \qquad \text{(for } a \neq 0\text{)}$$

are given by the formula

$$x = \frac{-b \pm \sqrt{b^2 - 4ac}}{2a}$$

The term $b^2 - 4ac$ in the quadratic formula is called the **discriminant** of the quadratic equation. If the discriminant is positive, the equation has two solutions, one coming from the formula with the sign $\pm$ replaced by $+$ and the other with $\pm$ replaced by $-$. If the discriminant is zero, the equation has only one solution since the formula reduces to $x = \dfrac{-b}{2a}$. If the discriminant is negative, the equation has no real solutions since negative numbers do not have real square roots.

The use of the quadratic formula is illustrated in the following examples.

EXAMPLE A.17 Solve the equation $x^2 + 3x + 1 = 0$.

SOLUTION
This is a quadratic equation with $a = 1$, $b = 3$, and $c = 1$. Using the quadratic formula, you get

$$x = \frac{-3 + \sqrt{5}}{2} \approx -0.38 \qquad \text{and} \qquad x = \frac{-3 - \sqrt{5}}{2} \approx -2.62$$

EXAMPLE A.18 Solve the equation $x^2 + 18x + 81 = 0$.

SOLUTION
This is a quadratic equation with $a = 1$, $b = 18$, and $c = 81$. Using the

quadratic formula, you find that the discriminant is zero and that the formula for x gives

$$x = \frac{-18 \pm \sqrt{0}}{2} = -\frac{18}{2} = -9.$$

EXAMPLE A.19 Solve the equation $x^2 + x + 1 = 0$.

SOLUTION
This is a quadratic equation with $a = 1$, $b = 1$, and $c = 1$. Using the quadratic formula, you get

$$x = \frac{-1 \pm \sqrt{-3}}{2}$$

Since there is no real square root of -3, it follows that the equation has no real solution.

Systems of Equations

A collection of equations that are to be solved simultaneously is called a **system of equations.** Some of the calculus problems in Chapter 9 involve the solution of systems of two (or more) equations in two (or more) unknowns. A typical example is to find the real numbers x and y that satisfy the system

$$2x + 3y = 5$$

$$x + 2y = 4$$

The procedure for solving a system of two equations in two unknowns is to (temporarily) eliminate one of the variables, thereby reducing the problem to a single equation in one variable, which you then solve for its variable. Once you have found the value of one of the variables, you can substitute it into either of the original equations and solve to get the value of the other variable.

The most common techniques for the elimination of variables are illustrated in the next two examples.

Elimination by Multiplication and Addition

EXAMPLE A.20 Solve the system

$$4x + 3y = 13$$

$$3x + 2y = 7$$

SOLUTION

To eliminate y, multiply both sides of the first equation by 2 and both sides of the second equation by -3 so that the system becomes

$$8x + 6y = 26$$

$$-9x - 6y = -21$$

and add the equations to get

$$-x + 0 = 5 \quad \text{or} \quad x = -5$$

To find y, you can substitute $x = -5$ into either of the original equations. If you choose the second equation, you get

$$3(-5) + 2y = 7 \quad 2y = 22 \quad \text{or} \quad y = 11$$

That is, the solution of the system is $x = -5$ and $y = 11$.

To check this answer, substitute $x = -5$ and $y = 11$ into each of the original equations. From the first equation you get

$$4(-5) + 3(11) = -20 + 33 = 13$$

and from the second equation you get

$$3(-5) + 2(11) = -15 + 22 = 7$$

as required.

Elimination by Substitution

EXAMPLE A.21 Solve the system

$$2y^2 - x^2 = 14$$

$$x - y = 1$$

SOLUTION

Solve the second equation for x to get

$$x = y + 1$$

and substitute this into the first equation to eliminate x. This gives

$$2y^2 - (y + 1)^2 = 14$$

$$2y^2 - (y^2 + 2y + 1) = 14$$

$$2y^2 - y^2 - 2y - 1 = 14$$

$$y^2 - 2y - 15 = 0$$

or

$$(y + 3)(y - 5) = 0$$

from which it follows that

$$y = -3 \quad \text{or} \quad y = 5$$

If $y = -3$, the second equation gives

$$x - (-3) = 1 \quad \text{or} \quad x = -2$$

and if $y = 5$, the second equation gives

$$x - 5 = 1 \quad \text{or} \quad x = 6$$

Hence the system has two solutions,

$$x = 6, y = 5 \quad \text{and} \quad x = -2, y = -3$$

To check these answers, substitute each pair x, y into the first equation. If $x = 6$ and $y = 5$, you get

$$2(5^2) - 6^2 = 50 - 36 = 14$$

and if $x = -2$ and $y = -3$, you get

$$2(-3)^2 - (-2)^2 = 18 - 4 = 14$$

as required.

Summation Notation

Sums of the form

$$a_1 + a_2 + \cdots + a_n$$

appear in Chapters 7 and 11. To describe such a sum, it suffices to specify the general term a_j and to indicate that n terms of this form are to be added, starting with the term in which $j = 1$ and ending with the term in which $j = n$. It is customary to use the Greek letter Σ (sigma) to denote summation and to express the sum compactly as follows.

Summation Notation

$$a_1 + a_2 + \cdots + a_n = \sum_{j=1}^{n} a_j$$

The use of summation notation is illustrated in the following examples.

EXAMPLE A.22 Use summation notation to express the following sums.

(a) $1 + 4 + 9 + 16 + 25 + 36 + 49 + 64$

(b) $(1 - x_1)^2 \Delta x + (1 - x_2)^2 \Delta x + \cdots + (1 - x_{15})^2 \Delta x$

SOLUTION

(a) This is a sum of 8 terms of the form j^2, starting with $j = 1$ and ending with $j = 8$. Hence,

$$1 + 4 + 9 + 16 + 25 + 36 + 49 + 64 = \sum_{j=1}^{8} j^2$$

(b) The jth term of this sum is $(1 - x_j)^2 \, \Delta x$. Hence,

$$(1 - x_1)^2 \, \Delta x + (1 - x_2)^2 \, \Delta x + \cdots + (1 - x_{15})^2 \, \Delta x$$

$$= \sum_{j=1}^{15} (1 - x_j)^2 \, \Delta x$$

EXAMPLE A.23   Evaluate the following sums.

(a) $\sum_{j=1}^{4} (j^2 + 1)$ 　　　　　　　　　　(b) $\sum_{j=1}^{3} (-2)^j$

SOLUTION

(a) $\sum_{j=1}^{4} (j^2 + 1) = (1^2 + 1) + (2^2 + 1) + (3^2 + 1) + (4^2 + 1)$

$$= 2 + 5 + 10 + 17 = 34$$

(b) $\sum_{j=1}^{3} (-2)^j = (-2)^1 + (-2)^2 + (-2)^3 = -2 + 4 - 8 = -6$

Problems

Intervals   In Problems 1 through 4, use inequalities to describe the given interval.

1.

2.

3.

4.

In Problems 5 through 8, represent the given interval as a line segment on a number line.

5. $x \geq 2$ 　　　　　　　　　　　　6. $-6 \leq x < 4$

7. $-2 < x \leq 0$ 　　　　　　　　　8. $x > 3$

Distance In Problems 9 through 12, find the distance on the number line between the given pair of real numbers.

9. 0 and -4 10. 2 and 5

11. -2 and 3   12. -3 and -1

Absolute value and intervals In Problems 13 through 18, find the interval or intervals consisting of all real numbers x that satisfy the given inequality.

13. $|x| \leq 3$ 14. $|x - 2| \leq 5$

15. $|x + 4| \leq 2$ 16. $|1 - x| < 3$

17. $|x + 2| \geq 5$ 18. $|x - 1| > 3$

Exponential notation In Problems 19 through 26, evaluate the given expression without using a calculator.

19. 5^3 20. 2^{-3}

21. $16^{1/2}$ 22. $36^{-1/2}$

23. $8^{2/3}$ 24. $27^{-4/3}$

25. $\left(\dfrac{1}{4}\right)^{1/2}$ 26. $\left(\dfrac{1}{4}\right)^{-3/2}$

In Problems 27 through 34, evaluate the given expression without using a calculator.

27. $\dfrac{2^5(2^2)}{2^8}$ 28. $\dfrac{3^4(3^3)}{(3^2)^3}$

29. $\dfrac{2^{4/3}(2^{5/3})}{2^5}$ 30. $\dfrac{5^{-3}(5^2)}{(5^{-2})^3}$

31. $\dfrac{2(16^{3/4})}{2^3}$ 32. $\dfrac{\sqrt{27}\,(\sqrt{3})^3}{9}$

33. $[\sqrt{8}\,(2^{5/2})]^{-1/2}$ 34. $[\sqrt{27}\,(3^{5/2})]^{1/2}$

In Problems 35 through 42, solve the given equation for n. (Assume $a > 0$ and $a \neq 1$.)

35. $a^3 a^7 = a^n$ 36. $\dfrac{a^5}{a^2} = a^n$

37. $a^4 a^{-3} = a^n$ 38. $a^2 a^n = \dfrac{1}{a}$

39. $(a^3)^n = a^{12}$ 40. $(a^n)^5 = \dfrac{1}{a^{10}}$

41. $a^{3/5} a^{-n} = \dfrac{1}{a^2}$ 42. $(a^n)^3 = \dfrac{1}{\sqrt{a}}$

Factoring   In Problems 43 through 48, factor the given expression.

43. $x^5 - 4x^4$ 44. $3x^3 - 12x^4$

45. $100 - 25(x - 3)$ 46. $60 - 20(4 - x)$

47. $8(x + 1)^3(x - 2)^2 + 6(x + 1)^2(x - 2)^3$

48. $12(x + 3)^5(x - 1)^3 - 8(x + 3)^6(x - 1)^2$

Simplification of   In Problems 49 through 52, simplify the given quotient by factoring the
quotients   numerator and then canceling.

49. $\dfrac{(x + 3)^3(x + 1) - (x + 3)^2(x + 1)^2}{(x + 3)(x + 1)}$

50. $\dfrac{3(x - 2)^2(x + 1)^2 - 2(x - 2)(x + 1)^3}{(x - 2)^4}$

51. $\dfrac{4(1 - x)^2(x + 3)^3 + 2(1 - x)(x + 3)^4}{(1 - x)^4}$

52. $\dfrac{6(x + 2)^5(1 - x)^4 - 4(x + 2)^6(1 - x)^3}{(x + 2)^8(1 - x)^2}$

Factoring polynomials   In Problems 53 through 66, factor the given polynomial using integer
with integer   coefficients.
coefficients

53. $x^2 + x - 2$ 54. $x^2 + 3x - 10$

55. $x^2 - 7x + 12$ 56. $x^2 + 8x + 12$

57. $x^2 - 2x + 1$ 58. $x^2 + 6x + 9$

59. $x^2 - 4$ 60. $64 - x^2$

61. $x^3 - 1$ 62. $x^3 - 27$

63. $x^7 - x^5$ 64. $x^3 + 2x^2 + x$

65. $2x^3 - 8x^2 - 10x$ 66. $x^4 + 5x^3 - 14x^2$

Solution of equations   In Problems 67 through 80, solve the given equation by factoring.
by factoring

67. $x^2 - 2x - 8 = 0$ 68. $x^2 - 4x + 3 = 0$

69. $x^2 + 10x + 25 = 0$ 70. $x^2 + 8x + 16 = 0$

71. $x^2 - 16 = 0$ 72. $x^2 - 25 = 0$

73. $2x^2 + 3x + 1 = 0$ 74. $x^2 - 2x + 1 = 0$

75. $4x^2 + 12x + 9 = 0$ 76. $6x^2 + 7x - 3 = 0$

77. $1 + \dfrac{4}{x} - \dfrac{5}{x^2} = 0$ 78. $\dfrac{9}{x^2} - \dfrac{6}{x} + 1 = 0$

79. $2 + \dfrac{2}{x} - \dfrac{4}{x^2} = 0$ 80. $\dfrac{3}{x^2} - \dfrac{5}{x} - 2 = 0$

Quadratic formula In Problems 81 through 86, use the quadratic formula to solve the given equation.

81. $2x^2 + 3x + 1 = 0$ 82. $-x^2 + 3x - 1 = 0$

83. $x^2 - 2x + 3 = 0$ 84. $x^2 - 2x + 1 = 0$

85. $4x^2 + 12x + 9 = 0$ 86. $x^2 + 12 = 0$

Systems of equations In Problems 87 through 92, solve the given system of equations.

87. $x + 5y = 13$
    $3x - 10y = -11$

88. $2x - 3y = 4$
    $3x - 5y = 2$

89. $5x - 4y = 12$
    $2x - 3y = 2$

90. $3x^2 - 9y = 0$
    $3y^2 - 9x = 0$

91. $2y^2 - x^2 = 1$
    $x - 2y = 3$

92. $2x^2 - y^2 = -7$
    $2x + y = 1$

Summation notation In Problems 93 through 96, evaluate the given sum.

93. $\displaystyle\sum_{j=1}^{4} (3j + 1)$ 94. $\displaystyle\sum_{j=1}^{5} j^2$

95. $\displaystyle\sum_{j=1}^{10} (-1)^j$ 96. $\displaystyle\sum_{j=1}^{5} 2^j$

In Problems 97 through 102, use summation notation to express the given sum.

97. $1 + \dfrac{1}{2} + \dfrac{1}{3} + \dfrac{1}{4} + \dfrac{1}{5} + \dfrac{1}{6}$

98. $3 + 6 + 9 + 12 + 15 + 18 + 21 + 24 + 27 + 30$

99. $2x_1 + 2x_2 + 2x_3 + 2x_4 + 2x_5 + 2x_6$

100. $1 - 1 + 1 - 1 + 1 - 1$

101. $1 - 2 + 3 - 4 + 5 - 6 + 7 - 8$

102. $x - x^2 + x^3 - x^4 + x^5$

B Limits and Continuity

In calculus, one infers what is or "should be" happening at a particular point from knowledge about what is happening at other points that approach it. Indeed, the development of the derivative itself, involving the approximation of tangents by secants, is based on such an inference. And, indirectly, so are all the subsequent rules and applications developed in this book.

Central to this process of inference is the mathematical concept of **limit.** The purpose of this section is to give you a better feel for this important concept. The approach will be intuitive rather than formal. The ideas outlined here form the basis for a more rigorous development of the laws and procedures of calculus and are at the heart of many of the branches of modern mathematics.

Here is a slightly imprecise definition of limit that will be sufficient for our purposes.

Limit

> If the function values $f(x)$ get closer and closer to some number L whenever the variable x gets closer and closer to some number a, we say that L is the limit of $f(x)$ as x approaches a, and we write
>
> $$\lim_{x \to a} f(x) = L$$

In geometric terms, $\lim_{x \to a} f(x) = L$ means that the height of the graph of $y = f(x)$ approaches L as x approaches a.

Limits describe the behavior of a function *near* a particular point and not necessarily *at* the point itself. This is illustrated in Figure B.1. For all three of the functions graphed in this figure, the limit of $f(x)$ as x approaches a is equal to L. Yet the functions behave differently at $x = a$ itself. In Figure B.1a, $f(a)$ is equal to the limit L; in Figure B.1b, $f(a)$ has been defined (artificially) to be different from L; and in Figure B.1c, $f(a)$ is not defined at all.

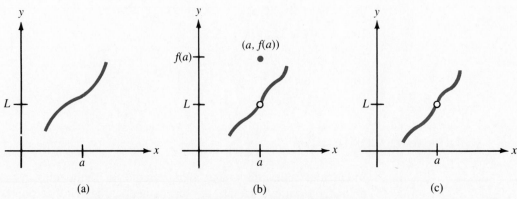

Figure B.1 Three functions for which $\lim_{x \to a} f(x) = L$.

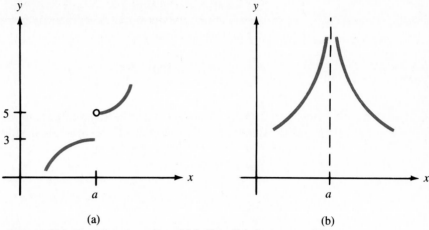

Figure B.2 Two functions for which $\lim\limits_{x \to a} f(x)$ does not exist.

Figure B.2 shows the graphs of two functions that do not have a limit as x approaches a. The function in Figure B.2a does not have a limit as x approaches a because $f(x)$ approaches 5 as x approaches a from the right, and it approaches a different value, 3, as x approaches a from the left. The function in Figure B.2b has no limit as x approaches a because the values of $f(x)$ increase without bound and do not approach any (finite) number.

Properties of Limits

Limits obey the following algebraic laws. These laws, which should seem plausible on the basis of our informal definition, are proved formally in more theoretical courses. They are important because they simplify the calculation of limits of algebraic functions.

The Limit of a Sum, Difference, or Product

If $\lim\limits_{x \to a} f(x)$ and $\lim\limits_{x \to a} g(x)$ exist, then

$$\lim_{x \to a} [f(x) + g(x)] = \lim_{x \to a} f(x) + \lim_{x \to a} g(x)$$
$$\lim_{x \to a} [f(x) - g(x)] = \lim_{x \to a} f(x) - \lim_{x \to a} g(x)$$
$$\lim_{x \to a} [f(x)g(x)] = \lim_{x \to a} f(x) \cdot \lim_{x \to a} g(x)$$

That is, the limit of a sum (or difference or product) is the sum (or difference or product) of the individual limits.

The Limit of a Quotient

If $\lim\limits_{x \to a} f(x)$ and $\lim\limits_{x \to a} g(x)$ exist and if $\lim\limits_{x \to a} g(x) \neq 0$, then

$$\lim_{x \to a} \frac{f(x)}{g(x)} = \frac{\lim\limits_{x \to a} f(x)}{\lim\limits_{x \to a} g(x)}$$

That is, if the limit of the denominator is not zero, the limit of a quotient is the quotient of the individual limits.

The Limit of a Power

If $\lim\limits_{x \to a} f(x) = L$ and p is any real number for which L^p is defined, then

$$\lim_{x \to a} [f(x)]^p = \left[\lim_{x \to a} f(x)\right]^p = L^p$$

That is, the limit of a power is the power of the limit.

The next two properties deal with the limits of two elementary linear functions, from which all other algebraic functions can be built.

The Limit of a Constant

If k is a constant,

$$\lim_{x \to a} k = k$$

That is, the limit of a constant is the constant itself.

In geometric terms, this says that the height of the graph of the constant function $f(x) = k$ approaches k as x approaches a. The situation

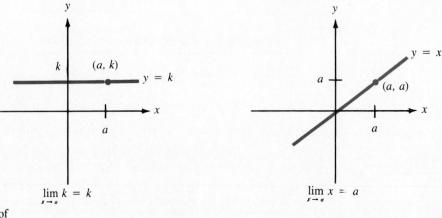

Figure B.3 The limits of two linear functions.

(a)

(b)

is illustrated in Figure B.3a. Of course, in this case, the height of the graph is actually equal to k when $x = a$ and for all other values of x as well.

The next property says that the height of the linear function $f(x) = x$ approaches a as x approaches a. The situation is illustrated in Figure B.3b.

The Limit of x

$$\lim_{x \to a} x = a$$

Computation of Limits

The following examples illustrate how the properties of limits can be used to calculate limits of algebraic functions. In the first example, you will see how to find the limit of a polynomial.

EXAMPLE B.1   Find $\lim\limits_{x \to 2} (x^2 - 3x + 1)$.

SOLUTION
Using the properties of limits, you get

$$\lim_{x \to 2} (x^2 - 3x + 1) = \left(\lim_{x \to 2} x\right)^2 - \lim_{x \to 2} 3 \cdot \lim_{x \to 2} x + \lim_{x \to 2} 1$$
$$= 2^2 - 3(2) + 1$$
$$= -1$$

In the next example, you will see how to find the limit of a rational function whose denominator does not approach zero.

EXAMPLE B.2   Find $\lim\limits_{x \to 0} \dfrac{3x^2 - 8}{x - 2}$.

SOLUTION
Since

$$\lim_{x \to 0} (x - 2) = -2 \neq 0$$

you can use the rule for the limit of a quotient to get

$$\lim_{x \to 0} \frac{3x^2 - 8}{x - 2} = \frac{\lim_{x \to 0} (3x^2 - 8)}{\lim_{x \to 0} (x - 2)} = \frac{-8}{-2} = 4$$

In the next example, the denominator of the given rational function approaches zero, while the numerator does not. When this happens you can conclude that the limit does not exist. The absolute value of such a quotient increases without bound and hence does not approach any (finite) number.

EXAMPLE B.3   Find $\lim\limits_{x \to 2} \dfrac{x + 1}{x - 2}$.

SOLUTION

The rule for the limit of a quotient does not apply in this case since the limit of the denominator is

$$\lim_{x \to 2} (x - 2) = 0$$

Since the limit of the numerator is

$$\lim_{x \to 2} (x + 1) = 3$$

which is not equal to zero, you can conclude that the limit of the quotient does not exist.

The graph of the function $f(x) = \dfrac{x + 1}{x - 2}$ has been sketched in Figure B.4 to give you a better idea of what is actually happening here.

Notice that $f(x)$ increases without bound as x approaches 2 from the right and decreases without bound as x approaches 2 from the left.

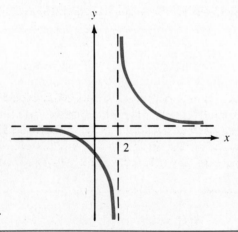

Figure B.4 The graph of the function $f(x) = \dfrac{x + 1}{x - 2}$.

In the next example, both the numerator and denominator of the given rational function approach zero. When this happens, you have to simplify the function algebraically in order to find the desired limit.

EXAMPLE B.4 Find $\lim\limits_{x \to 1} \dfrac{x^2 + x - 2}{x - 1}$.

SOLUTION

As x approaches 1, both the numerator and denominator approach zero, from which you can draw no conclusion about the size of the quotient.

To proceed, observe that the given function is not defined when $x = 1$, but that for all other values of x, you can simplify it by dividing numerator and denominator by $x - 1$ to get

$$\frac{x^2 + x - 2}{x - 1} = \frac{(x - 1)(x + 2)}{x - 1} = x + 2$$

(Since $x \neq 1$, you are not dividing by zero.) Now take the limit as x approaches (but is not equal to) 1 to get

$$\lim_{x \to 1} \frac{x^2 + x - 2}{x - 1} = \lim_{x \to 1} (x + 2) = 3$$

The graph of the function $f(x) = \dfrac{x^2 + x - 2}{x - 1}$ is sketched in Figure B.5. It is the straight line $y = x + 2$ with a hole at the point $(1, 3)$.

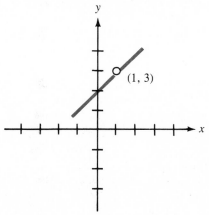

Figure B.5 The graph of the function $f(x) = \dfrac{x^2 + x - 2}{x - 1}$.

In general, when both the numerator and denominator of a quotient approach zero as x approaches a, your strategy will be to simplify the quotient algebraically (as you did in Example B.4 by canceling $x - 1$). In most cases, the simplified form of the quotient will be valid for all values of x except $x = a$. Since you are interested in the behavior of the quotient *near* $x = a$ and not *at* $x = a$, you may use the simplified form of the quotient to calculate the limit. Here is another example illustrating the technique.

EXAMPLE B.5 Find $\lim\limits_{x\to 1}\dfrac{\sqrt{x}-1}{x-1}$.

SOLUTION

Both the numerator and denominator approach zero as x approaches 1. To simplify the quotient, multiply the numerator and denominator by $\sqrt{x}+1$ to get

$$\frac{\sqrt{x}-1}{x-1} = \frac{(\sqrt{x}-1)(\sqrt{x}+1)}{(x-1)(\sqrt{x}+1)} = \frac{x-1}{(x-1)(\sqrt{x}+1)} = \frac{1}{\sqrt{x}+1}$$

and then take the limit to get

$$\lim_{x\to 1}\frac{\sqrt{x}-1}{x-1} = \lim_{x\to 1}\frac{1}{\sqrt{x}+1} = \frac{1}{2}$$

Continuity

The notion of continuity, which was introduced informally in Chapter 1, can be defined more precisely in terms of limits. Recall that a function was said to be **continuous** if its graph was an unbroken curve. Here is the more formal definition.

Continuity

A function f is continuous at $x = a$ if

(a) $f(a)$ is defined

(b) $\lim\limits_{x\to a} f(x)$ exists

and

(c) $\lim\limits_{x\to a} f(x) = f(a)$

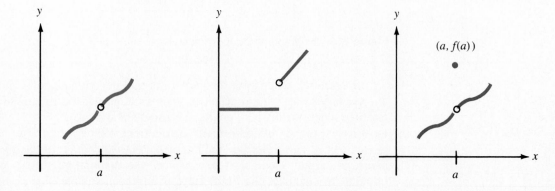

Figure B.6 Three functions with discontinuities at $x = a$.

Figure B.6 shows the graphs of three functions that are *not* continuous at $x = a$. The function in Figure B.6a is not continuous at $x = a$ because $f(a)$ is not defined. The function in Figure B.6b is not continuous at $x = a$ because $\lim_{x \to a} f(x)$ does not exist. And the function in Figure B.6c is not continuous at $x = a$ because $\lim_{x \to a} f(x) \neq f(a)$, even though $f(a)$ is defined and the limit exists.

A function whose graph is an unbroken curve on an open interval containing $x = a$ is continuous at $x = a$ because all three conditions in the definition of continuity are satisfied. Two such functions are sketched in Figure B.7.

You can use the properties of limits to show that a polynomial is continuous for every value of x and that a rational function is continuous for every value of x for which its denominator is not zero. Here are some examples.

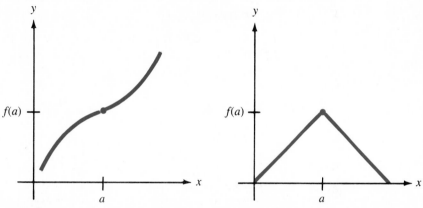

Figure B.7 Two continuous functions: $\lim_{x \to a} f(x) = f(a)$.

EXAMPLE B.7 Show that the polynomial $f(x) = 3x^2 - x + 5$ is continuous at $x = 1$.

SOLUTION
Verify that the three criteria for continuity are satisfied. Clearly $f(1)$ is defined. In fact, $f(1) = 7$. Moreover,

$$\lim_{x \to 1} f(x) = 3 \left(\lim_{x \to 1} x \right)^2 - \lim_{x \to 1} x + \lim_{x \to 1} 5 = 3 - 1 + 5 = 7 = f(1)$$

as required. Hence, f is continuous at $x = 1$.

EXAMPLE B.8 Show that the rational function $f(x) = \dfrac{x + 1}{x - 2}$ is continuous at $x = 3$.

SOLUTION
$$f(3) = \frac{3 + 1}{3 - 2} = 4$$

Moreover, since $\lim\limits_{x \to 3} (x - 2) = 1 \neq 0$, you can apply the rule for the limit of a quotient to get

$$\lim_{x \to 3} f(x) = \lim_{x \to 3} \frac{x + 1}{x - 2} = \frac{\lim\limits_{x \to 3} (x + 1)}{\lim\limits_{x \to 3} (x - 2)} = \frac{4}{1} = 4 = f(3)$$

as required. Hence, f is continuous at $x = 3$.

EXAMPLE B.9 Show that the rational function $f(x) = \dfrac{x + 1}{x - 2}$ is not continuous at $x = 2$.

SOLUTION
Since division by zero is impossible, $f(2)$ is undefined, violating the first of the criteria for continuity.

The Derivative

As you saw in Chapter 2, the derivative of a function is defined in terms of limits as follows, where x and Δx are related as in Figure B.8a.

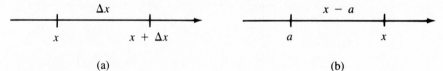

(a)   (b)

Figure B.8 Alternative notation for use in the definition of the derivative.

The Derivative

$$f'(x) = \lim_{\Delta x \to 0} \frac{f(x + \Delta x) - f(x)}{\Delta x}$$

Using the alternative notation illustrated in Figure B.8b, we can write this definition in the following form, which is more convenient for certain theoretical purposes.

Alternative Definition of the Derivative

$$f'(a) = \lim_{x \to a} \frac{f(x) - f(a)}{x - a}$$

Differentiability and Continuity

Using the alternative definition of the derivative, we can prove the following relationship between differentiability and continuity, which was mentioned without proof in Chapter 2, Section 1.

Differentiability Implies Continuity

> If $f'(a)$ exists, then $f(x)$ is continuous at $x = a$.

To prove this, we must show that

$$\lim_{x \to a} f(x) = f(a)$$

or, equivalently,

$$\lim_{x \to a} [f(x) - f(a)] = 0$$

Using the alternative definition of the derivative, this is easy. In particular,

$$\lim_{x \to a} [f(x) - f(a)] = \lim_{x \to a} \left[\frac{f(x) - f(a)}{x - a} (x - a) \right]$$

$$= \left[\lim_{x \to a} \frac{f(x) - f(a)}{x - a} \right] \left[\lim_{x \to a} (x - a) \right]$$

by the product rule for limits

$$= f'(a)(0) = 0$$

and the proof is complete.

Proofs of the Differentiation Rules

The rules for differentiation can be proved formally using the properties of limits and continuity. Here is a complete proof of the product rule from Chapter 2, Section 2. The proofs of the other rules are left as exercises for you to do. (See Problems 55, 56, and 57 at the end of this section.)

Proof of the Product Rule

EXAMPLE B.10   Prove that if f and g are differentiable at x, then

$$\frac{d}{dx} [f(x)g(x)] = f(x)g'(x) + g(x) f'(x)$$

SOLUTION

$$\frac{d}{dx} [f(x)g(x)] = \lim_{\Delta x \to 0} \frac{f(x + \Delta x)g(x + \Delta x) - f(x)g(x)}{\Delta x}$$

by the definition of the derivative

$$= \lim_{\Delta x \to 0} \left[\frac{f(x + \Delta x)g(x + \Delta x) - f(x + \Delta x)g(x)}{\Delta x} \right.$$

$$\left. + \frac{f(x + \Delta x)g(x) - f(x)g(x)}{\Delta x} \right]$$

$$= \lim_{\Delta x \to 0} f(x + \Delta x) \lim_{\Delta x \to 0} \frac{g(x + \Delta x) - g(x)}{\Delta x}$$

$$+ g(x) \lim_{\Delta x \to 0} \frac{f(x + \Delta x) - f(x)}{\Delta x}$$

by the sum rule and product rule for limits

$$= f(x) \lim_{\Delta x \to 0} \frac{g(x + \Delta x) - g(x)}{\Delta x}$$

$$+ g(x) \lim_{\Delta x \to 0} \frac{f(x + \Delta x) - f(x)}{\Delta x}$$

by the continuity of f, which follows from the differentiability of f

$$= f(x)g'(x) + g(x) f'(x)$$

by the definition of the derivative

Problems

In Problems 1 through 6, find $\lim_{x \to a} f(x)$ if it exists.

1.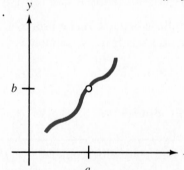

2.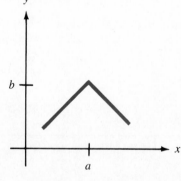

3.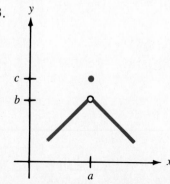

4.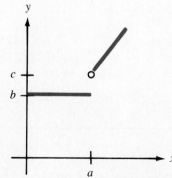

5.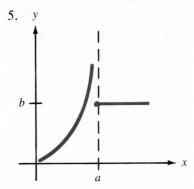

6.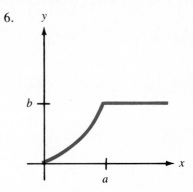

In Problems 7 through 28, find the indicated limit if it exists.

7. $\lim_{x \to 2} (3x^2 - 5x + 2)$

8. $\lim_{x \to -1} (x^3 - 2x^2 + x - 3)$

9. $\lim_{x \to 0} (x^5 - 6x^4 + 7)$

10. $\lim_{x \to 0} (1 - 5x^3)$

11. $\lim_{x \to 3} (x - 1)^2(x + 1)$

12. $\lim_{x \to -1} (x^2 + 1)(1 - 2x)^2$

13. $\lim_{x \to 2} \dfrac{x + 1}{x + 2}$

14. $\lim_{x \to 1} \dfrac{2x + 3}{x + 1}$

15. $\lim_{x \to 5} \dfrac{x + 3}{5 - x}$

16. $\lim_{x \to 3} \dfrac{2x + 3}{x - 3}$

17. $\lim_{x \to 1} \dfrac{x^2 - 1}{x - 1}$

18. $\lim_{x \to 3} \dfrac{9 - x^2}{x - 3}$

19. $\lim_{x \to 5} \dfrac{x^2 - 3x - 10}{x - 5}$

20. $\lim_{x \to 2} \dfrac{x^2 + x - 6}{x - 2}$

21. $\lim_{x \to 4} \dfrac{(x + 1)(x - 4)}{(x - 1)(x - 4)}$

22. $\lim_{x \to 0} \dfrac{x(x^2 - 1)}{x^2}$

23. $\lim_{x \to -2} \dfrac{x^2 - x - 6}{x^2 + 3x + 2}$

24. $\lim_{x \to 1} \dfrac{x^2 + 4x - 5}{x^2 - 1}$

25. $\lim_{x \to 4} \dfrac{\sqrt{x} - 2}{x - 4}$

26. $\lim_{x \to 9} \dfrac{\sqrt{x} - 3}{x - 9}$

27. $\lim_{x \to 1} \dfrac{x - 1}{\sqrt{x} - 1}$

28. $\lim_{x \to 9} \dfrac{x - 9}{\sqrt{x} - 3}$

In Problems 29 through 41, decide if the given function is continuous at the specified value of x.

29. $f(x) = 5x^2 - 6x + 1; x = 2$

30. $f(x) = x^3 - 2x^2 + x - 5; x = 0$

31. $f(x) = \dfrac{x + 2}{x + 1}; x = 1$

32. $f(x) = \dfrac{2x - 4}{3x - 2}; x = 2$

33. $f(x) = \dfrac{x + 1}{x - 1}; x = 1$

34. $f(x) = \dfrac{2x + 1}{3x - 6}; x = 2$

35. $f(x) = \dfrac{\sqrt{x} - 2}{x - 4}; x = 4$

36. $f(x) = \dfrac{\sqrt{x} - 2}{x - 4}; x = 2$

37. $f(x) = \begin{cases} x + 1 & \text{if } x \leq 2 \\ 2 & \text{if } x > 2 \end{cases} \quad ; \quad x = 2$

38. $f(x) = \begin{cases} 0 & \text{if } x < 1 \\ x - 1 & \text{if } x \geq 1 \end{cases} \quad ; \quad x = 1$

39. $f(x) = \begin{cases} x + 1 & \text{if } x < 0 \\ x - 1 & \text{if } x \geq 0 \end{cases} \quad ; \quad x = 0$

40. $f(x) = \begin{cases} x^2 + 1 & \text{if } x \leq 3 \\ 2x + 4 & \text{if } x > 3 \end{cases} \quad ; \quad x = 3$

41. $f(x) = \begin{cases} \dfrac{x^2 - 1}{x + 1} & \text{if } x < -1 \\ x^2 - 3 & \text{if } x \geq -1 \end{cases} \quad ; \quad x = -1$

In Problems 42 through 54 list all the values of x for which the given function is *not* continuous.

42. $f(x) = 3x^2 - 6x + 9$

43. $f(x) = x^5 - x^3$

44. $f(x) = \dfrac{x + 1}{x - 2}$

45. $f(x) = \dfrac{3x - 1}{2x - 6}$

46. $f(x) = \dfrac{3x + 3}{x + 1}$

47. $f(x) = \dfrac{x^2 - 1}{x + 1}$

48. $f(x) = \dfrac{3x - 2}{(x + 3)(x - 6)}$

49. $f(x) = \dfrac{x}{(x + 5)(x - 1)}$

50. $f(x) = \dfrac{x}{x^2 - x}$

51. $f(x) = \dfrac{x^2 - 2x + 1}{x^2 - x - 2}$

52. $f(x) = \begin{cases} 2x + 3 & \text{if } x \leq 1 \\ 6x - 1 & \text{if } x > 1 \end{cases}$

53. $f(x) = \begin{cases} x^2 & \text{if } x \leq 2 \\ 9 & \text{if } x > 2 \end{cases}$

54. $f(x) = \begin{cases} x - 1 & \text{if } x < 1 \\ 1 & \text{if } x = 1 \\ 1 - x & \text{if } x > 1 \end{cases}$

In Problems 55 through 57, prove the given differentiation rule, giving reasons for each major step.

55. The constant multiple rule: If f is a differentiable function and c is a constant, then

$$\frac{d}{dx}[cf(x)] = cf'(x)$$

56. Sum rule: If f and g are differentiable functions, then

$$\frac{d}{dx}[f(x) + g(x)] = f'(x) + g'(x)$$

57. Quotient rule: If f and g are differentiable functions, then

$$\frac{d}{dx}\left[\frac{f(x)}{g(x)}\right] = \frac{g(x)f'(x) - f(x)g'(x)}{[g(x)]^2}$$

provided $g(x) \neq 0$. *Hint*: The difference quotient is

$$\frac{1}{\Delta x}\left[\frac{f(x + \Delta x)}{g(x + \Delta x)} - \frac{f(x)}{g(x)}\right] = \frac{g(x)f(x + \Delta x) - f(x)g(x + \Delta x)}{g(x + \Delta x)g(x)\Delta x}$$

which you can rewrite in a more useful form by subtracting and adding $g(x)f(x)$ in the numerator.

TABLE I Powers of e

x	e^x	e^{-x}	x	e^x	e^{-x}	x	e^x	e^{-x}
0.00	1.0000	1.00000	0.50	1.6487	.60653	1.00	2.7183	.36788
0.01	1.0101	0.99005	0.51	1.6653	.60050	1.20	3.3201	.30119
0.02	1.0202	.98020	0.52	1.6820	.59452	1.30	3.6693	.27253
0.03	1.0305	.97045	0.53	1.6989	.58860	1.40	4.0552	.24660
0.04	1.0408	.96079	0.54	1.7160	.58275	1.50	4.4817	.22313
0.05	1.0513	.95123	0.55	1.7333	.57695	1.60	4.9530	.20190
0.06	1.0618	.94176	0.56	1.7507	.57121	1.70	5.4739	.18268
0.07	1.0725	.93239	0.57	1.7683	.56553	1.80	6.0496	.16530
0.08	1.0833	.92312	0.58	1.7860	.55990	1.90	6.6859	.14957
0.09	1.0942	.91393	0.59	1.8040	.55433	2.00	7.3891	.13534
0.10	1.1052	.90484	0.60	1.8221	.54881	3.00	20.086	.04979
0.11	1.1163	.89583	0.61	1.8404	.54335	4.00	54.598	.01832
0.12	1.1275	.88692	0.62	1.8589	.53794	5.00	148.41	.00674
0.13	1.1388	.87809	0.63	1.8776	.53259	6.00	403.43	.00248
0.14	1.1503	.86936	0.64	1.8965	.52729	7.00	1096.6	.00091
0.15	1.1618	.86071	0.65	1.9155	.52205	8.00	2981.0	.00034
0.16	1.1735	.85214	0.66	1.9348	.51685	9.00	8103.1	.00012
0.17	1.1853	.84366	0.67	1.9542	.51171	10.00	22026.5	.00005
0.18	1.1972	.83527	0.68	1.9739	.50662			
0.19	1.2092	.82696	0.69	1.9937	.50158			
0.20	1.2214	.81873	0.70	2.0138	.49659			
0.21	1.2337	.81058	0.71	2.0340	.49164			
0.22	1.2461	.80252	0.72	2.0544	.48675			
0.23	1.2586	.79453	0.73	2.0751	.48191			
0.24	1.2712	.78663	0.74	2.0959	.47711			
0.25	1.2840	.77880	0.75	2.1170	.47237			
0.26	1.2969	.77105	0.76	2.1383	.46767			
0.27	1.3100	.76338	0.77	2.1598	.46301			
0.28	1.3231	.75578	0.78	2.1815	.45841			
0.29	1.3364	.74826	0.79	2.2034	.45384			
0.30	1.3499	.74082	0.80	2.2255	.44933			
0.31	1.3634	.73345	0.81	2.2479	.44486			
0.32	1.3771	.72615	0.82	2.2705	.44043			
0.33	1.3910	.71892	0.83	2.2933	.43605			
0.34	1.4049	.71177	0.84	2.3164	.43171			
0.35	1.4191	.70469	0.85	2.3396	.42741			
0.36	1.4333	.69768	0.86	2.3632	.42316			
0.37	1.4477	.69073	0.87	2.3869	.41895			
0.38	1.4623	.68386	0.88	2.4109	.41478			
0.39	1.4770	.67706	0.89	2.4351	.41066			
0.40	1.4918	.67032	0.90	2.4596	.40657			
0.41	1.5068	.66365	0.91	2.4843	.40252			
0.42	1.5220	.65705	0.92	2.5093	.39852			
0.43	1.5373	.65051	0.93	2.5345	.39455			
0.44	1.5527	.64404	0.94	2.5600	.39063			
0.45	1.5683	.63763	0.95	2.5857	.38674			
0.46	1.5841	.63128	0.96	2.6117	.38298			
0.47	1.6000	.62500	0.97	2.6379	.37908			
0.48	1.6161	.61878	0.98	2.6645	.37531			
0.49	1.6323	.61263	0.99	2.6912	.37158			

Excerpted from *Handbook of Mathematical Tables and Formulas,* 5th ed., by R. S. Burington. Copyright © 1973 by McGraw-Hill, Inc. Used with permission of McGraw-Hill Book Company.

TABLE II The natural logarithm (base e)

x	$\ln x$	x	$\ln x$	x	$\ln x$	x	$\ln x$
.01	-4.60517	0.50	-0.69315	1.00	0.00000	1.5	0.40547
.02	-3.91202	.51	.67334	1.01	.00995	1.6	7000
.03	.50656	.52	.65393	1.02	.01980	1.7	0.53063
.04	.21888	.53	.63488	1.03	.02956	1.8	8779
		.54	.61619	1.04	.03922	1.9	0.64185
.05	-2.99573	.55	.59784	1.05	.04879	2.0	9315
.06	.81341	.56	.57982	1.06	.05827	2.1	0.74194
.07	.65926	.57	.56212	1.07	.06766	2.2	8846
.08	.52573	.58	.54473	1.08	.07696	2.3	0.83291
.09	.40795	.59	.52763	1.09	.08618	2.4	7547
0.10	-2.30259	0.60	-0.51083	1.10	.09531	2.5	0.91629
.11	.20727	.61	.49430	1.11	.10436	2.6	5551
.12	.12026	.62	.47804	1.12	.11333	2.7	9325
.13	.04022	.63	.46204	1.13	.12222	2.8	1.02962
.14	-1.96611	.64	.44629	1.14	.13103	2.9	6471
.15	.89712	.65	.43078	1.15	.13976	3.0	9861
.16	.83258	.66	.41552	1.16	.14842	4.0	1.38629
.17	.77196	.67	.40048	1.17	.15700	5.0	1.60944
.18	.71480	.68	.38566	1.18	.16551	10.0	2.30258
.19	.66073	.69	.37106	1.19	.17395		
0.20	-1.60944	0.70	-0.35667	1.20	.18232		
.21	.56065	.71	.34249	1.2i	.19062		
.22	.51413	.72	.32850	1.22	.19885		
.23	.46968	.73	.31471	1.23	.20701		
.24	.42712	.74	.30111	1.24	.21511		
.25	.38629	.75	.28768	1.25	.22314		
.26	.34707	.76	.27444	1.26	.23111		
.27	.30933	.77	.26136	1.27	.23902		
.28	.27297	.78	.24846	1.28	.24686		
.29	.23787	.79	.23572	1.29	.25464		
0.30	-1.20397	0.80	-0.22314	1.30	.26236		
.31	.17118	.81	.21072	1.31	.27003		
.32	.13943	.82	.19845	1.32	.27763		
.33	.10866	.83	.18633	1.33	.28518		
.34	.07881	.84	.17435	1.34	.29267		
.35	-1.04982	.85	-0.16252	1.35	.30010		
.36	.02165	.86	.15032	1.36	.30748		
.37	-0.99425	.87	.13926	1.37	.31481		
.38	.96758	.88	.12783	1.38	.32208		
.39	.94161	.89	.11653	1.39	.32930		
0.40	-0.91629	0.90	-0.10536	1.40	.33647		
.41	.89160	.91	.09431	1.41	.34359		
.42	.86750	.92	.08338	1.42	.35066		
.43	.84397	.93	.07257	1.43	.35767		
.44	.82098	.94	.06188	1.44	.36464		
.45	.79851	.95	.05129	1.45	.37156		
.46	.77653	.96	.04082	1.46	.37844		
.47	.75502	.97	.03046	1.47	.38526		
.48	.73397	.98	.02020	1.48	.39204		
.49	.71335	.99	.01005	1.49	.39878		

TABLE III Trigonometric functions

Degrees	Radians	sin	cos	tan	Degrees	Radians	sin	cos	tan
0	0.0000	0.0000	1.000	0.0000	45	0.7854	0.7071	0.7071	1.000
1	0.01745	0.01745	0.9998	0.01746	46	0.8028	0.7193	0.6947	1.036
2	0.03491	0.03490	0.9994	0.03492	47	0.8203	0.7314	0.6820	1.072
3	0.05236	0.05234	0.9986	0.05241	48	0.8378	0.7431	0.6691	1.111
4	0.06981	0.06976	0.9976	0.06993	49	0.8552	0.7547	0.6561	1.150
5	0.08727	0.08716	0.9962	0.08749	50	0.8727	0.7660	0.6428	1.192
6	0.1047	0.1045	0.9945	0.1051	51	0.8901	0.7772	0.6293	1.235
7	0.1222	0.1219	0.9926	0.1228	52	0.9076	0.7880	0.6157	1.280
8	0.1396	0.1392	0.9903	0.1405	53	0.9250	0.7986	0.6018	1.327
9	0.1571	0.1564	0.9877	0.1584	54	0.9425	0.8090	0.5878	1.376
10	0.1745	0.1736	0.9848	0.1763	55	0.9599	0.8192	0.5736	1.428
11	0.1920	0.1908	0.9816	0.1944	56	0.9774	0.8290	0.5592	1.483
12	0.2094	0.2079	0.9782	0.2126	57	0.9948	0.8387	0.5446	1.540
13	0.2269	0.2250	0.9744	0.2309	58	1.012	0.8480	0.5299	1.600
14	0.2444	0.2419	0.9703	0.2493	59	1.030	0.8572	0.5150	1.664
15	0.2618	0.2588	0.9659	0.2680	60	1.047	0.8660	0.5000	1.732
16	0.2792	0.2756	0.9613	0.2868	61	1.065	0.8746	0.4848	1.804
17	0.2967	0.2924	0.9563	0.3057	62	1.082	0.8830	0.4695	1.881
18	0.3142	0.3090	0.9511	0.3249	63	1.100	0.8910	0.4540	1.963
19	0.3316	0.3256	0.9455	0.3443	64	1.117	0.8988	0.4384	2.050
20	0.3491	0.3420	0.9397	0.3640	65	1.134	0.9063	0.4226	2.144
21	0.3665	0.3584	0.9336	0.3839	66	1.152	0.9136	0.4067	2.246
22	0.3840	0.3746	0.9272	0.4040	67	1.169	0.9205	0.3907	2.356
23	0.4014	0.3907	0.9205	0.4245	68	1.187	0.9272	0.3746	2.475
24	0.4189	0.4067	0.9136	0.4452	69	1.204	0.9336	0.3584	2.605
25	0.4363	0.4226	0.9063	0.4663	70	1.222	0.9397	0.3420	2.748
26	0.4538	0.4384	0.8988	0.4877	71	1.239	0.9455	0.3256	2.904
27	0.4712	0.4540	0.8910	0.5095	72	1.257	0.9511	0.3090	3.078
28	0.4887	0.4695	0.8830	0.5317	73	1.274	0.9563	0.2924	3.271
29	0.5062	0.4848	0.8746	0.5543	74	1.292	0.9613	0.2756	3.487
30	0.5236	0.5000	0.8660	0.5774	75	1.309	0.9659	0.2588	3.732
31	0.5410	0.5150	0.8572	0.6009	76	1.326	0.9703	0.2419	4.011
32	0.5585	0.5299	0.8480	0.6249	77	1.344	0.9744	0.2250	4.332
33	0.5760	0.5446	0.8387	0.6494	78	1.361	0.9782	0.2079	4.705
34	0.5934	0.5592	0.8290	0.6745	79	1.379	0.9816	0.1908	5.145
35	0.6109	0.5736	0.8192	0.7002	80	1.396	0.9848	0.1736	5.671
36	0.6283	0.5878	0.8090	0.7265	81	1.414	0.9877	0.1564	6.314
37	0.6458	0.6018	0.7986	0.7536	82	1.431	0.9903	0.1392	7.115
38	0.6632	0.6157	0.7880	0.7813	83	1.449	0.9926	0.1219	8.144
39	0.6807	0.6293	0.7772	0.8098	84	1.466	0.9945	0.1045	9.514
40	0.6981	0.6428	0.7660	0.8391	85	1.484	0.9962	0.08716	11.43
41	0.7156	0.6561	0.7547	0.8693	86	1.501	0.9976	0.06976	14.30
42	0.7330	0.6691	0.7431	0.9004	87	1.518	0.9986	0.05234	19.08
43	0.7505	0.6820	0.7314	0.9325	88	1.536	0.9994	0.03490	28.64
44	0.7679	0.6947	0.7193	0.9657	89	1.553	0.9998	0.01745	57.29
45	0.7854	0.7071	0.7071	1.000	90	1.571	1.000	0.0000	———

Answers to Odd-Numbered Problems and Review Problems

Chapter 1, Section 1 (page 8)

1. $f(1) = 6, f(0) = -2, f(-2) = 0$

3. $g(-1) = -2, g(1) = 2, g(2) = \dfrac{5}{2}$

5. $h(2) = 2\sqrt{3}, h(0) = 2, h(-4) = 2\sqrt{3}$

7. $f(1) = 1, f(5) = \dfrac{1}{27}, f(13) = \dfrac{1}{125}$

9. $f(1) = 0, f(2) = 2, f(3) = 2$

11. $f(-6) = 3, f(-5) = -4, f(16) = 4$

13. All real numbers x except $x = -2$.

15. All real numbers x for which $x \geq 5$.

17. All real numbers t.

19. All real numbers t for which $t \geq 2$.

21. All real numbers x for which $|x| > 3$.

23. All real numbers t except $t = 1$.

25. (a) $4,500   (b) $371

27. (a) $33\frac{1}{3}°$ Celsius   (b) Decreased by 7.5° Celsius

29. (a) All real numbers n except $n = 0$.
    (b) All positive integers n.
    (c) 7 minutes   (d) 12th trial
    (e) The time required will approach but never exceed 3 minutes.

31. (a) All real numbers x except $x = 300$.
    (b) All real numbers x for which $0 \leq x \leq 100$.
    (c) 120   (d) 300   (e) 60

33. (a) 192 feet (b) 80 feet   (c) 256 feet   (d) After 4 seconds.

35. $g[h(x)] = 3x^2 + 14x + 10$

37. $g[h(x)] = x^3 + 2x^2 + 4x + 2$

39. $g[h(x)] = \dfrac{1}{(x - 1)^2}$

41. $g[h(x)] = |x|$

43. $f(x - 2) = 2x^2 - 11x + 15$

45. $f(x - 1) = x^5 - 3x^2 + 6x - 3$

47. $f(x^2 + 3x - 1) = \sqrt{x^2 + 3x - 1}$

49. $f(x + 1) = \dfrac{x}{x + 1}$

51. $h(x) = 3x - 5,\ g(u) = \sqrt{u}$

53. $h(x) = x^2 + 1,\ g(u) = \dfrac{1}{u}$

55. $h(x) = x + 3,\ g(u) = \sqrt{u} - \dfrac{1}{(u + 1)^3}$

57. (a) $C[q(t)] = 625t^2 + 25t + 900$ (b) $6,600
    (c) After 4 hours.

Chapter 1, Section 2 (page 21)

1.

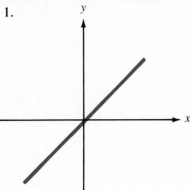

3.

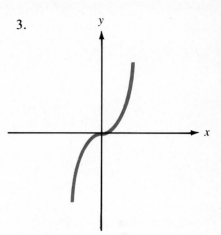

5.

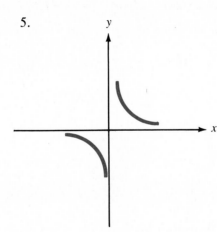

7.

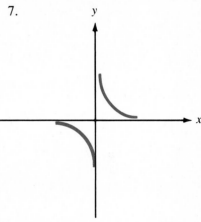

9.

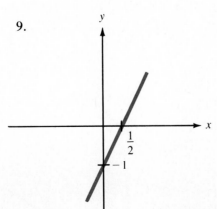

11.

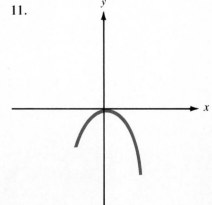

13.

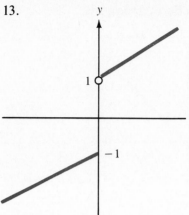

15.

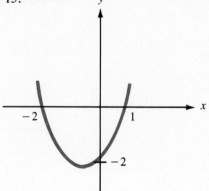

17.

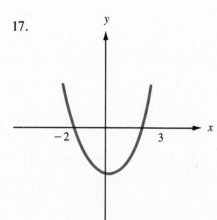

19.

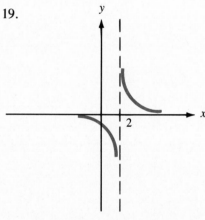

21.

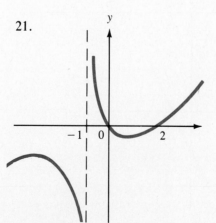

23.

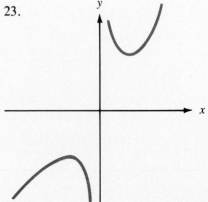

25. $P(x) = (x - 20)(120 - x)$

Optimal price = $70 per recorder

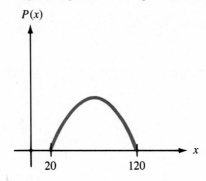

27. (a)

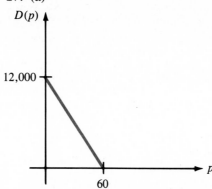

(c) $E(p)$

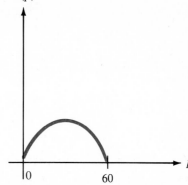

(b) $E(p) = -200p^2 + 12,000p = -200p(p - 60)$
(d) $E(0) = 0$ because price is zero.
$E(60) = 0$ because demand is zero.
(e) Optimal price = $30 per unit.

29. The graph has a practical interpretation
for $0 \leq x \leq 100$.

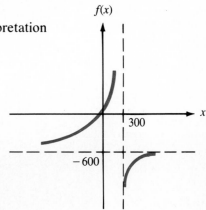

31. (a) 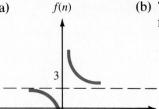   (b) The graph has a practical interpretation for $n = 1, 2, 3, \ldots$

(c) As n increases without bound, the height of the graph decreases and approaches 3. That is, as the number of trials increases, the time required for the rat to traverse the maze decreases, approaching a lower bound of 3 minutes.

33. 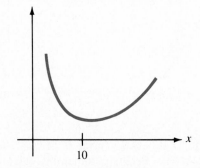   Optimal number of machines $= 10$.

35. $A(x) = x + 4 + \dfrac{16}{x}$

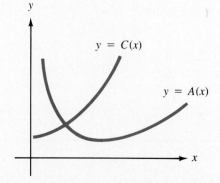

37. (a)

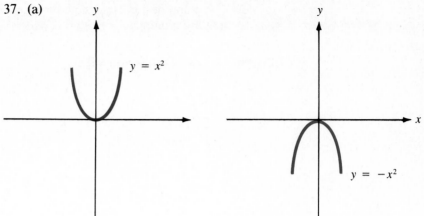

(b) The graph of g is obtained by reflecting the graph of f across the x axis.

39. (a)

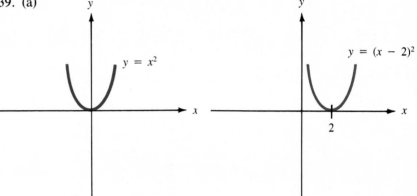

(b)

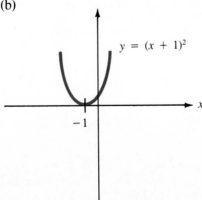

(c) The graph of g is obtained by translating the graph of f horizontally by c units.

Chapter 1, Section 3 (page 35)

1. $-\dfrac{7}{2}$ 3. -1 5. Undefined

7. $m = 5, b = 2$ 9. $m = -1, b = 2$ 11. $m = \dfrac{1}{2}, b = -3$

13. $m = 2, b = -3$ 15. $m = 0, b = 2$

17. $y = x - 2$ 19. $y = -\dfrac{1}{2}x + \dfrac{1}{2}$

21. $y = 5$
23. $y = -x + 1$

25. $y = x + 5$ 27. $x = 1$

29. (a) $C(x) = 0.14x + 20$ (b) \$27   (c) 180 miles

31. (a) $F(x) = -12.5x + 150$ (b) \$87.50

33. (a) $V(x) = -1{,}900x + 20{,}000$ (b) \$12,400

35. (a) $N(x) = 3x + 157$, where x is the number of days since the start of the program.
   (b) 289

37. (a) \$24; \$48
   (b) The function is not linear because the rate of change is not constant.

Chapter 1, Section 4 (page 45)

1. $\left(-\dfrac{1}{2}, \dfrac{7}{2}\right)$ 3. None

5. $(1, 0)$ 7. $(1, 1)$ and $(0, 0)$

9. None   11. $(-1, 2)$

13. $\left(\dfrac{1}{2}, 4\right)$ and $\left(-\dfrac{1}{2}, 4\right)$ 15. None

17. $\left(\dfrac{3 + \sqrt{5}}{2}, \dfrac{1 + \sqrt{5}}{2}\right)$ and $\left(\dfrac{3 - \sqrt{5}}{2}, \dfrac{1 - \sqrt{5}}{2}\right)$

19. (a) 4   (b) 7

21. Join the second club if fewer than 80 hours of tennis will be played and the first if more than 80 hours will be played.

23. $p = \$40$, $q = 360$ units

25. (a) $p = \$80$, $q = 70$ units

(c) $S(10) = 0$. Manufacturers will not supply any units unless the market price exceeds $10.

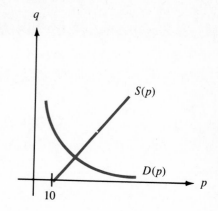

27. Of course!

Chapter 1, Section 5 (page 55)

1. $P(x) = 20(25 - x)(x - 3)$; optimal price $= \$14$

3. $R(x) = -\frac{1}{2}x(x - 155)$; optimal size $= 77$ or 78

5. $R(x) = 2(100 - x)(80 + x)$, where x is the number of days after July first; optimal harvest date $=$ July 11.

7. $f(x) = 2x + \dfrac{7,200}{x}$; optimal dimensions: 60 meters by 60 meters

9. $C(x) = 4x^2 + \dfrac{1,000}{x}$

11. $V(x) = 4x(9 - x)^2$

13. $C(r) = 0.08\pi\left(r^2 + \dfrac{2}{r}\right)$

15. (a) $C(x) = \begin{cases} 1.5x & \text{if } 0 < x < 50 \\ x & \text{if } x \ge 50 \end{cases}$

(b) $\$23.50$

17. $f(x) = \begin{cases} 20 & \text{if } 0 < x \le 1 \\ 37 & \text{if } 1 < x \le 2 \\ 54 & \text{if } 2 < x \le 3 \\ 71 & \text{if } 3 < x \le 4 \end{cases}$

19. (a) $f(x) = \begin{cases} 0.24x - 1,503 & \text{if } 15,000 < x \le 18,200 \\ 0.28x - 2,231 & \text{if } 18,200 < x \le 23,500 \\ 0.32x - 3,171 & \text{if } 23,500 < x \le 28,800 \end{cases}$

(b) $m_1 = 0.24$, $m_2 = 0.28$, $m_3 = 0.32$

21. $R(p) = kp$

23. $R(t) = k(M - t)$, where M is the temperature of the surrounding medium.

25. $R(x) = kx(n - x)$, where n is the total number of people involved.

27. $C(s) = \dfrac{k_1}{s} + k_2 s$

29. $D(t) = \sqrt{(60t)^2 + (300 - 30t)^2} = 30\sqrt{5t^2 - 20t + 100}$

31. $A(x) = 8x + \dfrac{100}{x} + 57$

Chapter 1, Review Problems (page 61)

1. (a) All real numbers x.
   (b) All real numbers x except $x = 1$ and $x = -2$.
   (c) All real numbers x for which $|x| \geq 3$.

2. (a) \$45   (b) \$1   (c) 9 months from now.
   (d) The price will approach \$40.

3. (a) $g[h(x)] = x^2 - 4x + 4$

   (b) $g[h(x)] = \dfrac{1}{2x + 5}$

   (c) $g[h(x)] = \sqrt{-2x - 3}$

4. (a) $f(x - 2) = x^2 - 5x + 10$

   (b) $f(x^2 + 1) = \sqrt{x^2 + 1} + \dfrac{2}{x^2}$

   (c) $f(x + 1) - f(x) = 2x + 1$

5. (a) $g(u) = u^5$ and $h(x) = x^2 + 3x + 4$

   (b) $g(u) = u^2 + \dfrac{5}{2(u + 1)^3}$ and $h(x) = 3x + 1$

6. (a) $Q[p(t)] = \sqrt{23.4 + 0.1t^2}$

   (b) 4.93 units   (c) 4 years from now.

7. $c = -4$

8. (a)   (b)

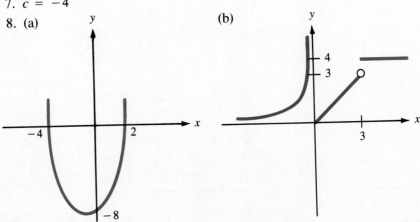

(c)

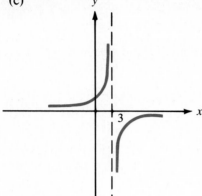

(d)

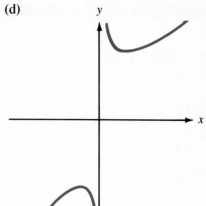

(e)

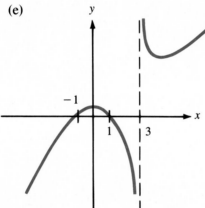

9. (a)

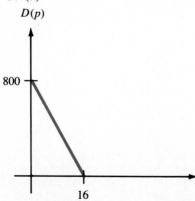

(b) $E(p) = -50p(p - 16)$

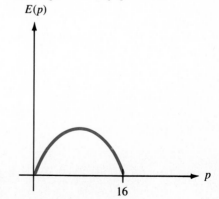

(c) Optimal price = $8 per unit.

10. (a) $f(x)$

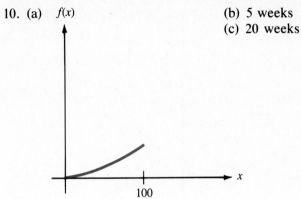

(b) 5 weeks

(c) 20 weeks

11. (a) $m = 3, b = 2$

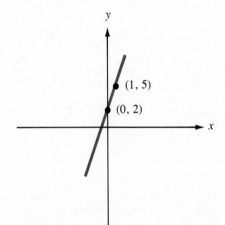

(b) $m = \dfrac{5}{4}, b = -5$

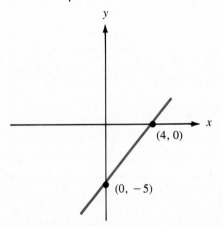

(c) $m = -\dfrac{3}{2}, b = 0$

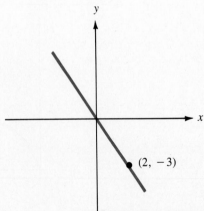

(d) $m = -\dfrac{2}{3}, b = 8$

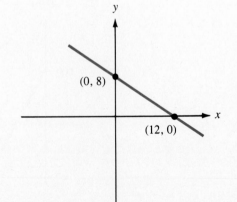

12. $y = 5x - 4$ 13. $y = -2x + 5$ 14. $y = 7x - 10$

15. (a) $P(x) = 2x + 93$, where x is the number of months since the beginning of the year.
    (b) 93 cents per gallon
    (c) \$1.11 per gallon

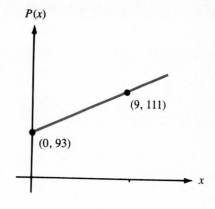

16. (a) $C(x) = 400x + 3,200$ where x is the number of months since the paper first appeared.
    (b) 5,200

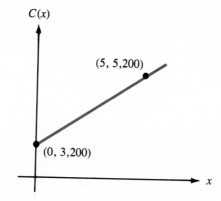

17. (a)

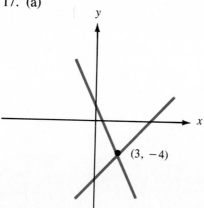

(b)

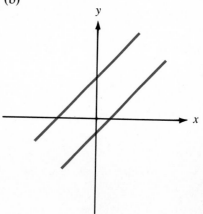

(c)

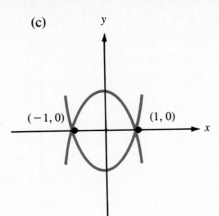

(d)

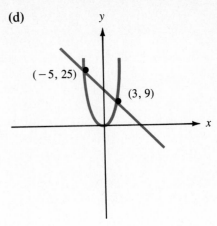

(e)

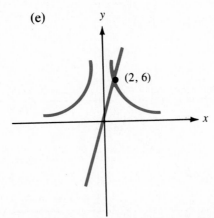

18. Call the first plumber if the work will take less than $1\frac{1}{2}$ hours and the second if the work will take more than $1\frac{1}{2}$ hours.

19. (a) 150   (b) \$1,500 profit   (c) 180

20. $P(x) = (50 - x)(x - 10)$
    Optimal price = \$30

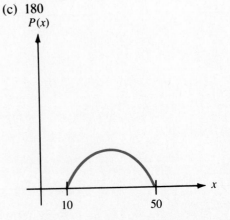

21. $P(x) = 2(100 - x)(x - 50)$
    Optimal price = \$75

22. $V(r) = 20r - \frac{3}{4}\pi r^3$

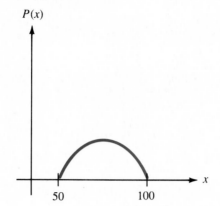

$P(x)$

50   100 x

23. $C(x) = 80x + \dfrac{11,520}{x}$

Optimal number of
machines = 12

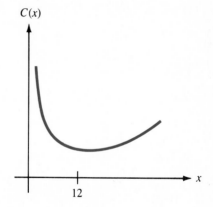

$C(x)$

12   x

24. $f(x) = \begin{cases} 0.21x - 1,150 & \text{if } 15,000 < x \le 18,200 \\ 0.25x - 1,878 & \text{if } 18,200 < x \le 23,500 \\ 0.29x - 2,818 & \text{if } 23,500 < x \le 28,800 \end{cases}$

25. $R(x) = k(n - x)$, where n is the total number of relevant facts in the
    subject's memory.

Chapter 2, Section 1 (page 77)

1. $f'(x) = 5$, $m = 5$

3. $f'(x) = 4x - 3$, $m = -3$

5. $f'(x) = -\dfrac{2}{x^2}$, $m = -8$

7. $f'(x) = \dfrac{1}{2\sqrt{x}}$, $m = \dfrac{1}{6}$

9. $y = 11x + 16$

11. $y = \dfrac{1}{2}x + 2$

13. (a) 3.31   (b) 3

15.   17.

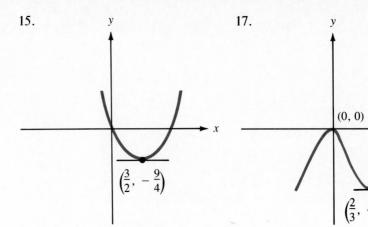

19. The graph of the function is rising for $a \leq x \leq b$.

21. (a) $\dfrac{d}{dx}(x^2) = 2x$, $\dfrac{d}{dx}(x^2 - 3) = 2x$. The graphs are "parallel."

   (b) $\dfrac{d}{dx}(x^2 + 5) = 2x$

23. (a) $\dfrac{d}{dx}(x^2) = 2x$, $\dfrac{d}{dx}(x^3) = 3x^2$

   (b) $\dfrac{d}{dx}(x^4) = 4x^3$, $\dfrac{d}{dx}(x^{27}) = 27x^{26}$

Chapter 2, Section 2 (page 86)

1. $\dfrac{dy}{dx} = 2x + 2$

3. $f'(x) = 9x^8 - 40x^7 + 1$

5. $\dfrac{dy}{dx} = -\dfrac{1}{x^2} - \dfrac{2}{x^3} + \dfrac{1}{2\sqrt{x^3}}$

7. $f'(x) = \dfrac{3}{2}\sqrt{x} - \dfrac{3}{2\sqrt{x^5}}$

9. $\dfrac{dy}{dx} = -\dfrac{1}{8}x - \dfrac{2}{x^2} - \dfrac{3}{2}\sqrt{x} - \dfrac{2}{3x^3} + \dfrac{1}{3}$

11. $f'(x) = 12x - 1$

13. $\dfrac{dy}{dx} = -300x - 20$

15. $f'(x) = \dfrac{1}{3}(5x^4 - 6x^2)$

17. $\dfrac{dy}{dx} = \dfrac{-3}{(x - 2)^2}$

19. $f'(x) = \dfrac{-x^2 - 2}{(x^2 - 2)^2}$

21. $\dfrac{dy}{dx} = \dfrac{-3}{(x + 5)^2}$

23. $f'(x) = \dfrac{11x^2 - 10x - 7}{(2x^2 + 5x - 1)^2}$

25. $\dfrac{dy}{dx} = -24x^2 + 44x + 7$

27. $y = -6x + 6$

29. $y = -\dfrac{1}{16}x + 2$

31. $y = 3x - 3$

33. $y = 6x - 2$

35. (a) $\dfrac{dy}{dx} = \dfrac{-4x + 9}{x^4}$ (b) $\dfrac{dy}{dx} = -3x^{-4}(2x - 3) + 2x^{-3}$

37.

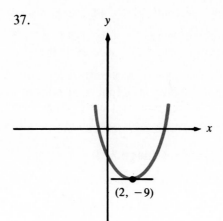

$(2, -9)$

39. $a = \dfrac{8}{9},\ b = -\dfrac{16}{3}$

41. $y = 6x$ and $y = -14x$

43. (a) $E(p) = -200p^2 + 12{,}000p$ (b) \$30 per unit

Chapter 2, Section 3 (page 94)

1. (a) $C'(t) = 200t + 400$
   (b) Increasing at the rate of 1,400 per year.
   (c) 1,500

3. (a) $f'(x) = -3x^2 + 12x + 15$
   (b) 24 radios per hour   (c) 26

5. (a) $P'(t) = \dfrac{6}{(t + 1)^2}$ thousand per year

   (b) 1,500 per year   (c) 1,000   (d) 60 per year
   (e) The rate of growth will approach zero.

7. 100 kilometers per hour

9. (a) \$241   (b) \$244

11. (a) \$248   (b) \$248.05

13. Daily output will increase by approximately 10 units.

15. (a) 20 people per month   (b) 0.39 percent per month

17. (a) $280 per year   (b) 17.95 percent per year

19. (a) $P(x) = \dfrac{100}{12 + x}$ (b) 7.69 percent per year

    (c) The percentage rate will approach zero.

21. (a) After 3 seconds   (b) 96 feet per second

23. (a) 32 feet per second   (b) 128 feet
    (c) 32 feet per second   (d) 96 feet per second

27. (a) Rate of change of cost with respect to output: dollars per unit.
    (b) Rate of change of output with respect to time: units per hour.
    (c) Rate of change of cost with respect to time: dollars per hour.

Chapter 2, Section 4 (page 104)

1. Cost will increase by approximately $50.08.

3. 200   5. 0.05 part per million

7. Daily output will increase by approximately 8 units.

9. 2.16 centimeters per second

11. Accurate to within $8.64\pi \approx 27.14$ cubic inches.

13. 0.2 unit   15. 4.066 percent

17. Area will increase by approximately 2 percent.

19. 2 percent   21. 0.5 percent

23. 2.4 percent

25. $9.03\pi \approx 28.37$ cubic inches

Chapter 2, Section 5 (page 114)

1. $\dfrac{dy}{dx} = 6(3x - 2)$ 3. $\dfrac{dy}{dx} = \dfrac{x + 1}{\sqrt{x^2 + 2x - 3}}$

5. $\dfrac{dy}{dx} = \dfrac{-4x}{(x^2 + 1)^3}$ 7. $\dfrac{dy}{dx} = -x(x^2 - 9)^{-3/2}$

9. $\dfrac{dy}{dx} = \dfrac{-2x}{(x^2 - 1)^2}$ 11. -160

13. $\dfrac{2}{3}$ 15. -16

17. $f'(x) = 8(2x + 1)^3$

19. $f'(x) = 8x^2(x^5 - 4x^3 - 7)^7(5x^2 - 12)$

21. $f'(x) = -\dfrac{10x - 6}{(5x^2 - 6x + 2)^2}$ 23. $f'(x) = \dfrac{-4x}{(4x^2 + 1)^{3/2}}$

25. $f'(x) = \dfrac{24x}{(1 - x^2)^5}$ 27. $f'(x) = \dfrac{15}{2}\dfrac{(1 + \sqrt{3x})^4}{\sqrt{3x}}$

29. $f'(x) = (x + 2)^2(2x - 1)^4(16x + 17)$

31. $f'(x) = \dfrac{-5}{2(3x + 1)^{1/2}(2x - 1)^{3/2}}$

33. $f'(x) = \dfrac{(x + 1)^4(9 - x)}{(1 - x)^5}$ 35. $f'(x) = \dfrac{5 - 6x}{(1 - 4x)^{3/2}}$

37. $y = 594x - 1,161$ 39. $y = -6x + 26$

41. (a) \$2,025 per year   (b) 10.125 percent per year

43. 0.31 part per million per year

45. Decreasing at the rate of 6 pounds per week.

47. 5.09 percent per year

Chapter 2, Section 6 (page 124)

1. $\dfrac{dy}{dx} = -\dfrac{x}{y}$ 3. $\dfrac{dy}{dx} = \dfrac{y - 3x^2}{3y^2 - x}$

5. $\dfrac{dy}{dx} = \dfrac{3 - 2y^2}{2y(1 + 2x)}$ 7. $\dfrac{dy}{dx} = \dfrac{1}{3(2x + y)^2} - 2$

9. $\dfrac{dy}{dx} = \dfrac{2y - 10x(x^2 + 3y^2)^4}{30y(x^2 + 3y^2)^4 - 2x}$

11. $\dfrac{1}{3}$ 13. $-\dfrac{1}{2}$ 15. $\dfrac{13}{12}$ 17. $\dfrac{8}{3}$

19. $\dfrac{dy}{dx} = \dfrac{2x - y}{x + 2} = \dfrac{x(x + 4)}{(x + 2)^2}$ 21. $\dfrac{dy}{dx} = \dfrac{y - 1}{1 - x} = \dfrac{-3}{(x - 1)^2}$

23. Increase input y by approximately 0.57 unit.

Chapter 2, Section 7, (page 131)

1. $f''(x) = 450x^8 - 120x^3$ 3. $\dfrac{d^2y}{dx^2} = \dfrac{-5}{4x^{3/2}} + \dfrac{18}{x^4} + \dfrac{1}{4x^{5/2}}$

5. $f''(x) = 180(3x + 1)^3$

7. $\dfrac{d^2y}{dx^2} = 80(x^2 + 5)^6(3x^2 + 1)$

9. $f''(x) = \dfrac{1}{(1 + x^2)^{3/2}}$

11. $\dfrac{d^2y}{dx^2} = \dfrac{4(3x^2 - 1)}{(1 + x^2)^3}$

13. $f''(x) = 16(2x + 1)^2(5x + 1)$

15. $\dfrac{d^2y}{dx^2} = \dfrac{2(1 - 2x)}{(x + 1)^4}$

17. $\dfrac{d^2y}{dx^2} = \dfrac{15}{4y^3}$

19. $\dfrac{d^2y}{dx^2} = -\dfrac{a}{b^2y^3}$

21. (a) 195 dollars per month
   (b) -16 dollars per month per month
   (c) The rate will decrease by approximately 8 dollars per month.
   (d) The rate will decrease by $8.75 per month.

23. The speed is decreasing at the rate of 6 meters per second per second.

25. (a) $A(t) = \dfrac{20}{3} - \dfrac{4}{3}t$
   (b) The speed is decreasing at the rate of $\dfrac{4}{3}$ kilometers per hour per hour.
   (c) The actual decrease in speed is 2 kilometers per hour.

27. $\dfrac{d^3y}{dx^3} = \dfrac{3}{8x^{5/2}} + \dfrac{3}{x^4}$

Chapter 2, Review Problems (page 134)

1. (a) $f'(x) = 2x - 3$　　(b) $f'(x) = \dfrac{-1}{(x - 2)^2}$

2. (a) $f'(x) = 24x^3 - 21x^2 + 2$

   (b) $f'(x) = 3x^2 + \dfrac{5}{3x^6} + \dfrac{1}{\sqrt{x}} + \dfrac{3}{x^2} - \dfrac{3}{x^4} + \dfrac{4}{x^3}$

   (c) $\dfrac{dy}{dx} = \dfrac{-14x}{(3x^2 + 1)^2}$

   (d) $\dfrac{dy}{dx} = 2(x + 1)(2x + 5)^2(5x + 8)$

   (e) $f'(x) = 20(5x^4 - 3x^2 + 2x + 1)^9(10x^3 - 3x + 1)$

   (f) $f'(x) = \dfrac{x}{\sqrt{x^2 + 1}}$

   (g) $f'(x) = 2\left(x + \dfrac{1}{x}\right)\left(1 - \dfrac{1}{x^2}\right) + \dfrac{15}{2(3x)^{3/2}}$

(h) $\dfrac{dy}{dx} = \dfrac{4(x + 1)}{(1 - x)^3}$

(i) $\dfrac{dy}{dx} = \dfrac{9(3x + 2)}{\sqrt{6x + 5}}$

(j) $f'(x) = \dfrac{3(3x + 1)^2(3x + 7)}{(1 - 3x)^5}$

(k) $f'(x) = \dfrac{-7}{2(1 - 2x)^{1/2}(3x + 2)^{3/2}}$

3. (a) $y = -x + 1$ (b) $y = -x - 1$

   (c) $y = x$ (d) $y = -\dfrac{2}{3}x + \dfrac{5}{3}$

4. (a) 1,652 people per week   (b) 1,514 people

5. (a) Output will increase by approximately 12,000 units.
   (b) Output will increase by 12,050 units.

6. 0.30 percent per month.

7. Output will decrease by approximately 5,000 units.

8. 1.055 percent   9. 10 percent

10. 1.5 percent

11. (a) $\dfrac{dy}{dx} = 3(30x + 11)$ (b) $\dfrac{dy}{dx} = \dfrac{-4}{(2x + 3)^3}$

12. (a) 2   (b) $\dfrac{3}{2}$

13. \$1,663.20 per hour   14. 16.27 percent per year

15. (a) $\dfrac{dy}{dx} = -\dfrac{5}{3}$ (b) $\dfrac{dy}{dx} = -\dfrac{2y}{x}$

   (c) $\dfrac{dy}{dx} = \dfrac{1 - 10(2x + 3y)^4}{15(2x + 3y)^4}$ (d) $\dfrac{dy}{dx} = -\dfrac{1 + 10y^3(1 - 2xy^3)^4}{4 + 30xy^2(1 - 2xy^3)^4}$

16. (a) $-\dfrac{2}{3}$ (b) -28

17. Decrease input y by approximately 0.172 unit.

18. (a) $f''(x) = 120x^3 - 24x + 10 + \dfrac{2}{x^3}$

   (b) $\dfrac{d^2y}{dx^2} = 24(3x^2 + 2)^2(21x^2 + 2)$

   (c) $f''(x) = \dfrac{2(x - 5)}{(x + 1)^4}$

19. $\dfrac{d^2y}{dx^2} = \dfrac{-9}{2y^3}$

20. (a) 27 units per hour   (b) 12 units per hour per hour
    (c) An increase of approximately 1.2 units per hour
    (d) An increase of 1.5 units per hour

21. (a) $\dfrac{d^4y}{dx^4} = 240x + 120 + \dfrac{24}{x^5}$

    (b) $f^{(4)}(x) = \dfrac{15\sqrt{3}}{16x^{7/2}} + \dfrac{180}{x^6}$

Chapter 3, Section 1 (page 149)

1. $f'(x) > 0$ for $-2 < x < 2$; $f'(x) < 0$ for $x < -2$ and $x > 2$.

3. $f'(x) > 0$ for $x < -4$ and $0 < x < 2$; $f'(x) < 0$ for $-4 < x < -2$, $-2 < x < 0$, and $x > 2$.

5.

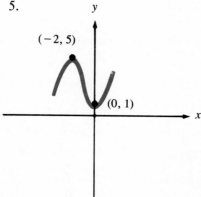

7.

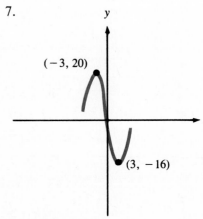

9.

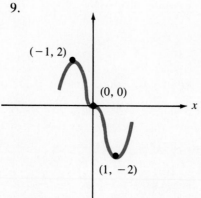

11.

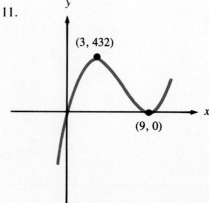

13.

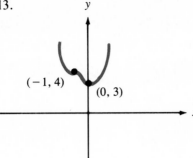

15.

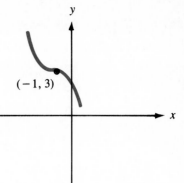

17.

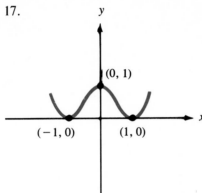

19.

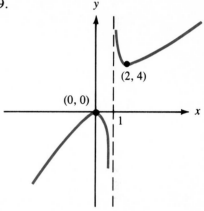

21.

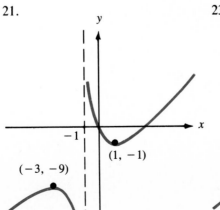

23.

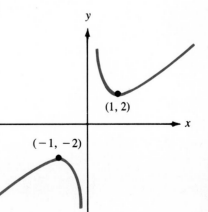

25.

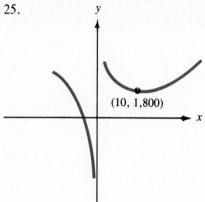

(10, 1,800)

27.

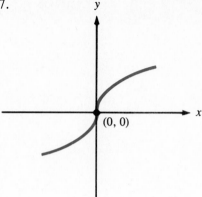

(0, 0)

29.

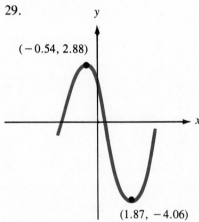

(−0.54, 2.88)

(1.87, −4.06)

31.

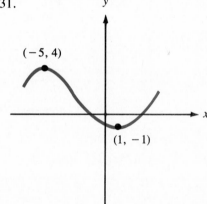

(−5, 4)

(1, −1)

33.

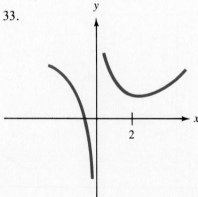

2

35. $a = -\dfrac{9}{25}$, $b = \dfrac{18}{5}$, $c = 3$

Chapter 3, Section 2 (page 160)

1. $f''(x) > 0$ for $x > 2$ and $f''(x) < 0$ for $x < 2$.

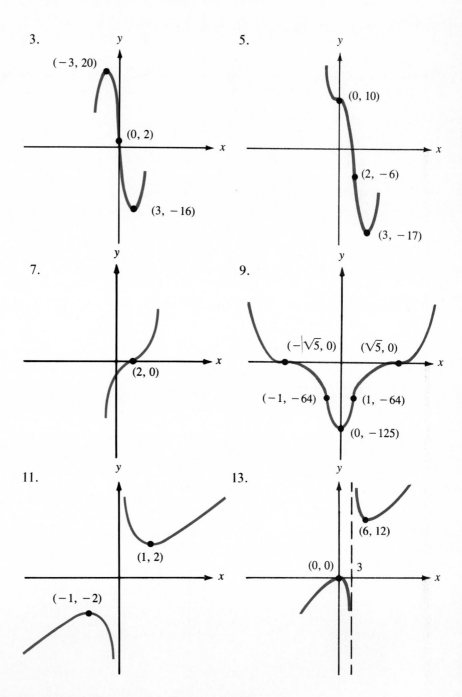

3.

(−3, 20)

(0, 2)

(3, −16)

5.

(0, 10)

(2, −6)

(3, −17)

7.

(2, 0)

9.

$(-\sqrt{5}, 0)$ $(\sqrt{5}, 0)$

(−1, −64)   (1, −64)

(0, −125)

11.

(1, 2)

(−1, −2)

13.

(6, 12)

(0, 0)   3

15.

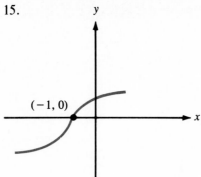

$(-1, 0)$

17.

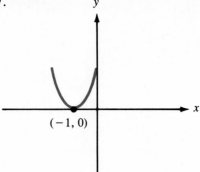

$(-1, 0)$

19.

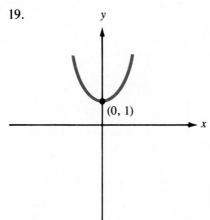

$(0, 1)$

21.

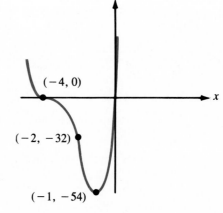

$(-4, 0)$

$(-2, -32)$

$(-1, -54)$

23.

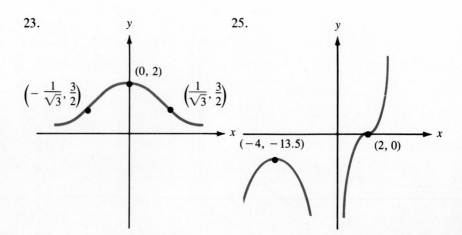

$\left(-\dfrac{1}{\sqrt{3}}, \dfrac{3}{2}\right)$

$(0, 2)$

$\left(\dfrac{1}{\sqrt{3}}, \dfrac{3}{2}\right)$

25.

$(-4, -13.5)$

$(2, 0)$

27.

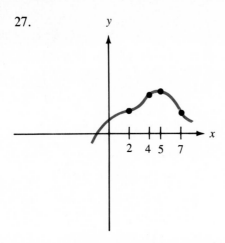

29. (a) Increasing for $x < 0$ and $x > 4$; decreasing for $0 < x < 4$.
    (b) Concave upward for $x > 2$; concave downward for $x < 2$.
    (c) Relative maximum when $x = 0$; relative minimum when $x = 4$; inflection point when $x = 2$.

31. Increasing for $x > 2$; decreasing for $x < 2$; concave upward for all x; relative minimum when $x = 2$.

33. Increasing for $x > 2$; decreasing for $x < 2$; concave upward for $x < -3$ and $x > -1$; concave downward for $-3 < x < -1$; relative minimum when $x = 2$; inflection points when $x = -3$ and $x = -1$.

Chapter 3 Section 3 (page 173)

Absolute Maximum	Absolute Minimum
1. $f(1) = 10$	$f(-2) = 1$
3. $f(0) = 2$	$f(2) = -\dfrac{40}{3}$
5. $f(-1) = 2$	$f(-2) = -56$
7. $f(-3) = 3{,}125$	$f(0) = -1{,}024$
9. $f(3) = \dfrac{10}{3}$	$f(1) = 2$
11. none	$f(1) = 2$
13. none	none
15. $f(0) = 1$	none

17. (a) Membership $= 46{,}400$ in 1974
    (b) Membership $= 12{,}100$ in 1981

19. $12.50 per radio.

21. At the central axis.

27. (a) $A(q) = 3q + 1 + \dfrac{48}{q}$ (b) $q = 4$ (c) $q = 4$

Chapter 3, Section 4 (page 188)

1. $14   3. 80

5. 10 days from now.

11. 2 meters by 2 meters by $\dfrac{4}{3}$ meters.

13. 12 inches by 12 inches by 3 inches.

15. Run the cable entirely under the water.

17. The mathematician.

19. $8 + 5\sqrt{2}$ centimeters by $4 + \dfrac{5\sqrt{2}}{2}$ centimeters

21. Radius = 1 inch, height = 4 inches

23. $h = 2r$

25. (a) 8   (b) $160   (c) $160

27. (a) 10:00 A.M.   (b) 8:00 A.M. and noon

29. Minimal slope $= \dfrac{9}{2}$ at $\left(\dfrac{1}{2}, \dfrac{5}{2}\right)$

31. (a) 3 years from now   (b) Now

33. 17   35. 11:00 A.M.

Chapter 3, Section 5 (page 206)

1. 200 cases   3. 4,000 maps

5. (a) $q = 5$
   (b) $q = 5$

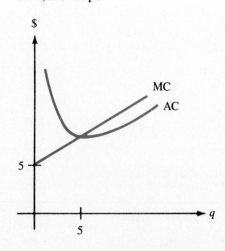

7. (a) $q = 8$

(b) $A(q)$ is increasing for $0 < q < 8$ and decreasing for $q > 8$.

(c) $

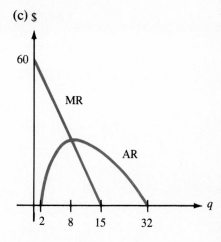

9. (a) $\eta = -\dfrac{0.1p}{60 - 0.1p}$

(b) $\eta = -0.5$

(c) $p = 300$

11. (a) Unit elasticity when $p = 125$; inelastic when $0 \le p < 125$; elastic when $125 < p \le 250$

(b) Increasing for $0 \le p < 125$; decreasing for $125 < p \le 250$. Relative maximum at $p = 125$.

(c) $R(p) = 500p - 2p^2$

(d)

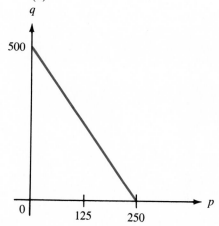

15. (a) $\eta = -\dfrac{30 - q}{q}$

(b) $\eta = -2$

(c) $\eta = -\dfrac{p}{60 - p}$

Chapter 3, Review Problems (page 211)

1. (a)

(2, 15)

(−1, −12)

(b)

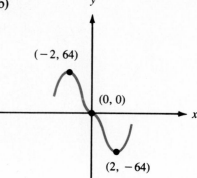

(−2, 64)

(0, 0)

(2, −64)

(c)

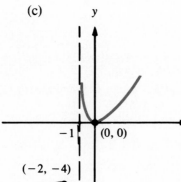

−1

(0, 0)

(−2, −4)

(d)

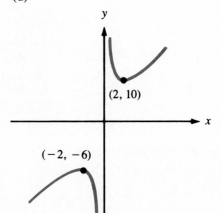

(2, 10)

(−2, −6)

2.

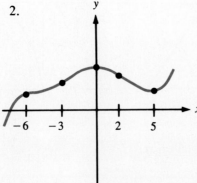

−6   −3   2   5

3. (a)

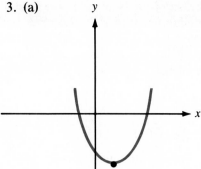

(b)

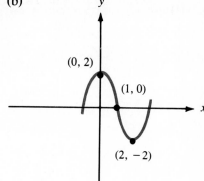

(c)

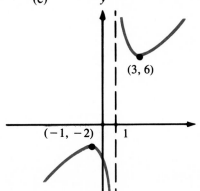

(d)

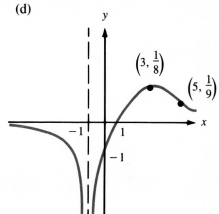

4.	Relative Maxima	Relative Minima
(a)	(2, 15)	(−1, −12)
(b)	(−2, −4)	(0, 0)
(c)	(−2, −6)	(2, 10)

5.   Absolute Maximum   Absolute Minimum

(a) $f(-3) = 40$ $f(-1) = -12$

(b) $f(2) = 6$ $f(3) = -37$

(c) $f\left(-\dfrac{1}{2}\right) = f(1) = \dfrac{1}{2}$ $f(0) = 0$

(d) none $f(2) = 10$

6. The maximum speed is 52 miles per hour at 1:00 P.M. and 7:00 P.M., and the minimum speed is 20 miles per hour at 5:00 P.M.

8. $75 per camera

9. Each plot should be 50 meters by 37.5 meters.

10. $h = \dfrac{3}{2}r$

11. Row all the way to the town.

12. After 2 hours and 20 minutes on the job.

13. 12   14. 77 or 78

15. (a) $D(p)$

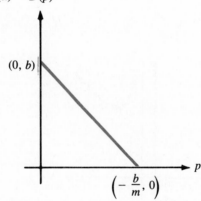

(b) $E(p) = p(mp + b)$

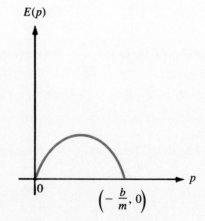

(c) Optimal price $= -\dfrac{b}{2m}$

16. $\sqrt{\dfrac{2bq}{s}}$ units

17. (a) $q = \sqrt{\dfrac{c}{a}}$

18. (a) Demand is elastic when $30 < p \le \sqrt{2{,}700}$; inelastic when $0 \le p < 30$; and of unit elasticity when $p = 30$.
    (b) Increasing for $0 \le p < 30$; decreasing for $30 < p \le \sqrt{2{,}700}$; relative maximum when $p = 30$.
    (c) $R(p) = 27p - 0.01p^3$
    (d)

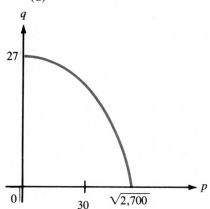

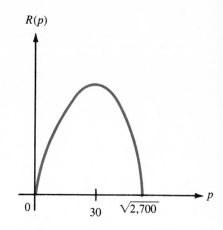

Chapter 4, Section 1 (page 225)

1. $e^2 \simeq 7.389$, $e^{-2} \simeq 0.135$, $e^{0.05} \simeq 1.051$, $e^{-0.05} \simeq 0.951$, $e^0 = 1$, $e \simeq 2.718$, $\sqrt{e} \simeq 1.649$, $\dfrac{1}{\sqrt{e}} \simeq 0.607$

3.

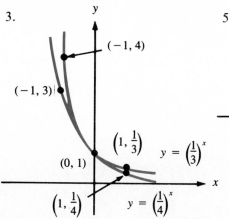

5.

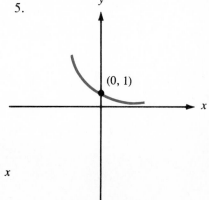

7.

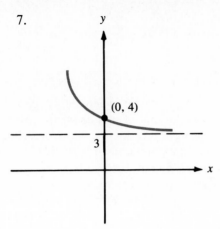

9.

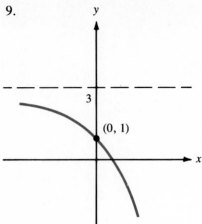

11.

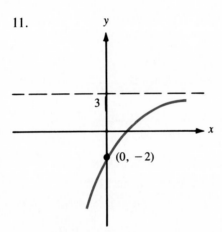

13.

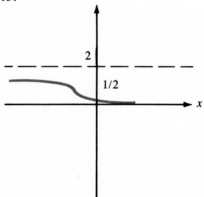

15.

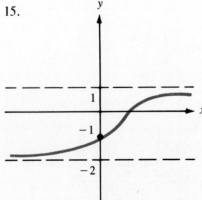

17. 8

19. $\dfrac{110}{3}$

21. (a) $1,967.15   (b) $2,001.60
    (c) $2,009.66   (d) $2,013.75

23. (a) $P = Be^{-rt}$ (b) $5,488.12

25.

n	1,000	10,000	25,000	50,000
$\left(1 + \dfrac{1}{n}\right)^n$	2.71692	2.71815	2.71823	2.71825

Chapter 4, Section 2 (page 232)

1. (a) 50 million   (b) 91.11 million

3. 202.5 million

5. 324 billion dollars

7. (a) 12,000 people per square mile
   (b) 5,959 people per square mile

9. 204.8 grams

11. (a)

(b) 0.7408
(c) 0.0888

13. (a)

(b) The height of the graph approaches A because the number of facts recalled approaches the total number of relevant facts in the person's memory.

15. (a) *V(t)*

(b) $5,200
(c) $1,049.61

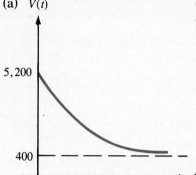

5,200

400

t

17. 18.75° Celsius

19. (a) *P(t)*

(b) 4 million
(c) 9.31 million
(d) The population will approach 10 million.

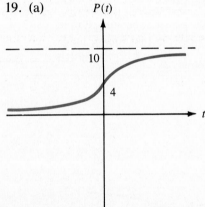

10

4

t

21. 4 hours

Chapter 4, Section 3 (page 243)

1. $\ln 1 = 0$, $\ln 2 \approx 0.693$, $\ln e = 1$, $\ln 5 \approx 1.609$, $\ln \frac{1}{5} \approx -1.609$, $\ln e^2 = 2$; $\ln 0$ and $\ln -2$ are undefined.

3. $\frac{1}{2}$ 5. 9   7. $\frac{19}{6}$ 9. 0.58

11. $e^{-b/2}$ 13. $e^2 \approx 7.39$ 15. $\frac{9}{25}$ 17. $\ln a$

19. e and $\frac{1}{e}$ 21. 5

23. 5.33 percent   25. In the year 2095

27. 5,614.06 years   31. $Q(t) = 6,000e^{0.02t}$

33. $Q(t) = 500 - 200e^{-0.133t}$ 35. 9,082 years old

39. $\log_a x = \dfrac{\ln x}{\ln a}$

Chapter 4, Section 4 (page 256)

1. $f'(x) = 5e^{5x}$

3. $f'(x) = 2(x + 1)e^{x^2 + 2x - 1}$

5. $f'(x) = -0.5e^{-0.05x}$

7. $f'(x) = (6x^2 + 20x + 33)e^{6x}$

9. $f'(x) = (1 - x)e^{-x}$

11. $f'(x) = -6e^x(1 - 3e^x)$

13. $f'(x) = \dfrac{3}{2\sqrt{3x}}\, e^{\sqrt{3x}}$

15. $f'(x) = \dfrac{3}{x}$

17. $f'(x) = \dfrac{2x + 5}{x^2 + 5x - 2}$

19. $f'(x) = 2x \ln x + x$

21. $f'(x) = \dfrac{1}{x^2}(1 - \ln x)$

23. $f'(x) = \dfrac{-2}{(x + 1)(x - 1)}$

25. $f'(x) = 2$

27. $\dfrac{dy}{dx} = \dfrac{y(1 - xy)}{x(xy - 3)}$

29. $\dfrac{dy}{dx} = \dfrac{y(2x^2y^3 + 1)}{x(1 - 3x^2y^3)}$

31. (a) 1.22 million per year
    (b) Constant rate of 2 percent per year

33. (a) \$1,082.68 per year
    (b) Constant rate of 40 percent per year

37. (a) 0.7805 billion per year   (b) 3.375 percent per year

39. (a) Approximately 406 copies   (b) 368 copies

41. \$15 per radio

43.   45.

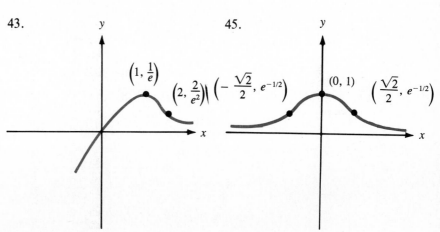

47.

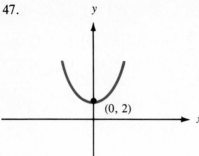

49.

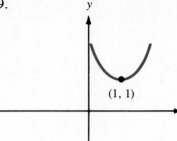

51.

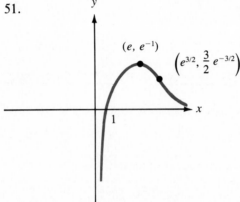

53. $f'(x) = -\dfrac{(3x + 61)(x + 2)^4}{(3x - 5)^7}$ 　　55. $f'(x) = 2^x \ln x$

57. $f'(x) = x^x(1 + \ln x)$ 　　59. $f'(x) = (2^x)^{\ln x} (\ln 2)(1 + \ln x)$

61. 207.94 years from now 　　65. 12.5 percent per year

Chapter 4, Section 5 (page 268)

1. (a) \$10,000.00 　　(b) \$13,425.32 　　(c) \$13,591.41

3. (a) 5.86 years 　　(b) 5.78 years

5. Doubling time $= \dfrac{\ln 2}{k \ln(1 + r/k)}$

7. 18.58 years

9. Tripling time $= \dfrac{\ln 3}{k \ln (1 + r/k)}$

11. (a) 15.39 years   (b) 15.27 years

13. (a) 6.14 percent   (b) 6.18 percent

15. 10.51 percent and 10.74 percent

17. 10 percent   19. 5.83 percent

21. (a) $6,095.65   (b) $6,023.88

23. $6,653.15   25. $209.18

27. $1,732.55

29. 69.44 years from now   31. 6.5 years from now

Chapter 4, Review Problems (page 273)

1. (a)

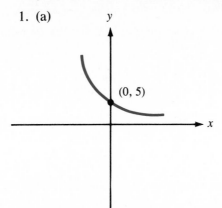

(b)

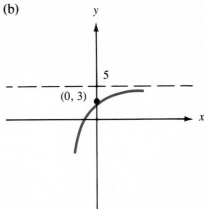

(c)

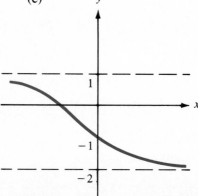

2. 32.5   3. $8,000

4. (a)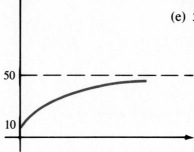

(b) 10,000

(c) 32,027

(d) $9,808

(e) 50,000

5. 60 units

6. (a)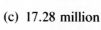

(b) 10 million

(c) 17.28 million

(d) The population will approach 30 million.

7. (a) 5   (b) 2   (c) 32

8. (a) 34.66   (b) 0   (c) $e^2 \approx 7.39$ (d) 1.86

9. 14.75 minutes

10. (a) $f'(x) = 6e^{3x+5}$ (b) $f'(x) = x(2 - x)e^{-x}$

(c) $f'(x) = \dfrac{x + 2}{x^2 + 4x + 1}$ (d) $f'(x) = 2(1 + \ln x)$

(e) $f'(x) = \dfrac{\ln 2x - 1}{(\ln 2x)^2}$

11. (a) $\dfrac{dy}{dx} = \dfrac{y(x + y)}{x(y - 2x)}$ (b) $\dfrac{dy}{dx} = \dfrac{y(1 - 3x^3y^2)}{x(2x^3y^2 + 1)}$

12. (a) 0.13 part per million per year
    (b) Constant rate of 3 percent per year

13. $140 per camera

14. (a)

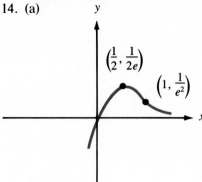

(b)

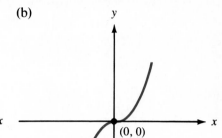

(c)

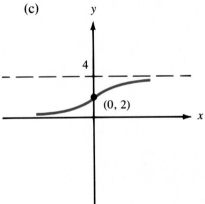

(d)

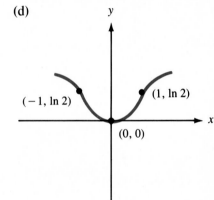

15. (a) $f'(x) = \dfrac{x(2x^2 + 3)}{(x^2 + 1)^{1/2}(x^2 + 2)^{1/2}}$

(b) $f'(x) = x^{x^2+1}(2 \ln x + 1)$

(c) $f'(x) = (x^x)(2^x)(1 + \ln 2x)$

16. (a) 11.57 years   (b) 11.45 years

17. 8.20 percent compounded continuously

18. (a) $1,080.81   (b) $1,070.52

19. $7,433.55

20. 51.02 years from now

Chapter 5, Section 1 (page 284)

1. $\frac{1}{6}x^6 + C$

3. $-\frac{1}{x} + C$

5. $5x + C$

7. $x^3 - \frac{5}{2}x^2 + 2x + C$

9. $2x^{3/2} + \frac{1}{x^2} + \ln|x| + C$

11. $2e^x + \ln x^6 + x \ln 2 + C$

13. $\frac{2}{5}x^{5/2} - \sqrt{x} + x\sqrt{2} + C$

15. $x + \ln x^2 - \frac{1}{x} + C$

17. $\frac{4}{45}x^5 - \frac{20}{27}x^3 + \frac{25}{9}x + C$

19. $\frac{2}{7}x^{7/2} - \frac{2}{3}x^{3/2} + C$

21. 10,128   23. \$1,000   25. \$436   27. \$2,300

29. $y = 2x^2 + x - 1$

31. $y = \frac{1}{4}x^4 + \frac{2}{x} + 2x - \frac{5}{4}$

33. $\frac{1}{3}e^{3x} + C$

35. $\frac{1}{12}(2x + 3)^6 + C$

37. $\frac{1}{2}\ln|2x + 1| + C$

Chapter 5, Section 2 (page 293)

1. $\frac{1}{12}(2x + 6)^6 + C$

3. $\frac{1}{6}(4x - 1)^{3/2} + C$

5. $-e^{1-x} + C$

7. $\frac{1}{2}e^{x^2} + C$

9. $\frac{1}{12}(x^2 + 1)^6 + C$

11. $\frac{4}{21}(x^3 + 1)^{7/4} + C$

13. $\frac{2}{5}\ln|x^5 + 1| + C$

15. $\frac{1}{26}(x^2 + 2x + 5)^{13} + C$

17. $\frac{3}{5}\ln|x^5 + 5x^4 + 10x + 12| + C$ 19. $-\frac{3}{2}\left(\frac{1}{x^2 - 2x + 6}\right) + C$

21. $\frac{1}{2}(\ln 5x)^2 + C$

23. $-\frac{1}{\ln x} + C$

25. $\frac{1}{2}[\ln(x^2 + 1)]^2 + C$

27. $x + \ln|x - 1| + C$

29. $-\frac{1}{4}(x - 5)^{-4} - (x - 5)^{-5} + C$

31. $\ln |x - 4| - \dfrac{7}{x - 4} + C$

33. $\dfrac{1}{5}(2x - 1)^{5/2} + \dfrac{4}{3}(2x - 1)^{3/2} + C$

35. $y = -\dfrac{1}{3} \ln |1 - 3x^2| + 5$

37. $849.61

39. $510.56 per acre

Chapter 5, Section 3 (page 299)

1. $-(x + 1)e^{-x} + C$

3. $-5(x + 5)e^{-x/5} + C$

5. $(2 - x)e^x + C$

7. $\dfrac{1}{2}x^2\left(\ln 2x - \dfrac{1}{2}\right) + C$

9. $\dfrac{2}{3}x(x - 6)^{3/2} - \dfrac{4}{15}(x - 6)^{5/2} + C$

11. $\dfrac{1}{9}x(x + 1)^9 - \dfrac{1}{90}(x + 1)^{10} + C$

13. $2x(x + 2)^{1/2} - \dfrac{4}{3}(x + 2)^{3/2} + C$

15. $-(x^2 + 2x + 2)e^{-x} + C$

17. $(x^3 - 3x^2 + 6x - 6)e^x + C$

19. $\dfrac{1}{3}x^3 \ln x - \dfrac{1}{9}x^3 + C$

21. $-\dfrac{1}{x}(\ln x + 1) + C$

23. $\dfrac{1}{2}(x^2 - 1)e^{x^2} + C$

25. $\dfrac{1}{36}x^4(x^4 + 5)^9 - \dfrac{1}{360}(x^4 + 5)^{10} + C$

27. 176.87

29. (b) $\left(\dfrac{1}{5}x^3 - \dfrac{3}{25}x^2 + \dfrac{6}{125}x - \dfrac{6}{625}\right)e^{5x} + C$

Chapter 5, Section 4 (page 304)

1. $-\dfrac{1}{3} \ln \left|\dfrac{x}{2x - 3}\right| + C$

3. $\ln |x + \sqrt{x^2 + 25}| + C$

5. $\dfrac{1}{4} \ln \left|\dfrac{2 + x}{2 - x}\right| + C$

7. $\dfrac{1}{2} \ln \left|\dfrac{x}{3x + 2}\right| + C$

9. $\left(\dfrac{1}{3}x^2 - \dfrac{2}{9}x + \dfrac{2}{27}\right)e^{3x} + C$

11. $-\dfrac{1}{2} \ln |2 - x^2| + C$

13. $x(\ln 2x)^2 - 2x \ln 2x + 2x + C$ 15. $\dfrac{1}{3\sqrt{5}} \ln \left| \dfrac{\sqrt{2x + 5} - \sqrt{5}}{\sqrt{2x + 5} + \sqrt{5}} \right| + C$

17. $\dfrac{1}{2}x + \dfrac{1}{2} \ln |2 - 3e^{-x}| + C$

Chapter 5, Review Problems (page 306)

1. $\dfrac{1}{6}x^6 - x^3 - \dfrac{1}{x} + C$

2. $\dfrac{3}{5}x^{5/3} - \ln |x| + 5x + \dfrac{2}{3}x^{3/2} + C$

3. $\dfrac{2}{9}(3x + 1)^{3/2} + C$ 4. $\dfrac{1}{3}(3x^2 + 2x + 5)^{3/2} + C$

5. $\dfrac{1}{12}(x^2 + 4x + 2)^6 + C$ 6. $\dfrac{1}{2} \ln |x^2 + 4x + 2| + C$

7. $-\dfrac{3}{4}\left(\dfrac{1}{2x^2 + 8x + 3}\right) + C$ 8. $\dfrac{1}{13}(x - 5)^{13} + C$

9. $\dfrac{1}{14}(x - 5)^{14} + \dfrac{5}{13}(x - 5)^{13} + C = \dfrac{x}{13}(x - 5)^{13} - \dfrac{1}{182}(x - 5)^{14} + C$

10. $\dfrac{5}{3}e^{3x} + C$ 11. $\left(\dfrac{5}{3}x - \dfrac{5}{9}\right)e^{3x} + C$

12. $-2(x + 2)e^{-x/2} + C$ 13. $\dfrac{1}{3}(x^3 - 1)e^{x^3} + C$

14. $10(2x - 19)e^{0.1x} + C$ 15. $\dfrac{1}{2}x^2 \ln 3x - \dfrac{1}{4}x^2 + C$

16. $x \ln 3x - x + C$ 17. $\dfrac{1}{2} (\ln 3x)^2 + C$

18. $-\dfrac{1}{x}(\ln 3x + 1) + C$

19. $\dfrac{1}{18}x^2(x^2 + 1)^9 - \dfrac{1}{180}(x^2 + 1)^{10} + C$

20. $(x^2 + 1)[\ln(x^2 + 1) - 1] + C$ 21. $y = \dfrac{1}{8}(x^2 + 1)^4 + 3$

22. 11,250   23. 10,945   24. \$2,265.80

25. $\dfrac{5}{8} \ln \left| \dfrac{2 + x}{2 - x} \right| + C$ 26. $\dfrac{2}{3} \ln \left| x + \sqrt{x^2 + \dfrac{16}{9}} \right| + C$

27. $-2(x^2 + 4x + 8)e^{-x/2} + C$

Chapter 6, Section 1 (page 313)

1. $\dfrac{9}{20}$ 3. 144   5. $\dfrac{8}{3} + \ln 3$ 7. $\dfrac{2}{9}$

9. $-\dfrac{16}{3}$ 11. $\dfrac{4}{3}$ 13. $\dfrac{7}{6}$ 15. e 17. $e^2 + 1$

19. $-3e^{-2} - e^2$ 21. $\dfrac{8}{3}$ 23. 152.85   25. 98 people

27. \$1,870   29. \$774   31. \$75   33. \$4,081,077.40

35. (b) 1   (c) $\dfrac{70}{3}$

Chapter 6, Section 2 (page 325)

1. $\dfrac{8}{3}$ 3. 15   5. $\dfrac{38}{3}$ 7. $\dfrac{4}{3}$ 9. $\dfrac{1}{2}$ 11. $\dfrac{7}{6}$ 13. $\dfrac{128}{3}$

15. (a) 0.0577   (b) 0.4512   (c) 0.5488

17. (a) 0.6321   (b) 0.3012

19. 33   21. $\dfrac{1}{6}$ 23. $\dfrac{1}{3}$ 25. $\dfrac{625}{12}$ 27. $\dfrac{3}{4}$

Chapter 6, Section 3 (page 335)

1. (a) 12 years   (b) \$1,008

3. (a) 9 years   (b) \$12,150

5. (a) 10 weeks   (b) \$14,857

7. \$1,000   9. \$314.78

11. \$1,920   13. \$284.30

Chapter 6, Review Problems (page 338)

1. 0   2. $\dfrac{17}{3}$ 3. 1,710   4. $1 - e^{-1}$ 5. $\dfrac{65}{8}$

6. $\dfrac{3}{5}$ 7. $2e^{-1}$ 8. $\dfrac{1}{2}$ 9. $\dfrac{1}{9}(2e^3 + 1)$

10. $55e^2 + 45$ 11. 126 people   12. \$76.80

13. 259.49 billion barrels per year

14. 36   15. $\dfrac{3}{4}$ 16. $\dfrac{9}{2}$ 17. $\dfrac{16}{3} + 8\ln 2$ 18. $\dfrac{3}{10}$ 19. $\dfrac{13}{2}$

20. (a) 22.10 percent   (b) 55.07 percent   (c) 44.93 percent

21. (a) 15   (b) $33,750

22. (a) 3 units   (b) $72   (c) $27

Chapter 7, Section 1 (page 349)

1. 30 meters

3. $\int_0^5 r(t)\ dt$

5. $7,040,000

7. $\int_0^{12} n(x)p(x)\ dx$

9. $75   11. $480   13. 7π 15. $\dfrac{1,532\pi}{15}$

17. $\dfrac{32\pi}{3}$ 19. 2π 21. $\dfrac{4}{3}\pi r^3$

Chapter 7, Section 2 (page 361)

1. 2   3. $\dfrac{4}{3}$ 5. 18.7° Celsius

7. 492.83 letters per hour

9. (a) $\dfrac{1}{N}\int_0^N S(t)\ dt$ (b) $\int_0^N S(t)\ dt$

11. $13,994.35   13. $27,124.92

15. $10,367.27   17. $4,511.88

19. The spy should take the 35,000 pounds. The present value of the pension is only 31,606 pounds.

21. 4,207

23. $P_0 f(N) + \int_0^N r(t)f(N - t)\ dt$

25. $\int_0^R 2\pi r S(r)\ dr$ 27. 116,039   29. 19,567 pounds

Chapter 7, Section 3 (page 371)

1. $\dfrac{1}{2}$ 3. ∞ 5. ∞ 7. $\dfrac{1}{10}$ 9. $\dfrac{5}{2}$

11. $\dfrac{1}{9}$ 13. ∞ 15. $\dfrac{2}{e}$ 17. $\dfrac{2}{9}$ 19. $5e^{10}$

21. ∞ 23. 2   25. $20,000   27. $150,000

31. 200   33. 50

Chapter 7, Section 4 (page 385)

1. (a) 1 (b) $\frac{1}{3}$ (c) $\frac{1}{3}$

3. (a) 1 (b) $\frac{3}{16}$ (c) $\frac{9}{16}$

5. (a) 1 (b) $\frac{7}{8}$ (c) $\frac{1}{8}$

7. (a) 1 (b) 0.3496 (c) 0.9817

9. $\frac{1}{3}$ 11. $\frac{1}{6}$ 13. 0.1353

15. $E(x) = \frac{7}{2}$; $Var(x) = \frac{3}{4}$

17. $E(x) = \frac{4}{3}$; $Var(x) = \frac{8}{9}$

19. $E(x) = 10$; $Var(x) = 100$

21. 10 minutes 23. 4 minutes

Chapter 7, Section 5 (page 398)

1. (a) 2.3438 (b) 2.3333

3. (a) 0.7828 (b) 0.7854

5. (a) 1.1515 (b) 1.1478

7. (a) 0.7430 (b) 0.7469

9. (a) 0.5090; $|E_4| \le 0.0313$ (b) 0.5004; $|E_4| \le 0.0026$

11. (a) 2.7967; $|E_{10}| \le 0.0017$ (b) 2.7974; $|E_{10}| \le 0.00002$

13. (a) 1.4907; $|E_4| \le 0.0849$ (b) 1.4637; $|E_4| \le 0.0045$

15. 0.2881

17. (a) $n = 164$ (b) $n = 18$

19. (a) $n = 36$ (b) $n = 6$

21. (a) $n = 179$ (b) $n = 8$

Chapter 7, Review Problems (page 400)

1. $20,000

2. $\int_0^{12} D(x)P(x)\,dx$

3. $\frac{234\pi}{15}$

4. 6.32π

5. $1.32 per pound   6. $7,377.37

7. $7,191.64   8. 62

9. 73,186

10. ∞ 11. 1   12. ∞ 13. $\dfrac{3}{5}$ 14. $\dfrac{1}{4}$ 15. $\dfrac{2}{3}$

16. $\dfrac{1}{4}$ 17. $\dfrac{1}{\ln 2}$ 18. $\dfrac{1}{3}$

19. 10,000   20. $120,000

21. The population will increase without bound.

22. (a) 1   (b) $\dfrac{1}{3}$ (c) $\dfrac{1}{3}$

23. (a) 1   (b) $\dfrac{1}{3}$

24. (a) 1   (b) 0.3694   (c) 0.3679

25. $E(x) = \dfrac{5}{2}$; $\text{Var}(x) = \dfrac{3}{4}$

26. $E(x) = 1$; $\text{Var}(x) = \dfrac{1}{2}$

27. $E(x) = 5$; $\text{Var}(x) = 25$

28. $\dfrac{2}{9}$

29. (a) 0.0498   (b) 2 minutes

30. (a) 1.1016; $|E_{10}| \leq 0.0133$ (b) 1.0987; $|E_{10}| \leq 0.0004$

31. (a) 17.5651; $|E_8| \leq 10.2372$ (b) 16.5386; $|E_8| \leq 1.0901$

32. (a) $n = 58$ (b) $n = 8$

33. (a) $n = 59$ (b) $n = 6$

Chapter 8, Section 1 (page 410)

1. $\dfrac{dQ}{dt} = kQ$ 3. $\dfrac{dQ}{dt} = 0.07Q$ 5. $\dfrac{dP}{dt} = 500$

7. $\dfrac{dQ}{dt} = k(M - Q)$, where M is the temperature of the surrounding medium and k is a positive constant.

9. $\dfrac{dQ}{dt} = k(N - Q)$, where N is the total number of relevant facts in the person's memory.

11. $\dfrac{dQ}{dt} = kQ(N - Q)$, where N is the total number of people involved.

17. $y = x^3 + \dfrac{5}{2}x^2 - 6x + C$ 19. $V = \ln (x + 1)^2 + C$

21. $P = 25t^2 + C_1t + C_2$ 23. $y = \dfrac{1}{5}e^{5x} + \dfrac{4}{5}$

25. $V = 2(t^2 + 1)^4 - 1$ 27. $A = 4e^{-t/2} + 3t - 2$

29. (a) $V(t) = 4,800e^{-t/5} - 4,800 + V_0$ (b) \$1,049.61

31. (a) 30,000   (b) 45,000

33. (a) $Q(t) = 0.03(36 + 16t - t^2)^{1/2} + 0.07$
    (b) 0.37 part per million at 3:00 P.M.

35. \$75

37. (a) $D(t) = -14t^2 + S_0t$ (b) 138.29 feet

Chapter 8, Section 2 (page 421)

1. $y = Ce^{3x}$ 3. $y = -\ln (C - x)$

5. $y = \pm\sqrt{x^2 + C}$ 7. $y = Ce^x - 10$

9. $y = \dfrac{1}{2 + Be^{-x}}$ 11. $y = \dfrac{2}{3 - 2x^4}$

13. $y = 8 - 2e^{-5x}$ 15. $Q(t) = 1,000e^{0.07t}$

19. $Q(t) = B(1 - e^{-kt})$, where B is the total number of relevant facts in the person's memory.

21. $Q(t) = 80 - 40e^{-0.014t}$

23. $Q(t) = S - (S - Q_0)e^{-kAt}$, where S is the concentration of the solute outside the cell and A is the area of the cell wall.

25. $Q(t) = 400 + 200e^{-t/40}$

27. $Q(t) = \dfrac{2,000}{1 + 3e^{-0.8t}}$

Chapter 8, Review Problems (page 426)

1. $y = \dfrac{1}{4}x^4 - x^3 + 5x + C$ 2. $y = Ce^{0.02x}$

3. $y = 80 - Ce^{-kx}$ 4. $y = \dfrac{1}{1 + Ce^{-x}}$

5. $y = x^5 - x^3 - 2x + 6$ 6. $y = 100e^{0.06x}$

7. $y = 3 - e^{-x}$ 8. $y = x^2 + 3x + 5$

9. $22,857   10. $549,504

11. 18.75 pounds   12. $Q(t) = \dfrac{10}{1 + 1.5e^{-0.06t}}$

Chapter 9, Section 1 (page 432)

1. The domain of f consists of all ordered pairs (x, y) of real numbers; $f(2, -1) = -3$ and $f(1, 2) = 16$.

3. The domain of f consists of all ordered pairs (x, y) of real numbers for which $|y| \geq |x|$; $f(4, 5) = 3$ and $f(-1, 2) = \sqrt{3} \approx 1.732$.

5. The domain of f consists of all ordered pairs (x, y) of real numbers for which $y > 0$ and $y \neq 1$; $f(-1, e^3) = -\dfrac{1}{3}$ and $f(\ln 9, e^2) = \ln 3 \approx 1.099$.

7. (a) $R(x_1, x_2) = 200x_1 - 10x_1^2 + 25x_1x_2 + 100x_2 - 10x_2^2$
   (b) $R(6, 5) = \$1,840$

9. (a) 160,000
   (b) The level will increase by 16,400 units.
   (c) The level will increase by 4,000 units.
   (d) The level will increase by 20,810 units.

Chapter 9, Section 2 (page 440)

1. $f_x = 2y^5 + 6xy + 2x; f_y = 10xy^4 + 3x^2$

3. $\dfrac{\partial z}{\partial x} = 15(3x + 2y)^4; \dfrac{\partial z}{\partial y} = 10(3x + 2y)^4$

5. $f_x = \dfrac{-3y}{2x^2}; f_y = \dfrac{3}{2x}$

7. $\dfrac{\partial z}{\partial x} = (xy + 1)e^{xy}; \dfrac{\partial z}{\partial y} = x^2e^{xy}$

9. $f_x = \dfrac{5y}{(y - x)^2}; f_y = \dfrac{-5x}{(y - x)^2}$

11. $\dfrac{\partial z}{\partial x} = \ln y; \dfrac{\partial z}{\partial y} = \dfrac{x}{y}$

13. Daily output will increase by approximately 10 units.

15. 12 cents

17. (a) Butter and margarine

   (b) $\dfrac{\partial Q_1}{\partial p_2} \geq 0$ and $\dfrac{\partial Q_2}{\partial p_1} \geq 0$ (c) No

19. $f_{xx} = 60x^2y^3$, $f_{yy} = 30x^4y$, $f_{xy} = f_{yx} = 60x^3y^2 + 2$

21. $f_{xx} = 2y(1 + 2x^2y)e^{x^2y}$, $f_{yy} = x^4e^{x^2y}$, $f_{xy} = f_{yx} = 2x(1 + x^2y)e^{x^2y}$

23. $f_{xx} = \dfrac{y^2}{(x^2 + y^2)^{3/2}}$, $f_{yy} = \dfrac{x^2}{(x^2 + y^2)^{3/2}}$, $f_{xy} = f_{yx} = \dfrac{-xy}{(x^2 + y^2)^{3/2}}$

25. $\dfrac{\partial^2 Q}{\partial K^2} \simeq$ change in marginal product of capital generated by an increase in capital investment of \$1,000.

27. (a) $\dfrac{\partial^2 Q}{\partial L^2} > 0$ for $L < L_0$ and $\dfrac{\partial^2 Q}{\partial L^2} < 0$ for $L > L_0$

Chapter 9, Section 3 (page 447)

1. $\dfrac{dz}{dt} = 7$

3. $\dfrac{dz}{dt} = \dfrac{1}{3}$

5. $\dfrac{dz}{dt} = -3t^{-4}$

7. $\dfrac{dz}{dt} = -162t^5$

9. 23   11. $-\dfrac{1}{3}$ 13. 5

15. The monthly demand will be increasing at the rate of 7 per month.

17. Approximately 61.6 additional units will be produced.

19. Daily profit will increase by approximately 24 cents.

21. 112 square yards   23. 0.5 percent   25. 1 percent

27. $\dfrac{d^2z}{dt^2} = a^2f_{xx} + 2abf_{xy} + b^2f_{yy}$

Chapter 9, Section 4 (page 458)

1.

3.

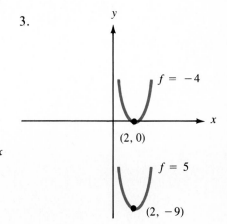

5.   7.

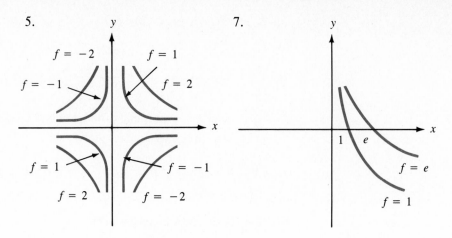

9. $\dfrac{y}{x}$ 11. $\dfrac{3 - 2xy}{x^2 + 6y^2 - 2}$ 13. $-\dfrac{y}{x}\ln y$ 15. $-\dfrac{1}{6}$ 17. 2

19. The level of unskilled labor should be reduced by approximately 2.4 hours.

21. Approximately 125 worker-hours

23. 1.2 units

Chapter 9, Section 5 (page 467)

	Relative Maxima	Relative Minima	Saddle Points
1.	(0, 0)	none	none
3.	none	none	(0, 0)
5.	(−2, −1)	(1, 1)	(−2, 1) and (1, −1)
7.	none	$\left(4, \dfrac{19}{2}\right)$	$\left(2, \dfrac{7}{2}\right)$

9. Price the first system at \$3,000 and the second at \$4,500.

11. $x = 200$ and $y = 300$

13. (a) Relative minimum   (b) Saddle point

Chapter 9, Section 6 (page 480)

1. $f\left(\dfrac{1}{2}, \dfrac{1}{2}\right) = \dfrac{1}{4}$ 3. $f(1, 1) = f(-1, -1) = 2$

5. $f(0, 2) = f(0, -2) = -4$

7. 40 meters by 80 meters

9. 3,456 cubic inches

11. Radius = 1 inch, height = 4 inches

13. $40,000 on labor; $80,000 on equipment.

15. Maximum output would increase by approximately 31.75 units.

17. (a) $4,000 on development and $6,000 on promotion
(b) $\lambda = 0$

Chapter 9, Section 7 (page 488)

1. $y = \frac{1}{4}x + \frac{3}{2}$

3. $y = 3$

5. $y = 0.78x + 1.06$

7. $y = -\frac{1}{2}x + 4$

9. (b) $y = 0.42x - 0.71$ (c) 1,306

11. (b) $y = 3.05x + 6.10$ (c) 42.7 percent

Chapter 9, Review Problems (page 491)

1. (a) $f_x = 6x^2y + 3y^2 - \dfrac{y}{x^2}, f_y = 2x^3 + 6xy + \dfrac{1}{x}$

(b) $f_x = 5y^2(xy^2 + 1)^4, f_y = 10xy(xy^2 + 1)^4$

(c) $f_x = y(1 + xy)e^{xy}, f_y = x(1 + xy)e^{xy}$

2. (a) $f_{xx} = 2, f_{yy} = 6y - 4x, f_{xy} = f_{yx} = -4y$

(b) $f_{xx} = (2 + 4x^2)e^{x^2+y^2}, f_{yy} = (2 + 4y^2)e^{x^2+y^2}$

$f_{xy} = f_{yx} = 4xye^{x^2+y^2}$

(c) $f_{xx} = 0, f_{yy} = \dfrac{-x}{y^2}, f_{xy} = f_{yx} = \dfrac{1}{y}$

3. Daily output will increase by approximately 16 units.

4. $\dfrac{\partial^2 Q}{\partial K \partial L} > 0$

5. (a) $\dfrac{dz}{dt} = -30t^4 + 24t^2$ (b) $\dfrac{dz}{dt} = 4t$

6. The demand will drop by approximately 46 cans per week.

7. 1.33 percent

8. Decreasing at the rate of approximately 18 pies per month.

9. (a) 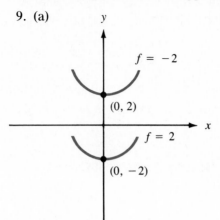 (b)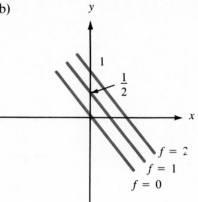

10. (a) $\dfrac{2}{3}$ (b) $-\dfrac{1}{2}$

11. The level of unskilled labor should be decreased by approximately 2 workers.

12. (a) Relative maximum at $(-2, 3)$; relative minimum at $(0, 9)$; saddle points at $(0, 3)$ and $(-2, 9)$.

 (b) Relative minimum at $\left(-\dfrac{23}{2}, 5\right)$; saddle point at $\left(\dfrac{1}{2}, 1\right)$.

13. Maximum value $= f(1, \sqrt{3}) = f(1, -\sqrt{3}) = 12$; minimum value $= f(-2, 0) = 3$.

15. $5,000 on development and $8,000 on promotion.

16. $4,000 on development and $7,000 on promotion.

17. Maximum profit will increase by approximately $235.

18. $y = \dfrac{4}{9}x + 1$

19. (b) $y = 11.54x + 44.45$ (c) $102,150

Chapter 10, Section 1 (page 497)

1. $\dfrac{7}{6}$ 3. -1 5. $4 \ln 2$ 7. 32

9. $\dfrac{1}{3}$ 11. $\dfrac{3}{2}e^4 + \dfrac{1}{2}$ 13. $\dfrac{1}{3} \ln 2$

15. $\dfrac{1}{2}$ 17. $-\dfrac{1}{4}e^2 + e - \dfrac{1}{4}$

Chapter 10, Section 2 (page 506)

1. $0 \le x \le 3$ and $x^2 \le y \le 3x$; $0 \le y \le 9$ and $\dfrac{y}{3} \le x \le \sqrt{y}$

3. $-1 \le x \le 2$ and $1 \le y \le 2$; $1 \le y \le 2$ and $-1 \le x \le 2$

5. $1 \le x \le e$ and $0 \le y \le \ln x$; $0 \le y \le 1$ and $e^y \le x \le e$

7. $\dfrac{3}{2}$ 9. $\dfrac{1}{2}$ 11. $\dfrac{44}{15}$ 13. $32 \ln 2 - 12$

15. 3   17. 1   19. $\dfrac{3}{2} \ln 5$ 21. $2e - 4$

23. $\displaystyle\int_0^4 \int_0^{\sqrt{4-y}} f(x, y)\, dx\, dy$

25. $\displaystyle\int_0^1 \int_{y^2}^{\sqrt[3]{y}} f(x, y)\, dx\, dy$

27. $\displaystyle\int_0^2 \int_1^{e^y} f(x, y)\, dx\, dy$

29. $\displaystyle\int_1^2 \int_{-\sqrt{y-1}}^{\sqrt{y-1}} f(x, y)\, dx\, dy$

31. $\displaystyle\int_0^1 \int_0^y f(x, y)\, dx\, dy + \int_1^2 \int_0^{2-y} f(x, y)\, dx\, dy$

33. $\displaystyle\int_0^4 \int_{-\sqrt{y}}^{\sqrt{y}} f(x, y)\, dx\, dy + \int_4^9 \int_{-\sqrt{y}}^{6-y} f(x, y)\, dx\, dy$

Chapter 10, Section 3 (page 516)

1. 18   3. $\dfrac{16}{3}$ 5. $\dfrac{4}{3}$ 7. 1   9. $\dfrac{19}{6}$

11. 8   13. $e - 1$ 15. $\dfrac{1}{2}e^2 - e + \dfrac{1}{2}$

17. $\dfrac{\displaystyle\iint_R f(x, y)\, dA}{\displaystyle\iint_R 1\, dA}$

19. \$188.61 per acre

21. 2   23. 0   25. $3(e - 2)$

27. 0.2285   29. 0.0803

31. 0.8452   33. 0.5269

Chapter 10, Review Problems (page 520)

1. -4 2. $\dfrac{1}{9}$ 3. $\dfrac{4}{5}\ln 2$ 4. $(e^{-2}-1)(e^{-1}-1) \simeq 0.5466$

5. $\dfrac{1}{8}(e^2-1)$ 6. 81 7. $\dfrac{9}{2}$ 8. 256

9. $e-2$ 10. $1{,}026$

11. $\displaystyle\int_0^2 \int_{y^2}^{\sqrt{8y}} f(x,\,y)\ dx\ dy$

12. $\displaystyle\int_1^{e^2} \int_{\ln x}^{2} f(x,\,y)\ dy\ dx$

13. $\displaystyle\int_0^4 \int_0^{\sqrt{y}} f(x,\,y)\ dx\ dy + \int_4^8 \int_0^{4-y/2} f(x,\,y)\ dx\ dy$

14. $\displaystyle\int_0^1 \int_{-\sqrt{y}}^{\sqrt{y}} f(x,\,y)\ dx\ dy + \int_1^5 \int_{-\sqrt{y}}^{1} f(x,\,y)\ dx\ dy$

$\displaystyle + \int_5^9 \int_{-\sqrt{y}}^{6-y} f(x,\,y)\ dx\ dy$

15. $\dfrac{9}{2}$ 16. $e-1$ 17. $\dfrac{5}{6}$ 18. $\dfrac{1}{3}$ 19. $\dfrac{1}{2}(3-e)$ 20. $\dfrac{1}{2}$

21. (a) $\displaystyle\iint_R f(x,\,y)\ dA$ (b) $\dfrac{\displaystyle\iint_R f(x,\,y)\ dA}{\displaystyle\iint_R 1\ dA}$

22. (a) 0.8625 (b) 0.9500

23. 0.4582

Chapter 11, Section 1 (page 531)

1. $\displaystyle\sum_{n=1}^{\infty} \dfrac{1}{3^n}$ 3. $\displaystyle\sum_{n=1}^{\infty} \dfrac{n}{n+1}$ 5. $\displaystyle\sum_{n=1}^{\infty} \dfrac{(-1)^{n+1}n^2}{n+1}$

7. $\dfrac{15}{16}$ 9. $-\dfrac{7}{12}$ 11. $\dfrac{1}{4}$ 13. $\dfrac{1}{2}$ 15. $\dfrac{2}{3}$

17. 2 19. Diverges   21. -1 23. $-\dfrac{2}{5}$ 25. Diverges

Chapter 11, Section 2 (page 540)

1. 5 3. 3 5. Diverges   7. $\dfrac{3}{20}$

9. 45   11. $\dfrac{3}{16}$ 13. 100   15. $\dfrac{1}{3}$ 17. $\dfrac{25}{99}$

19. 575 billion dollars   21. \$12,358.32

23. 27   25. 0.8202

Chapter 11, Section 3 (page 552)

1. $\displaystyle\sum_{n=0}^{\infty} \frac{3^n}{n!} x^n$

3. $\displaystyle\sum_{n=1}^{\infty} \frac{(-1)^{n+1}}{n} x^n$

5. $\displaystyle\sum_{n=0}^{\infty} \frac{1}{(2n)!} x^{2n}$

7. $\displaystyle\sum_{n=0}^{\infty} \left(\frac{n+1}{n!}\right) x^n$

9. $\displaystyle\sum_{n=0}^{\infty} \frac{2^n e^2}{n!} (x-1)^n$

11. $\displaystyle\sum_{n=0}^{\infty} (-1)^n (x-1)^n$

13. $\displaystyle\sum_{n=0}^{\infty} (x-1)^n$

15. $\displaystyle\sum_{n=0}^{\infty} 5^n x^{n+1}$ for $|x| < \dfrac{1}{5}$

17. $\displaystyle\sum_{n=0}^{\infty} \frac{x^n}{n!(-2)^n}$ for all x

19. 1.94936   21. 0.09531   23. 1.34984

25. 0.46127   27. 0.09967

Chapter 11, Section 4 (page 559)

1. 3.464   3. 0.755   5. -0.352

7. 4.791 and 0.209   9. 1.41421   11. 2.08008

Chapter 11, Review Problems (page 561)

1. $\dfrac{5}{6}$ 2. 1   3. $-\dfrac{1}{2}$ 4. $-\dfrac{1}{2}$

5. Diverges   6. 1.5415   7. 72   8. $\dfrac{17}{11}$

9. 54 feet   10. \$3,921.67   11. 18.16 units   12. 0.3297

13. $\displaystyle\sum_{n=0}^{\infty} (-1)^n \left(\frac{1}{3}\right)^{n+1} x^n$

14. $\displaystyle\sum_{n=0}^{\infty} \frac{(-3)^n}{n!} x^n$

15. $\displaystyle\sum_{n=0}^{\infty} \left(\frac{2n+1}{n!}\right) x^n$

16. $\displaystyle\sum_{n=0}^{\infty} (n+1)(x+2)^n$

17. $\sum_{n=0}^{\infty} (-1)^n (4^{n+1}) x^{2n+1}$ for $|x| < \dfrac{1}{2}$

18. 0.94869   19. 0.11919   20. 7.41620

21. -3.104 22. 0.258

Chapter 12, Section 1 (page 577)

1. $30°$ 3. $120°$ 5. $-120°$

7. 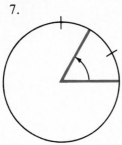   9. 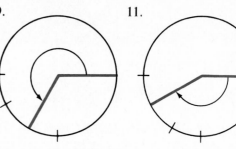   11.

13. $\dfrac{\pi}{12}$ 15. $-\dfrac{5\pi}{6}$ 17. 3π 19. $150°$

21. $270°$ 23. $-135°$ 25. $\dfrac{5\pi}{6}$ 27. $\dfrac{3\pi}{4}$ 29. $-\dfrac{2\pi}{3}$

31. 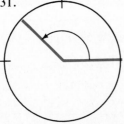   33. 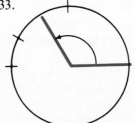   35.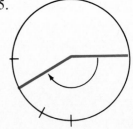

37. 0   39. 0   41. 0   43. 0

45. $\cos \dfrac{\pi}{3} = \dfrac{1}{2}$, $\sin \dfrac{\pi}{3} = \dfrac{\sqrt{3}}{2}$

θ	$\dfrac{7\pi}{6}$	$\dfrac{5\pi}{4}$	$\dfrac{4\pi}{3}$	$\dfrac{3\pi}{2}$	$\dfrac{5\pi}{3}$	$\dfrac{7\pi}{4}$	$\dfrac{11\pi}{6}$	2π
$\sin \theta$	$-\dfrac{1}{2}$	$-\dfrac{\sqrt{2}}{2}$	$-\dfrac{\sqrt{3}}{2}$	-1	$-\dfrac{\sqrt{3}}{2}$	$-\dfrac{\sqrt{2}}{2}$	$-\dfrac{1}{2}$	0
$\cos \theta$	$-\dfrac{\sqrt{3}}{2}$	$-\dfrac{\sqrt{2}}{2}$	$-\dfrac{1}{2}$	0	$\dfrac{1}{2}$	$\dfrac{\sqrt{2}}{2}$	$\dfrac{\sqrt{3}}{2}$	1

49. $-\dfrac{1}{2}$ 51. $\dfrac{1}{2}$ 53. $\dfrac{\sqrt{3}}{3}$ 55. 2   57. -1

59. $\dfrac{3}{4}$ 61. $\dfrac{3}{4}$ 63. $\dfrac{4}{5}$

65. $\theta = \dfrac{\pi}{2}$, $\theta = \dfrac{3\pi}{2}$, $\theta = \dfrac{\pi}{6}$, and $\theta = \dfrac{5\pi}{6}$

67. $\theta = \dfrac{\pi}{2}$, $\theta = \dfrac{\pi}{3}$, and $\theta = \dfrac{2\pi}{3}$

69. $\theta = \dfrac{\pi}{2}$ and $\theta = \dfrac{\pi}{4}$

71. $\theta = \dfrac{\pi}{2}$

73. $\theta = \dfrac{\pi}{6}$ and $\theta = \dfrac{5\pi}{6}$

Chapter 12, Section 2 (page 587)

1. $f'(\theta) = 3 \cos 3\theta$

3. $f'(\theta) = -2 \cos (1 - 2\theta)$

5. $f'(\theta) = -3\theta^2 \sin (\theta^3 + 1)$

7. $f'(\theta) = 2 \cos \left(\dfrac{\pi}{2} - \theta\right) \sin \left(\dfrac{\pi}{2} - \theta\right) = \sin (\pi - 2\theta) = \sin 2\theta$

9. $f'(\theta) = -6(1 + 3\theta) \sin (1 + 3\theta)^2$

11. $f'(\theta) = -e^{-\theta/2} \left(2\pi \sin 2\pi\theta + \dfrac{1}{2} \cos 2\pi\theta\right)$

13. $f'(\theta) = \dfrac{\cos \theta}{(1 + \sin \theta)^2}$

15. $f'(\theta) = -5\theta^4 \sec^2 (1 - \theta^5)$

17. $f'(\theta) = -4\pi \tan \left(\dfrac{\pi}{2} - 2\pi\theta\right) \sec^2 \left(\dfrac{\pi}{2} - 2\pi\theta\right)$

19. $f'(\theta) = 2 \dfrac{\cos \theta}{\sin \theta} = 2 \cot \theta$

Chapter 12, Section 3 (page 596)

1. 60 radians per hour   3. 0.15 radian per minute

5. $\dfrac{\pi}{2}$ radians   7. $\dfrac{\pi}{3}$ radians   9. 8 feet

Chapter 12, Review Problems (page 601)

1. (a) $\dfrac{2\pi}{3}$ radians; $120°$ (b) $-\dfrac{5\pi}{4}$ radians; $-225°$

2. 0.8727 radian   3. $14.3239°$

4. (a) $\dfrac{\sqrt{3}}{2}$ (b) $\dfrac{\sqrt{2}}{2}$ (c) 2 (d) $-\dfrac{1}{\sqrt{3}}$

5. $\dfrac{4}{3}$ 6. $\dfrac{3}{2}$

7. $\theta = \dfrac{\pi}{2}$ and $\theta = \dfrac{3\pi}{2}$ 8. $\theta = \dfrac{\pi}{3}$ and $\theta = \dfrac{2\pi}{3}$

9. $\theta = \dfrac{\pi}{6}$ and $\theta = \dfrac{5\pi}{6}$ 10. $\theta = \dfrac{\pi}{6}$ and $\theta = \dfrac{5\pi}{6}$

14. $f'(\theta) = 6(3\theta + 1) \cos (3\theta + 1)^2$

15. $f'(\theta) = -6 \cos (3\theta + 1) \sin (3\theta + 1)$

16. $f'(\theta) = 6\theta \sec^2 (3\theta^2 + 1)$

17. $f'(\theta) = 12\theta \tan (3\theta^2 + 1) \sec^2 (3\theta^2 + 1)$

18. $f'(\theta) = \dfrac{-1}{1 - \cos \theta}$

19. $f'(\theta) = -2 \tan \theta$

21. 0.012 radian per minute   22. $\dfrac{\pi}{3}$ radians   23. 27 feet

Appendix, Section A (page 620)

1. $1 < x \le 5$ 3. $x > -5$

5.   7.

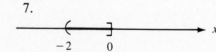

9. 4 11. 5

13. $-3 \le x \le 3$ 15. $-6 \le x \le -2$ 17. $x \ge 3$ or $x \le -7$

19. 125 21. 4 23. 4 25. $\dfrac{1}{2}$

27. $\dfrac{1}{2}$ 29. $\dfrac{1}{4}$ 31. 2 33. $\dfrac{1}{4}$

35. $n = 10$ 37. $n = 1$ 39. $n = 4$ 41. $n = \dfrac{13}{5}$

43. $x^4(x - 4)$

45. $25(7 - x)$

47. $2(x + 1)^2(x - 2)^2(7x - 2)$

49. $2(x + 3)$

51. $\dfrac{2(x + 3)^3(5 - x)}{(1 - x)^3}$

53. $(x + 2)(x - 1)$

55. $(x - 3)(x - 4)$

57. $(x - 1)^2$

59. $(x + 2)(x - 2)$

61. $(x - 1)(x^2 + x + 1)$

63. $x^5(x + 1)(x - 1)$

65. $2x(x - 5)(x + 1)$

67. $x = 4$ and $x = -2$

69. $x = -5$

71. $x = -4$ and $x = 4$

73. $x = -\dfrac{1}{2}$ and $x = -1$

75. $x = -\dfrac{3}{2}$

77. $x = -5$ and $x = 1$

79. $x = -2$ and $x = 1$

81. $x = -\dfrac{1}{2}$ and $x = -1$

83. No real solutions

85. $x = -\dfrac{3}{2}$

87. $x = 3, y = 2$

89. $x = 4, y = 2$

91. $x = -7, y = -5$ and $x = 1, y = -1$

93. 34

95. 0

97. $\displaystyle\sum_{j=1}^{6} \frac{1}{j}$ 99. $\displaystyle\sum_{j=1}^{6} 2x_j$ 101. $\displaystyle\sum_{j=1}^{8} (-1)^{j+1} j$

Appendix, Section B (page 634)

1. $\displaystyle\lim_{x \to a} f(x) = b$

3. $\displaystyle\lim_{x \to a} f(x) = b$

5. Limit does not exist.

7. 4

9. 7

11. 16

13. $\dfrac{3}{4}$

15. Limit does not exist.

17. 2

19. 7

21. $\dfrac{5}{3}$

23. 5

25. $\dfrac{1}{4}$

27. 2

29. Yes   31. Yes   33. No   35. No

37. No   39. No   41. Yes

43. None   45. $x = 3$

47. $x = -1$ 49. $x = -5$ and $x = 1$

51. $x = -1$ and $x = 2$ 53. $x = 2$

Index